Heath Mathematics

Walter E. Rucker • Clyde A. Dilley • David W. Lowry
Earl G. Ockenga

D. C. Heath and Company
Lexington, Massachusetts Toronto

Contributing Authors Edward Manfre, Philip Mirabelli: 38–39, 78–79, 88–89, 102–103, 106–107, 162–163, 164–165, 198–199, 222–223, 224–225, 254–255, 306–307, 398–399

Illustrations Leo Abbett; ANCO; Linda Strauss Edwards; Charlie Freeman/Sharon Kurlansky; Hal Frenck/Bookmakers; Alan Hering; George Hughes/Gwen Goldstein; Mark Kelly/Sharon Kurlansky; Kathy Parkinson; Julie Perron/Lisa Bassett; Andrew Shiff

Photography Erik Anderson: 272, 285, 312, 391; K. Anderson/Picture Cube: 157; Arthur d'Arazian/The Image Bank: 264; Jonathan Barkan: 125; Sharon Beals: 90–91, 172–173, 211; Fredrik D. Bodin: 97; Andrew Brilliant: 24, 32, 37, 48, 74, 82, 116, 126, 130, 142, 161, 168, 189, 209, 239, 248, 287, 365, 366, 452, 453; Gerald Brimacombe/The Image Bank: 28; Daniel Brody/Stock Boston: 316; Read D. Brugger/Picture Cube: 460; Stuart Cohen: 58; Elizabeth Crews/Stock Boston: 84; Donald Dietz/Stock Boston: 65; P. Ellin/Picture Cube: 76; Kevin Galvin: 19; Jonathan Goell: 338; Tad Goodale: 123, 138, 148, 202, 290; J. Howard/Stock Boston: 104; Jerry Howard/Margarite Bradley: iv, 66, 86, 88, 102, 137, 162, 164, 240, 292, 351, 401; Mark Kane: 181; Tom Magno: 108, 114, 204, 212–213, 218, 219, 233, 261, 379, 380, 456–457; Ned McCormick: title page; D. McCoy/Rainbow: 159; NASA: 20, 22, 47; Julie O'Neil: 53, 229, 347; Carol Palmer: 302; Tom Pantages: 289; Sarah Putnam/Picture Cube: 392; Wayne Ruple/Globe Photos, Inc.: 362–363; Walter Silver: 55, 80, 93, 100, 111, 155, 227, 250, 262, 332, 358, 368, 370, 371, 375; Deidra Delano Stead: 207, 268; Augustus Upitis/Shostal Associates: 323; Visual Horizons: 364, 365; Jerry Wachter/Focus on Sports: 121; Jonathan Wright/Bruce Coleman, Inc.: 57

Cover Bruce Terzian, Designer
Ned McCormick, Photographer

Published simultaneously in Canada.

Printed in the United States of America.

International Standard Book Number: 0-669-15929-8

4 5 6 7 8 9 0

Contents

Warmup

Use the Review Lessons to

- discover whether you need to review some whole-number skills.
- review and practice whole-number skills.
- apply your skills on challenging problems.

PRETEST

You can use this pretest to find which skills you need to review and practice.

A. Write in words.

1. 72,000 2. 42,600,000 3. 7,040,020

B. Round 459,701,985 to the nearest

1. million 2. hundred 3. ten thousand

C.

1.	2.	3.	4.	5.
$\begin{array}{r} 4 \\ 8 \\ 7 \\ +8 \\ \hline \end{array}$	$\begin{array}{r} 56 \\ 48 \\ +73 \\ \hline \end{array}$	$\begin{array}{r} 754 \\ 237 \\ +691 \\ \hline \end{array}$	$\begin{array}{r} 8654 \\ +3917 \\ \hline \end{array}$	$\begin{array}{r} 842 \\ 960 \\ 637 \\ +295 \\ \hline \end{array}$

D.

1.	2.	3.	4.	5.
$\begin{array}{r} 734 \\ -456 \\ \hline \end{array}$	$\begin{array}{r} 950 \\ -162 \\ \hline \end{array}$	$\begin{array}{r} 8082 \\ -7367 \\ \hline \end{array}$	$\begin{array}{r} 700 \\ -383 \\ \hline \end{array}$	$\begin{array}{r} 9000 \\ -5692 \\ \hline \end{array}$

E.

1.	2.	3.	4.	5.
$\begin{array}{r} 187 \\ \times 4 \\ \hline \end{array}$	$\begin{array}{r} 308 \\ \times 7 \\ \hline \end{array}$	$\begin{array}{r} 82 \\ \times 37 \\ \hline \end{array}$	$\begin{array}{r} 461 \\ \times 25 \\ \hline \end{array}$	$\begin{array}{r} 826 \\ \times 476 \\ \hline \end{array}$

F.

1.	2.	3.	4.	5.
$\begin{array}{r} 437 \\ \times 600 \\ \hline \end{array}$	$\begin{array}{r} 592 \\ \times 380 \\ \hline \end{array}$	$\begin{array}{r} 746 \\ \times 409 \\ \hline \end{array}$	$\begin{array}{r} 8314 \\ \times 270 \\ \hline \end{array}$	$\begin{array}{r} 6947 \\ \times 802 \\ \hline \end{array}$

G. 1. $7\overline{)453}$ 2. $3\overline{)802}$ 3. $47\overline{)1753}$ 4. $24\overline{)8741}$

H. 1. $27\overline{)8153}$ 2. $53\overline{)37,461}$ 3. $438\overline{)89,404}$ 4. $561\overline{)33,704}$

REVIEW LESSON A **Place value**

REMEMBER: Commas help you read large numbers. Look at the place-value chart.

Trillions			Billions			Millions			Thousands			Ones		
Hundreds	Tens	Ones	Hundreds	Tens	Ones	Hundreds	Tens	Ones	Hundreds	Tens	Ones	Hundreds	Tens	Ones
				9	6	1	4	6	0	8	5	2	7	3

Standard numeral: 9 6, 1 4 6, 0 8 5, 2 7 3

Read as "ninety-six billion, one hundred forty-six million, eighty-five thousand, two hundred seventy-three."

PRACTICE WITH SCRAMBLED ANSWERS

Give the place value of the underlined digit.

1. 146,235 (4 ten thousands)
2. 7,216,305
3. 296,003,165
4. 573,806
5. 34,872,129
6. 7,238,422,000
7. 95,061
8. 6,291,082
9. 8,152,883,420,000
10. 861,597
11. 1,666,660,000
12. 56,842,073,000,000

You should be able to find your answers in this list.

8 tens 9 tens 8 hundreds 4 ten thousands 9 ten thousands 2 hundred thousands 6 millions
3 ten millions 6 hundred millions 7 billions 1 hundred billion 6 trillions

PRACTICE

Write in words.

13. 62,000,000
14. 6,200,000
15. 620,000
16. 62,000
17. 6,020,000
18. 60,000,200
19. 600,020
20. 602,000

Write in standard form.

21. two million
22. two thousand
23. two million, two thousand
24. twenty thousand

25. two million, twenty thousand

26. two million, twenty-two thousand

27. ten thousand

28. ten million

29. ten

30. ten million, ten thousand

31. ten thousand, ten

32. ten million, ten

33. five million

34. six hundred thousand

35. ten million, ten thousand, ten

36. forty-seven thousand

37. five million, six hundred thousand

38. five million, forty-seven thousand

39. five million, six hundred forty-seven thousand

Write the number in standard form.

40. The average distance of the earth from the sun is ninety-three million miles.

41. The speed of light is about one hundred eighty-six thousand miles per second.

SPEED DRILL

Try to write these numbers in words in less than 1 minute.

42. 321,000

43. 7,007,007

Challenge!

Solve.

1. Build as long a string of math words as you can.

Rules

- The last letter of a math word must match the first letter of the next word.
- You cannot use a word more than once.

2. Kokomo is as far north of Indianapolis as Ottenbein is west of Kokomo. Indianapolis is in what direction from Ottenbein?

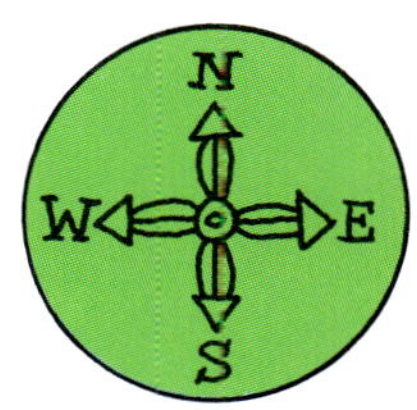

3. If 2 girls can fill 2 notebooks in 2 weeks and 3 boys can fill 3 notebooks in 3 weeks, how many notebooks can 6 girls and 6 boys fill in 6 weeks?

REVIEW LESSON B Rounding

REMEMBER: These rules tell you how to round.

1. Find the place to which you are rounding.
2. Look at the digit to the right.
3. If that digit is 5 or greater, round up.
   If that digit is less than 5, round down.

EXAMPLES: Round to the nearest ten thousand.

4<u>1</u>8,350 — 8 is greater than 5. Round up: 420,000

1,3<u>6</u>4,291 — 4 is less than 5. Round down: 1,360,000

3<u>2</u>5,000 — Round up when this digit is 5: 330,000

PRACTICE WITH SCRAMBLED ANSWERS

Round to the nearest hundred.

1. 742 **2.** 616 **3.** 1846 **4.** 1395 **5.** 6982

Round to the nearest thousand.

6. 8432 **7.** 8500 **8.** 19,874 **9.** 27,499 **10.** 399,876

Round to the nearest ten thousand.

11. 58,342 **12.** 13,099 **13.** 715,036 **14.** 746,428

Round to the nearest million.

15. 45,421,000 **16.** 46,800,000 **17.** 29,790,000 **18.** 15,455,800

You should be able to find your answers in this list.

600 700 1400 1800 7000 8000 9000 10,000
20,000 27,000 60,000 400,000 720,000 750,000
15,000,000 30,000,000 45,000,000 47,000,000

PRACTICE

Round to the nearest hundred.

19. 570 **20.** 820 **21.** 977 **22.** 1485 **23.** 99,977

24. 608 **25.** 13,429 **26.** 791 **27.** 20,950 **28.** 43,349

Round to the nearest thousand.

29. 57,631 **30.** 486,731 **31.** 89,744

32. 75,500 **33.** 99,499 **34.** 99,500

Round to the nearest hundred thousand.

35. 268,421 **36.** 99,463 **37.** 6,873,211

38. 8,949,999 **39.** 8,950,000 **40.** 8,950,601

Round to the nearest million.

41. 5,762,000 **42.** 6,432,000 **43.** 49,621,444

44. 987,431 **45.** 99,643,000 ★**46.** 374,521

Problem Solving

47. What is the smallest whole number which rounded to the nearest hundred is equal to 700?

48. What is the largest whole number which rounded to the nearest hundred is equal to 700?

49. What is the largest whole number which rounded to the nearest ten is equal to 700?

SPEED DRILL
Try to round these numbers to the nearest thousand in less than 1 minute.

50. 86,473 **51.** 29,751 **52.** 103,416

53. 987,444 **54.** 879,500 **55.** 59,741

Challenge!

Solve.

1. Here are four views of the same cube. The cube has the numbers 1–6. What numbers are on opposite sides of the cube?

2. A month has 5 Mondays. The month neither begins nor ends on Monday. On what day of the week does the month begin?

3. Look on these two pages. Find the smallest number with a 4 in the thousands place. Round the number to the nearest hundred. Did you get 374,500?

REVIEW LESSON C **Addition**

REMEMBER: When the sum of the digits in a column is 10 or greater, you must regroup.

	Add ones. Regroup.	Add tens. Regroup.	Add hundreds. Regroup.
858 +796	1 858 +796 4	1 1 858 +796 54	1 1 858 +796 1654

PRACTICE WITH SCRAMBLED ANSWERS

Add.

1. 78 + 56
2. 93 + 48
3. 207 + 509
4. 529 + 298
5. 834 + 957
6. 999 + 999
7. 5173 + 8861
8. 48 + 21 + 68
9. 37 + 43 + 58
10. 58 + 36 + 429
11. 392 + 517 + 636
12. 8476 + 1295 + 863
13. 17 + 23 + 14 + 56
14. 248 + 367 + 421 + 588
15. 831 + 1526 + 395 + 4288

You should be able to find your answers in this list.

110	134	137	138	141
523	716	827	1545	1624
1791	1998	7040	10,634	14,034

PRACTICE

Add.

16. $57 + 86$

17. $984 + 367$

18. $421 + 689$

19. $86 + 34 + 57$

20. $95 + 37 + 88$

21. $864 + 321 + 977$

22. $431 + 207 + 980$

23. $8766 + 5904$

24. $9741 + 8773 + 2914$

ADDITION SHORTCUT

To add 99, you can add 100 and subtract 1.

$375 + 99$

$375 + 100 = 475$, $475 - 1 = 474$

Use the shortcut to add these.

25. $543 + 99$

26. $678 + 99$

27. $548 + 99$

28. $956 + 99$

29. $693 + 99$

30. $496 + 99$

SPEED DRILL

Try to find the three errors in less than 1 minute.

31. $629 + 498 = 1127$

32. $4106 + 3986 = 7092$

33. $14{,}261 + 7{,}559 = 21{,}820$

34. $6926 + 629 = 7655$

35. $7169 + 887 = 8056$

36. $85{,}326 + 9{,}099 = 94{,}335$

Challenge!

Solve.

1.

Sonya and Bill weigh 150 kg.

Bill and Rita weigh 140 kg.

Sonya and Rita weigh 130 kg.

How much does each person weigh?

2. Cristina and Jeff bowled three games. The scores were:

game 1—152 to 122
game 2—132 to 112
game 3—150 to 140

Cristina's total score was 424. Which games did she win?

3. Look on these two pages. Find the largest number less than 1000. Find the smallest number greater than 1000. Add. Did you get a total of 2126?

REVIEW LESSON D Subtraction

REMEMBER: If you cannot subtract in a column, you must regroup.

	Regroup a ten. Subtract ones.	Regroup a hundred. Subtract tens.	Subtract hundreds.
563 −278	5 15 5~~6~~3 −278 5	4 15 ~~5~~~~6~~3 −278 85	4 15 ~~5~~~~6~~3 −278 285

REMEMBER: Sometimes you have to regroup more than once before you can subtract.

	Regroup. Then subtract.
7000 −437	6 9 9 ~~7~~~~0~~~~0~~0 −437 6563

PRACTICE WITH SCRAMBLED ANSWERS

Subtract.

1. 96 − 58
2. 182 − 67
3. 521 − 496
4. 4731 − 2914
5. 8630 − 975
6. 304 − 142
7. 800 − 417
8. 900 − 571
9. 2001 − 768
10. 3004 − 1876
11. 6000 − 1591
12. 9004 − 7567
13. 10,000 − 4,631
14. 9031 − 6755
15. 10,000 − 467

You should be able to find your answers in this list.

25	38	115	162	329
383	1128	1233	1437	1817
2276	4409	5369	7655	9533

PRACTICE

Subtract.

16. $49 - 31$ **17.** $103 - 47$ **18.** $573 - 428$

19. $763 - 598$ **20.** $604 - 319$ **21.** $298 - 76$

22. $703 - 477$ **23.** $5276 - 4197$ **24.** $2873 - 944$

25. $3004 - 1776$ **26.** $5002 - 4977$ **27.** $8000 - 6731$

SUBTRACTION SHORTCUT

To subtract 199, subtract 200, then add 1.

$$\begin{array}{r} 503 \\ -199 \\ \hline \end{array} \quad \begin{array}{r} 503 \\ -200 \\ \hline 303 \\ +\ 1 \\ \hline 304 \end{array}$$

Use the shortcut to subtract these.

28. $400 - 199$ **29.** $500 - 199$ **30.** $701 - 499$

31. $650 - 399$ **32.** $875 - 199$ **33.** $4234 - 899$

SPEED DRILL

Try to write these differences in less than 1 minute.

34. $704 - 196$ **35.** $538 - 277$ **36.** $644 - 398$

37. $8002 - 766$ **38.** $9003 - 4567$ **39.** $8703 - 4436$

Challenge!

Solve.

1.

The sum of the digits shown on the watch is 10. How many minutes before the sum of the digits will be 20?

2. You bought a horse for $60 on Monday. On Tuesday you sold it for $70. You bought the horse back for $80 on Wednesday. On Thursday you sold the horse again for $90. Did you make or lose money? How much money did you make or lose?

3. Carolyn, Brad, Jennifer, and Paul collected a total of 14 bumper stickers. Each had more than one sticker. Jennifer had the most. Carolyn had more than Paul. Brad had the fewest. How many bumper stickers did each person have?

REVIEW LESSON E **Multiplication**

REMEMBER: Multiply by the ones first.

	Multiply by 6 ones.	Multiply by 5 tens.	Add.
$\begin{array}{r} 329 \\ \times 56 \\ \hline \end{array}$	$\begin{array}{r} 329 \\ \times 56 \\ \hline 1974 \end{array}$	$\begin{array}{r} 329 \\ \times 56 \\ \hline 1974 \\ 1645 \\ \hline \end{array}$	$\begin{array}{r} 329 \\ \times 56 \\ \hline 1974 \\ 1645 \\ \hline 18{,}424 \end{array}$

PRACTICE WITH SCRAMBLED ANSWERS

Multiply.

1. $\begin{array}{r} 37 \\ \times 8 \\ \hline \end{array}$ 2. $\begin{array}{r} 56 \\ \times 9 \\ \hline \end{array}$ 3. $\begin{array}{r} 81 \\ \times 4 \\ \hline \end{array}$ 4. $\begin{array}{r} 448 \\ \times 7 \\ \hline \end{array}$ 5. $\begin{array}{r} 237 \\ \times 9 \\ \hline \end{array}$

6. $\begin{array}{r} 721 \\ \times 6 \\ \hline \end{array}$ 7. $\begin{array}{r} 834 \\ \times 4 \\ \hline \end{array}$ 8. $\begin{array}{r} 93 \\ \times 76 \\ \hline \end{array}$ 9. $\begin{array}{r} 44 \\ \times 53 \\ \hline \end{array}$ 10. $\begin{array}{r} 41 \\ \times 71 \\ \hline \end{array}$

11. $\begin{array}{r} 63 \\ \times 28 \\ \hline \end{array}$ 12. $\begin{array}{r} 804 \\ \times 77 \\ \hline \end{array}$ 13. $\begin{array}{r} 235 \\ \times 462 \\ \hline \end{array}$ 14. $\begin{array}{r} 730 \\ \times 349 \\ \hline \end{array}$ 15. $\begin{array}{r} 1046 \\ \times 78 \\ \hline \end{array}$

You should be able to find your answers in this list.

296	324	504	1764	2133
2332	2911	3136	3336	4326
7068	61,908	81,588	108,570	254,770

PRACTICE

Multiply.

16. $\begin{array}{r} 58 \\ \times 3 \\ \hline \end{array}$ 17. $\begin{array}{r} 86 \\ \times 5 \\ \hline \end{array}$ 18. $\begin{array}{r} 94 \\ \times 7 \\ \hline \end{array}$ 19. $\begin{array}{r} 521 \\ \times 6 \\ \hline \end{array}$ 20. $\begin{array}{r} 344 \\ \times 8 \\ \hline \end{array}$

21. $\begin{array}{r} 48 \\ \times 27 \\ \hline \end{array}$ 22. $\begin{array}{r} 65 \\ \times 44 \\ \hline \end{array}$ 23. $\begin{array}{r} 39 \\ \times 56 \\ \hline \end{array}$ 24. $\begin{array}{r} 82 \\ \times 12 \\ \hline \end{array}$ 25. $\begin{array}{r} 46 \\ \times 75 \\ \hline \end{array}$

26. 321×63

27. 456×21

28. 876×32

29. 596×83

30. 404×67

31. 721×83

32. 496×21

33. 387×39

34. 844×67

MULTIPLICATION SHORTCUT

Here is a way to multiply two 2-digit numbers that requires less paper work.

Multiply ones. Cross-multiply and add. Multiply tens and add.

Use the shortcut to multiply these.

35. 12×32

36. 42×25

37. 53×37

38. 28×43

39. 56×27

40. 25×56

SPEED DRILL

Try to write these products in less than 1 minute.

41. 57×26

42. 638×41

43. 103×78

Challenge!

Solve.

1. In a game, Brian's score was 10 points more than Dave's. Calvin's score was 6 points less than Brian's. Mike's score was 8 points more than Calvin's.

   Who had the highest score?

   Who had the lowest score?

2. 

   Nine of these signs have 30 sides.

   How many of them are No-Passing signs?

3. Jill and Brenda panned for gold. Jill said to Brenda, "Give me one of your nuggets and I'll have as many as you have." Brenda said, "Give me one of your nuggets and I'll have twice as many as you." How many nuggets did each girl have?

REVIEW LESSON F **Zeros in multiplication**

REMEMBER: Start each partial product directly under the digit by which you are multiplying.

$$\begin{array}{r} 329 \\ \times 806 \\ \hline 1974 \\ 2632 \\ \hline 265{,}174 \end{array} \qquad \begin{array}{r} 385 \\ \times 700 \\ \hline 269{,}500 \end{array} \qquad \begin{array}{r} 721 \\ \times 340 \\ \hline 28840 \\ 2163 \\ \hline 245{,}140 \end{array}$$

partial product

PRACTICE WITH SCRAMBLED ANSWERS

Multiply.

1. 397 × 40
2. 885 × 700
3. 629 × 800
4. 438 × 120
5. 681 × 470
6. 926 × 380
7. 234 × 102
8. 680 × 405
9. 702 × 904
10. 216 × 503
11. 6225 × 420
12. 8341 × 602
13. 5876 × 130
14. 8200 × 300
15. 5003 × 601

You should be able to find your answers in this list.

15,880	23,868	52,560	108,648	275,400
320,070	351,880	503,200	619,500	634,608
763,880	2,460,000	2,614,500	3,006,803	5,021,282

PRACTICE

Multiply.

16. 823 × 104
17. 971 × 700
18. 502 × 502
19. 216 × 101
20. 328 × 640
21. 672 × 300
22. 598 × 200
23. 675 × 50
24. 831 × 250
25. 936 × 205

26. 755×300

27. 291×302

28. 291×320

29. 476×303

30. 523×607

31. 298×404

32. 507×321

33. 298×460

34. 347×207

35. 5981×601

36. 2004×304

37. 3040×206

38. 3089×750

39. 9012×634

40. 7060×408

Problem Solving

41. The speed of light is about 186,000 miles per second. What is the speed of light in miles per hour?

42. How many miles does light travel in a day?

43. How far does light travel in a year?

44. How far away is a star if its light takes 6 years to reach the earth?

SPEED DRILL

Try to write these products in less than 1 minute.

45. 261×201

46. 378×502

47. 4010×606

Challenge!

Solve.

1. What is the price of a parrot if three times the price is $50 more than two times the price?

2. A class had a contest to see who could guess the number of pennies in a jar.

   Kevin guessed 450 pennies.
   Lance guessed 580 pennies.
   Carla guessed 318 pennies.

   One guess was off by 91, one was off by 39, and one was off by 171.

   How many pennies are in the jar?

3. If you bought 12 of these boxes of nails, how many boxes would be left?

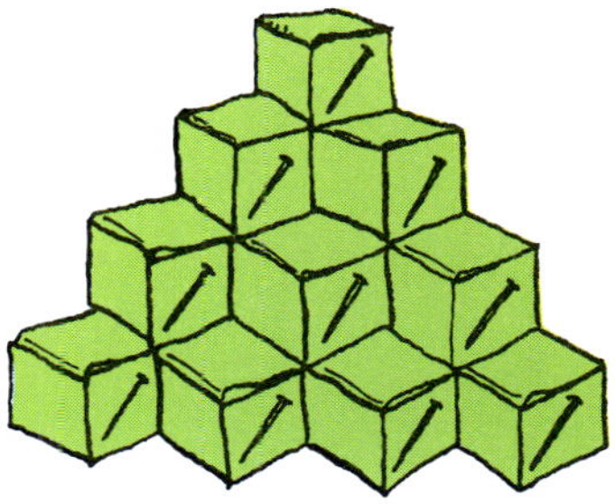

REVIEW LESSON G Division

REMEMBER: When you divide, be sure to put each digit of the answer in the correct place.

Not enough hundreds.	Divide 4 into 26 tens. Multiply and subtract.	Divide 4 into 29 ones. Multiply and subtract.
4)269	4)269 = 6 −24 2	4)269 = 67 R1 −24 29 −28 1

REMEMBER: When the answer is a 2-digit number, you may have to estimate the digits in the quotient.

Think of the divisor as rounded to 40.	Divide 40 into 112 tens. Multiply and subtract.	Divide 40 into 367 ones. Multiply and subtract.
(40) 38)1127	(40) 38)1127 = 2 −76 36	(40) 38)1127 = 29 R25 −76 367 −342 25

PRACTICE WITH SCRAMBLED ANSWERS

Divide.

1. 8)673 **2.** 5)931 **3.** 6)2887 **4.** 4)1953

5. 42)3125 **6.** 36)1735 **7.** 81)2816 **8.** 67)9287

9. 38)8988 **10.** 21)9755 **11.** 801)63,301 **12.** 394)93,397

You should be able to find your answers in this list.

34 R62	48 R7	74 R17	79 R22
84 R1	138 R41	186 R1	236 R20
237 R19	464 R11	481 R1	488 R1

PRACTICE
Divide.

13. $9\overline{)756}$ 14. $8\overline{)7841}$ 15. $7\overline{)5963}$

16. $5\overline{)3275}$ 17. $8\overline{)3414}$ 18. $6\overline{)2984}$

19. $4\overline{)2737}$ 20. $19\overline{)5634}$ 21. $41\overline{)3516}$

22. $53\overline{)2988}$ 23. $72\overline{)56,441}$ 24. $36\overline{)19,411}$

25. $32\overline{)67,734}$ 26. $51\overline{)21,143}$ 27. $77\overline{)98,341}$

28. $62\overline{)83,144}$ 29. $76\overline{)31,442}$ 30. $216\overline{)57,134}$

Problem Solving

The speed of sound at sea level is about 742 mph.

31. What is the speed of sound in feet per hour?

32. What is the speed of sound in feet per minute?

33. How far away is a flash of lightning if the sound takes 60 seconds to reach you?

34. How far away is a flash of lightning if the sound takes 15 seconds to reach you?

SPEED DRILL
Try to write these quotients in less than 1 minute.

35. $6\overline{)7293}$ 36. $5\overline{)6218}$ 37. $26\overline{)5413}$

Challenge!

Solve.

1. There are six bills (ones, fives, and tens) in the billfold. There are more fives than tens. There are more tens than ones. How much money is in the billfold?

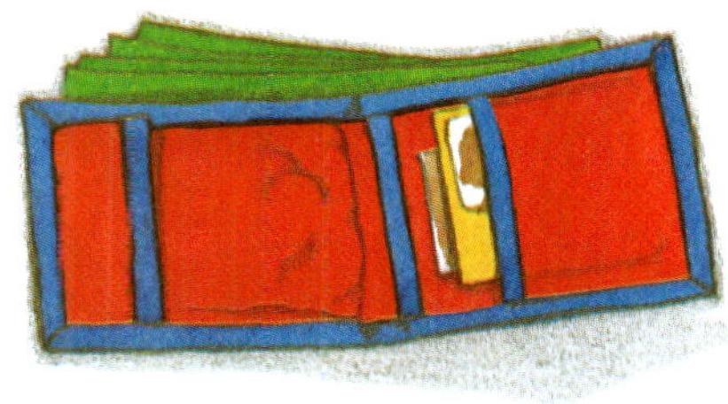

2. 2 whistles and 3 books cost \$19. 3 whistles and 2 books cost \$16. How much does one whistle cost?

3. Elmer counted 9 pigs and cows.
He counted 10 cows and horses.
He counted 16 animals in all.
How many cows did he count?

REVIEW LESSON H Zeros in division

REMEMBER: If you estimate the quotient first, you are not likely to leave out a zero.

Estimate.

Not enough thousands.	The first digit is in the hundreds place.	Write 0's in the remaining places.	
$27\overline{)8267}$	$\begin{array}{r} 3 \\ 27\overline{)8267} \end{array}$	$\begin{array}{r} 300 \\ 27\overline{)8267} \end{array}$	The quotient is near 300.

Now divide.

$$\begin{array}{r} 306 \text{ R5} \\ 27\overline{)8267} \\ -81 \\ \hline 167 \\ -162 \\ \hline 5 \end{array}$$

Your estimate told you that the quotient was near 300. Knowing that, you would not have left out the zero.

PRACTICE WITH SCRAMBLED ANSWERS

Estimate. Then divide.

1. $2\overline{)61}$
2. $4\overline{)825}$
3. $43\overline{)895}$
4. $67\overline{)694}$
5. $28\overline{)1703}$
6. $52\overline{)3671}$
7. $52\overline{)36,714}$
8. $3\overline{)1226}$
9. $4\overline{)16,024}$
10. $5\overline{)20,004}$
11. $173\overline{)6956}$
12. $284\overline{)22,721}$

You should be able to find your answers in this list.

10 R24	20 R35	30 R1	40 R36
60 R23	70 R31	80 R1	206 R1
408 R2	706 R2	4000 R4	4006

PRACTICE

Estimate. Then divide.

13. $6\overline{)242}$
14. $5\overline{)351}$
15. $8\overline{)326}$
16. $7\overline{)6315}$
17. $4\overline{)1212}$
18. $3\overline{)1816}$
19. $9\overline{)7214}$
20. $6\overline{)54,027}$

21. 9)36,047 22. 12)36,085 23. 23)11,549

24. 56)22,739 25. 44)13,392 26. 57)17,083

27. 41)20,767 28. 35)21,092 29. 87)8896

30. 7)35,022 31. 8)9630 32. 5)16,023

33. 6)13,214 34. 9)28,814 35. 32)16,014

36. 72)21,693 37. 51)10,259 38. 8)16,029

Problem Solving

At 55 mph on the highway, the crankshaft in a car's engine is turning at 2700 revolutions per minute.

39. What is the rate in revolutions per second?

40. What is the rate in revolutions per hour?

41. Each cylinder fires once each revolution. How many times does a 4-cylinder engine fire per minute if it is turning at 2700 rpm?

42. How many times does a 6-cylinder engine fire in a minute if it is turning at 2700 rpm?

SPEED DRILL

Try to write these quotients in less than 1 minute.

43. 8)32,011 44. 5)125,125

45. 12)6071 46. 28)3375

Challenge!

Solve.

1. 

Hot dog	Salad	Milk
49¢	46¢	38¢

Karen bought two items. She gave the cashier a $1 bill and got back 3 coins in change. What did she buy?

2. lure $3
reel $29
rod $15
tackle box $12

Cindy, Tim, and Lance each bought 3 different items. Cindy said she spent $47. Tim said he spent $30. Lance said he spent $45. Who miscalculated?

3. Two dimes, some nickels, one quarter, and seven pennies total 77¢. How many nickels are there?

POSTTEST

A good score on this test means you are ready to start Chapter 1.

A. Write in words.

1. 207,000
2. 36,400,000
3. 8,050,010

B. Round 327,401,985 to the nearest

1. million.
2. hundred.
3. ten thousand.

C.

1. $6473 + 7409$
2. $46 + 93 + 85$
3. $619 + 278 + 653$
4. $827 + 68 + 298$
5. $506 + 948 + 674 + 871$

D.

1. $851 - 283$
2. $9073 - 4624$
3. $8412 - 766$
4. $600 - 249$
5. $7012 - 6747$

E.

1. 581×4
2. 6027×8
3. 814×74
4. 627×481
5. 903×397

F.

1. 827×800
2. 659×460
3. 724×209
4. 5816×780
5. 2943×508

G.

1. $7\overline{)4580}$
2. $38\overline{)2328}$
3. $45\overline{)7642}$
4. $436\overline{)36{,}200}$

H.

1. $26\overline{)9107}$
2. $72\overline{)43{,}511}$
3. $122\overline{)61{,}223}$
4. $387\overline{)15{,}486}$

1

Decimals

Karen Campel's time in the white-water kayak race is printed on the photograph. The winning time was 1:27.9 minutes. How much faster was the winning time?

Decimals

The space shuttle *Columbia* roared into orbit for the first time on April 12, 1981. *Columbia* remained in orbit for 54.5 hours before reentering the earth's atmosphere and landing in the southern California desert.

During its descent *Columbia* was traveling 7 times the speed of sound, or about 1.4167 miles each second.

To read this reentry speed, we can use a place-value chart.

Thousands	Hundreds	Tens	Ones	Tenths	Hundredths	Thousandths	Ten-thousandths	Hundred-thousandths
			1.	**4**	**1**	**6**	**7**	

1. Read the whole-number part
   "one" **1.4167**

2. Read the decimal point as "and"
   "and" **1.4167**

3. Read the decimal part as you would read a whole number
   "four thousand one hundred sixty-seven" **1.4167**

4. Read the value of the last place on the right
   "ten-thousandths" **1.4167** ↑

One and four thousand one hundred sixty-seven ten-thousandths

EXERCISES

For 295.781603, give the digit that is in the

1. tens place
2. tenths place
3. hundreds place
4. hundredths place
5. ten-thousandths place
6. thousandths place
7. millionths place
8. hundred-thousandths place

Read aloud. Then write in words.

9. 0.52
10. 0.036
11. 5.72
12. 0.0591
13. 7.324
14. 6.0938
15. 42.16590
16. 7.28304
17. 6203.149
18. 620.3149
19. 62.03149
20. 6.203149

Write the standard numeral.

21. eight and nine tenths
22. fifteen and six hundredths
23. forty and one hundred twenty-five thousandths
24. ninety-four ten-thousandths
25. two hundred seventy-eight and seven hundred sixty-five thousandths
26. one hundred fifty-two and three hundred eighty-two millionths

Challenge!

Give the standard numeral.

27. 2.8 million

The 2 is in the millions place.
2.8 million = 2,800,000

28. 3.4 million
29. 15.6 million
30. 4.9 billion
31. 22.3 billion

Match.

32. 3,280,000
33. 32,800,000,000
34. 32,800,000
35. 3,280,000,000

a. 32.8 million
b. 32.8 billion
c. 3.28 billion
d. 3.28 million

Rounding decimals

The space shuttle *Columbia* can carry cargo as well as people into space. The cargo bay measures 18.146 meters in length.

To write a rule for rounding decimals, we can use a place-value chart.

Tens	Ones	Tenths	Hundredths	Thousandths
1	**8.**	**1**	**4**	**6**

Numbers like this can be rounded in several ways. To round to the nearest whole number, follow these steps.

Step 1. Find the digit in the ones place.

$\downarrow$ (under 8)

18.146

Step 2. Look at the digit that is one place to the right.

$\downarrow$ (under 1)

18.146

- If it is 5 or greater, round up.
- If it is less than 5, write the digit in the ones place.

18 is the nearest whole number.

Rounded to the nearest tenth, 18.146 is 18.1.
Rounded to the nearest hundredth, 18.146 is 18.15.

EXERCISES

Round each decimal to the nearest one (or whole number).

1. 26.3	**2.** 8.9	**3.** 35.6	**4.** 52.5
5. 74.09	**6.** 24.83	**7.** 76.45	**8.** 89.368
9. 99.72	**10.** 153.495	**11.** 279.89	**12.** 399.634

Round to the nearest tenth.

13. 6.73	**14.** 5.28	**15.** 12.06	**16.** 23.029
17. 65.38	**18.** 51.46	**19.** 79.82	**20.** 184.60
21. 196.193	**22.** 220.095	**23.** 358.551	**24.** 296.742

Round to the nearest hundredth.

25. 79.835	**26.** 16.374	**27.** 29.182	**28.** 74.365
29. 32.298	**30.** 38.409	**31.** 86.635	**32.** 72.382
33. 263.1141	**34.** 591.2963	**35.** 748.5294	**36.** 599.6994

Round 369.4278351 to the nearest

37. tenth	**38.** thousandth	**39.** hundredth
40. ten-thousandth	**41.** millionth	**42.** hundred-thousandth

Challenge!

43. Study the clues to find this number. □.□□□

Clue 1

If you round the number to the nearest whole number, you get 6.

Clue 2

If you round the number to the nearest tenth, you get 6.0.

Clue 3

If you round the number to the nearest hundredth, you get 6.00.

Clue 4

It is the smallest such number.

Comparing decimals

Cascade Junior High School holds a coin-stacking contest each year. The table shows the time it took each of 5 contestants to stack 50 pennies.

Contestant	*Time (seconds)*
John Bradley	43.53
Bev Dorr	43.17
Vern Eastman	45.39
Charlie Fitch	44.07
Tara Kilby	43.57

Who stacked the coins in a shorter time, John Bradley or Bev Dorr?

To answer the question, compare the decimals by starting at the left and comparing the digits that are in the same place.

same

43.53 > **43.17**

>

So, Bev Dorr stacked the coins in a shorter time.

EXERCISES

1. Who stacked the coins in a shorter time, Vern Eastman or Tara Kilby?

2. Charlie Fitch's time is shorter than only one other person's time. Whose time is that?

3. Is Tara Kilby's time longer or shorter than John Bradley's time?

4. Rank the times from shortest time to longest time. Who had the shortest time?

<, >, **or** =?

5. 0.39 ● 0.39

6. 12.4 ● 12.1

7. 15.237 ● 15.273

8. 3.8 ● 38

9. 0.097 ● 0.097

10. 6.99 ● 7

11. 10.94 ● 11

12. 39.34 ● 393.40

13. 26.8 ● 26.800

14. 3.8 ● 3.9

15. 5.06 ● 5.02

16. 12.34 ● 12.034

17. 9.09 ● 10

18. 82.05 ● 82.049

19. 75.283 ● 75.29

20. 0.12 ● 0.099

21. 15.03 ● 14.13

22. 8.235 ● 12.68

23. 3.27 ● 0.327

24. 86.73 ● 867.3

25. 2.999 ● 3

26. 100 ● 99.99

27. 5.09 ● 5.090

28. 0.003 ● 0.020

In the tallest-stack event, each contestant gets 3 tries to build the tallest stack of pennies. The table shows the time it took contestants to build their tallest stacks.

Use the table to complete.

Contestant	*Number of pennies in stack*	*Time*
Juan Amigna	99	2 minutes 1.43 seconds
Bev Dorr	99	2 minutes 1.56 seconds
Jennifer Hartshorn	101	2 minutes 12.13 seconds
Bill McNelly	99	2 minutes 1.46 seconds
John Roberts	101	2 minutes 12.06 seconds

29. Who stacked 101 pennies in a shorter time?

30. Who stacked 99 pennies in the shortest time?

31. A contestant with a taller stack ranks higher than a contestant with a shorter stack. When 2 contestants build stacks with the same number of pennies, the contestant with the shorter time ranks higher. Rank the contestants.

Adding and subtracting decimals

Jane made two runs down the slalom course. Her times were 35.76 seconds and 38.19 seconds. In a slalom race the two times are added and the skier with the fastest total time wins.

$$\begin{array}{r} 35.76 \\ +38.19 \\ \hline \end{array}$$

We can estimate the sum by rounding each addend to the nearest whole number: 36 + 38 = 74.

Is the answer near the estimate?

Alice had a total time of 76.02 seconds. You can subtract to find out how much better Jane's time was. You can estimate the difference by subtracting 74 from 76.

$$\begin{array}{r} 76.02 \\ -73.95 \\ \hline \end{array}$$

76 − 74 = 2

Is the answer near the estimate?

EXERCISES

First estimate the sum. Then add.

1. $\begin{array}{r} 3.7 \\ +2.8 \\ \hline \end{array}$
2. $\begin{array}{r} 5.94 \\ +3.8 \\ \hline \end{array}$
3. $\begin{array}{r} 65.4 \\ +78.9 \\ \hline \end{array}$
4. $\begin{array}{r} 6.88 \\ +9.758 \\ \hline \end{array}$
5. $\begin{array}{r} 5.9384 \\ +2.6578 \\ \hline \end{array}$

6. 5.9 + 6.8 + 3.7

7. 9.08 + 3.65 + 8.17

8. 4.3 + 6.58 + 9.4

9. 88.60 + 5.79 + 30.70

10. 0.2735 + 0.9115 + 0.2673

First estimate the difference. Then subtract.

11. 5.3 − 2.7

12. 9.0 − 3.5

13. 1.83 − 0.91

14. 2.834 − 1.799

15. 0.544 − 0.270

16. 6.3482 − 1.7063

17. 8.07 − 2.9

18. 5.4 − 3.68

19. 4 − 2.38

20. 2.3 − 1.657

Give each sum or difference. Remember to line up the decimal points.

21. 5.63 + 2.84
22. 4.38 − 1.59
23. 7.98 + 3.59
24. 0.084 − 0.009
25. 5.93 + 24.3
26. 15.6 − 8.34
27. 9.06 + 10.4
28. 17 − 9.83
29. 8.4 + 3.9 + 6.2
30. 4 − 1.382
31. 6.84 + 2.056
32. 6.831 − 2.4

Problem Solving

Time in seconds

Name	*First Run*	*Second Run*
Brown	35.26	36.19
Davis	35.74	38.28
Franklin	36.29	35.48
Jones	38.02	37.54
Langley	36.13	38.29
Manning	37.18	36.42

33. Find each skier's total time.

34. How much less was Franklin's total time than Langley's?

35. What was the difference between the fastest and slowest total times?

36. Rank the six skiers from fastest to slowest. (Use total times.)

37. What time would Davis have had to get on her second run to win the slalom?

Challenge!

Solve.

38. The letters M, E, and Y stand for three digits. Find the digit that each letter stands for.

$$\begin{array}{r} M.E \\ +\,M.Y \\ \hline EY.E \end{array}$$

Multiplying decimals

Below are two methods for finding the product $\begin{array}{r} 13.4 \\ \times 2.8 \\ \hline \end{array}$

ESTIMATION METHOD

Step 1.
Multiply as whole numbers.

$$\begin{array}{r} 13.4 \\ \times 2.8 \\ \hline 1072 \\ 268 \\ \hline 3752 \end{array}$$

Step 2.
Estimate the product and place the decimal point.

$$\begin{array}{r} 13.4 \\ \times 2.8 \\ \hline 1072 \\ 268 \\ \hline 37.52 \end{array}$$

We can estimate the product by rounding each factor to the nearest whole number: 13 × 3 = 39.

Since the product is near 39, the decimal point goes between the 7 and the 5.

SHORTCUT METHOD

Step 1.
Multiply as whole numbers.

$$\begin{array}{r} 13.4 \\ \times 2.8 \\ \hline 1072 \\ 268 \\ \hline 3752 \end{array}$$

Step 2.
Count the digits to the right of the decimal points.

$$\begin{array}{r} 13.\underset{\smile}{4} \\ \times 2.\underset{\smile}{8} \\ \hline 1072 \\ 268 \\ \hline 3752 \end{array}$$

Step 3.
Count off the same number of digits to place the decimal point in the product.

$$\begin{array}{r} 13.4 \\ \times 2.8 \\ \hline 1072 \\ 268 \\ \hline 37.\underset{\smile\smile}{52} \end{array}$$

EXERCISES

Write each product with the decimal point in the correct place.

1. $\begin{array}{r} 7.3 \\ \times 5.8 \\ \hline 4234 \end{array}$ **2.** $\begin{array}{r} 9.01 \\ \times 5.2 \\ \hline 46852 \end{array}$ **3.** $\begin{array}{r} 3.4 \\ \times 0.16 \\ \hline 544 \end{array}$ **4.** $\begin{array}{r} 200 \\ \times 7.1 \\ \hline 14200 \end{array}$ **5.** $\begin{array}{r} 6.9 \\ \times 0.15 \\ \hline 1035 \end{array}$

6. $\begin{array}{r} 8.6 \\ \times 0.07 \\ \hline 602 \end{array}$ **7.** $\begin{array}{r} 3.01 \\ \times 0.23 \\ \hline 6923 \end{array}$ **8.** $\begin{array}{r} 0.47 \\ \times 1.8 \\ \hline 846 \end{array}$ **9.** $\begin{array}{r} 4.7 \\ \times 1.8 \\ \hline 846 \end{array}$ **10.** $\begin{array}{r} 4.7 \\ \times 0.18 \\ \hline 846 \end{array}$

First <u>estimate</u> the product. Then multiply.

11. $\begin{array}{r} 5.2 \\ \times 2.1 \\ \hline \end{array}$ **12.** $\begin{array}{r} 9.4 \\ \times 0.32 \\ \hline \end{array}$ **13.** $\begin{array}{r} 0.78 \\ \times 0.43 \\ \hline \end{array}$ **14.** $\begin{array}{r} 6.9 \\ \times 0.18 \\ \hline \end{array}$ **15.** $\begin{array}{r} 4.9 \\ \times 36 \\ \hline \end{array}$

16. $\begin{array}{r} 24.2 \\ \times 2.8 \\ \hline \end{array}$ **17.** $\begin{array}{r} 3.91 \\ \times 3.5 \\ \hline \end{array}$ **18.** $\begin{array}{r} 46.8 \\ \times 0.24 \\ \hline \end{array}$ **19.** $\begin{array}{r} 3.29 \\ \times 0.73 \\ \hline \end{array}$ **20.** $\begin{array}{r} 74.2 \\ \times 8.5 \\ \hline \end{array}$

21. $\begin{array}{r} 59.3 \\ \times 1.65 \\ \hline \end{array}$ **22.** $\begin{array}{r} 8.26 \\ \times 1.03 \\ \hline \end{array}$ **23.** $\begin{array}{r} 7.52 \\ \times 0.242 \\ \hline \end{array}$ **24.** $\begin{array}{r} 983 \\ \times 3.11 \\ \hline \end{array}$ **25.** $\begin{array}{r} 6.54 \\ \times 42.1 \\ \hline \end{array}$

Multiple choice. [*Hint:* <u>Estimate</u>!]

26. Four members of a relay team skated 100 meters each in 10.9 seconds, 11.3 seconds, 11.6 seconds, and 10.8 seconds. What was their total-team time?

a. 400.8 s **b.** 11 s
c. 4.5 s **d.** 44.6 s

27. A four-member relay team skated a 400-meter race. The average time for each skater was 11.2 seconds. What was the team time in seconds?

a. 4.48 **b.** 4480
c. 448 **d.** 44.8

28. In the 1500-meter race in the 1980 Winter Olympics, Eric Heiden's time was 1 minute 55.44 seconds. This was 3.84 seconds better than the previous record. What was the previous record?

a. 1 min 49.38 s **b.** 1 min 51.60 s
c. 1 min 59.28 s **d.** 4 min 3.84 s

29. In the 1000-meter race, Eric Heiden beat the previous record of 1 minute 19.32 seconds by 4.14 seconds. What was his time?

a. 1 min 15.18 s **b.** 1 min 23.46 s
c. 5 min 15.18 s **d.** 1 min 29.32 s

Zeros in multiplication

Sometimes you have to write some extra zeros before you place the decimal point in the product.

Step 1. Multiply.

$$\begin{array}{r} 0.04 \\ \times 0.2 \\ \hline 8 \end{array}$$

Step 2. Need 3 digits to the right of the decimal point. So, write 2 extra zeros.

$$\begin{array}{r} 0.04 \\ \times 0.2 \\ \hline 008 \end{array}$$

Step 3. Place the decimal point in the product.

$$\begin{array}{r} 0.04 \\ \times 0.2 \\ \hline 0.008 \end{array}$$

EXERCISES

Give each product.

1. 0.4 ×0.2	**2.** 0.4 ×0.3	**3.** 0.02 ×0.4	**4.** 0.02 ×0.5	**5.** 0.02 ×0.9
6. 0.12 ×0.3	**7.** 0.13 ×0.4	**8.** 2.8 ×0.3	**9.** 0.28 ×0.3	**10.** 0.028 ×0.3
11. 0.004 ×0.02	**12.** 0.007 ×0.03	**13.** 0.013 ×0.14	**14.** 0.029 ×0.35	**15.** 0.036 ×0.57
16. 3.56 ×0.42	**17.** 7.04 ×0.31	**18.** 0.067 ×0.08	**19.** 4.07 ×0.28	**20.** 5.13 ×4.02
21. 28.7 ×8.06	**22.** 51.7 ×0.0004	**23.** 82.7 ×50.2	**24.** 6.07 ×6.07	**25.** 0.412 ×0.044

Multiple choice.
Choose the correct answer by estimating.

26. 53.77 + 82.31
a. 126.28 **b.** 136.08 **c.** 147.08 **d.** 251.28

27. 703.11 − 498.26
a. 304.85 **b.** 204.85 **c.** 145.85 **d.** 263.85

28. 50 − 7.24
a. 52.76 **b.** 43.76 **c.** 57.24 **d.** 42.76

29. (25.8 + 36.5) + 41.1

Work inside parentheses first.

a. 103.4 **b.** 92.4 **c.** 123.4 **d.** 96.4

30. (56.2 − 13.9) + 21.7
a. 64.0 **b.** 56.6 **c.** 54.0 **d.** 60.6

31. 29.7 × 4.2
a. 96.64 **b.** 132.74 **c.** 124.74 **d.** 112.64

32. (3.1 + 5.3) × 2.7
a. 35.28 **b.** 22.68 **c.** 16.98 **d.** 43.58

33. (3.14 × 5.1) × 5.1
a. 81.6714 **b.** 93.2124 **c.** 15.2673 **d.** 60.5134

34. 14.4 × (1.2 + 5.6)
a. 22.8 **b.** 71.92 **c.** 97.92 **d.** 123.12

35. 58.1 − (17.9 × 1.9)
a. 10.79 **b.** 24.09 **c.** 33.59 **d.** 45.29

Challenge!

Give the missing input or output.

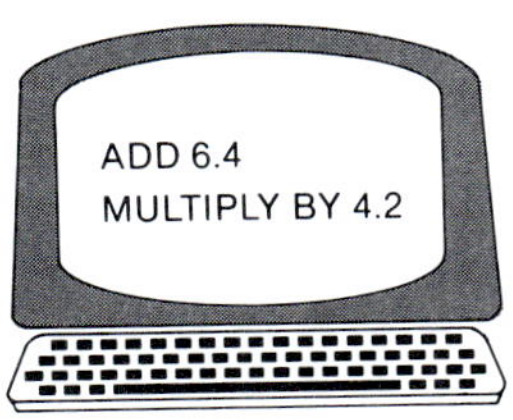

	Input	Output
	3.4	41.16
36.	15.4	?
37.	20.63	?
38.	?	53.76

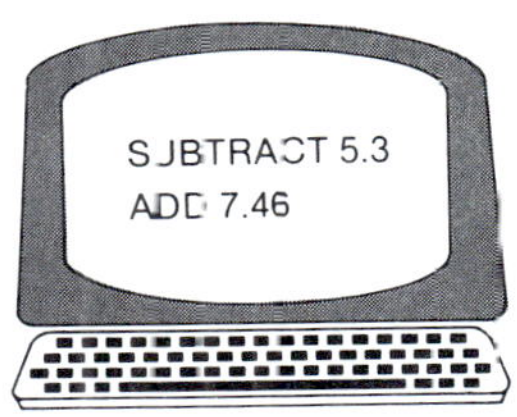

	Input	Output
	14.25	16.41
39.	18.9	?
40.	?	31.53
41.	?	42.6

Dividing a decimal by a whole number

You divide a decimal by a whole number just as you divide a whole number by a whole number. Study these examples.

EXERCISES

Give each quotient.

1. $2\overline{)5.8}$	**2.** $3\overline{)9.3}$	**3.** $5\overline{)0.25}$	**4.** $7\overline{)1.47}$	**5.** $6\overline{)0.456}$
6. $4\overline{)16.4}$	**7.** $2\overline{)1.7}$	**8.** $4\overline{)2.7}$	**9.** $8\overline{)2.0}$	**10.** $5\overline{)32.4}$
11. $24\overline{)55.2}$	**12.** $45\overline{)283.5}$	**13.** $36\overline{)27.0}$	**14.** $25\overline{)82.4}$	**15.** $81\overline{)52.65}$

Divide. Round each quotient to the nearest hundredth.

16. $3\overline{)94.1}$	**17.** $7\overline{)9.56}$	**18.** $6\overline{)42.83}$	**19.** $9\overline{)6.178}$	**20.** $6\overline{)59.378}$
21. $19\overline{)58.3}$	**22.** $32\overline{)5.11}$	**23.** $42\overline{)17.2}$	**24.** $183\overline{)26.8}$	**25.** $205\overline{)396.8}$

Problem Solving Using a graph

Can you give the value of the bills on which these portraits appear?

One thousand junior high school students were asked this question in a survey. The results are shown in the graph.

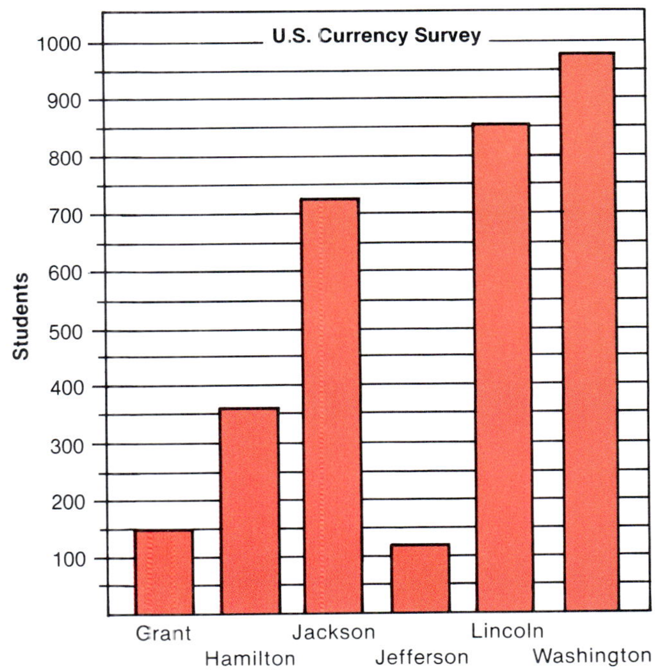

1. How many more students knew the value of Jackson's bill than knew the value of Hamilton's bill?

2. Whose portrait appears on the bill that was known by 125 students?

3. Which bills were known by more than $\frac{1}{2}$ of the students surveyed?

Use the clues to match the portrait with the value of the bill.

4. 700 more students knew the five-dollar bill than knew the fifty-dollar bill.

5. Fewer students knew the two-dollar bill than knew the one-dollar bill. These were the least- and best-known bills in the survey.

6. The ten-dollar bill
   a. was known by fewer than 400 students.
   b. was known by more students than knew Grant's bill.

7. The twenty-dollar bill
   a. was known by more students than knew Hamilton's bill.
   b. was known by fewer students than knew Lincoln's bill.

Dividing by a decimal

Below are two methods for finding this quotient: **5.4)9.18**

ESTIMATION METHOD

Step 1. Divide as whole numbers.

```
      17
5.4)9.18
   -54
    378
   -378
      0
```

Step 2. Estimate the quotient and place the decimal point.

```
      1.7
5.4)9.18
   -54
    378
   -378
      0
```

We can estimate the quotient by rounding the divisor and the dividend to the nearest whole numbers: 9 ÷ 5 is a little less than 2.

Since the quotient is a little less than 2, the decimal point goes between the 1 and the 7.

SHORTCUT METHOD

You can multiply both the divisor and the dividend by the same number without changing the quotient.

Step 1. To divide by a whole number, multiply both divisor and dividend by 10.

5.4)9.18

To multiply by 10, move both decimal points one place to the right.

Step 2. Place the decimal point for the quotient directly above the decimal point in the dividend.

5.4)9.18

Step 3. Divide.

```
      1.7
5.4)9.18
   -54
    378
   -378
      0
```

EXERCISES

Divide.

1. $0.06\overline{)0.0912}$
2. $0.8\overline{)19.68}$
3. $0.005\overline{)0.412}$
4. $0.04\overline{)1.408}$
5. $1.2\overline{)3.840}$
6. $3.4\overline{)53.04}$
7. $0.56\overline{)23.968}$
8. $0.74\overline{)18.722}$
9. $3.8\overline{)171.38}$
10. $0.025\overline{)5.0325}$
11. $0.46\overline{)22.862}$
12. $0.52\overline{)24.336}$
13. $13.5\overline{)434.7}$
14. $0.274\overline{)93.434}$
15. $82.9\overline{)5015.45}$

Divide. Round each quotient to the nearest hundredth.

16. $0.3\overline{)5.5}$
17. $0.6\overline{)3.61}$
18. $0.03\overline{)0.287}$
19. $0.09\overline{)0.537}$
20. $0.47\overline{)5.666}$
21. $0.035\overline{)63.8}$
22. $2.31\overline{)12.9}$
23. $4.52\overline{)1.46}$
24. $53.6\overline{)8.68}$

Problem Solving

Round each answer to the nearest hundredth.

25. Diane scored 263 points in 15 basketball games. How many points did she average per game?
26. Karen ran 9.2 kilometers in 46.92 minutes. What was the average number of kilometers per minute?
27. Jim swam 5.5 laps of the pool in 3.82 minutes. How many minutes per lap did he average?
28. John ran 40 meters in 4.4 seconds. At that rate, how long would it take him to run 100 meters?

Challenge!

Give each missing number.

29. 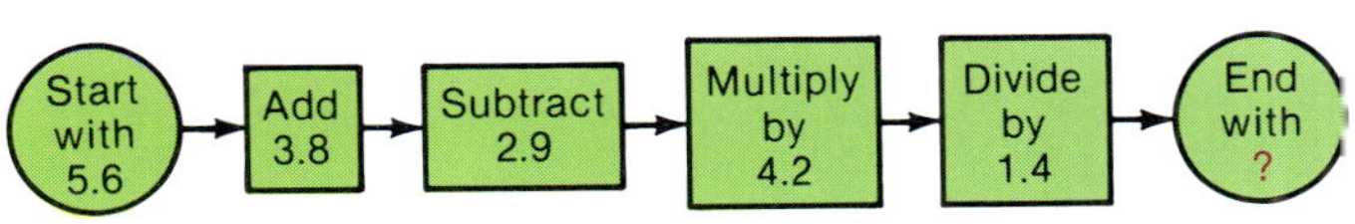

30. 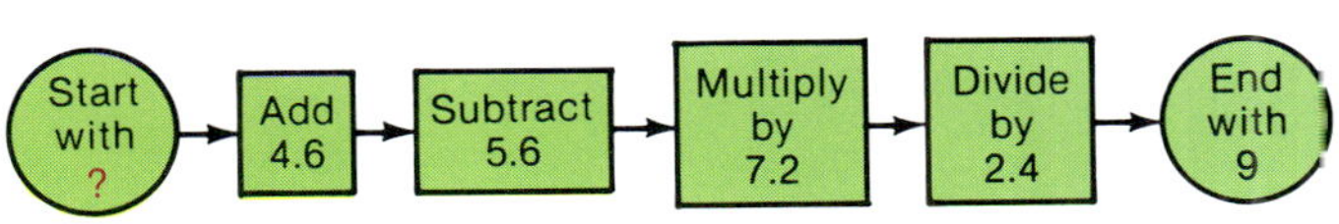

KEEPING SKILLS SHARP

Estimate to find the incorrect answer.

1. a. $47.2 + 7.31 = 54.51$
   b. $102.8 - 54.09 = 48.71$
   c. $39.02 + 111.1 = 150.12$
   d. $60.4 - 3.21 = 28.3$

Add.

2. $\begin{array}{r} 6.73 \\ +2.98 \\ \hline \end{array}$

3. $\begin{array}{r} 57.4 \\ +28.8 \\ \hline \end{array}$

4. $\begin{array}{r} 93.17 \\ 46.29 \\ +5.67 \\ \hline \end{array}$

Subtract.

5. $\begin{array}{r} 60.4 \\ -21.5 \\ \hline \end{array}$

6. $\begin{array}{r} 77.43 \\ -28.76 \\ \hline \end{array}$

7. $\begin{array}{r} 50.04 \\ -26.57 \\ \hline \end{array}$

8. $\begin{array}{r} 6.72 \\ -1.835 \\ \hline \end{array}$

Estimating

Calculator errors can be the result of pressing the wrong buttons or pressing the buttons in the wrong order. You can find calculator errors by <u>estimating</u>.

For example, if you round each number below as shown, you can tell that the calculator answer is wrong.

(48.9 × 2.1) + 58.75 = ?

50 × 2 + 60

The answer should be near 160.

Nine of these calculator answers are wrong. Find them by <u>estimating</u>.

1. 9.6 + 13.9 = 23.5
2. 36.74 − 10.31 = 26.43
3. 24.2 × 4 = 9.68
4. 50 ÷ 2.5 = 12.5
5. 75.16 × 9.7 = 729.052
6. 153.6 + 72.4 = 81.2
7. 30.6 ÷ 4.8 = 6.375
8. 204.6 − 129.8 = 174.8
9. (5.2 + 3.9) × 8 = 72.8
10. (19.8 − 4.9) × 4 = 9.1
11. (60 ÷ 0.5) + 17.8 = 29.8
12. (56 + 3.9) − 14.8 = 45.1
13. (2.8 × 4.2) + 5.9 = 593.84
14. (52 × 3.6) − 25.4 = 161.8
15. (12.6 − 3.5) × 15.6 = 251.16
16. (14.6 ÷ 7.3) × 9.8 = 196

Challenge!

Give each key ([×] or [÷]) you would push to get the answer.

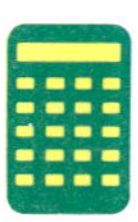

17. 19.2 □ 2.4 □ 1.2 [=] 9.6
18. 24.5 □ 1.75 □ 3.5 [=] 12.25
19. 12.5 □ 2.5 □ 1.5 [=] 7.5
20. 14.4 □ 3.6 □ 1.2 [=] 4.8
21. 7.56 □ 1.2 □ 6.3 [=] 1.44
22. 51.2 □ 6.4 □ 1.6 [=] 12.8
23. 17.28 □ 4.8 □ 3.6 [=] 12.96
24. 21.16 □ 9.2 □ 2.3 [=] 84.64

SKILL GAME

Rules

1. Player X (or O) chooses a red number and a blue number and estimates the product of the two numbers.
2. The player marks the number on the game board that is closest to the estimate.
3. Players take turns until one wins by getting 5 marks in any row, column, or diagonal.
4. One player may challenge the other player's choice by performing the exact multiplication and showing there is a closer choice that could be made. If the challenger is right, he/she gets to mark the square. If the challenger is wrong, he/she loses a turn.

Red	Blue
2.9	7.93
5.1	9.89
41.2	15.3
89.6	39.78
99.6	51.3

Baseball Game Board

44.37	28.681	115.362	148.77	985.044
407.468	1638.936	710.528	326.716	886.144
3564.288	4596.48	50.439	1370.88	3962.088
5109.48	789.828	1523.88	22.997	261.63
40.443	78.03	2113.56	630.36	202.878 (marked X)

EXAMPLE. Choose 5.1 and 39.78.
To estimate the product:

39.78	Round to	40
×5.1	Round to	×5
		200

Mark 202.878 on the game board, since it is closest to 200, the estimated product.

Problem Solving Estimating in different situations

Work in groups. Read each estimation case and discuss what kind of rounding and estimation you would do, if any, to make the decision.

1. You are helping out your family with the shopping. You have only a $20 bill, and so far you have put these items in your cart. You need to decide whether you will have enough money to pay for them. There is a 5% sales tax on all nonfood items.

2. You are managing a popular music group, The Toes. Fans have bought only 6,972 tickets so far for the concert two weeks away. For previous Toes concerts, only 1/10 of the tickets sold were bought in the two weeks before the performance. Tickets cost $25 each. Expenses for the trip and concert will come to about $83,500. In addition, renting the concert hall will cost $25,000 plus 1/5 of the ticket sales. You need to decide whether or not to go ahead with the concert.

3. You are in charge of purchasing for a large clothing store. For one department, you buy plain sweatshirts that the store then imprints with different messages. It is time to order more. You like to order the sweatshirts about once a month. Each size comes in boxes of 48 shirts to a box. You need to decide how many boxes of each size to order.

Last Year's Sweatshirt Sales	
Small	691
Medium	1245
Large	877

4. You are driving to an important meeting in Chanceville, but you may need more gas. The last time you measured your car's fuel consumption, it was "32 point something" miles per gallon. The owner's manual for the car says the tank holds ten gallons. You need to decide whether or not to stop for gas at the next exit, which may be your last chance before Chanceville.

5. You are planning to tile the floor of a school cafeteria with rectangular tiles that are 9 inches by 12 inches. The rectangular cafeteria is 178 feet long and 44 feet wide. The tiles come in crates of 500 tiles. You need to decide how many crates to order.

6. Discuss your results and methods with another group.
   a. For which cases was there the most disagreement? Why?
   b. Was a calculator useful for any of these cases? Why or why not?

Exponents

Study the three ways of writing the number of blocks.

16

$\mathbf{4 \times 4}$

$\mathbf{4^2}$

Read "4^2" as "4 squared" or "4 to the second power."

64

$\mathbf{4 \times 4 \times 4}$

$\mathbf{4^3}$

Read "4^3" as "4 cubed" or "4 to the third power."

Notice that instead of writing 4 as a factor several times, you can use an exponent. The exponent tells you how many times the base, 4, was used as a factor.

Study these examples.

$\mathbf{2 \times 2 \times 2 \times 2 \times 2 = 2^5}$

2 is the base.
5 is the exponent.

Read "2^5" as "2 to the fifth power."

$\mathbf{3 = 3^1}$

Read "3^1" as "3 to the first power."

Special Agreement

Any number (not zero) raised to the 0 power is equal to 1.

$\mathbf{8^0 = 1}$ $\mathbf{5^0 = 1}$ $\mathbf{10^0 = 1}$

EXERCISES

Write in exponential form.

1. $2 \times 2 \times 2$
2. $4 \times 4 \times 4 \times 4$
3. $5 \times 5 \times 5 \times 5 \times 5 \times 5$
4. $8 \times 8 \times 8 \times 8$
5. $1 \times 1 \times 1 \times 1 \times 1$
6. 9×9
7. $4 \times 4 \times 4$
8. $7 \times 7 \times 7 \times 7 \times 7$
9. 2×2
10. $8 \times 8 \times 8$
11. 3×3
12. $6 \times 6 \times 6 \times 6 \times 6$
13. $10 \times 10 \times 10 \times 10$
14. $3 \times 3 \times 3 \times 3 \times 3 \times 3$
15. 10×10
16. $11 \times 11 \times 11 \times 11$
17. $1 \times 1 \times 1 \times 1 \times 1 \times 1$
18. $11 \times 11 \times 11$
19. $2 \times 2 \times 2 \times 2 \times 2$
20. $10 \times 10 \times 10$
21. $12 \times 12 \times 12$

Write the standard numeral.

22. 6^0
23. 2^4
24. 7^2
25. 3^0
26. 6^4
27. 8^4
28. 3^3
29. 3^1
30. 1^5
31. 7^3
32. 3^5
33. 2^5
34. 1^6
35. 6^3
36. 9^4
37. 2^3
38. 9^3
39. 1^8
40. 4^5
41. 2^6
42. 8^0
43. 3^2
44. 2^8
45. 3^4
46. 10^0
47. 10^1
48. 10^2
49. 10^3
50. 10^4
51. 10^5
52. eight squared
53. four to the fifth power
54. six to the first power
55. two to the eighth power
56. one to the seventh power
57. five to the fourth power
58. seven to the zero power
59. three to the sixth power
60. nine cubed
61. eight to the zero power
62. twelve squared
63. fifteen cubed

Challenge!

Solve. [*Hint:* Look for a pattern.]

64. $2^1 = 2$

$2^2 = 4$

$2^3 = 8$

$2^4 = 16$

$2^5 = 32$

.
.
.

$2^{40} = 1{,}099{,}511{,}627{,}77?$

What is the missing digit?

65. $1^3 = ?$

$1^3 + 2^3 = ?$

$1^3 + 2^3 + 3^3 = ?$

$1^3 + 2^3 + 3^3 + 4^3 = ?$

$1^3 + 2^3 + 3^3 + 4^3 + 5^3 = ?$

$1^3 + 2^3 + 3^3 + 4^3 + 5^3 + 6^3 = ?$

Multiplying and dividing by 10, 100, or 1000

If you multiply a number by 10, 100, or 1000, the product is greater than the number.

SHORTCUT
MENTAL MATH

Remember that the product must be greater. So, to multiply by 10, move the decimal point 1 place to the right.

× 10 $5.92 \times 10 = 5\ 9.2$

× 100 moves the decimal point 2 places to the right.

$12.384 \times 100 = 12\ 38.4$

× 1000 moves the decimal point 3 places to the right.

$42.79 \times 1000 = 42\ 790$

If you divide a number by 10, 100, or 1000, the quotient is less than the number.

SHORTCUT
MENTAL MATH

Remember that the quotient must be less. So, to divide by 10, move the decimal point 1 place to the left.

÷ 10 $64.5 \div 10 = 6.4\ 5$

÷ 100 moves the decimal point 2 places to the left.

$65.2 \div 100 = 0.65\ 2$

÷ 1000 moves the decimal point 3 places to the left.

$75.83 \div 1000 = 0.075\ 83$

Notice that in each example the number of zeros in the divisor tells how many places to move the decimal point.

EXERCISES

Give each product.

1. 67×1000
2. 67×100
3. 67×10
4. 8.63×100
5. 5.942×1000
6. 7.638×10
7. 59.76×100
8. 317.4×10
9. 29.38×1000
10. 0.0356×10
11. 0.004×100
12. 0.03285×1000

Give each quotient.

13. $7594 \div 1000$
14. $7594 \div 100$
15. $7594 \div 10$
16. $5.9 \div 10$
17. $5.9 \div 100$
18. $5.9 \div 1000$
19. $382 \div 10$
20. $1.6 \div 100$
21. $7903 \div 1000$
22. $0.04 \div 10$
23. $67.93 \div 1000$
24. $0.8 \div 100$

Simplify.

25. 38×10
26. $6.8 \div 10$
27. 2.91×100
28. $0.56 \div 10$
29. 153×10
30. $12.52 \div 100$
31. 8.159×1000
32. 3946×10
33. $0.71 \div 1000$
34. $5274 \div 1000$
35. $6.74 \div 10$
36. 4.238×1000
37. 75×100
38. $9.4 \div 100$
39. 429×1000
40. $13.78 \div 1000$
41. 7.361×100
42. $5.714 \div 100$

Challenge!

Here is how to write a decimal as an expanded numeral.

823.65

$$\boxed{800} + \boxed{20} + \boxed{3} + \boxed{0.6} + \boxed{0.05}$$

expanded numeral ⟶ $(8 \times 100) + (2 \times 10) + (3 \times 1) + (6 \times 0.1) + (5 \times 0.01)$

Give the expanded numeral.

43. 9.74
44. 28.653
45. 792.153
46. 946.382
47. 3.6598
48. 3659.8
49. 42.953
50. 4.2953

Multiplying and dividing by powers of 10

> *When 10 is the base, the exponent tells the number of zeros in the standard numeral.*

Study these examples.

$10^1 = 10$

$10^2 = 10 \times 10 = 100$

$10^3 = 10 \times 10 \times 10 = 1000$

When you are multiplying by a power of 10, the exponent tells you how many places to move the decimal point to the right.

$5.634 \times 10^2 = ?$

$5.634 \times 10^2 = 5\,63.4$ *Move right 2 places.*

When you are dividing by a power of 10, the exponent tells you how many places to move the decimal point to the left.

$923.4 \div 10^3 = ?$

$923.4 \div 10^3 = 0.923\,4$ *Move left 3 places.*

EXERCISES

Give the standard numeral.

1. 10^2
2. 10^4
3. 10^3
4. 10^5
5. 10^6
6. 10^8
7. 10^1

Write as a power of 10.

8. 1000
9. 10,000
10. 1,000,000
11. 100,000
12. 10,000,000
13. 100,000,000

Give each product.

14. 3.56×10
15. 0.758×100
16. 0.08652×1000
17. $0.00825 \times 10{,}000$
18. 9.62×10^2
19. 17.84×10^1
20. 0.2971×10^2
21. 0.03825×10^3
22. 0.0563×10^5
23. 1.7×10^4
24. 0.816×10^6
25. 0.0936×10^8
26. 0.031×10^4
27. 6.529×10^7

Give each quotient.

28. $29.6 \div 10$
29. $7542 \div 100$
30. $618.3 \div 1000$
31. $51.6 \div 10{,}000$
32. $2.9 \div 10^1$
33. $74.8 \div 10^3$
34. $319.5 \div 10^2$
35. $764.82 \div 10^5$
36. $5361.7 \div 10^4$
37. $28.14 \div 10^6$
38. $89.6 \div 10^8$
39. $713.4 \div 10^7$

Challenge!

Here is how to use powers of 10 to write a standard numeral as an expanded numeral:

$$397 = (3 \times 10^2) + (9 \times 10^1) + (7 \times 10^0)$$

Write each standard numeral as an expanded numeral.

40. 72
41. 87
42. 256
43. 423
44. 781
45. 5293
46. 2817
47. 64,318
48. 93,629

KEEPING SKILLS SHARP

Estimate to find the incorrect answer.

1. a. $3.67 \times 5.1 = 187.2$

   b. $179.52 \div 3.2 = 56.1$

Multiply.

2. 924×0.8
3. 359×0.06
4. 27.4×1.2
5. 82.06×0.35
6. 6.581×3.04

Divide. Round the quotient to the nearest hundredth.

7. $0.3\overline{)382}$
8. $0.06\overline{)0.394}$
9. $1.9\overline{)5.29}$
10. $3.8\overline{)89.64}$
11. $0.47\overline{)3}$
12. $8.4\overline{)2.9}$

Large numbers and scientific notation

You can think of the planet Earth as a spinning "spaceship" that takes us around the sun once every year. Each day it travels about 2,570,000 km. Here is that number written in **scientific notation:**

a number between 1 and 10 $\longrightarrow$ $\mathbf{2.57 \times 10^6}$ $\longleftarrow$ *a power of 10*

Here is how to write a standard numeral in scientific notation.

EXAMPLE 1. **739,000,000**

Step 1. Locate the decimal point for a number between 1 and 10. **7.39000000**

Step 2. To find the exponent of 10, count the digits that are to the right of the decimal point. **7.39000000**

Step 3. Write the number as the product of a number between 1 and 10 and a power of 10. $\mathbf{7.39 \times 10^8}$

A number in scientific notation can be written as a standard numeral by multiplying.

EXAMPLE 2. $\mathbf{8.11 \times 10^5}$

$$8.11 \times 10^5 = 8\,11000$$
$$= 811{,}000$$

Move the decimal point 5 places to the right.

EXERCISES

Write the standard numeral.

1. 5.3×10^2
2. 2.6×10^3
3. 7.4×10^1
4. 9.3×10^5
5. 8.92×10^3
6. 5.66×10^4
7. 4.93×10^5
8. 7.4×10^2
9. 9.31×10^6
10. 1.6×10^5
11. 2.38×10^3
12. 3.02×10^4
13. 5.381×10^7
14. 3.014×10^8
15. 8.342×10^6
16. 5.671×10^9

Write in scientific notation.

17. 89
18. 70
19. 520
20. 880
21. 9600
22. 15,000
23. 362,000
24. 58,000,000
25. 291,000,000
26. 42,800,000,000
27. 4,740,000,000,000
28. 90,000,000,000,000

Write each number in scientific notation.

29. At the equator, the earth is spinning at about 1600 kilometers per hour.
30. The earth stays about 150,000,000 kilometers from the sun.
31. The age of the earth is about 4,500,000,000 years.
32. The earth weighs about 5,990,000,000,000,000,000,000 metric tons.
33. The diameter of the sun is about 1,380,000 kilometers.
34. The temperature at the center of the sun is about 20,000,000 degrees Celsius.
35. The earth receives approximately 170,000,000,000,000 kilowatts of power per hour from the sun.
36. The star closest to the earth is about 40,700,000,000,000 kilometers away.

Small numbers and scientific notation

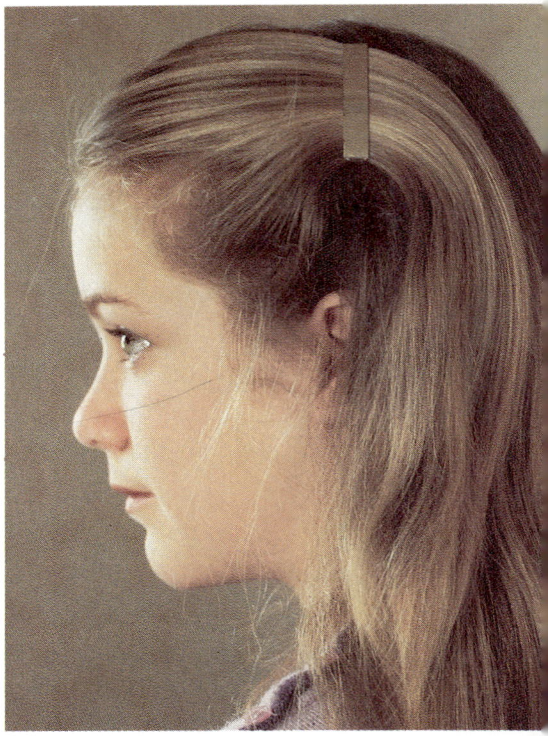

Negative exponents are used to write numbers less than 1 in scientific notation.

$$0.1 = 10^{-1}$$
$$0.01 = 10^{-2}$$
$$0.001 = 10^{-3}$$
$$0.0001 = 10^{-4}$$
$$0.00001 = 10^{-5}$$

Read as "10 to the negative 1 power."

The thickness of a human hair is about 0.000093 meter. Here is the number written in scientific notation:

A number between 1 and 10 ⟶ 9.3×10^{-5} ⟵ *a power of 10*

Here is how to write a standard numeral in scientific notation.

EXAMPLE 1. **0.00076**

Step 1. Locate the decimal point for a number between 1 and 10. — 0.00076

Step 2. To find the negative exponent of 10, count the digits that are between the decimal points. — 0.00076 (4) *There are 4 digits.*

Step 3. Write the number as the product of a number between 1 and 10 and a power of 10. — 7.6×10^{-4}

A number in scientific notation can be written as a standard numeral by multiplying.

EXAMPLE 2. 6.49×10^{-3}

$$6.49 \times 10^{-3} = 0.00649$$

Move the decimal point 3 places to the left.

EXERCISES

Write the standard numeral.

1. 3.4×10^{-3}	**2.** 2.9×10^{-2}	**3.** 7.4×10^{-1}	**4.** 6.9×10^{-5}
5. 2.81×10^{-2}	**6.** 4.293×10^{-4}	**7.** 3.6×10^{-3}	**8.** 6.08×10^{-5}
careful! **9.** 4.003×10^{-3}	**10.** 2.44×10^{-4}	**11.** 8.33×10^{-5}	**12.** 8.43×10^{-2}
13. 5.1×10^{3}	**14.** 8.16×10^{-4}	**15.** 7.09×10^{2}	**16.** 3.2×10^{1}
17. 2.683×10^{-5}	**18.** 4.93×10^{4}	**19.** 7.004×10^{-6}	**20.** 5.96×10^{3}

Write in scientific notation.

21. 0.0052	**22.** 0.059	**23.** 0.00076	**24.** 0.000017
25. 0.0137	**26.** 0.000043	**27.** 0.00846	**28.** 0.00094
29. 0.0664	**30.** 0.0047	**31.** 0.0000318	**32.** 0.0381
careful! **33.** 18,000	**34.** 23,421	**35.** 0.06890	**36.** 0.00117
37. 326,000	**38.** 0.0051	**39.** 436,800	**40.** 3,520,000

Computers are programmed to use scientific notation with large and small numbers. If you input PRINT 40000 * 60000, you might expect your computer to output 2400000000. Instead, it will probably print something like this:

2.4E+09

This means that 2.4 is to be multiplied by 10 with an exponent of $^{+}9$ to get the standard numeral.

$$2.4 \times 10^9 = 2{,}400{,}000{,}000$$

Translate from computer-style scientific notation to standard numerals.

41. 2.4E+Ø2	**42.** 3.ØE−Ø4	**43.** 1.65E+12	**44.** 2.Ø6E−Ø5
45. 9.8E−Ø3	**46.** 1.943E+Ø3	**47.** 7.ØØ82E−Ø3	**48.** 4.33E+Ø9

Write these numbers in computer-style scientific notation.

49. 0.0063	**50.** 9,832,000	**51.** 19,035,000,000	**52.** 0.036
53. 425,000,000,000	**54.** 0.0000032	**55.** 0.00459	**56.** 706,300,000

Problem Solving More than one way to solve

1. Study and understand.
2. Plan and do.
3. Answer and check.

Sometimes there is more than one way to solve a problem.

Margo's Deluxe photo album holds 8 photos per page. Her "Slim Jim" album holds 6 photos per page. She has 20 pages of each album filled. How many photos does Margo have?

Jennifer's way to solve

$$\begin{array}{r} 20 \\ \times 8 \\ \hline 160 \end{array}$$ photos in Deluxe album

$$\begin{array}{r} 20 \\ \times 6 \\ \hline 120 \end{array}$$ photos in "Slim Jim" album

$$\begin{array}{r} 160 \\ +120 \\ \hline 280 \end{array}$$ photos altogether

Paul's way to solve

$$\begin{array}{r} 8 \\ +6 \\ \hline 14 \end{array}$$ photos on each page of both albums

$$\begin{array}{r} 14 \\ \times 20 \\ \hline 280 \end{array}$$ photos on the 20 pages of the two albums

EXERCISES

One way to solve each problem is given. Find another way to solve it.

1. Margo bought 3 rolls of 12-exposure film and 3 rolls of 20-exposure film. How many pictures can she take?

$$\begin{array}{r} 12 \\ \times 3 \\ \hline 36 \end{array} \qquad \begin{array}{r} 20 \\ \times 3 \\ \hline 60 \end{array} \qquad \begin{array}{r} 36 \\ +60 \\ \hline 96 \end{array}\ \text{pictures}$$

2. Tim bought a 12-exposure roll of film for $2.40 and 12 flashbulbs for $3.00. How much will it cost to take one flash picture?

$$\begin{array}{r} \$2.40 \\ +\ \$3.00 \\ \hline \$5.40 \end{array} \qquad \begin{array}{r} \$\ .45 \\ 12\,)\overline{\$5.40} \\ -48 \\ \hline 60 \\ -60 \\ \hline 0 \end{array}$$

Solve.

3. A 12-exposure roll of film costs $2.40 and a 20-exposure roll of film costs $3.10. Robin bought 2 rolls of each. How much money did she spend?

4. Jim bought a 20-exposure roll of film for $3.10 and a carton of flash cubes for $4.80. He gave the clerk a $20 bill. How much change should Jim get back?

5. Dana had $50. She bought an Instant Pix camera for $29.89 and a carrying case for $5.29. How much money did she have left?

6. Dana wants to use her new camera to take a picture of each of the 30 students in her homeroom. A 10-picture cartridge of film for the camera costs $7.30. How much money will Dana spend for the film?

7. Juan had 2 rolls of 12-exposure film and 1 roll of 20-exposure film developed and printed. How much change did he get back from a $20 bill?

8. How much less does it cost to develop and print 2 rolls of 24-exposure film than 4 rolls of 12-exposure film?

9. Which is less expensive to develop and print, 2 rolls of 24-exposure film or 1 roll of 12-exposure film and 1 roll of 36-exposure film? How much less expensive?

Price List for Developing and Printing

Exposures	*Cost*
12	$3.69
20	$4.89
24	$5.59
36	$7.29

Chapter Checkup

Write the standard numeral. [pages 20–21]

1. eight tenths
2. six and four hundredths
3. one hundred and forty-two thousandths
4. nine and sixteen thousandths

Round 42.9856028 to the nearest [pages 22–23]

5. tenth.
6. ten-thousandth.
7. hundredth.

< or >? [pages 24–25]

8. 38.296 ● 38.290
9. 1.536 ● 15.32
10. 0.1 ● 0.098
11. 0.499 ● 0.5

Add or subtract. [pages 26–27]

12. $\begin{array}{r} 9.7 \\ 3.62 \\ +7.8 \\ \hline \end{array}$

13. $\begin{array}{r} 19.08 \\ -5.63 \\ \hline \end{array}$

14. 28.41 + 5.9

15. 182.6 − 13.97

Multiply. [pages 28–31]

16. $\begin{array}{r} 6.21 \\ \times 8 \\ \hline \end{array}$

17. $\begin{array}{r} 72.51 \\ \times 4.3 \\ \hline \end{array}$

18. $\begin{array}{r} 0.86 \\ \times 5.5 \\ \hline \end{array}$

19. $\begin{array}{r} 0.04 \\ \times 0.02 \\ \hline \end{array}$

Divide. Round quotients to the nearest hundredth.
[pages 32, 34–35]

20. $5\overline{)3.2}$
21. $0.08\overline{)2.3}$
22. $2.4\overline{)52}$
23. $0.58\overline{)3.77}$

Write the standard numeral. [pages 40–49]

24. 4^3
25. 10^2
26. 0.86×10^4
27. $0.4 \div 10^2$
28. 9.52×10^3
29. 7.4×10^2
30. 5.67×10^{-3}
31. 4.2×10^{-4}

Solve. [pages 33, 38–39, 50–51]

32. One of the world's most valuable stamps is an 1856 British Guiana 1¢ stamp. It sold in 1917 for $32,250. The present owner paid $280,000. How much did the price increase from 1917 to the present?

33. A stamp dealer bought 24 stamps for $160. He sold 17 of them for $12.45 each and the remaining stamps for $10.75 each. What was the total profit?

Chapter Project

DOING A SIMPLER PROBLEM

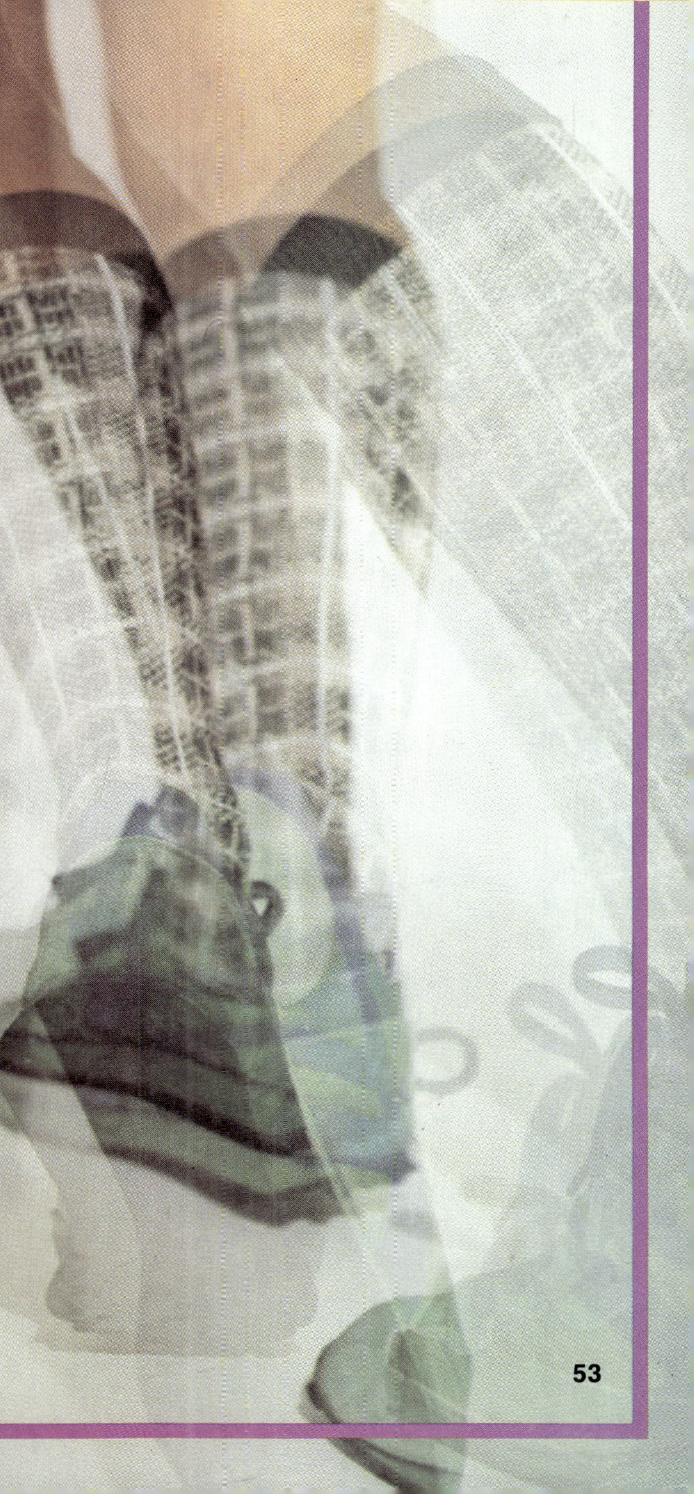

1. How long does it take to walk **50** meters at a normal pace? Give your answer to the nearest tenth second.
2. Compute how long it would take to walk a kilometer. Give your answer in minutes and seconds.
3. Find out how many kilometers it is from Miami, Florida, to Seattle, Washington, and compute how long it would take you to walk from Miami to Seattle.
4. If you walked **8** hours a day, how many days would it take you?

Chapter Review

thousandths place

200.016

two hundred *and* *sixteen thousandths*

Round to the nearest tenth.

tenths place

57.25

5 or greater round up

57.3

$$\begin{array}{r} 5.260 \\ +2.784 \\ \hline 8.044 \end{array} \qquad \begin{array}{r} 4.590 \\ -1.782 \\ \hline 2.808 \end{array}$$

$$\begin{array}{rl} 2.15 & \leftarrow \text{2 digits} \\ \times 3.4 & \leftarrow \text{1 digit} \\ \hline 860 & \\ 645 & \\ \hline 7.310 & \leftarrow \text{3 digits} \end{array}$$

$$\begin{array}{r} 0.27 \\ 4.3 \overline{)1.1\,61} \\ -8\,6 \\ \hline 3\,01 \\ -3\,01 \\ \hline 0 \end{array}$$

Write in words.

1. 0.4
2. 0.06
3. 0.007
4. 50.123
5. 700.015
6. 85.1375

Round to the nearest tenth.

7. 4.62
8. 20.75
9. 19.97
10. 0.162
11. 1.985
12. 4.4499

Add or subtract.

13. $\begin{array}{r} 1.67 \\ +3.29 \\ \hline \end{array}$

14. $\begin{array}{r} 14.26 \\ +2.984 \\ \hline \end{array}$

15. $\begin{array}{r} 2.58 \\ 3.77 \\ +4.81 \\ \hline \end{array}$

16. $\begin{array}{r} 2.35 \\ -1.462 \\ \hline \end{array}$

17. $\begin{array}{r} 60.04 \\ -7.56 \\ \hline \end{array}$

18. $\begin{array}{r} 325.106 \\ -148.15 \\ \hline \end{array}$

Multiply.

19. $\begin{array}{r} 5.7 \\ \times 3.4 \\ \hline \end{array}$

20. $\begin{array}{r} 0.62 \\ \times 3.4 \\ \hline \end{array}$

21. $\begin{array}{r} 0.04 \\ \times 0.3 \\ \hline \end{array}$

Divide.

22. $0.4\overline{)8.1}$
23. $0.08\overline{)3.3}$
24. $2.5\overline{)74}$

Chapter Challenge

FINDING A PATTERN

Suppose that you were offered a job that paid 1¢ the first day, 2¢ the second day, 4¢ the third day, and so on.

Copy and complete this table.

1.

Day	Salary (in cents)	
	Standard numeral	*Power of 2*
1	1	2^0
2	2	2^1
3	4	2^2
4	8	?
5	?	?
6	?	?
7	?	?

2. How much would you earn the first week (7 days)?
3. Guess how much you would earn during the second week.
4. Compute your earnings for the second week. Was your guess too small or too large?
5. How many days would you have to work to earn a total of $1000?
6. *Fact:* $2^{20} = 1{,}048{,}576$.
   Use the above fact to compute your earnings for
   (a) your 20th day.
   (b) your 22nd day.

Cumulative Checkup

STANDARDIZED FORMAT

Choose the correct letter.

1. The standard numeral for six hundred and fifty-four thousandths is

a. 0.654
b. 600.054
c. 0.0654
d. none of these

2. 86.5 rounded to the nearest whole number is

a. 86
b. 85
c. 86.5
d. none of these

3. 3.7249 rounded to the nearest hundredth is

a. 3.73
b. 3.725
c. 3.72
d. none of these

4. Add.

9.28 + 63.975

a. 72.155
b. 63.903
c. 73.255
d. none of these

5. Subtract.

4.036 − 0.2574

a. 3.7786
b. 1.462
c. 3.8786
d. none of these

6. Multiply.

6.5×2.46

a. 1.599
b. 15.99
c. 16.09
d. none of these

7. Divide.

14.224 ÷ 2.54

a. 56
b. 5.58
c. 5.6
d. none of these

8. Divide and round to the nearest hundredth.

26.359 ÷ 0.021

a. 1255.19
b. 1255.20
c. 125.52
d. none of these

9. 970,000 = ?

a. 0.97×10
b. 9.7×10^4
c. 9.7×10^5
d. none of these

10. 0.00047 = ?

a. 4.7×10^4
b. 4.7×10^3
c. 4.7×10^5
d. none of these

11. A 12-ounce box of fish costs $1.62. What is the cost of a 6-ounce serving?

a. 81¢ **b.** 84¢
c. $9.72 **d.** none of these

12. An 8-ounce box of beans costs 84¢. What is the cost of 4 boxes?

a. 42¢ **b.** 44¢
c. $3.36 **d.** none of these

2

Integers and Equations

The wind makes the air feel colder than the thermometer indicates. When the photograph was taken, the windchill was 39° lower than the thermometer reading. What was the thermometer reading at the time?

Positive and negative integers

We can use **positive** and **negative integers** to measure quantities that have direction.

On one passing play the Richmond Eagles gained 17 yards. A gain of 17 yards can be represented by

$$^{+}17$$

Read as "positive 17."

On one running play they lost 6 yards. A loss of 6 yards can be represented by

$$^{-}6$$

Read as "negative 6."

Any time we work with **directed** quantities, we can use positive integers for quantities with one direction and negative integers for quantities with the opposite direction.

Positive and negative integers are shown on the number line below.

$^{-}7 \quad ^{-}6 \quad ^{-}5 \quad ^{-}4 \quad ^{-}3 \quad ^{-}2 \quad ^{-}1 \quad 0 \quad ^{+}1 \quad ^{+}2 \quad ^{+}3 \quad ^{+}4 \quad ^{+}5 \quad ^{+}6 \quad ^{+}7$

The integers to the right of 0 are positive.
The integers to the left of 0 are negative.
Zero is neither positive nor negative.

Integers can also represent electrical charges. Imagine containers that have mixtures of positive and negative charges in them. The charges form as many positive-negative pairs as possible. The pairs have no charge. Count the unpaired charges to find the charge in the container.

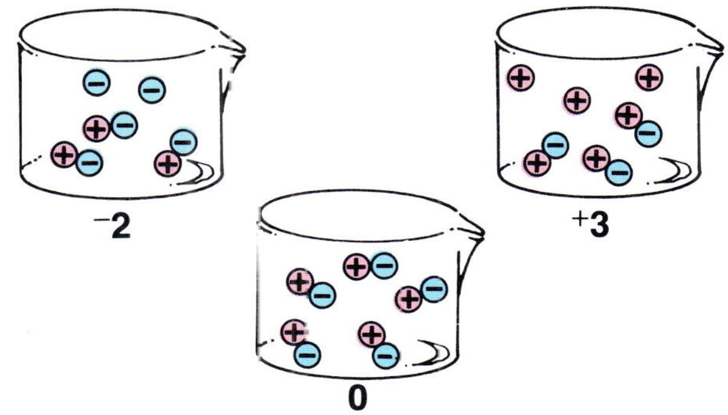

EXERCISES

If $^{-}8°$ represents 8° below zero, then what integer would represent

1. 7 degrees above zero?
2. 15 degrees below zero?

If $^{+}2$ represents being 2 hours ahead of schedule, then what integer would represent

3. 4 hours ahead of schedule?
4. 6 hours behind schedule?
5. neither ahead of nor behind schedule?

Use integers to represent the charges in these containers.

6. 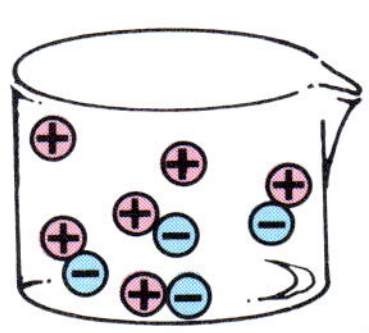

7. 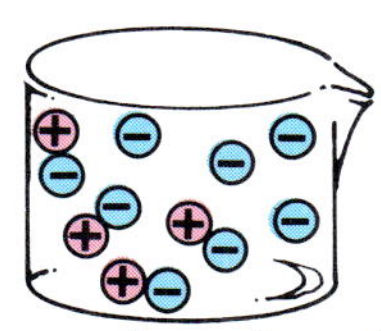

8. 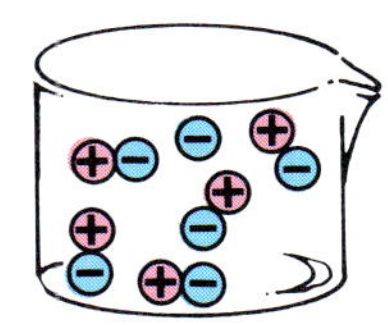

The number line can be used to show the order of positive and negative integers. An integer is less than every integer to its right and greater than every integer to its left.

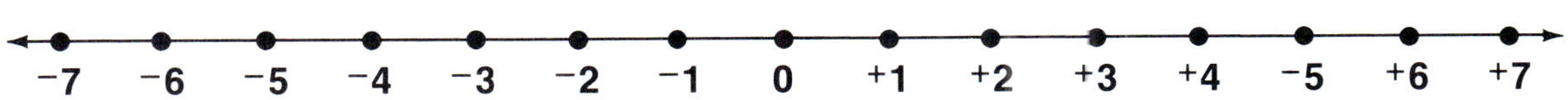

$<$ or $>$?

9. $^{-}2$ (?) $^{-}3$
10. $^{+}2$ (?) $^{+}3$
11. $^{-}3$ (?) $^{+}2$
12. $^{+}3$ (?) $^{-}2$
13. $^{+}1$ (?) 0
14. $^{-}4$ (?) $^{-}2$
15. $^{+}5$ (?) $^{-}2$
16. 0 (?) $^{-}2$
17. $^{-}9$ (?) $^{-}7$
18. $^{+}2$ (?) $^{+}7$
19. $^{-}13$ (?) $^{-}14$
20. $^{-}16$ (?) $^{-}12$
21. $^{-}10$ (?) $^{+}10$
22. $^{-}6$ (?) $^{-}10$
23. 0 (?) $^{+}7$
24. $^{-}15$ (?) $^{-}20$

Adding integers

To understand how to add, you can think about putting charges together.

EXAMPLE 1. $^{+}4 + {}^{+}2 = ?$

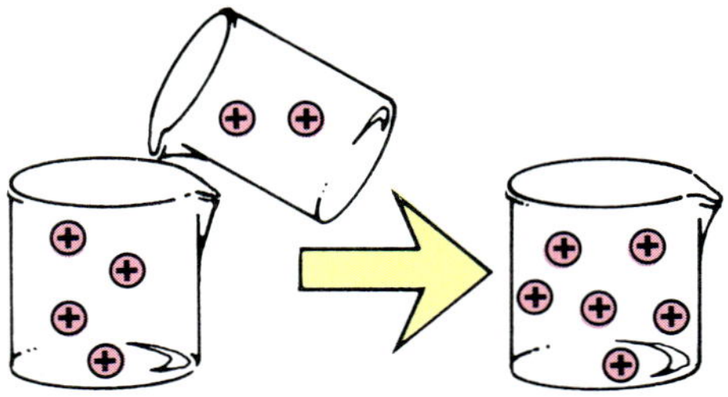

$^{+}4 + {}^{+}2 = {}^{+}6$

EXAMPLE 2. $^{-}2 + {}^{-}3 = ?$

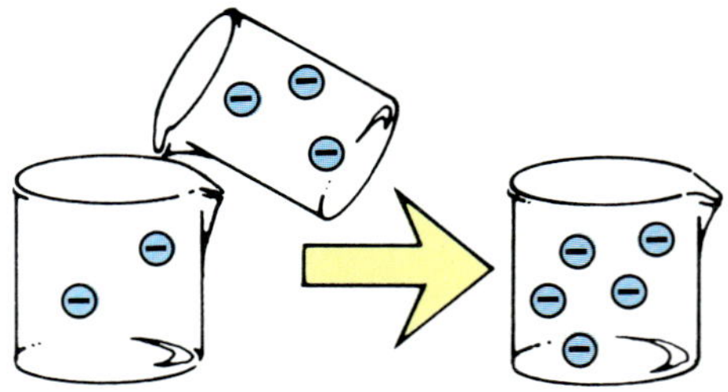

$^{-}2 + {}^{-}3 = {}^{-}5$

EXAMPLE 3. $^{+}5 + {}^{-}3 = ?$

$^{+}5 + {}^{-}3 = {}^{+}2$

EXAMPLE 4. $^{-}5 + {}^{+}2 = ?$

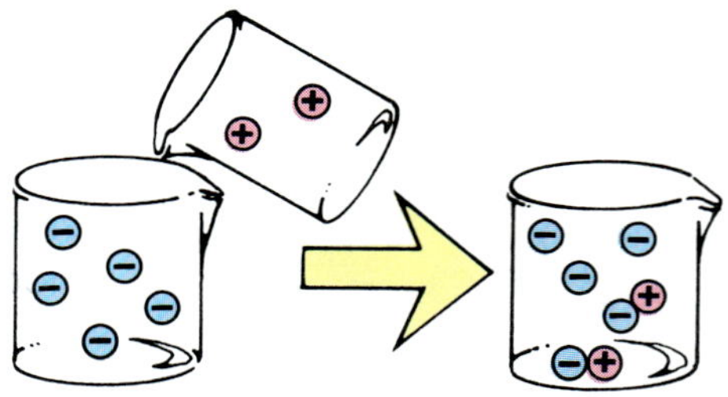

$^{-}5 + {}^{+}2 = {}^{-}3$

EXAMPLE 5. $^{+}4 + {}^{-}4 = ?$

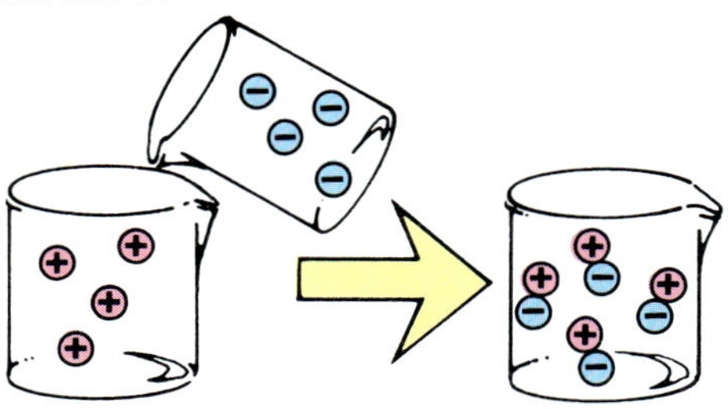

$^{+}4 + {}^{-}4 = 0$

$^{+}4$ and $^{-}4$ are called **opposites,** because their sum is 0.

$^{-}16$ is the opposite of $^{+}16$. 0 is its own opposite.

EXERCISES

Give each sum.

1. $^{+}5 + {}^{+}2$	**2.** $^{-}5 + {}^{-}2$	**3.** $^{+}5 + {}^{-}2$	**4.** $^{-}5 + {}^{+}2$
5. $^{+}6 + {}^{+}8$	**6.** $^{-}6 + {}^{-}8$	**7.** $^{+}6 + {}^{-}8$	**8.** $^{-}6 + {}^{+}8$
9. $^{+}7 + {}^{+}5$	**10.** $^{-}7 + {}^{-}5$	**11.** $^{+}7 + {}^{-}5$	**12.** $^{-}7 + {}^{+}5$
13. $^{+}4 + {}^{+}6$	**14.** $^{-}4 + {}^{-}6$	**15.** $^{+}4 + {}^{-}6$	**16.** $^{-}4 + {}^{+}6$
17. $^{+}9 + 0$	**18.** $^{-}9 + 0$	**19.** $^{+}4 + 0$	**20.** $^{-}4 + 0$
21. $^{+}18 + {}^{+}12$	**22.** $^{-}18 + {}^{-}12$	**23.** $^{+}18 + {}^{-}12$	**24.** $^{-}18 + {}^{+}12$
25. $^{-}8 + {}^{+}7$	**26.** $^{-}8 + {}^{-}7$	**27.** $^{+}8 + {}^{-}7$	**28.** $^{+}8 + {}^{+}7$
29. $^{-}16 + {}^{-}15$	**30.** $^{+}16 + {}^{-}15$	**31.** $^{-}16 + {}^{+}15$	**32.** $^{+}16 + {}^{+}15$
33. $^{-}12 + {}^{+}6$	**34.** $^{-}19 + {}^{-}6$	**35.** $^{+}8 + {}^{-}13$	**36.** $^{+}20 + {}^{+}12$
37. $^{+}35 + {}^{-}10$	**38.** $^{-}22 + {}^{-}8$	**39.** $^{+}14 + {}^{-}14$	**40.** $^{+}15 + {}^{-}24$
41. $^{-}38 + {}^{+}69$	**42.** $0 + {}^{-}75$	**43.** $^{+}85 + {}^{-}34$	**44.** $^{-}24 + {}^{+}25$

True or false?

45. The sum of two positive integers is always positive.

46. The sum of two negative integers is always positive.

47. The sum of a positive integer and a negative integer is always positive.

48. The sum of a positive integer and a negative integer is always negative.

49. The sum of a positive integer and a negative integer may be positive, negative, or 0.

50. The sum of two opposites is 0.

Give the opposite of each integer.

51. $^{+}6$ **52.** $^{-}23$ **53.** $^{+}62$ **54.** $^{-}37$ **55.** 0 **56.** $^{+}7082$

Subtracting integers

You can understand how to subtract by thinking about taking charges out of a container.

EXAMPLE 1. $^{+}4 - {}^{+}1 = ?$

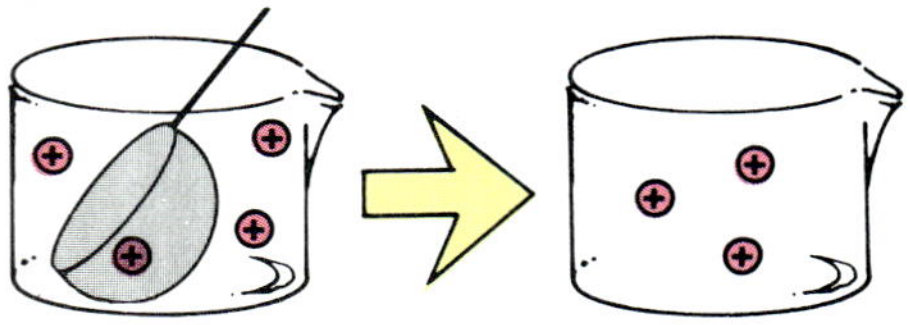

$^{+}4 - {}^{+}1 = {}^{+}3$

EXAMPLE 2. $^{-}3 - {}^{-}2 = ?$

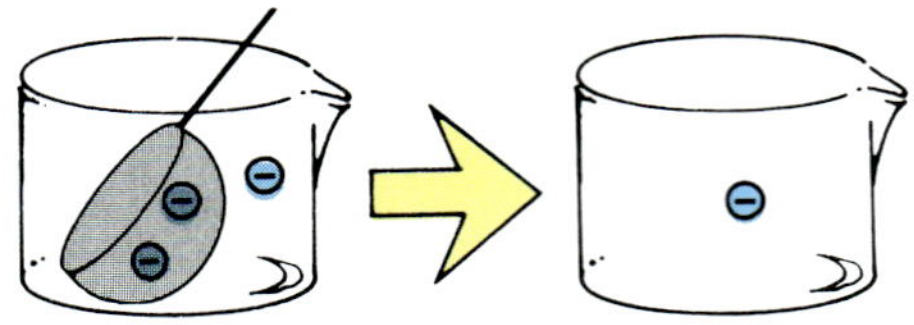

$^{-}3 - {}^{-}2 = {}^{-}1$

EXAMPLE 3. $^{+}2 - {}^{+}5 = ?$

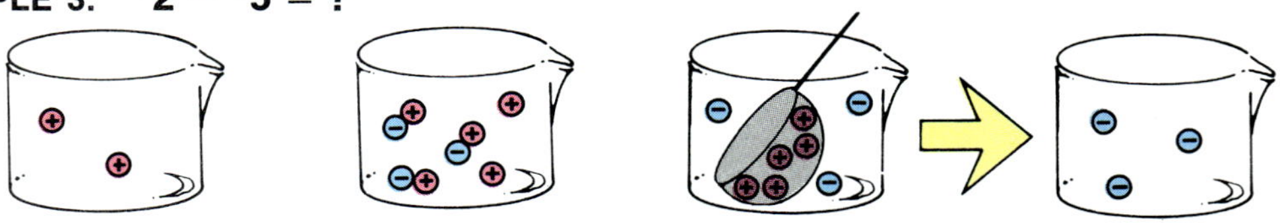

We do not have 5 ⊕ to take out.

Draw at least 3 ⊕⊖ pairs.

$^{+}2 - {}^{+}5 = {}^{-}3$

EXAMPLE 4. $^{+}1 - {}^{-}4 = ?$

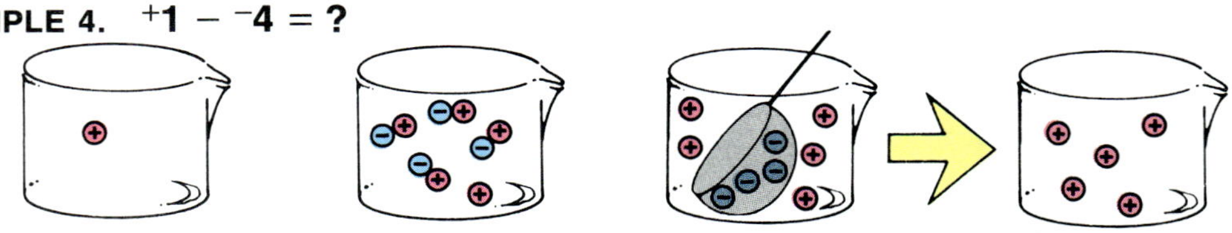

We do not have 4 ⊖ to take out.

Draw at least 4 ⊕⊖ pairs.

$^{+}1 - {}^{-}4 = {}^{+}5$

Addition and subtraction are closely related. Study these examples.

$^{+}6 - {}^{+}3 = {}^{+}3$	$^{-}5 - {}^{-}4 = {}^{-}1$	$^{+}3 - {}^{+}7 = {}^{-}4$	$^{+}3 - {}^{-}8 = {}^{+}11$
$^{+}6 + {}^{-}3 = {}^{+}3$	$^{-}5 + {}^{+}4 = {}^{-}1$	$^{+}3 + {}^{-}7 = {}^{-}4$	$^{+}3 + {}^{+}8 = {}^{+}11$

These examples give us a way to subtract.

To subtract an integer, add its opposite.

$^{-}7 - {}^{+}4 = {}^{-}7 + {}^{-}4 = {}^{-}11$

EXERCISES

Copy and complete.

1. ${}^+9 - {}^+3 = {}^+9 + {}^-3$
   $= ?$

2. ${}^-6 - {}^+2 = {}^-6 + {}^-2$
   $= ?$

3. ${}^+8 - {}^-4 = {}^+8 + {}^+4$
   $= ?$

4. ${}^-7 - {}^-5 = {}^-7 + {}^+5$
   $= ?$

5. ${}^-3 - {}^+7 = {}^-3 + ?$
   $= ?$

6. ${}^+9 - {}^-4 = {}^+9 + ?$
   $= ?$

Give each difference.

7. ${}^+5 - {}^+8$
8. ${}^+6 - {}^-2$
9. ${}^-9 - {}^-5$
10. ${}^-7 - {}^+4$
11. ${}^+3 - {}^-8$
12. ${}^+4 - 0$
13. ${}^+8 - {}^-3$
14. ${}^-8 - {}^+3$
15. ${}^-6 - {}^-2$
16. $0 - {}^-5$
17. ${}^-9 - {}^+3$
18. ${}^+6 - {}^-8$

Simplify. Remember to work inside the grouping symbols first.

19. ${}^-16 + ({}^-3 + {}^+5)$
20. ${}^+24 + ({}^-6 + {}^-2)$
21. ${}^+8 + ({}^-6 + {}^-5)$
22. ${}^+10 - ({}^-8 - {}^-5)$
23. ${}^+12 - ({}^-8 + {}^+2)$
24. ${}^-20 - ({}^+8 - {}^-10)$
25. ${}^-6 - ({}^+3 + {}^-6)$
26. ${}^-15 + ({}^-6 + {}^-8)$

Challenge!

Find the missing input or output.

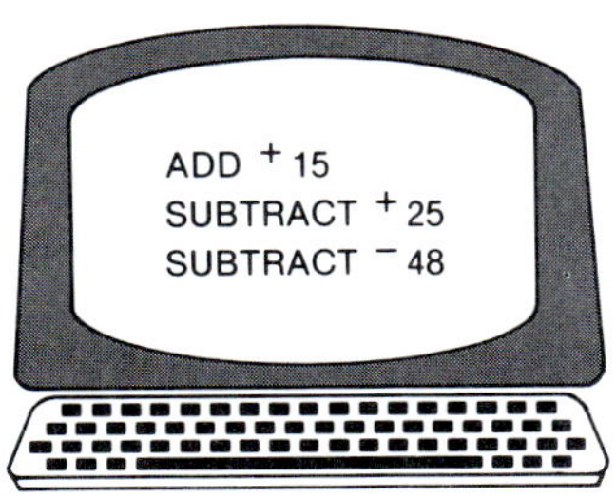

	Input	Output
	${}^+15$	${}^+53$
27.	${}^-6$	?
28.	?	0
29.	0	?

KEEPING SKILLS SHARP

Round to the nearest tenth.

1. 7.07
2. 0.63
3. 2.96
4. 3.24
5. 9.98
6. 6.829
7. 4.348

Round to the nearest hundredth.

8. 16.342
9. 9.468
10. 0.235
11. 28.5268
12. 50.397
13. 8.995
14. 14.2061

Round to the nearest thousandth.

15. 5.2834
16. 8.3608
17. 11.2675
18. 9.4296
19. 0.3594
20. 8.8997
21. 0.00068

Multiplying integers

To understand multiplication of integers, think of multiplying by $^{+}3$ as putting in three sets of charges, and think of multiplying by $^{-}3$ as taking out three sets of charges.

EXAMPLE 1. $^{+}3 \times {}^{+}2 = ?$

Put in 3 sets of $^{+}2$ charges.

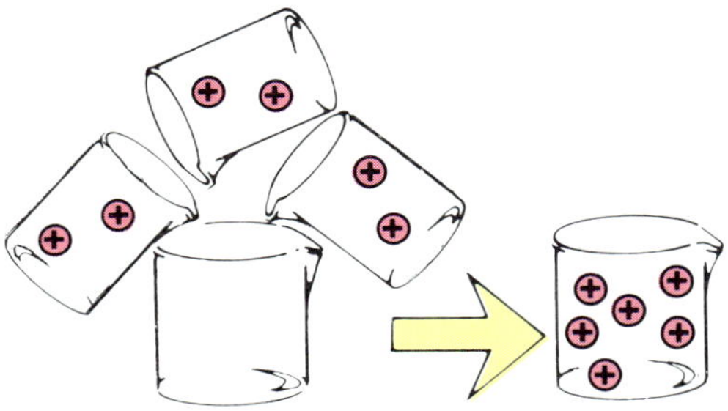

$^{+}3 \times {}^{+}2 = {}^{+}6$

EXAMPLE 2. $^{+}3 \times {}^{-}2 = ?$

Put in 3 sets of $^{-}2$ charges.

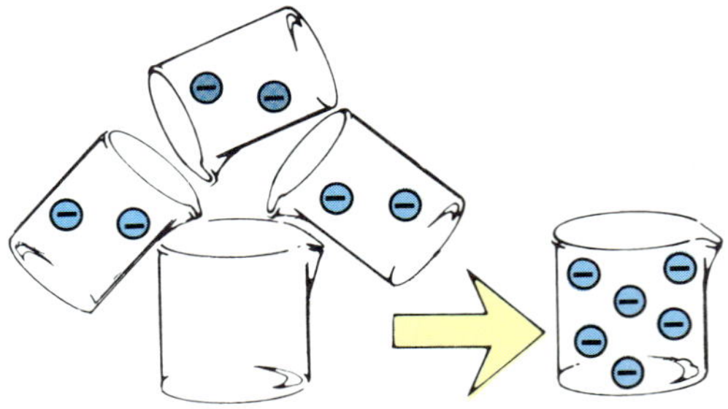

$^{+}3 \times {}^{-}2 = {}^{-}6$

EXAMPLE 3. $^{-}3 \times {}^{+}2 = ?$

We need to take out 3 sets of $^{+}2$ charges. So, draw at least 6 ⊕⊖ pairs.

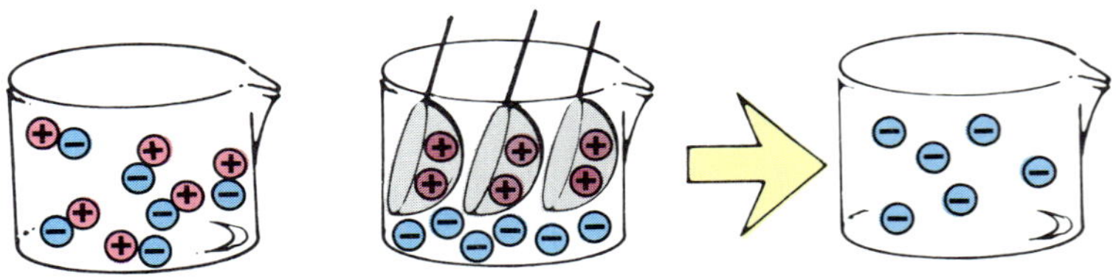

$^{-}3 \times {}^{+}2 = {}^{-}6$

EXAMPLE 4. $^{-}3 \times {}^{-}2 = ?$

We need to take out 3 sets of $^{-}2$ charges. So, draw at least 6 ⊕⊖ pairs.

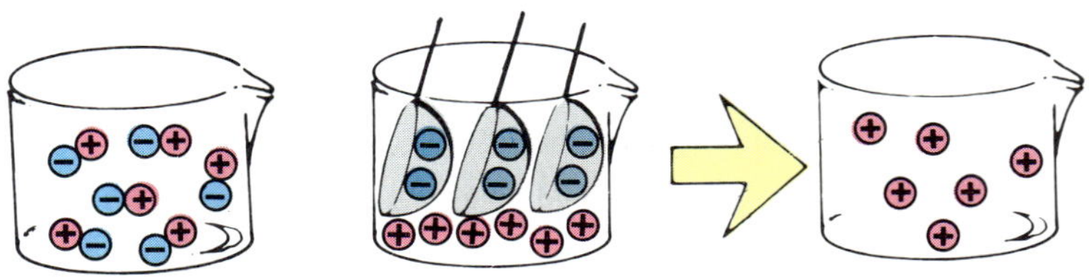

$^{-}3 \times {}^{-}2 = {}^{+}6$

These examples give us a way to multiply.

> The product of two integers with the *same* sign is *positive*.
> The product of two integers with *different* signs is *negative*.
> The product of any integer and 0 is 0.

EXERCISES

Give the product.

1. $^{+}8 \times {}^{+}7$ **2.** $^{-}8 \times {}^{-}7$ **3.** $^{+}8 \times {}^{-}7$ **4.** $^{-}8 \times {}^{+}7$

5. $^{+}5 \times {}^{+}9$ **6.** $^{-}5 \times {}^{-}9$ **7.** $^{+}5 \times {}^{-}9$ **8.** $^{-}5 \times {}^{+}9$

9. $^{-}7 \times {}^{+}3$ **10.** $^{+}5 \times 0$ **11.** $^{-}8 \times 0$ **12.** $^{-}7 \times {}^{-}9$

13. $^{-}4 \times {}^{+}3$ **14.** $^{-}3 \times {}^{-}20$ **15.** $^{-}1 \times {}^{+}1$ **16.** $^{-}32 \times {}^{-}2$

17. $^{+}1 \times {}^{-}9$ **18.** $^{-}12 \times {}^{-}3$ **19.** $^{-}39 \times {}^{+}3$ **20.** $^{+}13 \times {}^{+}5$

21. $0 \times {}^{-}6$ **22.** $^{-}42 \times {}^{+}2$ **23.** $^{-}50 \times {}^{-}4$ **24.** $^{-}23 \times {}^{+}10$

25. $^{-}15 \times {}^{-}3$ **26.** $^{-}25 \times 0$ **27.** $^{+}70 \times {}^{-}6$ **28.** $^{-}80 \times {}^{+}9$

$<$, $=$, **or** $>$**?**

29. $^{+}3 \times {}^{-}5$ (?) $^{+}15$ **30.** $^{-}6 \times 0$ (?) $^{+}6$ **31.** $^{-}8 \times {}^{+}6$ (?) $^{-}45$

32. $^{-}8 \times {}^{-}5$ (?) $^{+}38$ **33.** $^{-}7 \times {}^{+}4$ (?) $^{-}25$ **34.** $^{+}6 \times {}^{-}5$ (?) $^{-}32$

35. $^{-}4 \times {}^{-}9$ (?) $^{+}36$ **36.** $^{+}9 \times {}^{-}9$ (?) $^{-}100$ **37.** $^{-}7 \times {}^{-}6$ (?) $^{-}42$

Challenge!

38. You got into a hotel elevator. You went up 3 floors, down 6 floors, and up 8 floors. You were then at the top floor.
You went down 7 floors, up 2 floors, and down 9 floors. You were then on the first floor. Where did you get on?

Dividing integers

EXAMPLE 1. $^+36 \div {}^+9 = ?$

$^+9 \times ? = {}^+36$

So, $^+36 \div {}^+9 = {}^+4$

EXAMPLE 2. $^-36 \div {}^+9 = ?$

$^+9 \times ? = {}^-36$

So, $^-36 \div {}^+9 = {}^-4$

EXAMPLE 3. $^-36 \div {}^-9 = ?$

$^-9 \times ? = {}^-36$

So, $^-36 \div {}^-9 = {}^+4$

EXAMPLE 4. $^+36 \div {}^-9 = ?$

$^-9 \times ? = {}^+36$

So, $^+36 \div {}^-9 = {}^-4$

These examples give us a way to divide.

The quotient of two integers with the *same* sign is *positive*.
The quotient of two integers with *different* signs is *negative*.
The quotient of 0 divided by any nonzero integer is 0.

EXERCISES
Give each quotient.

1. $^-18 \div {}^-3$
2. $^-18 \div {}^+3$
3. $^+18 \div {}^-3$
4. $^+18 \div {}^+3$
5. $^+24 \div {}^+6$
6. $^+24 \div {}^-6$
7. $^-24 \div {}^-6$
8. $^-24 \div {}^+6$
9. $^-56 \div {}^+8$
10. $^-56 \div {}^-8$
11. $^+56 \div {}^-8$
12. $^+56 \div {}^+8$

13. $^{+}12 \div {}^{-}4$

14. $^{-}16 \div {}^{-}2$

15. $0 \div {}^{+}3$

16. $^{-}30 \div {}^{-}5$

17. $^{-}20 \div {}^{+}5$

18. $^{-}21 \div {}^{+}3$

19. $^{+}24 \div {}^{+}8$

20. $^{+}36 \div {}^{-}9$

21. $^{+}27 \div {}^{-}3$

22. $^{-}36 \div {}^{-}6$

23. $^{+}72 \div {}^{-}8$

24. $^{+}42 \div {}^{-}7$

25. $^{+}81 \div {}^{+}9$

26. $^{-}28 \div {}^{-}7$

27. $^{-}45 \div {}^{-}9$

28. $^{-}54 \div {}^{+}6$

29. $^{-}32 \div {}^{+}4$

30. $^{-}64 \div {}^{-}8$

31. $^{+}28 \div {}^{-}4$

32. $^{-}100 \div {}^{-}10$

33. $^{+}72 \div {}^{+}9$

34. $^{+}15 \div {}^{-}1$

35. $^{-}88 \div {}^{-}4$

36. $^{+}66 \div {}^{-}2$

Simplify.

37. $(9 - {}^{-}2) + 6$

38. $9 - ({}^{-}2 + 6)$

39. $5 + ({}^{-}3 + 3)$

40. $(3 \times {}^{-}2) + 5$

41. $3 \times ({}^{-}2 + 5)$

42. $4 + ({}^{-}3 \times 2)$

43. $(15 \div {}^{-}3) + {}^{-}2$

44. $15 \div ({}^{-}3 + {}^{-}2)$

45. $12 \div (4 - 2)$

46. $(8 + 2) - (3 + 1)$

47. $({}^{-}5 - 3) + (6 + {}^{-}4)$

48. $({}^{-}2 - {}^{-}9) + (7 - {}^{-}4)$

49. $({}^{-}15 \div 3) + (21 \div {}^{-}7)$

50. $({}^{-}24 \div {}^{-}6) \times (5 - {}^{-}1)$

51. $(3 \times {}^{-}5) - (12 \div {}^{-}4)$

Challenge!

Find the missing input or output.

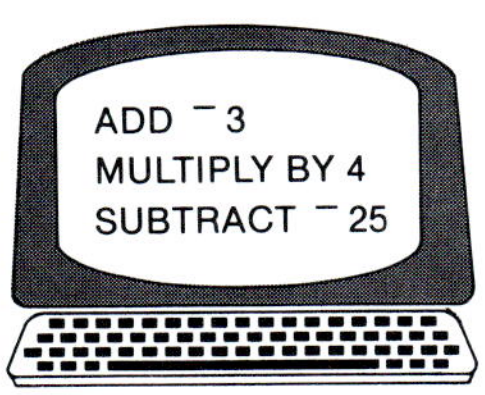

	Input	Output
	16	77
52.	0	?
53.	?	101
54.	?	$^{-}3$

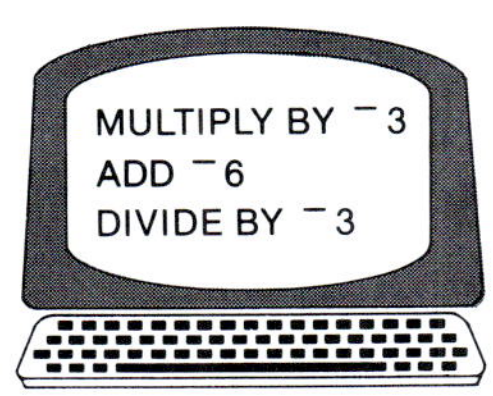

	Input	Output
	$^{-}5$	$^{-}3$
55.	0	?
56.	?	6
57.	?	$^{-}22$

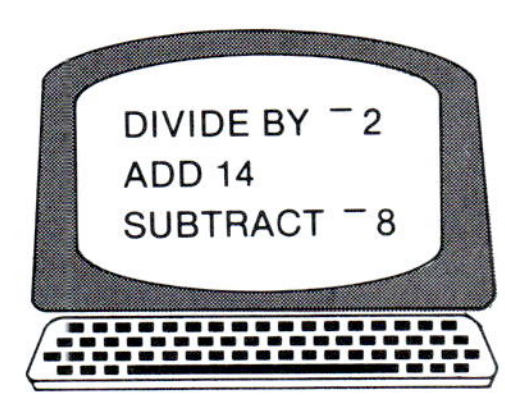

	Input	Output
	$^{-}4$	24
58.	12	?
59.	?	26
60.	?	42

Problem Solving Doing systematic trials

1. Study and understand.
2. Plan and do.
3. Answer and check.

At an ultra-light-airplane show, two student tickets and one adult ticket cost $8.50. Two adult tickets and one student ticket cost $9.50. How much did a student ticket cost?

		First Trial	*Second Trial*	*Third Trial*
Guess	Student ticket	$2	$3	$2.50
	Adult ticket ($8.50 − student)	$4.50	$2.50	$3.50
Check	2 Adult + 1 student	$11.00 *Too much! Guess more costly student ticket.*	$8.00 *Too little! Guess less costly student ticket.*	$9.50 *Just right!*

Answer: A student ticket cost $2.50.

Solve by doing systematic trials.

1. Two hot-air-balloon models and 1 ultra-light-airplane model cost $39. One hot-air-balloon and 2 ultra-light models cost $42. What did a hot-air-balloon model cost?

2. The ultra-light flying club sold stickers. Three small stickers and 1 large sticker sold for 54¢. One small sticker and 3 large stickers sold for 66¢. How much did each size of sticker sell for?

Absolute value of an integer

You can see on the number line that

$^{-}6$ is located 6 units from zero.

5 is located 5 units from zero.

The number of units an integer is from zero on the number line is the **absolute value** of the integer.

The absolute value of $^{-}6$ equals 6.
We write

$$|{}^{-}6| = 6$$

The absolute value of 5 equals 5.
We write

$$|5| = 5$$

An expression with symbols for absolute values can be simplified. Study these examples.

$$|{}^{-}6| - |5| = 6 - 5 = 1$$

$$|{}^{-}6 - 5| = |{}^{-}11| = 11$$

EXERCISES

Simplify.

1. $|{}^{-}2|$
2. $|8|$
3. $|{}^{-}7|$
4. $|9|$
5. $|0|$
6. $|{}^{-}13|$
7. $|{}^{-}8| + |2|$
8. $|{}^{-}8 + 2|$
9. $|{}^{-}5| + |{}^{-}1|$
10. $|{}^{-}5 + {}^{-}1|$
11. $|3| - |{}^{-}4|$
12. $|3 - {}^{-}4|$
13. $|{}^{-}10| - |{}^{-}3|$
14. ${}^{-}10 - {}^{-}3|$
15. $|{}^{-}5| \times |4|$
16. $|{}^{-}5 \times 4|$
17. $|18| \div |{}^{-}9|$
18. $|18 \div {}^{-}9|$

Properties of addition and multiplication

This lesson reviews some properties of addition and multiplication.

ADDITION

Adding 0 Property
The sum of any number and 0 is the number. $279 + 0 = 279$

Commutative Property of Addition
Changing the order of the addends does not change the sum.
$^{-}258 + 312 = 312 + {}^{-}258$

Associative Property of Addition
Changing the grouping of the addends does not change the sum.
$(82 + 65) + {}^{-}35 = 82 + (65 + {}^{-}35)$

MULTIPLICATION

Multiplying by 1 Property
The product of any number and 1 is the number. $^{-}374 \times 1 = {}^{-}374$

Commutative Property of Multiplication
Changing the order of the factors does not change the product.
$125 \times {}^{-}64 = {}^{-}64 \times 125$

Associative Property of Multiplication
Changing the grouping of the factors does not change the product.
$(56 \times 25) \times {}^{-}4 = 56 \times (25 \times {}^{-}4)$

Distributive Property

product sum → sum of products

$^{-}36 \times (83 + 17) = ({}^{-}36 \times 83) + ({}^{-}36 \times 17)$

EXERCISES

Give two examples of each of the following properties.

1. commutative property of addition
2. associative property of addition
3. commutative property of multiplication
4. associative property of multiplication
5. distributive property
6. adding 0 property

Use one of the properties to help you compute each answer in your head.

7. $(39 \times 5) \times 20$
[*Hint:* $39 \times (5 \times 20)$]

8. $(18 \times 7) + (18 \times 3)$
[*Hint:* $18 \times (7 + 3)$]

9. $(186 + 31) + 69$
[*Hint:* $186 + (31 + 69)$]

10. $(^{-}65 \times 4) + (^{-}65 \times 6)$
[*Hint:* $^{-}65 \times (4 + 6)$]

11. $(723 + 65) + 35$

12. $(78 \times {^{-}25}) + (78 \times {^{-}75})$

13. $(154 \times 2) \times 500$

14. $^{-}758 \times 1$

15. $193 + 0$

16. $(723 \times 25) + (723 \times {^{-}24})$

17. $(63 + {^{-}26}) + 26$

18. $(^{-}82 \times 2) + (^{-}82 \times 50)$

19. $(97 + 110) + {^{-}10}$

20. $(79 \times 105) + (79 \times {^{-}5})$

21. $(86 \times {^{-}25}) \times {^{-}4}$

22. $30 \times (20 + 7)$

Challenge!

Here is how to use the distributive property to multiply.

$$\begin{aligned} 57 \times 99 &= 57 \times (100 + {^{-}1}) \\ &= (57 \times 100) + (57 \times {^{-}1}) \\ &= 5700 + {^{-}57} \\ &= 5643 \end{aligned}$$

Use the distributive property to compute each product in your head.

23. 37×99
24. 45×99
25. 68×99
26. 51×99
27. 25×99
28. 42×99
29. 36×101
30. 33×101
31. 64×101
32. 53×101
33. 48×101
34. 24×102

KEEPING SKILLS SHARP

Estimate to find the incorrect answer.

1. a. 4.07×2.7 = 12.699

b. 3.11×2.06 = 6.4066

Multiply.

2. 3.67×3.6

3. 4.98×4.5

4. 62.1×5.3

5. 3.06×0.82

6. 784×5.5

7. 96.3×0.73

8. 8.17×90

Order of operations

Who is right? Carla and Jon were both asked to simplify the expression $6 + 4 \times 7$. Notice that they got different answers. They got different answers because they performed the operations in different orders. Carla multiplied first and then added. Jon added first and then multiplied.

To be sure that an expression has only one value, mathematicians have agreed on these rules for the **order of operations**:

(1) First, do the operations within the grouping symbols (parentheses).

(2) Next, work from left to right doing the multiplications and divisions.

(3) Last, work from left to right doing the additions and subtractions.

Study the rules. Who was right, Carla or Jon?

Here are some examples. The order of operations is shown by the numbered arrows.

$$\mathbf{8 \overset{1}{\div} 4 \overset{2}{\times} 2 = 4}$$

$$\mathbf{10 \overset{1}{-} 6 \overset{2}{+} {}^{-}3 = 1}$$

$$\mathbf{7 \overset{2}{+} 3 \overset{1}{\times} 8 \overset{3}{-} 4 = 27}$$

$$\mathbf{4 \overset{1}{\times} 2 \overset{3}{+} 20 \overset{2}{\div} {}^{-}4 = 3}$$

$$\mathbf{5 \overset{3}{+} (6 \overset{1}{+} 2) \overset{2}{\div} 4 = 7}$$

$$\mathbf{(7 \overset{1}{+} 6) \overset{2}{\times} {}^{-}2 = {}^{-}26}$$

EXERCISES

Simplify each expression.

1. $5 \times 6 - 4$
2. $16 + 8 \div 4$
3. $20 \div 5 - 3$
4. $^{-}18 \div 3 \times 3$
5. $32 - 8 - 6$
6. $20 - 8 \div 4$
7. $15 - 4 + 6$
8. $24 \div {}^{-}4 \times 2$
9. $36 + 14 - 10$
10. $^{-}72 \times (8 - 3)$
11. $(96 + 24) \div 6$
12. $16 \times (15 - 5)$
13. $^{-}16 + 8 \div 4 + 4$
14. $(^{-}16 + 8) \div 4 + 4$
15. $^{-}16 + 8 \div (4 + 4)$
16. $12 + 10 \times 5 - 1$
17. $(12 + 10) \times 5 - 1$
18. $12 + 10 \times (5 - 1)$
19. $6 \times 3 + {}^{-}4 \times 3$
20. $(6 + {}^{-}4) \times 3$
21. $6 \times (3 + {}^{-}4) \times 3$
22. $18 - (15 + 6) \div 3$
23. $18 - 15 + 6 \div 3$
24. $(18 - 15) + 6 \div 3$

★ 25. $36 \div (2 + 4) \times 4 - 5$

★ 26. $36 \div 2 + 4 \times 4 - 2$

★ 27. $36 \div 2 + 4 \times (4 - 2)$

★ 28. Four different expressions are described below.

a. All the operations are additions.

b. All the operations are subtractions.

c. All the operations are multiplications.

d. All the operations are divisions.

In which of these expressions will you get the same answer no matter in what order you do the operations?

Challenge!

What number was entered first?

29. ? [×] 15 [÷] 24 [=] 75
30. ? [÷] 18 [×] 25 [=] 175
31. ? [×] 27 [+] 150 [=] 1095
32. ? [÷] 36 [−] 38 [=] 40
33. ? [×] 98 [−] 254 [=] 1118
34. ? [÷] 12 [+] 315 [=] 1323
35. ? [×] 17 [×] 23 [=] 6256
36. ? [÷] 24 [÷] 18 [=] 8

Variables and expressions

These students are on a walkathon to raise money for a muscular-dystrophy drive. Each student has a sponsor that has agreed to pay a certain amount for each kilometer walked. How much will they raise?

Bill's sponsor is paying him \$2 a kilometer. If we let n be the number of kilometers that he walks, the number of dollars that he will raise is given by the expression $2n$. $2n$ is a short way to write $2 \times n$.

The letter n is called a **variable.** To find the number of dollars that he would raise if he walked 15 kilometers, we can **substitute** 15 for n and then simplify the expression.

$$2 \times 15 = 30$$

Another way to write this is

$$2(15) = 30$$

How many dollars will Bill raise if he walks 18 kilometers? 25 kilometers?

Betty's sponsor is paying her \$3 a kilometer. She is also giving \$5 of her own money. If m is the number of kilometers she walks, the number of dollars she will raise is given by the expression $3m + 5$. If she walks 11 kilometers, the number of dollars raised will be

$$3(11) + 5 = 38$$

How many dollars will Betty raise if she walks 15 kilometers? 20 kilometers?

EXERCISES

Substitute 12 for n in the expression and simplify.

1. $n + 8$
2. $n + 15$
3. $n - 2$
4. $n - 6$
5. $2n$
6. $5n$
7. $3n$
8. $6n$
9. $2n + 1$
10. $3n + 5$
11. $4 + 3n$
12. $5 + 6n$
13. $32 - 2n$
14. $3n - 4$
15. $48 - 4n$

Substitute $^-5$ for m in the expression and simplify.

16. $m + 6$
17. $m - 3$
18. $3m$
19. $7m$
20. $m + 1$
21. $2m + 10$
22. $17 + 2m$
23. $3m - 4$
24. $12 - 4m$

To find how many dollars each of the following students raised, substitute the number of kilometers walked in the given expression.

25. Ruth: $n + 18$
    kilometers walked: 12
26. Andy: $n + 5$
    kilometers walked: 17
27. Robert: $2n + 1$
    kilometers walked: 11
28. Marcia: $3n + 6$
    kilometers walked: 8
29. David: $4n + 2$
    kilometers walked: 6
30. Susan: $3n + 5$
    kilometers walked: 9

Let n be the number of kilometers walked. Write an expression for the number of dollars you would raise if your sponsors agreed to pay you

31. \$2 a kilometer.
32. \$3 a kilometer.
33. \$1 a kilometer plus \$3 for going on the walkathon.
34. \$3 a kilometer plus \$1 for going on the walkathon.
35. \$2 a kilometer plus \$3 for going on the walkathon.
36. \$3 a kilometer plus \$2 for going on the walkathon.

KEEPING SKILLS SHARP

Estimate to find the incorrect answer.

1. a. $8.82 \div 4.3 = 2.05$
   b. $19.15 \div 6.3 = 0.304$

Divide.

2. $2.1\overline{)20.223}$
3. $1.1\overline{)3.916}$
4. $0.03\overline{)16.110}$
5. $3.2\overline{)2.1312}$
6. $50\overline{)458.50}$
7. $6.3\overline{)505.89}$
8. $78.4\overline{)337.12}$
9. $9.21\overline{)66.312}$
10. $0.803\overline{)39.347}$
11. $6.25\overline{)331.25}$
12. $3.08\overline{)5.0512}$
13. $0.246\overline{)7.3062}$

Formulas

Think about riding a bicycle at a constant speed of 20 kilometers an hour for 5 hours. This table gives the distance you would have traveled at the end of each hour. Notice that the distance traveled in kilometers is 20 times the number of hours.

Distance (kilometers)	*Time (hours)*
20	1
40	2
60	3
80	4
100	5

We can use this **formula** to show the relationship:

$$d = 20t,$$

where d is the distance in kilometers and t is the time in hours. To find how many kilometers are traveled in 3.2 hours, we can **substitute** 3.2 for t.

$$d = 20\,(3.2)$$
$$= 64$$

EXERCISES

Use the formula $d = 20t$ to compute the distance in kilometers for each of the following times.

1. 1.5 hours **2.** 3.6 hours **3.** 4.3 hours **4.** 2.9 hours

Sound travels at about 335 meters per second. The formula $d = 335t$ gives the distance (d) in meters that sound travels in t seconds.

Use the formula to compute the distance that sound travels for each of the following times.

5. 2 seconds
6. 8 seconds
7. 10 seconds
8. 8.2 seconds
9. 5.4 seconds
10. 9.5 seconds
11. 1 minute
12. 2.5 minutes

13. Suppose that 5.2 seconds after you observe a flash of lightning you hear the thunder. How far away was the lightning?

14. Suppose that you are 1 kilometer away from a fireworks show. Will it take more or less than 3 seconds for you to hear a boom?

The formula $d = 4.88t^2$ gives the approximate distance (d) in meters that an object will fall in t seconds.

Use the formula to compute how far an object will fall in the following times.

15. 10 seconds
16. 5 seconds
17. 3 seconds
18. 1.5 seconds

19. Does an object fall exactly twice the distance in twice the time?

20. Suppose that you dropped a pebble from a bridge and saw the splash 3.5 seconds later. How high above the water is the bridge?

Did you know that you can determine the temperature by counting the chirps of a cricket? The formula $T = \frac{n}{8} + 5$ tells the temperature (T) in degrees Celsius for the number of chirps (n) in one minute.

Use the formula to compute the temperature when the number of chirps is each of the following.

21. 80
22. 64
23. 92
24. 74
25. 102

Problem Solving

26. Suppose you counted 32 chirps in 30 seconds. How many degrees Celsius is the temperature?

★ **27.** How many times should a cricket chirp at a temperature of 25°C?

Making and testing formulas

Formulas are often useful for describing patterns. In this lesson, you will be making up and testing your own formulas.

Work in groups. You will need some squared paper, and you may need some cubes or blocks.

Series A

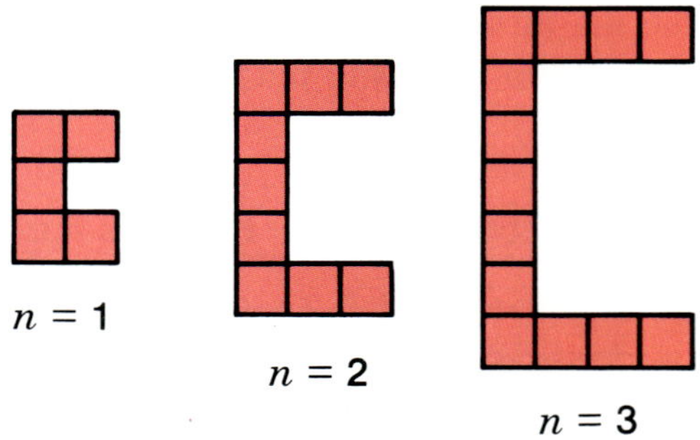

PART A

1. Using the pattern of the figures in Series A, draw the figure for $n = 4$.
2. Draw the figure for $n = 6$.
3. Find the area and perimeter of the figures for $n = 2$ and $n = 3$.

n	Area (square units)	Perimeter (units)
1	5	12
2		
3		

4. Make up a formula for finding the area of any figure in the series.
5. Check to see if your formula works. Use it to predict the area of a figure in the series that has not been drawn yet. Then draw the figure to check. Does your formula work?
6. Make up a formula for finding the perimeter of any figure in the series. Check your formula. Does it work?
7. Extend the pattern of the figures backward and draw what you think the figure for $n = 0$ would look like.
8. What do your formulas for area and perimeter predict for this figure? Do they check?
9. Compare your results for Part A with those of another group.

Series B

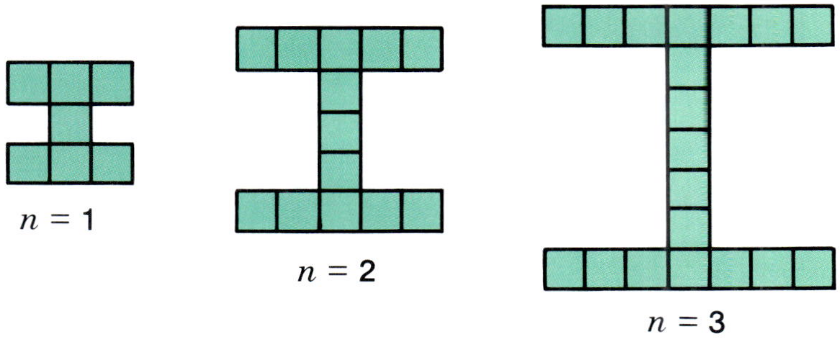

PART B

10. Using the pattern of the figures in Series B, make up formulas for finding the area and perimeter of any figure in the series. Test your formulas.

11. Draw what you think the figure for $n = 0$ would look like. Does your formula work for $n = 0$?

12. Suppose we change the way we numbered the figures in Series B. Let's say for the first figure we let $n = 2$ instead of 1 and for the second figure we let $n = 3$ instead of 2. Will your formulas for area and perimeter need to change? If so, what will your new formulas be?

PART C

13. Make up a formula for the number of cubes in any figure in Series C. Check to see that your formula works.

14. Compare the strategies you used to make up and test your formula with strategies used by other groups.

Series C

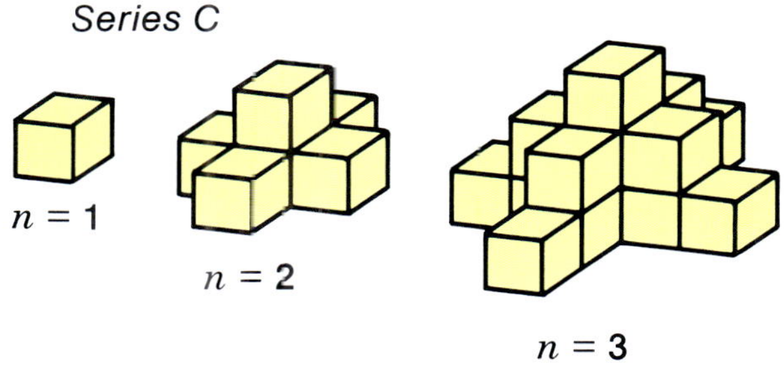

PART D

15. Use one of your formulas to predict the perimeter of the 100th figure in one of the series. Discuss how sure you are that your prediction is correct.

16. Make up your own series of figures. Try to find a formula for the area or perimeter of any figure in the series.

Solving addition and subtraction equations

Robert has 57 more stamps than David. Robert has 511 stamps in all. How many does David have?

Let n be the number of David's stamps. Then $n + 57$ is the number of Robert's stamps.

Next write an equation for the problem:

$$\overbrace{\underset{\substack{\uparrow \\ \textit{David's number}}}{n} + 57}^{\textit{Robert's number}} = \overset{\textit{Robert's number}}{511}$$

We **solve** the equation by finding the number that can be substituted for n to make the equation true.

Addition and subtraction are **inverse** operations. Each operation undoes the other. To undo the addition and get n by itself on one side of the equation, subtract 57 from both sides.

$$n + 57 - 57 = 511 - 57$$

Simplify both sides. $n = 454$

Check the solution by substituting 454 for n in the first equation.

Check: $454 + 57 = 511$

(Does the solution also check in the problem?)

So, David has 454 stamps.

Jan has 109 fewer stamps than Carl. She has 358 stamps in all. How many stamps does Carl have?

Let n be the number of Carl's stamps. Then $n - 109$ is the number of Jan's stamps.

Equation:

$$\overbrace{\underset{\substack{\uparrow \\ \textit{Carl's number}}}{n} - 109}^{\textit{Jan's number}} = \overset{\textit{Jan's number}}{358}$$

To undo the subtraction and get n by itself on one side of the equation, add 109 to both sides.

$$n - 109 + 109 = 358 + 109$$

Simplify both sides.

$$n = 467$$

Check: $467 - 109 = 358$

So, Carl has 467 stamps.

EXERCISES

Solve each equation. Check the solution.

1. $n + 54 = 79$	2. $n + 38 = 109$	3. $117 = n + 53$ (Same as $n + 53 = 117$)
4. $134 = x + 118$	5. $y + 56 = 40$	6. $z + 74 = 206$
7. $n - 15 = 38$	8. $n - 23 = 49$	9. $76 = n - 42$
10. $x - 65 = {}^{-}5$	11. $75 = z - 39$	12. $w - 71 = 90$
13. $40 + n = 125$	14. $g - 58 = 102$	15. $139 = 92 + n$
16. ${}^{-}100 = r - 85$	17. $36 + t = -2$	18. $u - 83 = {}^{-}10$
19. ${}^{-}58 + s = 90$	20. $t - 57 = 0$	21. $175 = 50 + r$
22. $t + 29 = 0$	23. $n - 43 = {}^{-}43$	24. ${}^{-}26 = x - 67$
25. ${}^{-}103 + b = {}^{-}60$	26. $34 = r - 10$	27. ${}^{-}59 = 3 + t$
28. $x + 2.8 = 4.1$	29. $n - 1.6 = 3.02$	30. $5.1 + a = 6$
31. $13 = y + 7.2$	32. $t - 1 = 5.8$	33. $9 = b - 1.7$
34. $r + 0.64 = 2$	35. $s - 4 = 17.5$	36. $10 = 4.13 + n$

Problem Solving

Decide whether Equation A or Equation B would be used to solve each problem. Then solve the problem.

Equation A: $n + 19 = 15$

Equation B: $n - 19 = 15$

37. 19 less than a number n is 15. What is the number?

38. 19 more than a number n is 15. What is the number?

39. The sum of a number n and 19 is 15. What is the number?

40. A number n decreased by 19 is 15. What is the number?

41. From last night to this morning the temperature dropped 19 degrees. The temperature this morning was 15 degrees. What was the temperature last night?

42. From this morning to this afternoon the temperature rose 19 degrees. The temperature this afternoon was 15 degrees. What was the temperature this morning?

Solving multiplication and division equations

Jan bought 3 identical fish for \$4.50. How much did she pay for each fish?

Let p be the price of 1 fish. Then $3p$ is the price of 3 fish.

price of 3

Equation: $3p = 4.50$

Multiplication and division are inverse operations. Each operation undoes the other. To undo the multiplication, divide both sides by 3.

The bars indicate division.

$$\frac{3p}{3} = \frac{4.50}{3}$$

Simplify both sides. $p = 1.50$

Check: $3 \times 1.50 = 4.50$

She paid \$1.50 for each fish.

There were 8 fish tanks in the pet shop. The average number of fish in a tank was 35. How many fish were there in all?

Let f be the number of fish there were in all. Then $\frac{f}{8}$ is the average number of fish in a tank.

average number of fish in a tank

Equation: $\frac{f}{8} = 35$

To undo the division, multiply both sides by 8.

$$8 \times \frac{f}{8} = 8 \times 35$$

Simplify both sides. $f = 280$

Check: $\frac{280}{8} = 35$

There were 280 fish in all.

EXERCISES

Solve each equation. Check the solution.

Same as $5r = 85$

1. $2a = 28$
2. $^-3b = 57$
3. $85 = 5r$
4. $^-108 = 9s$
5. $8c = 8$
6. $^-4d = {}^-20$
7. $17 = {}^-17n$
8. $^-80 = 5t$
9. $\frac{a}{3} = 5$
10. $\frac{n}{^-3} = 6$
11. $15 = \frac{x}{2}$
12. $^-20 = \frac{g}{4}$
13. $\frac{r}{^-2} = {}^-7$
14. $\frac{b}{5} = 5$
15. $^-9 = \frac{n}{9}$
16. $6 = \frac{t}{^-1}$
17. $^-4t = 64$
18. $\frac{y}{^-5} = 10$
19. $24 = \frac{m}{12}$
20. $81 = 3x$
21. $5m = {}^-470$
22. $\frac{r}{6} = {}^-20$
23. $^-7y = {}^-84$
24. $^-72 = \frac{c}{^-9}$
25. $12s = 360$
26. $^-32 = \frac{y}{^-8}$
27. $\frac{t}{10} = 10$
28. $0 = 16x$
29. $6a = 1.92$
30. $34g = 91.8$
31. $\frac{n}{7} = 1.2$
32. $\frac{a}{18} = 0.16$
33. $2.24 = 7c$
34. $22.8 = 3r$
35. $3.2 = \frac{x}{15}$
36. $1.7 = \frac{b}{16}$
37. $1.6a = 32$
38. $0.5n = 24$
39. $\frac{t}{1.6} = 32$
40. $\frac{r}{0.5} = 24$

Problem Solving

Decide whether Equation A or Equation B would be used to solve each problem. Then solve the problem.

Equation A. $6n = 84$

Equation B. $\frac{n}{6} = 84$

41. 6 times a number n is 84. What is the number?
42. A number n divided by 6 is 84. What is the number?
43. If you multiply a number n by 6, you get 84. What is the number?
44. The product of a number n and 6 is 84. What is the number?
45. There were 6 rabbit cages in the pet shop. Each cage had the same number of rabbits. If there were 84 rabbits in all, how many rabbits were in each cage?
46. During 6 years, the Garcias spent an average of $84 a year for dog food. How much did they spend for dog food during the 6 years?
47. Beth feeds her kittens 6 ounces of cat food each day. How many days will an 84-ounce bag of cat food last?
48. Bill's 6 puppies weigh a total of 84 pounds. What is their average weight?

Problem Solving Writing an equation

To solve each problem:

- Write an equation.
- Solve the equation.
- Check the solution in the problem.

Grace hiked all of Raccoon Trail in the morning and 7 miles along another trail in the afternoon. She hiked a total of 22 miles. How many miles long was Raccoon Trail?

Let m be the length of Raccoon Trail. Then $m + 7$ is the total number of miles hiked.

Equation: $\underbrace{m + 7}_{\text{total miles}} = \underbrace{22}_{\text{total miles}}$

Subtract 7 from each side. $m + 7 - 7 = 22 - 7$

Simplify. $m = 15$

Check: $15 + 7 = 22$

So, Raccoon Trail was 15 miles long.

The weight of Ann's pack was $\frac{1}{4}$ her own weight. Her pack weighed 27 pounds. How much did Ann weigh?

Let w be Ann's weight.

Then $\frac{w}{4}$ is the weight of Ann's pack.

Equation: $\underbrace{\frac{w}{4}}_{\text{weight of Ann's pack in pounds}} = \underbrace{27}_{\text{weight of Ann's pack in pounds}}$

Multiply both sides by 4. $4 \times \frac{w}{4} = 4 \times 27$

Simplify. $w = 108$

Check: $\frac{108}{4} = 27$

So, Ann weighed 108 pounds.

EXERCISES

Use an equation to solve each problem.

1. Grace bought some packages of Quick Freeze Hiking Food. She sold 12 of them to Ann. She had 24 packages left. How many packages did she buy?

2. The first rest area for hikers on Red Mountain was $\frac{1}{2}$ the way up the mountain. If they had hiked 6 miles when they reached the rest area, what was the total hiking distance up Red Mountain?

3. Grace's pack weighed 6 more pounds than Ann's pack. Grace's pack weighed 22 pounds. How many pounds did Ann's weigh?

4. Today there are 8200 more pine trees in the forest than there were 5 years ago. There are 23,100 trees today. How many were there 5 years ago?

5. On last year's trip, the highest pass the hikers crossed had an altitude of 3025 meters. The highest pass on this year's trip was 54 meters lower. What was the altitude of the lower pass?

6. The hikers passed a Forest Service tree farm. It had 52 rows of white pines, with the same number of trees in each row. There were 5512 white pines in all. How many trees were there in each row?

7. Roberta bought all the food for the weekend camping trip with 3 friends. She paid $55.00. How much money should her friends each give Roberta to pay for their share of the food?

8. Maria drove Roberta and their two friends on the camping trip. If she used 11 gallons of gas and paid $15.00, how much did 1 gallon of gas cost? Round your answer to the nearest cent.

Challenge!

Find a way to push each marked key once to get the answer.

9.

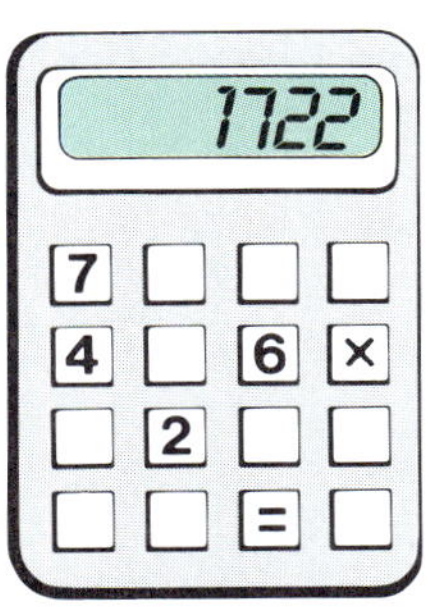

10.

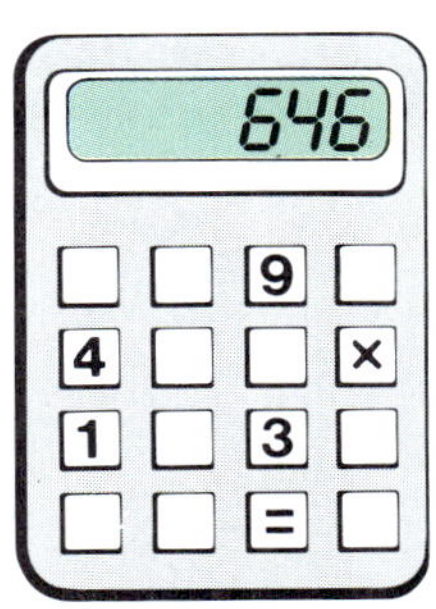

11.

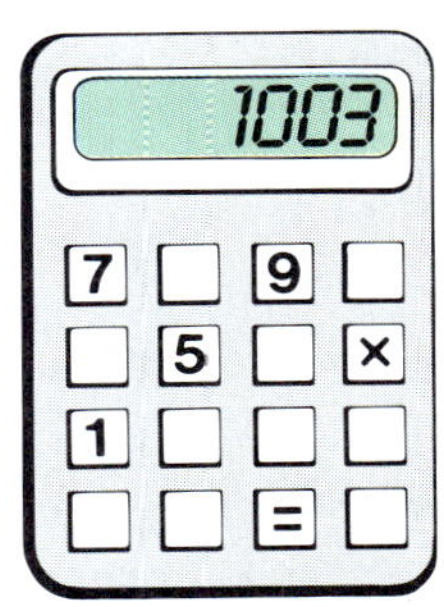

Solving two-step equations

In this lesson you will use what you have learned about solving simple equations to solve two-step equations. When solving an equation, you want to get the variable by itself on one side of the equation. You do this by undoing all the operations that were done on the variable. Study the examples.

EXAMPLE 1.

3 times my age plus 8 is 53. How old am I?

$$3n + 8 = 53$$

Step 1. Subtract 8 from both sides of the equation. Then simplify both sides.

$$3n + 8 - 8 = 53 - 8$$

$$3n = 45$$

Step 2. Divide both sides of the equation by 3. Then simplify both sides.

$$\frac{3n}{3} = \frac{45}{3}$$

$$n = 15$$

Check: $3(15) + 8 = 53$

He is 15 years old.

EXAMPLE 2.

$$27 = 5n - 3$$

$$27 + 3 = 5n - 3 + 3$$

$$30 = 5n$$

$$\frac{30}{5} = \frac{5n}{5}$$

$$6 = n$$

Check: $27 = 5(6) - 3$

EXAMPLE 3.

$$12 + 7n = {}^-44$$

$$12 - 12 + 7n = {}^-44 - 12$$

$$7n = {}^-56$$

$$\frac{7n}{7} = \frac{{}^-56}{7}$$

$$n = {}^-8$$

Check: $12 + 7({}^-8) = {}^-44$

EXERCISES

Solve and check.

1. $3n + 4 = 22$
2. $5n - 9 = 6$
3. $^-12 + {}^-4t = 0$
4. $6n + 8 = 38$
5. $^-6n - 8 = 10$
6. $9b - 36 = 0$
7. $7m + 6 = 6$
8. $8n - 12 = 36$
9. $23 = 13 + 2a$
10. $12 = 9a - 15$
11. $^-4c + 5 = 9$
12. $12 + 4x = 40$
13. $15 + 2d = 31$
14. $52 = 12 + 4x$
15. $6x - 2 = 40$
16. $38 + 5y = 38$
17. $19 + 3c = 40$
18. $4h - 42 = {}^-18$
19. $7k - 45 = 4$
20. $^-12r - 48 = 12$
21. $4j - 36 = 0$
22. $7 + 5d = 42$
23. $25 = 25 + 9r$
24. $12s - 24 = 0$

Find each number by writing an equation and solving it.

25. 3 more than 6 times a number n is 45. What is the number?
26. 5 less than 4 times a number n is 31. What is the number?
27. If you multiply a number n by 8 and subtract 15 from the product, you get 65. What is the number?
28. The sum of 6 times a number n and 15 is 9. What is the number?

Challenge!

Find the missing input or output.

	Input	Output
	14	39
29.	236	?
30.	?	414
31.	?	866

	Input	Output
	43	128
32.	13	?
33.	?	460
34.	?	316

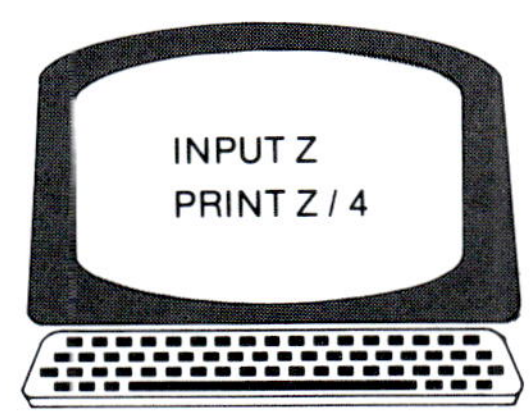

	Input	Output
	44	11
35.	128	?
36.	?	256
37.	?	$^-142$

Solving equations

Study the following number puzzle and solution.
Bill said to Jasmine, "I am thinking of a number. If I divide it by 4 and subtract 2, I get 3. Can you guess my number?"
Sample solution:

a. Jasmine thought, "Let n be his number. Then n divided by 4 minus 2 equals 3."

$$\frac{n}{4} - 2 = 3$$

b. Jasmine realized that she could work backward. "If he got 3 after subtracting 2, he must have had 2 more than 3." So Jasmine "undid" the subtraction of 2 by adding 2 to both sides of her equation.

$$\frac{n}{4} - 2 + 2 = 3 + 2$$

$$\frac{n}{4} = 5$$

c. "If his number divided by 4 equals 5, then it must be 5 multiplied by 4." So Jasmine "undid" the division by 4. She multiplied by 4 on both sides.

$$\frac{n}{4} \times 4 = 5 \times 4$$

d. Jasmine thought, "If I simplify this equation, I get 20. Bill's number must be 20."

$$n = 20$$

e. "I will check this number to be sure."

$$\frac{20}{4} - 2 = 3$$

Work in groups. Get a calculator.

1. Make up a puzzle like the one Bill told Jasmine:
   - **a.** Write down a mystery answer.
   - **b.** Divide your mystery answer by a number that evenly divides it.
   - **c.** Add or subtract some other number to your result from step b.

2. Ask the other people in your group to find your mystery answer, using the following steps:
   - **a.** Tell them your final result and the steps you took to get it.
   - **b.** Have them write your problem as an equation using a variable.
   - **c.** Ask them to solve the equation.

3. Repeat this activity, using at least one negative integer.

4. a. Can you think of an equation in which the mystery answer and the number it is divided by are positive, yet the final result is negative?
   b. Can you think of an equation in which the mystery number is a mixed number or a decimal?

5. Do a similar activity, but use multiplication instead of division in your equation.

Solve and then compare all the answers to 6 and 7. Look for a pattern within each set and discuss with your group.

6. What number when
   a. divided by 3 and decreased by 2 equals $^{-}2$?
   b. divided by 3 and decreased by 2 equals $^{-}1$?
   c. divided by 3 and decreased by 2 equals 0?

7. What number when
   a. tripled and increased by 12 equals 24?
   b. tripled and decreased by 12 equals 0?
   c. multiplied by $^{-}3$ and increased by 12 equals 0?
   d. multiplied by $^{-}3$ and decreased by 12 equals $^{-}24$?

Solve each mystery. Discuss with your group strategies you used.

8. Fifteen pirates divided the gold in a treasure chest. Then they all contributed 5 gold pieces to a worthy cause. If each pirate had 53 gold pieces left, how many pieces of gold were in the treasure chest to begin with?

9. Jack said to Jill: "I remember that I had more than $50 this morning. I divided my money into 2 or 3 equal parts—I don't remember which. Then I went shopping with one of these parts. I bought a shirt for $22 but didn't have quite enough, so I borrowed $3 from you." Exactly how much money did Jack have this morning?

10. Make up a mystery to fit one of these equations; then have your group solve the mystery.

    a. $\frac{x}{3} + 7 = 28$ b. $3x - 9 = 15$ c. $2x + 12 = {}^{-}8$

Problem Solving

Writing an equation

To solve each problem:

- Write an equation.
- Solve the equation.
- Check the solution in the problem.

A produce manager accepted delivery of 15 bushels of apples. The delivery charge was \$8. The total bill was \$79.25. What was the cost of each bushel?

Let c be the cost of each bushel. Then $15c$ is the cost of 15 bushels. So, the cost of 15 bushels and the delivery charge equals the total bill.

Equation: $\overbrace{15c + 8}^{\text{total bill in dollars}} = \overbrace{79.25}^{\text{total bill in dollars}}$

Subtract 8. $15c + 8 - 8 = 79.25 - 8$

Simplify. $15c = 71.25$

Divide by 15. $\frac{15c}{15} = \frac{71.25}{15}$

Simplify. $c = 4.75$

Check. $15(4.75) + 8 = 79.25$

The cost of each bushel was \$4.75.

EXERCISES

Use an equation to solve each problem.

1. Last year an apple orchard yielded 475 bushels per acre. That was 38 bushels per acre less than it yielded this year. How many bushels per acre did it yield this year?
2. One particular apple tree yielded 20 bushels of apples. The average number of apples per bushel is about 160. About how many apples were on this tree?

3. The owner of an apple orchard paid a beekeeper \$130 for 7 hives of bees to pollinate the apple blossoms on his trees. The owner was charged \$15 a day and a \$40 delivery-and-pick-up fee. How many days did the owner have the bees?

4. Each tree in an apple orchard was expected to yield 15 bushels of apples. At harvest the total yield was found to be 47,200 bushels, which was 80 bushels short of what was expected. How many trees were in the orchard?

5. An apple orchard has 421 trees. That is 54 trees more than it had 5 years ago. How many trees did it have 5 years ago?

6. The Ruiz family went to a pick-your-own apple orchard. They picked 8 bushels of apples and were charged \$25.60. What was the cost of each bushel?

7. One morning insecticide was sprayed on 420 trees in an apple orchard. This number was $\frac{1}{2}$ of the number of trees sprayed in the afternoon. How many trees were sprayed in the afternoon?

★ 8. Sarah was paid 75¢ a bushel for picking apples. After she spent \$2 of what she had earned, she had \$25 left. How many bushels did she pick?

Chapter Checkup

Give each sum. [pages 60–61, 63, 67]

1. $^-3 + ^+8$
2. $^+5 + ^+7$
3. $^-8 + ^-5$
4. $^+7 + ^-10$

Give each difference. [pages 62–63, 67]

5. $^-3 - ^+4$
6. $^+5 - ^-6$
7. $^+8 - ^+9$
8. $^-6 - ^-1$

Give each product. [pages 64–65, 67]

9. $^-5 \times ^+8$
10. $^-3 \times ^-6$
11. $^+7 \times ^+4$
12. $^+9 \times ^-2$

Give each quotient. [pages 66–68]

13. $^+20 \div ^-4$
14. $^-15 \div ^+5$
15. $^-24 \div ^-6$
16. $^+49 \div ^+7$

Solve. [pages 80–83, 86–89]

17. $x + 37 = 95$
18. $74 = y - 19$
19. $y - 8 = ^-5$
20. $5 + n = 0$
21. $85 = 5t$
22. $^-4a = 20$
23. $^-6 = 2x$
24. $12r = 60$
25. $\frac{t}{2} = 3$
26. $3 = \frac{r}{^-4}$
27. $\frac{n}{4} = 0$
28. $^-1 = \frac{w}{10}$
29. $3t + 18 = 57$
30. $15 = 3 + 6t$
31. $\frac{w}{3} + 9 = 13$
32. $87 = 5n - 13$
33. $^-2a + 7 = 11$
34. $7 + ^-2a = ^-11$
35. $\frac{n}{4} - 2 = ^-7$
36. $5 = 1 + \frac{w}{5}$
37. $10t + 13 = 63$

First write an equation. Then solve. [pages 68, 84–85, 90–91]

38. 23 less than a number is 55. What is the number?

39. If you divide a number by 7, you get $^-21$. What is the number?

40. Bill has 4 times as many stamps as Maria. If Bill has 228 stamps, how many stamps does Maria have?

41. Anne jogged 2 kilometers more than 3 times as far as Jill. If Anne jogged 11 kilometers, how many kilometers did Jill jog?

Chapter Project

ADDING AND SUBTRACTING INTEGERS

Make a slide rule for adding integers. Use two strips of paper. Mark them so the integers are equally spaced. To find $^-3 + 2$, follow these steps:

Step 1. Slide scale A so that its 0 mark is directly under the mark for $^-3$ on scale B.

Step 2. Locate the mark for 2 on scale A. Read the answer on scale B directly above the mark for 2 on scale A.

$^-3 + 2 = {}^-1$

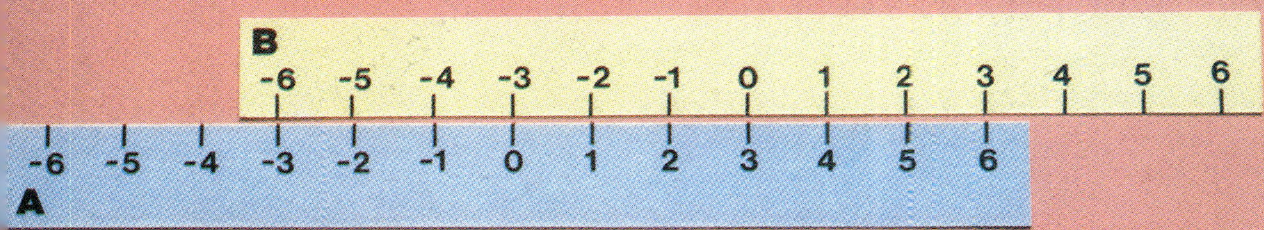

Use your slide rule to find each sum.

1. $^-4 + 2$	**2.** $^-5 + 4$	**3.** $^-1 + 2$	**4.** $^-1 + 5$
5. $^-2 + {}^-3$	**6.** $^-3 + {}^-1$	**7.** $^-4 + 3$	**8.** $5 + {}^-4$
9. $3 + 2$	**10.** $4 + {}^-4$	**11.** $6 + {}^-5$	**12.** $^-2 + 5$

Find a way to use your slide rule to give each difference.

13. $3 - {}^-1$	**14.** $4 - {}^-1$	**15.** $3 - {}^-2$	**16.** $1 - {}^-3$
17. $^-1 - {}^-2$	**18.** $^-3 - 1$	**19.** $^-2 - {}^-2$	**20.** $^-4 - {}^-2$
21. $0 - 5$	**22.** $4 - 3$	**23.** $0 - {}^-4$	**24.** $1 - 3$

Chapter Review

$^{-}7 + ^{+}2 = ^{-}5$

To subtract a number, add its opposite.

$^{-}8 - ^{+}3 = ^{-}8 + ^{-}3$
$= ^{-}11$

If the signs are the same, the product or quotient is positive.

$^{+}6 \times ^{+}3 = ^{+}18$
$^{-}16 \div ^{-}2 = ^{+}8$

If the signs are different, the product or quotient is negative.

$^{-}9 \times ^{+}4 = ^{-}36$
$^{+}21 \div ^{-}7 = ^{-}3$

$n + 9 = 6$
$n + 9 - 9 = 6 - 9$
$n = ^{-}3$

$3a - 7 = 20$
$3a - 7 + 7 = 20 + 7$
$3a = 27$
$\frac{3a}{3} = \frac{27}{3}$
$a = 9$

Give each sum.

1. $^{+}4 + ^{-}6$ 2. $^{-}3 + ^{-}2$ 3. $^{-}5 + ^{+}8$
4. $^{+}7 + ^{-}7$ 5. $^{-}5 + ^{-}6$ 6. $^{-}6 + ^{+}2$

Give each difference.

7. $^{-}4 - ^{+}1$ 8. $^{-}6 - ^{-}2$ 9. $^{+}9 - ^{+}8$
10. $^{+}8 - ^{-}5$ 11. $^{-}8 - ^{-}8$ 12. $^{-}3 - ^{+}7$

Give each product.

13. $^{+}5 \times ^{+}7$ 14. $^{-}3 \times ^{-}9$ 15. $^{+}6 \times ^{-}4$
16. $^{-}8 \times 0$ 17. $^{-}5 \times ^{+}4$ 18. $^{-}6 \times ^{-}7$

Give each quotient.

19. $^{-}42 \div ^{+}6$ 20. $^{+}48 \div ^{+}8$ 21. $^{+}56 \div ^{-}7$
22. $^{-}27 \div ^{-}9$ 23. $^{+}40 \div ^{+}5$ 24. $^{-}54 \div ^{+}9$

Solve.

25. $x + 7 = 18$ 26. $29 + y = 63$
27. $w - 12 = 43$ 28. $^{-}3 = t - 12$
29. $3n = 27$ 30. $65 = -5r$
31. $\frac{b}{4} = 10$ 32. $^{-}5 = \frac{d}{^{-}3}$

Solve.

33. $3a + 5 = 23$ 34. $64 = 10 + 6c$
35. $\frac{b}{4} - 6 = 2$ 36. $9 = 4 + \frac{d}{3}$
37. $^{-}2a - 3 = 7$ 38. $^{-}7 = 4r - 3$
39. $4 + \frac{c}{5} = 8$ 40. $\frac{n}{^{-}2} - 3 = 10$

Chapter Challenge

GRAPHING INEQUALITIES

Mathematical sentences that contain these symbols are called *inequalities.*

$<$	$>$	$\leq$	$\geq$	$\neq$
is less than	is greater than	is less than or equal to	is greater than or equal to	is not equal to

There are many integers that are solutions to this inequality:

$$2y < {}^-8$$

In fact, every integer less than $^-4$ is a solution.

You can show all of the solutions on a number-line graph.

Here are other examples.

$2y \leq {}^-8$

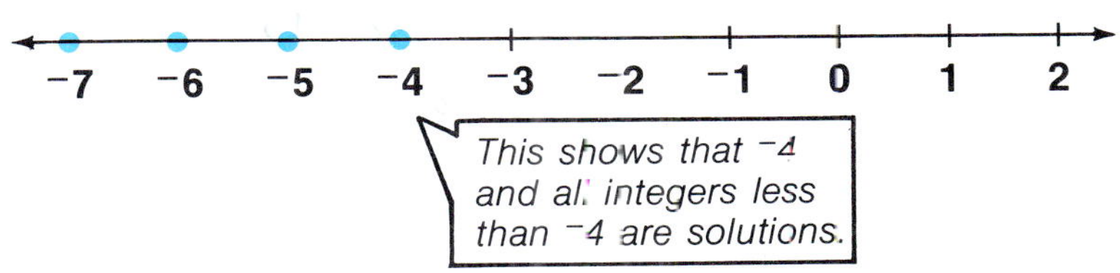

$m + {}^-2 \neq 1$

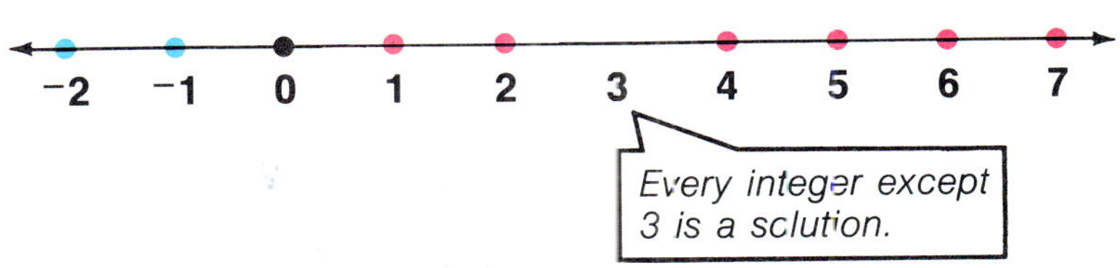

EXERCISES

Make graphs to show which integers are solutions to these inequalities.

1. $g - 3 < {}^-1$
2. $r + 4 > {}^-1$
3. $t + 3 \geq 2$
4. $x + {}^-2 < 2$
5. $3a \leq 9$
6. $2b > {}^-4$
7. $\frac{c}{5} < 0$
8. $\frac{m}{2} \neq {}^-2$
9. $2t + 1 \leq 7$

Cumulative Checkup

STANDARDIZED FORMAT

Choose the correct letter.

1. 5.0349 rounded to the nearest hundredth is

a. 5.03
b. 5.04
c. 5.035
d. none of these

2. Which of these numbers is greater than 0.2?

a. 0.09
b. 0.10
c. 0.199
d. none of these

3. Add.

48.3 + 7.91

a. 56.21
b. 127.4
c. 12.74
d. none of these

4. Subtract.

72.4 − 3.77

a. 34.7
b. 3.47
c. 68.63
d. none of these

5. Multiply.

4.21 × 0.07

a. 0.2947
b. 2.947
c. 0.02947
d. none of these

6. Divide.

$5.6\overline{)11.368}$

a. 20.3
b. 2.3
c. 2.03
d. none of these

7. $^{-}8 + 4 = \underline{?}$

a. 4
b. $^{-}4$
c. $^{-}12$
d. none of these

8. $^{-}2 \times {}^{-}6 = \underline{?}$

a. $^{-}12$
b. 12
c. 3
d. none of these

9. The solution of $y - 18 = 34$ is

a. 16
b. 24
c. 52
d. none of these

10. The solution of $6w + 18 = 48$ is

a. 30
b. 5
c. 11
d. none of these

11. Ellen had 3 times as many records as Frank. Ellen had 12 records. How many records did Frank have?

a. 4 **b.** 9
c. 36 **d.** none of these

12. Alan bought some \$8 records and one \$12 tape. He spent \$68. How many records did he buy?

a. 5 **b.** 6
c. 7 **d.** none of these

3

Measurement

The flask contains 231 milliliters of chemical. Susan is pouring 159 milliliters into the graduated cylinder. How much will be left in the flask?

Measuring length

The **meter** (m) is the basic unit for measuring length in the metric system. The length of the leather strip is about 1 meter.

These units are used to measure length in the metric system:

1 kilometer (km) = 1000 meters
1 hectometer (hm)* = 100 meters
1 dekameter (dam)* = 10 meters
1 meter (m) = 1 meter
1 decimeter (dm)* = 0.1 meter
1 centimeter (cm) = 0.01 meter
1 millimeter (mm) = 0.001 meter

* These units are seldom used.

Notice that both the metric system and our place-value system are based on the number 10. Because of this, we use decimals to write metric measurements.

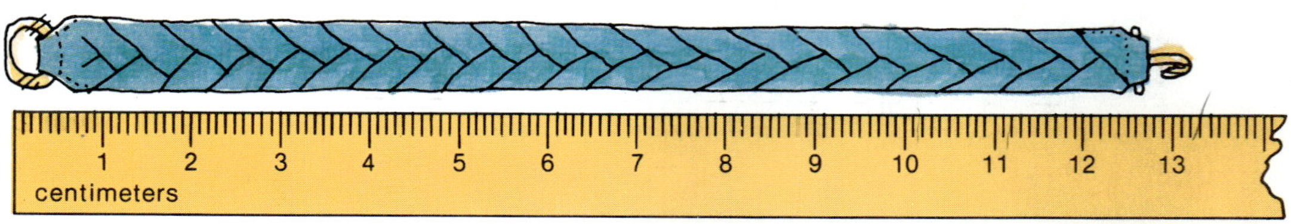

The length of the bracelet is 1 dm + 3 cm + 2 mm. Here are three ways that we can write the length of the bracelet:

1.32 dm
13.2 cm
132 mm

EXERCISES

Draw segments of these lengths.

1. 1 cm **2.** 5.2 cm **3.** 11.4 cm **4.** 19 mm **5.** 50 mm
6. 123 mm **7.** 1.5 dm **8.** 1.43 dm **9.** 14.3 cm **10.** 143 mm

11. **a.** Draw a segment 1.46 dm long.
b. How many cm long is the segment?
c. How many mm long is it?

12. **a.** Draw a segment 12.5 cm long.
b. How many dm long is the segment?
c. How many mm long is it?

13. **a.** Draw a segment that is 165 mm long.
b. How many cm long is the segment?
c. How many dm long is it?

14. **a.** On a chalkboard draw a segment that is 1.35 m long.
b. How many dm long is the segment?
c. How many cm long is it?
d. How many mm long is it?

Multiple choice. Pick the answer that seems reasonable.

15. The thickness of a dime is **a.** 1 cm **b.** 1 m **c.** 1 mm **d.** 1 dm

16. The diameter (length across) of a nickel is
a. 2.1 mm **b.** 2.1 m **c.** 2.1 dm **d.** 2.1 cm

17. The length of a new pencil is **a.** 1.89 dm **b.** 18.9 mm **c.** 1.89 m **d.** 189 cm

18. The height above the floor of a basketball hoop is
a. 3.05 dm **b.** 3.05 cm **c.** 3.05 m **d.** 3.05 km

Copy and complete this table. Estimate first!

	Length	*Estimate of length*	*Measurement of length*	*Difference between estimate and measurement*
19.	width of desk	?	?	?
20.	height of desk	?	?	?
21.	length of room	?	?	?
22.	width of room	?	?	?
23.	height of ceiling	?	?	?
24.	width of door	?	?	?

Changing units in the metric system

Since the metric system is based on the number 10, we can easily change from one unit to another by multiplying or dividing by a power of 10.

Study these examples.

EXAMPLE 1. Change 12.6 cm to mm.

Step 1. Recall: 1 cm = 10 mm

Step 2. Think: Since we are changing to a smaller unit, we should get a larger number. So, we should multiply.

Step 3. Compute: $12.6 \times 10 = 126$
So, 12.6 cm = 126 mm.

EXAMPLE 2. Change 15.6 dm to m.

Step 1. Recall: 10 dm = 1 m

Step 2. Think: Since we are changing to a larger unit, we should get a smaller number. So, we should divide.

Step 3. Compute: $15.6 \div 10 = 1.56$
So, 15.6 dm = 1.56 m.

EXAMPLE 3. Change 215 cm to m.

Step 1. Recall: 100 cm = 1 m

Step 2. Think: We should get a smaller number, so we divide.

Step 3. Compute: $215 \div 100 = 2.15$
So, 215 cm = 2.15 m.

EXAMPLE 4. Change 14 km to m.

Step 1. Recall: 1 km = 1000 m

Step 2. Think: We should get a larger number, so we multiply.

Step 3. Compute: $14 \times 1000 = 14{,}000$
So, 14 km = 14,000 m.

EXERCISES

1. Change 4.2 cm to mm.
2. Change 397 mm to m.
3. Change 7563.2 m to km.
4. Change 14.2 m to cm.

Copy and complete.

	km	*hm*	*dam*	*m*	*dm*	*cm*	*mm*
5.				100			
6.	3.2						
7.							3962
8.						4281	

Copy and complete.

9. 7 m = ? dm
10. 3 m = ? cm
11. 5 m = ? mm
12. 8 km = ? m
13. 3.9 km = ? m
14. 563 m = ? km
15. 672 cm = ? m
16. 43.2 cm = ? m
17. 8452 mm = ? m
18. 57.3 mm = ? cm
19. 3.9 m = ? mm
20. 3.9 mm = ? m
21. 349.4 cm = ? m
22. 583.1 dm = ? m
23. 42.71 km = ? m
24. 35.7 cm = ? m
25. 426 mm = ? m
26. 53.2 m = ? mm
27. 8 km = ? mm
28. 476.3 mm = ? km
29. 9 mm = ? km

Problem Solving

30. A distance record for throwing a fresh egg was 964.6 decimeters. How many meters was that?

31. One distance record for throwing a Frisbee was 128 meters. How many centimeters was that?

32. A height record for kite flying was 10.830 kilometers. How many meters was that?

33. The longest paper chain (made in less than a day) was made by 75 high school students in San Jose, California. It measured 21,762 meters. How many kilometers was that?

Exploring errors in measurement

No measurements are exact. In this lesson, you will explore what causes some errors in measurements and what happens when you add and subtract measurements.

Work in groups. Get a meterstick marked in centimeters only and a meterstick marked in millimeters.

PART A

Have each person measure the length of a desk top as accurately as possible using the meterstick marked in centimeters.

Now have each person measure the desk top again as accurately as possible, but this time using the meterstick marked in millimeters.

Record the measurements in a chart like the one shown.

Name	Length of Desk Top	
	as accurately as possible using cm	as accurately as possible using mm

Discuss.

1. In which column of your chart are more of the measurements the same? In which column are the measurements more accurate?
2. Think about the way you measured the desk top. What caused errors, or inaccuracies, in your measurement? Which of these causes of error could have been avoided? Which could not have been avoided? Why?

PART B

3. Measure the length of two sides of your chalkboard, door, or window. Measure using a meterstick marked in centimeters. Fill in the first column of the chart on page 103.

Side	Measured Length	Possible Range of True Length	
		Greatest	Smallest
1			
2			

4. The **greatest possible error** for your measurement is 0.5 cm. This means the true length can range from as much as 0.5 cm more than the measured length to 0.5 cm less than the measured length. Why do you think this is so? If you had measured using the meterstick marked in mm, what would be the greatest possible error? What if you had used the meter as your unit of measure?

5. For each measurement, find the greatest that the true length could be. Then find the smallest it could be. Complete the chart.

Discuss.

6. Find the sum of the two lengths you measured. Figure out how large the sum of the true lengths could be. Then figure out how small it could be. How far off from the true sum might the sum of your measurements be? How do you know?

7. Find the greatest possible error **(g.p.e.)** for the difference in the two lengths. Is your answer reasonable? How do you know?

8. Compare the g.p.e. of the difference with the g.p.e. of the sum. Does this result make sense? Why or why not?

9. Suppose you were measuring the length of 5 tables placed end-to-end. Should you measure each one and add, or should you measure all 5 together? Why?

10. Compare the way your group measured with the method used by another group. Which method do you think is more accurate? Why?

Perimeter

The **perimeter** of a figure is the distance around it.

Triangle

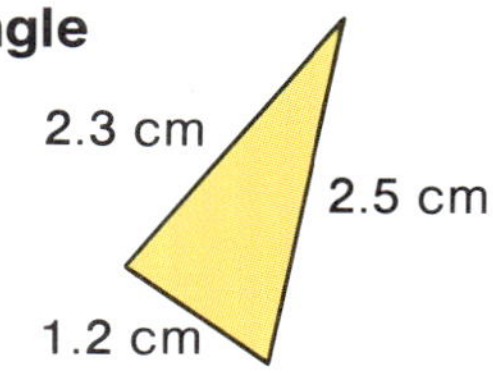

$$\begin{array}{r} 2.3 \text{ cm} \\ 2.5 \text{ cm} \\ +1.2 \text{ cm} \\ \hline 6.0 \text{ cm} \end{array}$$

The perimeter is 6.0 cm.

Square

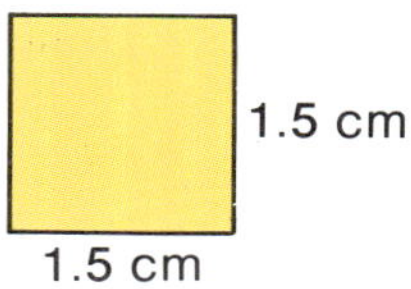

$$\begin{array}{r} 1.5 \text{ cm} \\ 1.5 \text{ cm} \\ 1.5 \text{ cm} \\ +1.5 \text{ cm} \\ \hline 6.0 \text{ cm} \end{array}$$

The perimeter is 6.0 cm.

The perimeter (P) of a square is 4 times the length of one side (s).

Formula

$P = 4s$
$P = 4(1.5 \text{ cm})$
$P = 6.0 \text{ cm}$

Rectangle

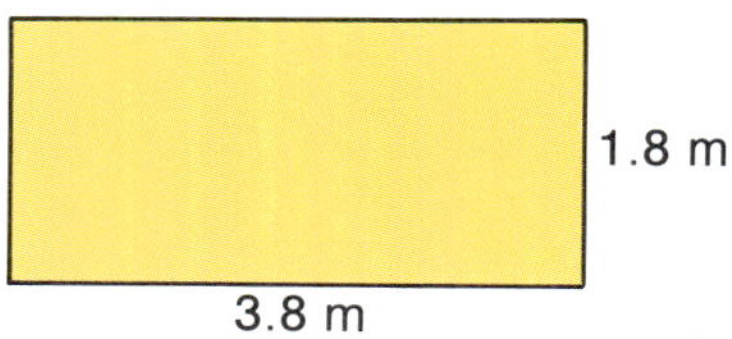

$$\begin{array}{r} 1.8 \text{ m} \\ 3.8 \text{ m} \\ 1.8 \text{ m} \\ +3.8 \text{ m} \\ \hline 11.2 \text{ m} \end{array}$$

The perimeter is 11.2 m.

The perimeter (P) of a rectangle is 2 times the sum of the length (l) and the width (w).

Formula

$P = 2(l + w)$
$P = 2(1.8 \text{ m} + 3.8 \text{ m})$
$P = 11.2 \text{ m}$

EXERCISES

Find each perimeter.

1.

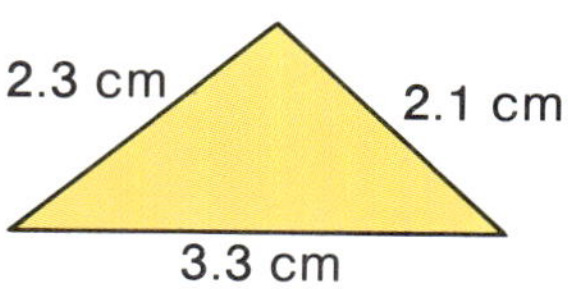

2.

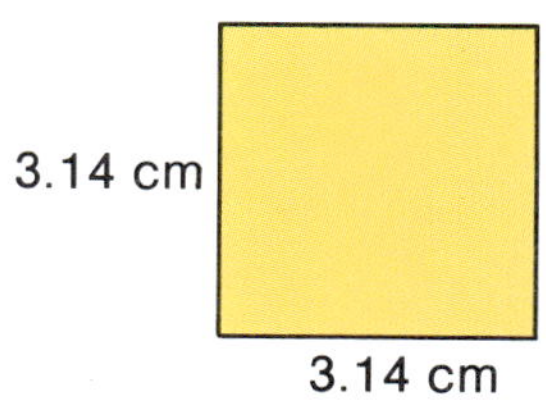

3.

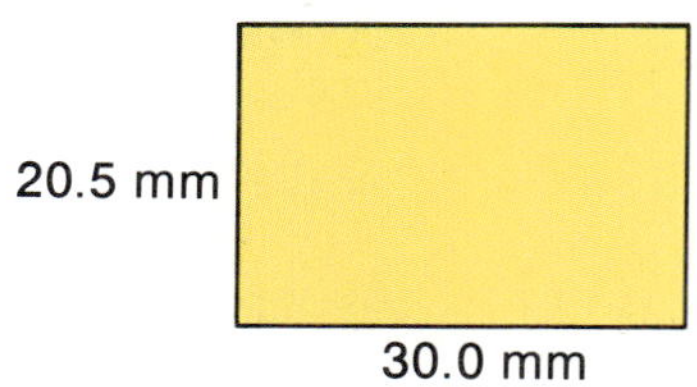

4. 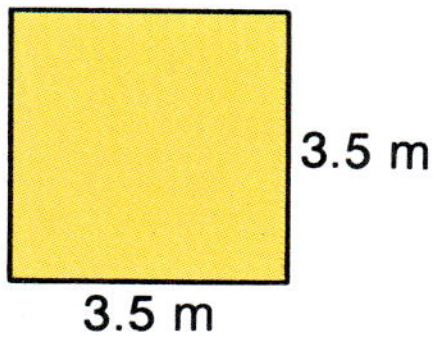

5. 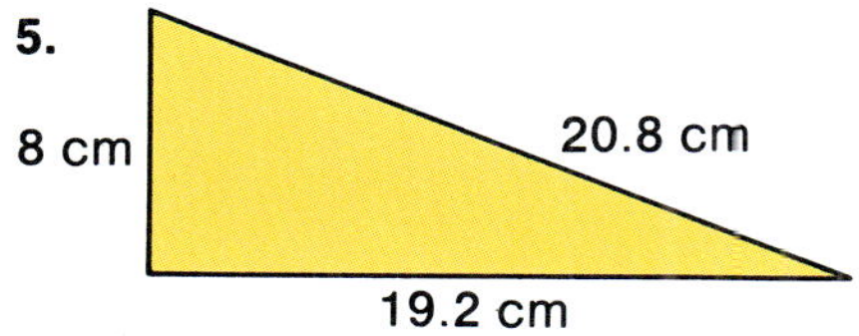

6. 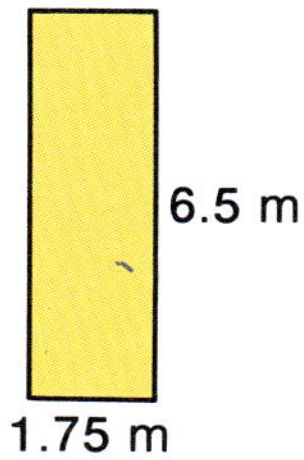

7. 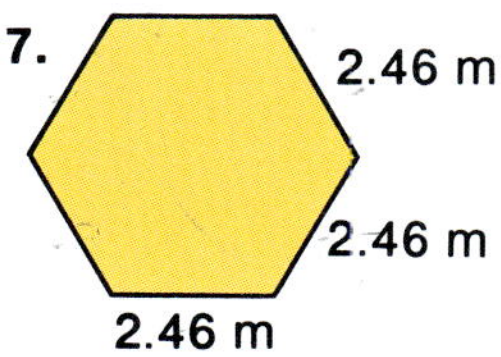

8. 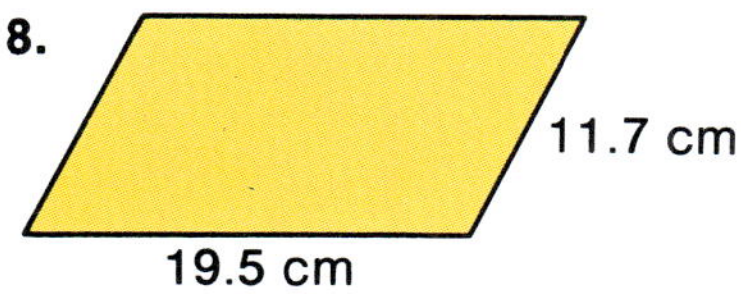

9. 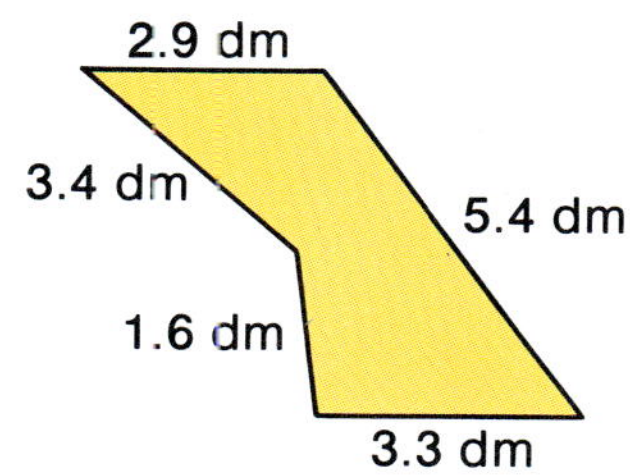

First measure each side to the nearest millimeter. Then compute the perimeter.

10. 

11. 

12. 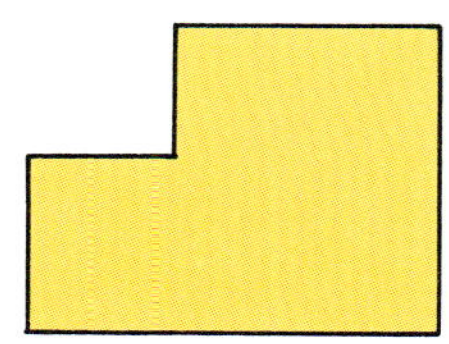

Problem Solving

13. A patio 6.3 m by 9.6 m is to be edged with brick. If each brick is to be 1.5 dm long, how many bricks will be needed?

14. A rectangular lot that measures 56 m by 120 m is to be fenced in with fencing that costs $4.30 a running meter. How much will the fencing cost?

15. The perimeter of a rectangular parking lot is 580 m. The lot is 150 m long. What is its width?

★ **16.** Each side of a square measured to the nearest cm is 12 cm. What is the g.p.e. of the measurement of each side? What is the perimeter? What is the g.p.e. of the perimeter?

Circumference and chords

Work in groups. Get a meterstick, a can or some other cylinder, and a calculator.

The distance around a circle is called its **circumference.**

1. Here is one way to find the circumference of the can:
   a. Tape some paper onto a flat surface. Mark a point on the circumference of the can. Carefully mark the distance on the paper that the can moves when you roll it one complete turn.
   b. Measure this distance accurately. Why is this distance equal to the circumference?

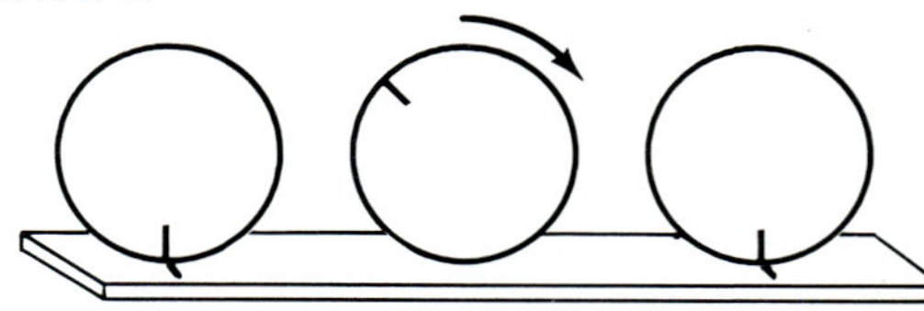

A **chord** is any segment with both endpoints on the circle. A circle's **diameter** is any of its longest chords.

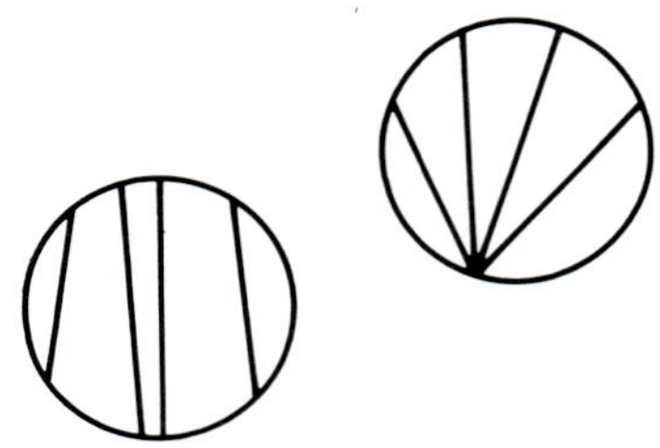

2. Use the can to draw a circle. To measure the circle's diameter:
   a. Mark a point on the circle. Then carefully measure the longest possible chord from that point. This chord is a diameter.
   b. Check by measuring the longest chord from another point.

3. Compare the can's circumference with its diameter.
   a. Record your group's and another group's answers in a chart like this one.

Can	C	d	$\pi = C/d$

π (read as "pi") is the circumference (C) of a circle divided by its diameter (d).

   b. About how many times longer is any circle's circumference than its diameter?
   c. How can you use all the calculations of π to find the best guess of π?
   d. Mathematicians have calculated the value of π to thousands of decimal places. Look up *pi* in your Glossary to get a close approximation of its value. Compare this with your best guess.

4. Suppose you know a tire's diameter is 35 cm.
   a. How can you find its circumference without measuring?
   b. Make up a rule to find the circumference when the diameter is known.
   c. Compare your rule to the method you used in problem 1.

5. Use the can to trace a circle on paper.

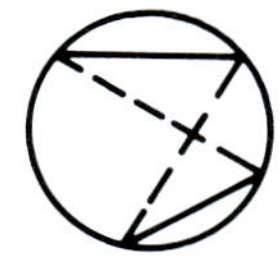

   a. Draw two chords of equal length that do not intersect. Connect opposite ends of the chords. Notice that you have made two triangles. Measure the sides of each triangle.
   b. Repeat this activity several times with different chords. Make up a rule to describe your results. Discuss.

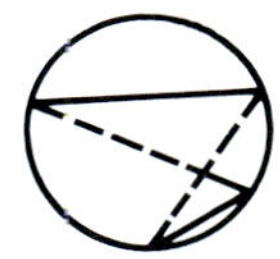

6. Draw a circle. Draw two chords that do not intersect but make one chord twice as long as the other. Connect opposite ends of the chords.
   a. Make a chart like the one shown. Measure only one side of one triangle. Record the information in your chart. Can you guess the length of the corresponding side of the other triangle? See if your guess is correct.

	Small Triangle	Large Triangle
Side 1		
Side 2		
Side 3		

   b. Measure all the other sides of the triangles. Record. What do you notice?
   c. Cut out the two triangles. Compare their sizes and shapes.

7. What would happen if you did an activity similar to exercise 6, but you began with one chord three times as long as the other? Try this activity to see if your predictions are true. Make a rule that says what you have discovered.

8. Did you use a method of computation (mental math, paper and pencil, calculator) or estimation for the activities in this lesson? Why? Discuss.

Liquid volume

The **liter (L)** is a unit for measuring liquid volume in the metric system.

1000 milliliters (mL) = 1 liter (L)

1 milliliter = 0.001 liter

1 L = 1000 cubic centimeters

1 mL = 1 cubic centimeter

1 liter

10 cm

10 cm

10 cm

The units for measuring liquid volume are related in the same way as the units for measuring length. Therefore, you can change from one unit of liquid volume to another in the same way as you do with units of length.

EXAMPLE. Change 325 mL to L.

Step 1. Recall: 1000 mL = 1 L

Step 2. Think: Since we are changing to a larger unit, we should get a smaller number. So we divide.

Step 3. Compute: $325 \div 1000 = 0.325$

So, 325 mL = 0.325 L.

EXERCISES

Complete.

1. 456 mL = ? L
2. 925 mL = ? L
3. 3625 mL = ? L
4. 8 L = ? mL
5. 0.375 L = ? mL
6. 2.5 L = ? mL
7. 382 L = ? mL
8. 8000 mL = ? L
9. 500 mL = ? L
10. 250 mL = ? L

Liter or milliliter? Which unit would most likely be used to measure the liquid volume of

11. a small milk carton?
12. a drinking glass?
13. a large milk carton?
14. a soft-drink can?
15. an eyedropper?
16. a soup spoon?

Choose the most reasonable liquid volume.

17.

a. 300 mL
b. 30 mL
c. 3 mL

18.

a. 0.1 L
b. 1 L
c. 10 L

19.

a. 4 mL
b. 4 L
c. 40 L

20.

a. 0.6 L
b. 6 L
c. 60 L

KEEPING SKILLS SHARP

Give each product.

1. 23×100
2. 8.6×10
3. 0.53×100
4. 4.6×1000
5. 0.074×1000
6. 9.1×100
7. 2.62×1000
8. 59.3×10
9. 0.745×100
10. 83.5×1000

Give each quotient.

11. $68 \div 10$
12. $12.6 \div 100$
13. $32.18 \div 1000$
14. $594.6 \div 100$
15. $38.72 \div 10$
16. $74.12 \div 1000$
17. $83.4 \div 100$
18. $0.86 \div 10$
19. $52.3 \div 100$
20. $9.54 \div 1000$

Measuring weight (mass)

The **gram (g)** is a unit for measuring weight (mass) in the metric system.

1 kilogram (kg) = 1000 grams (g)

1 milligram (mg) = 0.001 gram

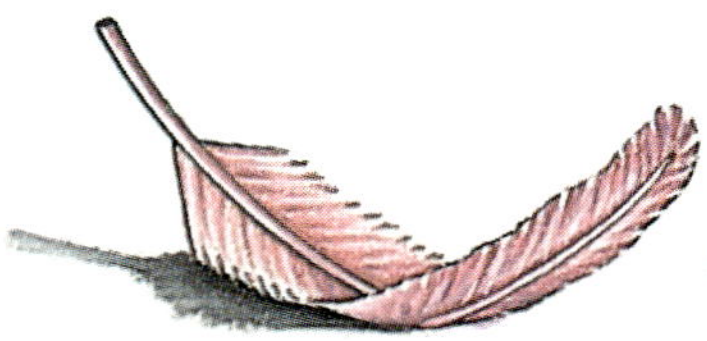

50 milligrams

400 grams

2 kilograms

EXERCISES

Complete.

1. 1 g = ? mg
2. 3 g = ? mg
3. 22 g = ? mg
4. 0.4 g = ? mg
5. 3.2 g = ? mg
6. 3000 mg = ? g
7. 1500 mg = ? g
8. 750 mg = ? g
9. 1 kg = ? g
10. 3.5 kg = ? g
11. 0.43 kg = ? g
12. 5000 g = ? kg
13. 4500 g = ? kg
14. 250 g = ? kg
15. 426.5 g = ? kg

Milligram, gram, or kilogram? Which unit would most likely be used to measure the weight of

16. a baseball?
17. a desk?
18. a straight pin?
19. a small pill?
20. a large piece of chalk?
21. an orange?
22. a bicycle?
23. an apple seed?
24. a sandwich?
25. your body?

Choose the most reasonable weight.

26. a baseball
- **a.** 142 mg
- **b.** 142 g
- **c.** 142 kg

27. a nickel
- **a.** 5 mg
- **b.** 5 g
- **c.** 5 kg

28. a bicycle
- **a.** 13 mg
- **b.** 13 g
- **c.** 13 kg

29. a small car
- **a.** 1100 mg
- **b.** 1100 g
- **c.** 1100 kg

30. an apple
- **a.** 105 mg
- **b.** 105 g
- **c.** 105 kg

31. a brick
- **a.** 0.9 mg
- **b.** 0.9 g
- **c.** 0.9 kg

32. Estimate the weight of this textbook. Then weigh it. How close was your estimate?

33. Estimate the weights of some objects in your classroom. Then weigh them. How close was each estimate?

In the metric system, the **Celsius** thermometer is used to measure temperature.

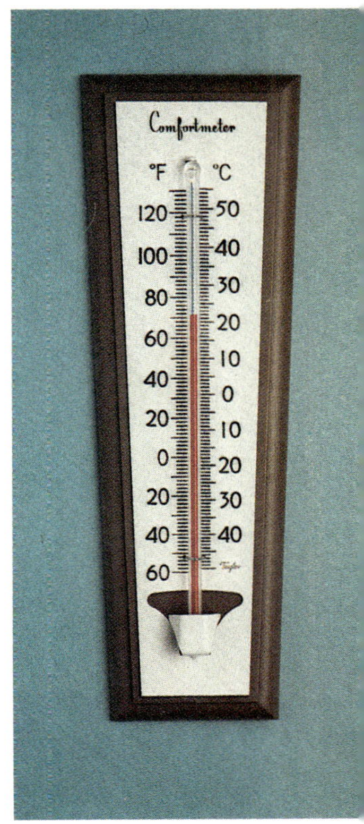

Choose the most reasonable temperature for

34. a very cold day. **a.** $^{-}10$°C **b.** 20°C **c.** 30°C

35. your classroom. **a.** 2°C **b.** 15°C **c.** 52°C

36. hot soup. **a.** 20°C **b.** 40°C **c.** 80°C

Problem Solving

37. One cold winter day, John recorded a temperature of $^{-}8$°C in the shade and 5°C in the sun. How much warmer was it in the sun?

38. One fall evening, the temperature dropped 20° from 6 P.M. to midnight. If the temperature was 19°C at 6 P.M., what was the temperature at midnight?

39. On the dashboard of Mary's car the thermometer read $^{-}7$°C when she first got in and 18°C twenty minutes later. How many degrees did the temperature in the car increase?

Computers in careers

Earlier you learned that a programmer often writes a flow chart before writing a program. Here is an example.

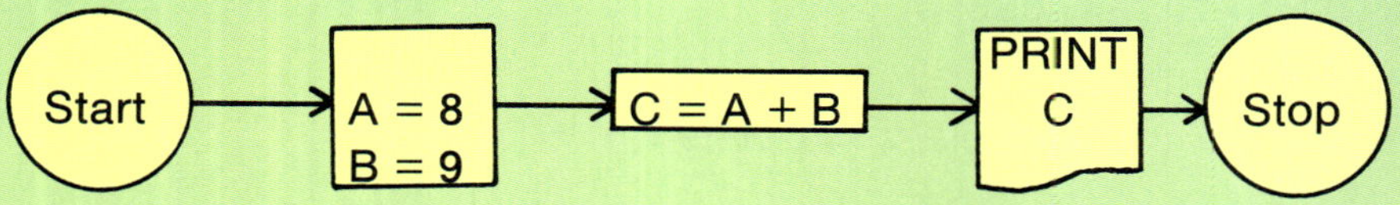

Once a flow chart is written for a computer, a detailed set of instructions, a program, must be written in a language that the computer understands. One simple computer language is called BASIC. Here is a program for the flow chart.

```
10   A = 8
20   B = 9
70   C = A + B
80   PRINT C
90   END
```

In this program there are 5 statements; each is on a separate line. The last statement must be "END."

Each line is numbered. Any number between 0 and 9999 may be used. As you can see, numbers may be skipped, but the numbers must be in order. The same number cannot be used more than once.

This program tells the computer to store the number 8 in a location in its memory referred to as A, to store 9 in a location referred to as B, and to store the sum in location C. Finally, in line 80, the computer is told to print the number stored in C. The last line tells the computer to stop.

These symbols can be used in BASIC programs:

+	(Add)	/	(Divide)
−	(Subtract)	↑	(Raise to a power)
*	(Multiply)	( )	(Grouping symbols)

The computer also recognizes the words PRINT and END. You will learn other words later.

Here are some sample program statements. Notice that the line number can be any number you choose.

```
15   X = A * B - 5
```
(Multiply A and B and then subtract 5.)

```
483   G = (A - B)/D
```
(Subtract B from A and divide that answer by D.)

```
37   R = M ↑ 3 - K
```
(Raise M to the third power and then subtract K.)

EXERCISES

Follow each program. Tell what the computer would print.

1.
```
1Ø   G = 1Ø
2Ø   H = 7
3Ø   I = 2
4Ø   K = G * H + I
5Ø   PRINT K
6Ø   END
```

2.
```
2ØØ   A = 7
3ØØ   B = 1
4ØØ   C = A ↑ 2 + B
5ØØ   PRINT C
6ØØ   END
```

3.
```
1ØØØ   X = 56
2ØØØ   Y = 2
3ØØØ   Z = (X - Y)/6
4ØØØ   PRINT Z
5ØØØ   END
```

4.
```
11Ø   A = 5
12Ø   B = 4
125   C = (A + B) * 2
13Ø   PRINT C
14Ø   END
```

5.
```
3ØØ   A = 3
4ØØ   B = 2
5ØØ   C = (A + B) ↑ 3
6ØØ   PRINT C
7ØØ   END
```

6.
```
4ØØØ   A = 8
5ØØØ   B = 4
6ØØØ   C = A ↑ 2 - B ↑ 2
7ØØØ   PRINT C
8ØØØ   END
```

7. Here is a flow chart for computing the circumference of a circle having a diameter of 3.

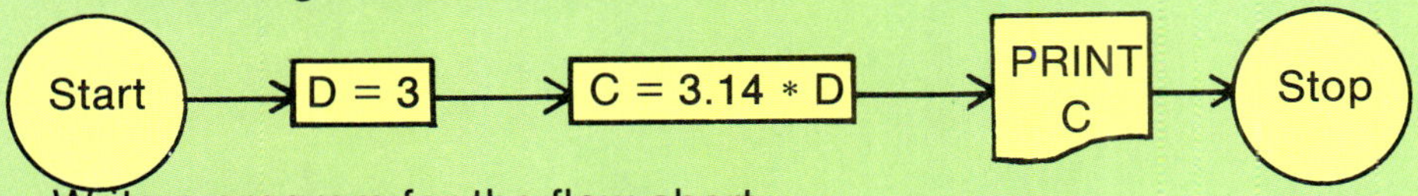

Write a program for the flow chart.

8. Write a program for computing the circumference of a circle having a radius of 8.

Length—customary units

12 inches (in. or ″) = 1 foot (ft or ′)

3 feet = 1 yard (yd)

36 inches = 1 yard

1760 yards = 1 mile (mi)

5280 feet = 1 mile

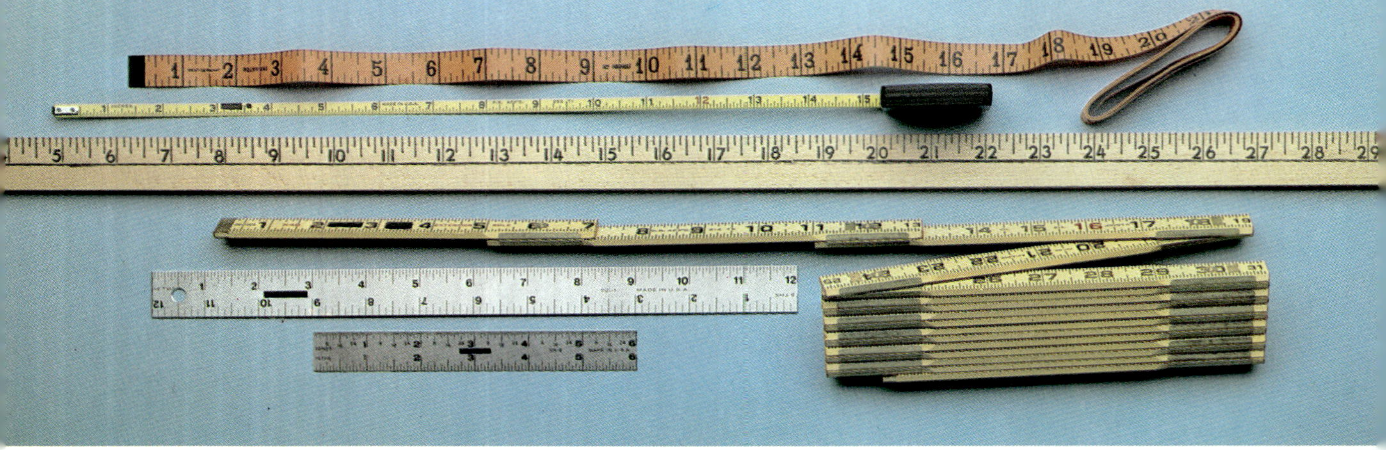

These examples show how to change from one customary unit of length to another.

EXAMPLE 1. Change 16 ft to in.

Step 1. Recall: 1 ft = 12 in.

Step 2. Think: Since we are changing from a larger unit to a smaller one, we should get a larger number. So, we should multiply.

Step 3. Compute:

$$\begin{array}{r} 16 \\ \underline{\times 12} \\ 32 \\ \underline{16} \\ 192 \end{array}$$

So, 16 ft = 192 in.

EXAMPLE 2. Change 76 ft to yd.

Step 1. Recall: 3 ft = 1 yd

Step 2. Think: Since we are changing to a larger unit, we should get a smaller number. So, we should divide.

Step 3. Compute:

$$\begin{array}{r} 25 \\ 3\overline{)76} \\ \underline{-6} \\ 16 \\ \underline{-15} \\ 1 \end{array}$$

So, 76 ft is $25\frac{1}{3}$ yd.
It is also 25 yd 1 ft.

EXERCISES

Copy and complete.

1. 8 yd = ? ft
2. 108 in. = ? ft
3. 3 yd = ? in.
4. 27 ft = ? yd
5. 4 ft = ? in.
6. 108 in. = ? yd
7. 1 yd 5 in. = ? in.
8. 51 ft = ? yd
9. 15 yd = ? ft
10. 4 mi = ? yd
11. 3 ft 4 in. = ? in.
12. 19 in. = ? ft 7 in.
13. 9 yd 2 ft = ? ft
14. 32 ft = ? yd 2 ft
15. 3 mi = ? yd
16. 117 in. = ? ft 9 in.
17. 1 mi 440 ft = ? ft
18. 12 yd 1 ft = ? ft
19. 6 ft 5 in. = ? in.
20. 104 ft = ? yd 2 ft
21. 127 in. = ? ft 7 in.
22. 14 yd 1 ft = ? ft
23. 65 ft = ? yd 2 ft
24. 1 yd 8 in. = ? in.
25. 154 in. = ? ft 10 in.
26. 15 ft 11 in. = ? in.
27. 75 in. = ? yd 3 in.

Problem Solving

28. A **fathom** is used to measure the depth of water.
    1 fathom = 6 ft
    How many fathoms deep is water that is 282 feet deep?
29. The deepest spot in any of the oceans is the Mariana Trench. It is 6033 fathoms deep. How many feet deep is it?
30. A **furlong** is a measure of length used in horse racing.
    1 furlong = 220 yd
    How many furlongs are in 1 mile?
31. A **hand** is a unit for measuring the height of a horse.
    1 hand = 4 in.
    How many inches tall is a horse whose height is 17 hands?

Challenge!

Solve.

32. Sea distances are usually measured in **nautical miles.**
    1 nautical mile is about 6076.1 feet.
    A **knot** is a measure of speed at sea.
    1 knot is 1 nautical mile per hour.
    How many miles per hour is 1 knot?

Liquid volume and weight—customary units

2 cups (c) = 1 pint (pt)

2 pints = 1 quart (qt)

2 quarts = 1 half-gallon

2 half-gallons = 1 gallon (gal)

EXERCISES

Complete.

1. 10 c = ? pt
2. 18 pt = ? qt
3. 8 qt = ? gal
4. 5 qt = ? c
5. 14 pt = ? c
6. 3 gal = ? qt
7. 1 gal = ? pt
8. 4 qt = ? half-gallons
9. 16 c = ? qt
10. 20 pt = ? qt
11. 3 qt = ? c
12. 5 gal = ? half-gallons
13. 2 pt 1 c = ? c
14. 3 qt 1 pt = ? pt
15. 2 gal 3 qt = ? qt
16. 15 pt = ? qt 1 pt
17. 19 qt = ? gal 3 qt
18. 34 pt = ? gal 1 qt

In the metric system, dry material such as grain is measured with the same units used for liquids. In the customary system, different units are used for dry measure.

2 pints (pt) = 1 quart (qt)
8 quarts = 1 peck (pk)
4 pecks = 1 bushel (bu)

Complete.

19. 10 pt = ? qt
20. 48 qt = ? pk
21. 24 pk = ? bu
22. 1 bu = ? qt
23. 3 bu = ? pk
24. 27 pk = ? bu 3 pk

Here are some units for measuring weight.

16 ounces (oz) = 1 pound (lb)
2000 pounds (lb) = 1 ton (T)

1 lb

1 T

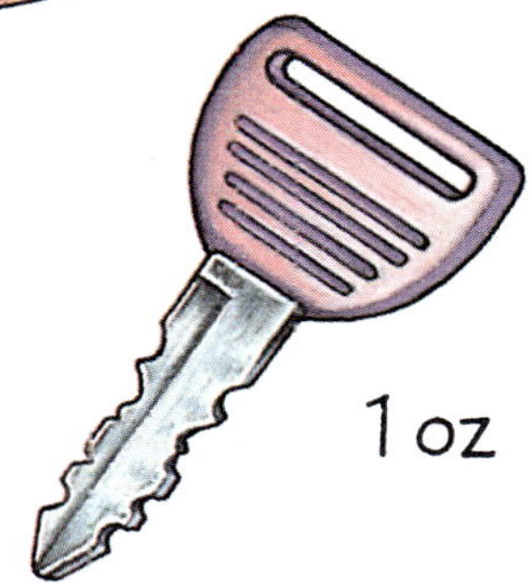

1 oz

Complete.

25. 5 lb = ? oz

26. 2 T = ? lb

27. 144 oz = ? lb

28. 27 lb = ? oz

29. 10,000 lb = ? T

30. 384 oz = ? lb

31. 1 lb 8 oz = ? oz

32. 2 lb 5 oz = ? oz

33. 3 lb 11 oz = ? oz

34. 20 oz = ? lb 4 oz

35. 50 oz = ? lb 2 oz

36. 108 oz = ? lb 12 oz

Problem Solving

37. A bread recipe calls for 2 lb of whole-wheat flour. If you have 1 lb 8 oz of flour, how much more do you need?

38. An empty truck weighs 11,000 lb. When full, it weighs 8 tons. What is the weight of its cargo in pounds?

Challenge!

Solve.

39. How can you get exactly 11 gallons of water using an 8-gallon bucket and a 5-gallon bucket?

40. How can you get exactly 6 gallons of water using an 8-gallon bucket and a 5-gallon bucket?

Computing with customary units

EXAMPLE 1. Add 1 ft 9 in. and 2 ft 5 in.

Step 1. Add inches and regroup.

$$\begin{array}{rr} \scriptstyle 1\text{ ft} & \\ 1\text{ ft} & 9\text{ in.} \\ +2\text{ ft} & 5\text{ in.} \\ \hline & 2\text{ in.} \end{array}$$

Step 2. Add feet.

$$\begin{array}{rr} \scriptstyle 1\text{ ft} & \\ 1\text{ ft} & 9\text{ in.} \\ +2\text{ ft} & 5\text{ in.} \\ \hline 4\text{ ft} & 2\text{ in.} \end{array}$$

EXAMPLE 2. Subtract 1 gal 3 qt from 3 gal 1 qt.

Step 1. Regroup.

$$\begin{array}{rr} \scriptstyle 2 & \scriptstyle 5 \\ \cancel{3}\text{ gal} & \cancel{1}\text{ qt} \\ -1\text{ gal} & 3\text{ qt} \\ \hline \end{array}$$

Step 2. Subtract quarts.

$$\begin{array}{rr} \scriptstyle 2 & \scriptstyle 5 \\ \cancel{3}\text{ gal} & \cancel{1}\text{ qt} \\ -1\text{ gal} & 3\text{ qt} \\ \hline & 2\text{ qt} \end{array}$$

Step 3. Subtract gallons.

$$\begin{array}{rr} \scriptstyle 2 & \scriptstyle 5 \\ \cancel{3}\text{ gal} & \cancel{1}\text{ qt} \\ -1\text{ gal} & 3\text{ qt} \\ \hline 1\text{ gal} & 2\text{ qt} \end{array}$$

EXAMPLE 3. Multiply 1 lb 8 oz by 3.

Step 1. Multiply ounces and regroup.

$$\begin{array}{rr} \scriptstyle 1\text{ lb} & \\ 1\text{ lb} & 8\text{ oz} \\ & \times 3 \\ \hline & 8\text{ oz} \end{array}$$

Step 2. Multiply pounds and add.

$$\begin{array}{rr} \scriptstyle 1\text{ lb} & \\ 1\text{ lb} & 8\text{ oz} \\ & \times 3 \\ \hline 4\text{ lb} & 8\text{ oz} \end{array}$$

EXAMPLE 4. Divide 38 yd 2 ft by 4.

Step 1. Divide yards.

$$\begin{array}{rrr} & 9\text{ yd} & \\ \hline 4\,) & 38\text{ yd} & 2\text{ ft} \\ & -36\text{ yd} & \\ \hline & 2\text{ yd} & \end{array}$$

Step 2. Regroup.

$$\begin{array}{rrr} & 9\text{ yd} & \\ \hline 4\,) & 38\text{ yd} & \cancel{2\text{ ft}} \\ & -36\text{ yd} & \\ \hline & \cancel{2\text{ yd}} & 8\text{ ft} \end{array}$$

Step 3. Divide feet.

$$\begin{array}{rrr} & 9\text{ yd} & 2\text{ ft} \\ \hline 4\,) & 38\text{ yd} & \cancel{2\text{ ft}} \\ & -36\text{ yd} & \\ \hline & \cancel{2\text{ yd}} & 8\text{ ft} \\ & & -8\text{ ft} \\ \hline & & 0 \end{array}$$

EXERCISES

Complete.

1. 3 ft 4 in. = $\underline{40}$ in.
2. 8 yd 2 ft = $\underline{\ ?\ }$ ft
3. 3 qt 1 pt = $\underline{\ ?\ }$ pt
4. 2 gal 3 qt = $\underline{\ ?\ }$ qt
5. 8 pt 1 c = $\underline{\ ?\ }$ c
6. 5 lb 2 oz = $\underline{\ ?\ }$ oz
7. 29 in. = 2 ft $\underline{\ ?\ }$ in.
8. 29 ft = $\underline{\ ?\ }$ yd 2 ft

Add.

9. 2 ft 8 in. + 1 ft 10 in.
10. 2 yd 2 ft + 5 yd 2 ft
11. 3 gal 3 qt + 1 gal 3 qt

Subtract.

12. 2 ft 3 in. − 1 ft 8 in.
13. 6 yd 1 ft − 3 yd 2 ft
14. 6 yd − 3 yd 1 ft

Multiply.

15. 1 ft 7 in. × 2
16. 4 qt 1 pt × 6
17. 7 lb 6 oz × 4

Divide.

18. $4\overline{)9 \text{ ft } 4 \text{ in.}}$
19. $2\overline{)7 \text{ yd } 1 \text{ ft}}$
20. $2\overline{)15 \text{ gal } 2 \text{ qt}}$

Find each perimeter or circumference. Use $\frac{22}{7}$ as an approximation for π.

21.

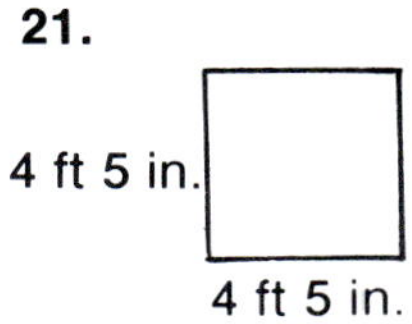

22.

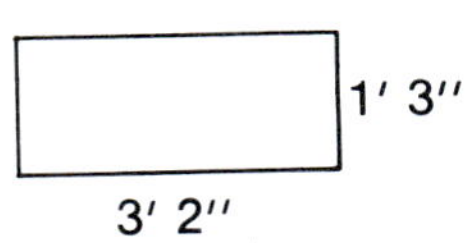

23.

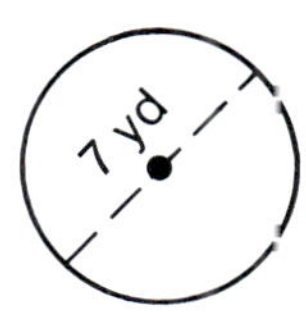

Challenge!

Solve.

24. The perimeter of a rectangular patio is 36 yd. The width is 6 yd 2 ft. What is its length?

KEEPING SKILLS SHARP

Solve and check.

1. $4a - 3 = 13$
2. $3y + 2 = {}^-16$
3. ${}^-5c - {}^-3 = {}^-22$
4. ${}^-42 = 6w + 6$
5. $12j + {}^-5 = {}^-5$
6. $\frac{w}{3} - 4 = {}^-2$
7. $7b + 2 = 16$
8. ${}^-11 = \frac{r}{6} - {}^-2$
9. $20 = 8d - 4$
10. $9x + 7 = 79$
11. ${}^-2a - 5 = {}^-13$
12. $12z + 9 = {}^-3$
13. $\frac{k}{3} - {}^-5 = 8$
14. $18 = \frac{b}{5} - 2$
15. $3f - 7 = 2$
16. $45 + 2s = 55$
17. ${}^-27 + {}^-4w = 9$
18. $\frac{y}{8} + 6 = 0$
19. $84 + 2d = 72$
20. $100 = \frac{z}{6} - 2$

Chapter Checkup

Copy and complete. [pages 98–101, 108–111]

1. 38.6 m = ? dm
2. 2.43 km = ? m
3. 5620 m = ? km
4. 56 cm = ? mm
5. 384 mm = ? m
6. 43.2 m = ? mm
7. 5.1 L = ? mL
8. 357.2 mL = ? L
9. 63.2 kg = ? g
10. 350 g = ? kg
11. 750 mg = ? g
12. 45 g = ? mg

Find the perimeter or circumference of each figure. Use 3.14 as an approximation for π. [pages 104–107]

13.

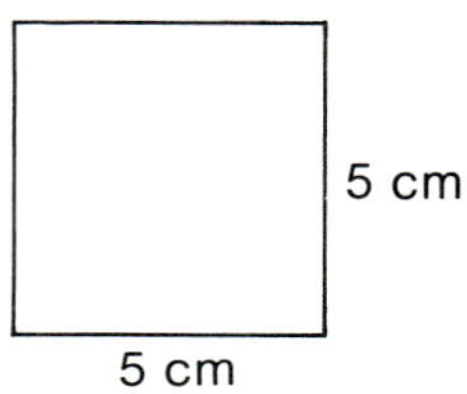

14.

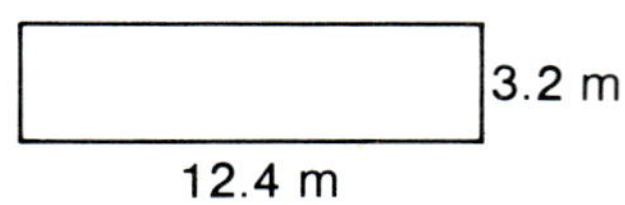

15.

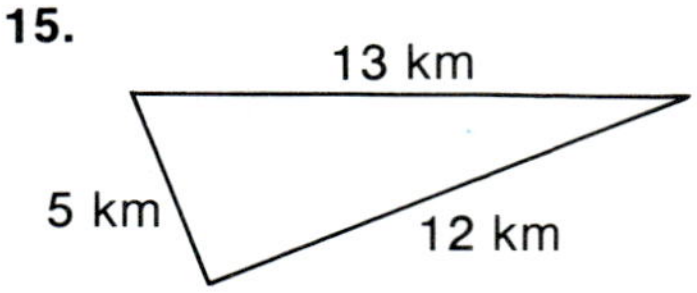

16.

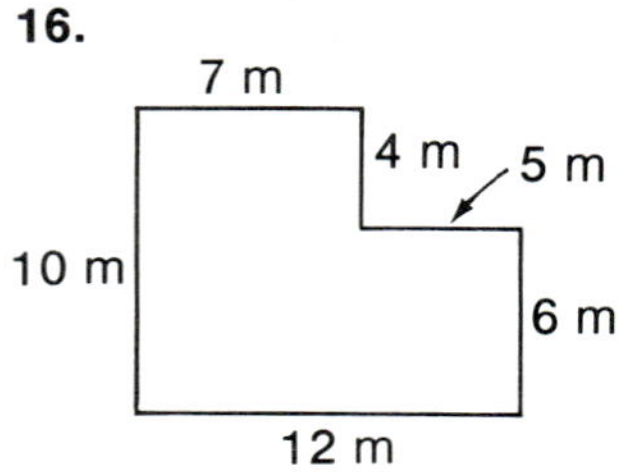

17.

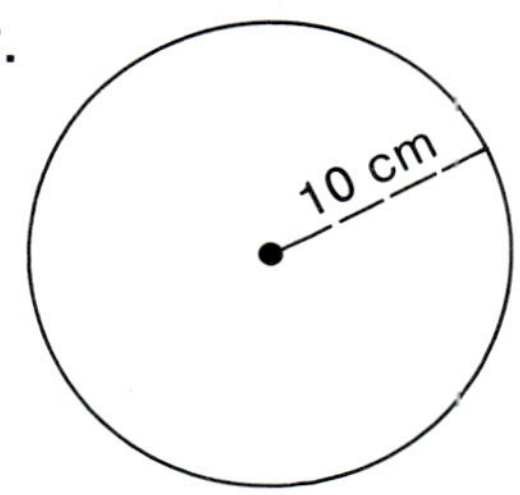

18.

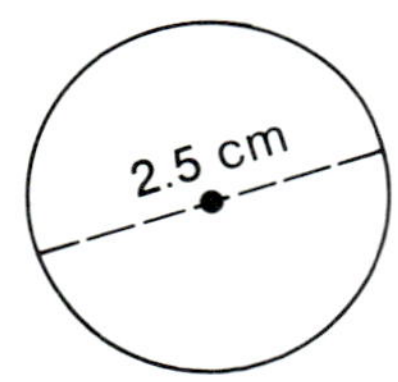

Copy and complete. [pages 114–117]

19. 8 ft = ? in.
20. 6 yd = ? ft
21. 54 in. = ? ft 6 in.
22. 7 lb = ? oz
23. 3.2 T = ? lb
24. 4 gal = ? qt
25. 15 pt = ? qt 1 pt
26. 17 pt = ? gal 1 pt
27. 3 qt = ? c

Compute. [pages 118–119]

28. 4 ft 6 in. × 3
29. 2 ft 8 in. + 1 ft 7 in.
30. 4 gal 1 qt − 1 gal 2 qt

Chapter Project

ESTIMATING SIZE

Think about the diameter of a basketball and a basketball hoop.

The diameter of the basketball hoop is clearly greater than the diameter of the basketball.

Estimate: How many times the diameter of the ball is the diameter of the hoop?

1.5 times? 1.7 times? 2 times? 2.4 times?

Measure: Find a way to measure the diameter of a basketball and the diameter of a basketball hoop.

Compute: How many times the diameter of the ball is the diameter of the hoop?

Was your estimate reasonably close?

Chapter Review

smaller unit ↓

13.4 m = ? mm

↑ larger number

So, multiply by 1000.

13.4 m = 13,400 mm

Complete.

1. 536 mm = ? m
2. 4.2 km = ? m
3. 17.3 cm = ? mm
4. 750 mL = ? L
5. 0.4 L = ? mL
6. 20 kg = ? g
7. 450 g = ? kg
8. 1230 mg = ? g

The perimeter of a figure is the distance around it.

The circumference of a circle is the distance around it.

$C = \pi d$

Give each perimeter (or circumference). Use 3.14 as an approximation for π.

9.

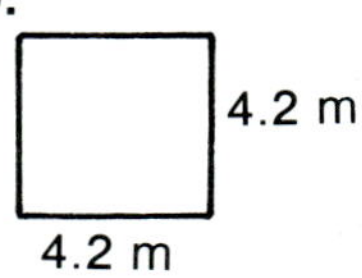

10.

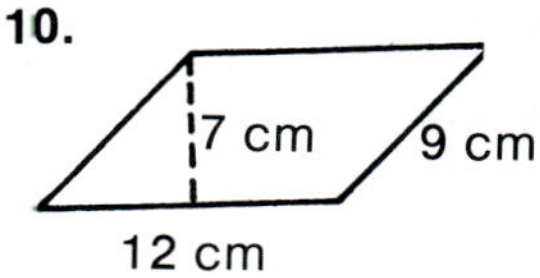

11.

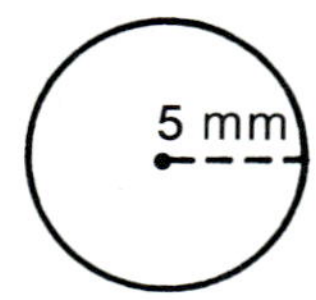

larger unit ↓

48 in. = ? ft

↑ smaller number

So, divide by 12.

48 in. = 4 ft

Copy and complete.

12. 8 ft = ? in.
13. 6 yd = ? ft
14. 1 mi = ? yd
15. 6 ft 3 in. = ? in.
16. 3 gal = ? qt
17. 4 qt = ? pt
18. 12 c = ? pt
19. 8 lb = ? oz

```
   2      19
   3̶ ft    7̶ in.
 − 1 ft   8 in.
 ─────────────
   1 ft  11 in.

   1 pt 1 c
         ×3
 ──────────
   4 pt 1 c
```

Compute.

20.
```
   1 ft 8 in.
 + 2 ft 9 in.
```

21.
```
   6 ft
 − 2 ft 7 in.
```

22.
```
 4 pt 1 c
       ×2
```

23. $2\overline{)3 \text{ yd } 8 \text{ in.}}$

Chapter Challenge

RELATIVE ERROR

Relative error

The length of the shorter piece of macramé is 12 cm to the nearest centimeter. The length of the longer one is 180 cm to the nearest centimeter.
In both cases the greatest possible error is 0.5 cm.

When compared to 12 cm (the shorter length) the g.p.e. is quite large.
When compared to 180 cm (the longer length) the g.p.e. is quite small.

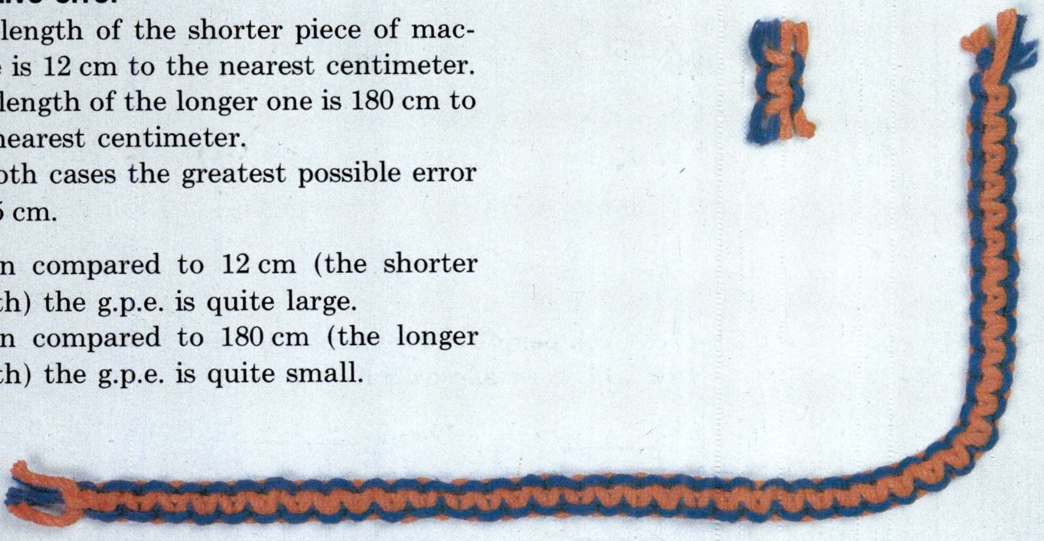

To find the **relative error** of a measurement, we can divide the greatest possible error by the measurement.

Relative error of 12-cm measurement

$$12\overline{)0.5000}\quad 0.0417$$

Relative error of 180-cm measurement

$$180\overline{)0.5000}\quad 0.0028$$

Notice that the smaller measurement has the greater relative error.

For each of these measurements, compute the relative error to the nearest ten-thousandth.

1. 15 cm to the nearest centimeter
2. 18 m to the nearest meter
3. 6.3 m to the nearest 0.1 m
4. 43.5 km to the nearest 0.1 km
5. 2.54 kg to the nearest 0.01 kg
6. 3.80 L to the nearest 0.01 L
7. Measure your handspan to the nearest centimeter. Find the relative error of your measurement.
8. Measure your height to the nearest centimeter. Find the relative error of your measurement.

Cumulative Checkup

STANDARDIZED FORMAT

Choose the correct letter.

1. Add.

$5.77 + 3.2$

a. 5.45
b. 8.97
c. 6.09
d. none of these

2. Subtract.

$2.001 - 0.693$

a. 1.308
b. 1.408
c. 1.418
d. none of these

3. Multiply.

$$\begin{array}{r} 0.04 \\ \times 0.03 \\ \hline \end{array}$$

a. 0.12
b. 0.012
c. 0.0012
d. none of these

4. Divide.

$4.2\overline{)13.02}$

a. 3.01
b. 3.1
c. 0.31
d. none of these

5. $3.75 \times 10^3 = \underline{?}$

a. 37.5
b. 375
c. 0.00375
d. none of these

6. $25 - (4 + 2) \times 3 = \underline{?}$

a. 69
b. 7
c. 57
d. none of these

7. Solve.

$4x - 6 = 22$

a. 7
b. 4
c. $^-4$
d. none of these

8. Solve.

$\frac{a}{2} + 6 = 44$

a. 76
b. 25
c. 19
d. none of these

9. $47.2 \text{ m} = \underline{?} \text{ cm}$

a. 47,200
b. 4720
c. 0.472
d. none of these

10. What is the perimeter?

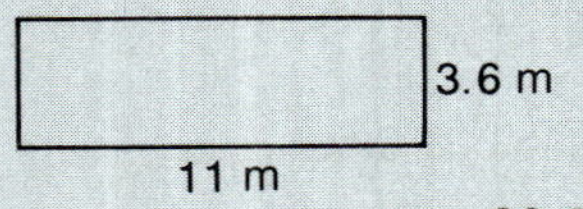

a. 29.2 m
b. 39.6 m
c. 14.6 m
d. none of these

11. Steve earned $34 for working 5 hours. How much did he earn per hour?

a. $7 **b.** $6.40
c. $170 **d.** none of these

12. A dime weighs 2.26 g. A penny weighs 3.1 g. How many more g does the penny weigh than the dime?

a. 0.84 **b.** 0.96
c. 19.5 **d.** none of these

4

Fractions—Addition and Subtraction

Find the five circles on the manhole cover. What fraction of the raised bars are inside the smallest circle?

Divisibility

Since there is no remainder in this problem, we say that 2538 is **divisible by** 6.

Below are some shortcuts for telling whether a whole number is divisible by certain numbers.

A whole number is divisible by

2 if its last digit is divisible by 2.

Example:

36,018 is divisible by 2.

3 if the sum of the digits is divisible by 3.

Example:

63,501 is divisible by 3.
6 + 3 + 5 + 0 + 1 = 15

4 if its last two digits are divisible by 4.

Example:

513,732 is divisible by 4.

5 if its last digit is divisible by 5, that is, if the last digit is 0 or 5.

Example:

37,145 is divisible by 5.

6 if it is divisible by both 2 and 3.

Example:

36,702 is divisible by 6.

8 if its last three digits are divisible by 8.

Example:

74,112 is divisible by 8.

9 if the sum of its digits is divisible by 9.

Example:

35,784 is divisible by 9.
3 + 5 + 7 + 8 + 4 = 27

10 if its last digit is divisible by 10, that is, if the last digit is 0.

Example:

27,850 is divisible by 10.

EXERCISES

1. Copy and complete.

Whole Number	*Divisible by*							
	2?	3?	4?	5?	6?	8?	9?	10?
60	Yes	?	?	?	?	?	?	?
196	?	?	?	?	?	?	?	?
620	?	?	?	?	?	?	?	?
152	?	?	?	?	?	?	?	?
394	?	?	?	?	?	?	?	?

Give the greatest whole number that divides both the numerator and the denominator.

2. $\frac{3}{6}$ (*numerator*: 3; *denominator*: 6)
3. $\frac{14}{16}$
4. $\frac{36}{6}$
5. $\frac{8}{4}$
6. $\frac{45}{18}$
7. $\frac{9}{12}$
8. $\frac{42}{36}$
9. $\frac{12}{18}$
10. $\frac{36}{28}$
11. $\frac{16}{20}$
12. $\frac{14}{21}$
13. $\frac{20}{50}$
14. $\frac{20}{16}$
15. $\frac{48}{42}$
16. $\frac{10}{15}$
17. $\frac{24}{16}$
18. $\frac{48}{56}$
19. $\frac{45}{81}$
20. $\frac{40}{45}$
21. $\frac{30}{50}$
22. $\frac{12}{8}$
23. $\frac{56}{40}$
24. $\frac{32}{40}$
25. $\frac{55}{45}$

True or false?

26. If a number is divisible by 9, then it is divisible by 3.
27. If a number is divisible by 3, then it is divisible by 9.
28. All numbers divisible by 10 are also divisible by 5.
29. Zero and all other numbers divisible by 2 are even numbers.
30. All numbers divisible by 3 are odd numbers.
31. Some numbers divisible by 5 are odd numbers.

Challenge!

Solve.

32. What is the largest 4-digit number that is divisible by 6?
33. What is the smallest 5-digit number that is divisible by 8?

Prime numbers and composite numbers

Check the list of numbers and their factors. The factors of 2 are 1 and 2,

since $1 \times 2 = 2$.

The factors of 4 are 1, 2, and 4,

since $1 \times 4 = 4$
and $2 \times 2 = 4$.

Numbers	*Factors*
1	1
2	1, 2
3	1, 3
4	1, 2, 4
5	1, 5
6	1, 2, 3, 6
7	1, 7
8	1, 2, 4, 8
9	1, 3, 9
10	1, 2, 5, 10
11	1, 11
12	1, 2, 3, 4, 6, 12

Whole numbers that have exactly two factors are called **prime numbers.** 2 is a prime number.

Whole numbers (other than 0) that have more than two factors are called **composite numbers.** 4 is a composite number.

Every composite number can be factored into a product of prime numbers. The numbers in the bottom row of the factor tree are prime numbers.

Composite number → **72**

$$2 \times 36$$

$$2 \times 6 \times 6$$

$$2 \times 2 \times 3 \times 2 \times 3$$ ← **Prime numbers**

To express a composite number as a product of prime numbers is to give the **prime factorization** of the composite number.

Here is how to write the prime factorization of 72 using exponents.

$$72 = 2^3 \times 3^2$$

EXERCISES

Give the prime factorization. Use exponents for factors that occur more than once.

1. 10
2. 24
3. 27
4. 36
5. 48
6. 54
7. 72
8. 84
9. 90
10. 96
11. 100
12. 124
13. 144
14. 200
15. 240
16. 256
17. 288
18. 300

Give the composite number.

19. 2^3
20. $2^2 \times 3$
21. $3^2 \times 5$
22. 2^5
23. $5^2 \times 7 \times 11$
24. $3^2 \times 5 \times 7$
25. $7^3 \times 5$
26. $3^4 \times 5$
27. $2^4 \times 5^2 \times 13$
28. $5^2 \times 17$
29. 3×13^2
30. 11×19^2

Give the prime factorization of the numerator and denominator of each fraction.

31. $\frac{9}{12}$ • $\frac{3 \times 3}{2 \times 2 \times 3}$
32. $\frac{16}{24}$
33. $\frac{48}{64}$
34. $\frac{96}{228}$
35. $\frac{18}{39}$
36. $\frac{6}{28}$
37. $\frac{27}{125}$
38. $\frac{56}{169}$

Problem Solving

39. What is the smallest composite number that is the product of the squares of two different prime numbers?

40. What two whole numbers are neither prime nor composite?

41. What is the smallest composite number that is the product of a prime number squared and a different prime number cubed?

42. Although there is no largest prime number, the largest prime that is presently known, $2^{21,701} - 1$, was found with the aid of a computer by two 18-year-olds. If you wrote $2^{21,701}$ as a product of 2's, how many 2's would you write?

Challenge!

Give the prime factorization.

43. 41,160
44. 6,126,120
45. 112,000

Factors and common factors

Listed below are the factors of 12 and 20.

factors of 12: 1, 2, 3, 4, 6, 12

factors of 20: 1, 2, 4, 5, 10, 20

Notice that 1, 2, and 4 are factors, or divisors, of both 12 and 20. They are called **common factors** of 12 and 20. The **greatest common factor (GCF)** of 12 and 20 is 4.

You can use the prime factorization of two numbers to find their greatest common factor (GCF).

EXAMPLE 1.

$12 = \boxed{2 \times 2} \times 3$

$20 = \boxed{2 \times 2} \times 5$

The GCF of 12 and 20 is 4.

EXAMPLE 2.

$24 = 2 \times 2 \times \boxed{2 \times 3}$

$18 = \boxed{2 \times 3} \times 3$

The GCF of 24 and 18 is 6.

SHORTCUT *To find the GCF, multiply the prime factors that are common to both factorizations.*

EXERCISES
Give the GCF.

1. 8, 12	**2.** 9, 6	**3.** 16, 10	**4.** 12, 15	**5.** 4, 16
6. 12, 18	**7.** 16, 40	**8.** 21, 14	**9.** 18, 27	**10.** 24, 18
11. 36, 9	**12.** 14, 18	**13.** 16, 12	**14.** 30, 24	**15.** 30, 28
16. 50, 30	**17.** 48, 40	**18.** 22, 28	**19.** 20, 30	**20.** 24, 6

Give the GCF of the numerator and denominator of each fraction.

21. $\frac{3}{6}$ **22.** $\frac{24}{16}$ **23.** $\frac{18}{20}$ **24.** $\frac{12}{18}$

25. $\frac{30}{18}$ **26.** $\frac{4}{12}$ **27.** $\frac{21}{35}$ **28.** $\frac{45}{36}$

29. $\frac{36}{21}$ **30.** $\frac{32}{60}$ **31.** $\frac{60}{50}$ **32.** $\frac{24}{48}$

33. $\frac{26}{39}$ **34.** $\frac{19}{57}$ **35.** $\frac{20}{15}$ **36.** $\frac{60}{72}$

Whole numbers that have 2 as a factor are called **even numbers.** 0, 2, 4, 6, . . . are even numbers.

Whole numbers that are not even numbers are called **odd numbers.** 1, 3, 5, 7, . . . are odd numbers.

Problem Solving

37. What is the smallest odd number that is greater than 1000?

38. What is the largest even 3-digit number?

39. Is the sum of two odd numbers an even number or an odd number?

40. Is the product of an even number and an odd number an even number?

Here are the first three square numbers:

1 1×1 **4** 2×2 **9** 3×3

41. Give the first ten square numbers.

42. Do square numbers have an even or an odd number of factors?

KEEPING SKILLS SHARP

Write the standard numeral.

1. 6^4

2. 10^0

3. 2^{16}

4. 3^5

5. 4^3

6. 7^2

7. 8^0

8. 9^3

9. 10^6

10. 11^1

Write using exponents.

11. $3 \times 3 \times 3$

12. 4×4

13. $7 \times 7 \times 7 \times 7$

14. $10 \times 10 \times 10$

15. 12×12

16. $1 \times 1 \times 1 \times 1 \times 1$

17. 13×13

18. $2 \times 2 \times 2 \times 2 \times 2$

19. $11 \times 11 \times 11 \times 11$

20. $20 \times 20 \times 20$

Equivalent fractions

The numerator and denominator of a fraction are the **terms** of the fraction.

terms → $\frac{2}{4}$ ← *numerator* (2), ← *denominator* (4)

Equivalent fractions name the same number.

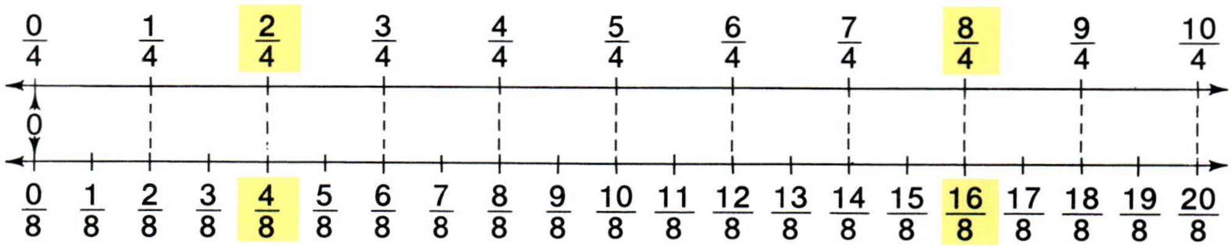

From the number lines we see that

$$\frac{2}{4} = \frac{4}{8} \qquad \text{and} \qquad \frac{8}{4} = \frac{16}{8}$$

You can get an equivalent fraction in **higher terms** by multiplying both numerator and denominator by the same whole number (greater than 1).

EXAMPLES.

$$\frac{2}{4} = \frac{4}{8} \quad (\times 2) \qquad \frac{1}{5} = \frac{3}{15} \quad (\times 3) \qquad \frac{2}{3} = \frac{8}{12} \quad (\times 4)$$

You can get an equivalent fraction in **lower terms** by dividing both numerator and denominator by the same whole number (greater than 1).

EXAMPLES.

$$\frac{4}{8} = \frac{2}{4} \quad (\div 2) \qquad \frac{6}{9} = \frac{2}{3} \quad (\div 3) \qquad \frac{12}{20} = \frac{3}{5} \quad (\div 4)$$

When you divide both terms of a fraction by their greatest common factor (GCF), you get the equivalent fraction in **lowest terms.**

The GCF of 8 and 12 is 4.

$$\frac{8}{12} \overset{\div 4}{=} \frac{2}{3} \quad (\div 4)$$

Here is how to find the fraction that is equivalent to $\frac{3}{6}$ and has a denominator of 12.

Step 1. The denominator was multiplied by 2. $\frac{3}{6} = \frac{?}{12}$ (×2)

Step 2. So, multiply the numerator by 2. $\frac{3}{6} = \frac{6}{12}$ (×2)

EXERCISES

Copy and complete.

1. $\frac{1}{2} = \frac{?}{6}$
2. $\frac{1}{3} = \frac{?}{6}$
3. $\frac{1}{4} = \frac{?}{16}$
4. $\frac{3}{2} = \frac{6}{?}$
5. $\frac{2}{7} = \frac{?}{21}$
6. $\frac{5}{8} = \frac{?}{40}$
7. $\frac{4}{3} = \frac{12}{?}$
8. $\frac{5}{2} = \frac{10}{?}$
9. $\frac{5}{6} = \frac{?}{30}$
10. $\frac{4}{3} = \frac{?}{15}$

Write in lowest terms.

11. $\frac{6}{8}$
12. $\frac{4}{16}$
13. $\frac{9}{12}$
14. $\frac{16}{14}$
15. $\frac{24}{30}$
16. $\frac{14}{18}$
17. $\frac{18}{32}$
18. $\frac{8}{12}$
19. $\frac{27}{36}$
20. $\frac{25}{15}$
21. $\frac{16}{24}$
22. $\frac{14}{21}$
23. $\frac{14}{12}$
24. $\frac{36}{45}$

A fraction is near $\frac{1}{2}$ if the denominator is about 2 times the numerator.

Complete these fractions so they are a little less than $\frac{1}{2}$.

25. $\frac{\square}{37}$
26. $\frac{\square}{51}$
27. $\frac{\square}{99}$
28. $\frac{\square}{69}$
29. $\frac{\square}{101}$

Challenge!

Write in lowest terms.

30. $\frac{180}{630}$
31. $\frac{210}{576}$
32. $\frac{315}{1050}$
33. $\frac{2475}{8085}$

Fractions and decimals

Every fraction is equivalent to a decimal. A fraction can be changed to a decimal by dividing its numerator by its denominator.

EXAMPLE 1. $\frac{7}{8} = ?$

$$\begin{array}{r} 0.875 \\ 8\overline{)7.000} \\ -6\,4 \\ \hline 60 \\ -56 \\ \hline 40 \\ -40 \\ \hline 0 \end{array}$$

So, $\frac{7}{8} = 0.875$

Since the decimal above terminates (comes to an end), it is called a **terminating decimal.**

Some fractions are equivalent to decimals that do not terminate. They are called **repeating decimals.**

EXAMPLE 2. $\frac{7}{11} = ?$

$$\begin{array}{r} 0.6363 \\ 11\overline{)7.0000} \\ -6\,6 \\ \hline 40 \\ -33 \\ \hline 70 \\ -66 \\ \hline 40 \\ -33 \\ \hline 7 \end{array}$$

So, $\frac{7}{11} = 0.6363\ldots$ *The digits repeat.*

or $\frac{7}{11} = 0.\overline{63}$ *You can write a bar over the repeating digits.*

or $\frac{7}{11} = 0.63\frac{7}{11}$ *You can write a repeating decimal as a mixed decimal.*

Every fraction is equivalent to a terminating or a repeating decimal.

These examples show how to change a terminating decimal to a fraction.

8 tenths

$0.8 = \frac{8}{10} = \frac{4}{5}$

75 hundredths

$0.75 = \frac{75}{100} = \frac{3}{4}$

875 thousandths

$0.875 = \frac{875}{1000} = \frac{7}{8}$

EXERCISES

Write each fraction as a decimal. Tell whether the decimal is terminating or repeating.

1. $\frac{1}{2}$	**2.** $\frac{1}{6}$	**3.** $\frac{1}{8}$	**4.** $\frac{5}{3}$	**5.** $\frac{3}{5}$	**6.** $\frac{2}{3}$
7. $\frac{5}{8}$	**8.** $\frac{5}{4}$	**9.** $\frac{1}{3}$	**10.** $\frac{4}{9}$	**11.** $\frac{3}{8}$	**12.** $\frac{3}{2}$

Complete.

13. If $\frac{1}{5} = 0.2$ then $\frac{2}{5} = \underline{?}$ $\frac{3}{5} = \underline{?}$ $\frac{4}{5} = \underline{?}$ $\frac{5}{5} = \underline{?}$ $\frac{6}{5} = \underline{?}$

14. If $\frac{1}{4} = 0.25$ then $\frac{2}{4} = \underline{?}$ $\frac{3}{4} = \underline{?}$ $\frac{4}{4} = \underline{?}$ $\frac{5}{4} = \underline{?}$ $\frac{6}{4} = \underline{?}$

Write each decimal as an equivalent fraction in lowest terms.

15. 0.28	**16.** 0.35	**17.** 0.6	**18.** 0.06	**19.** 0.006
20. 0.375	**21.** 1.3	**22.** 4.2	**23.** 1.25	**24.** 0.625

$<$, $=$, or $>$?

25. 0.3 ● $\frac{1}{3}$	**26.** $\frac{5}{9}$ ● 0.56	**27.** $\frac{9}{4}$ ● 2.25	**28.** $\frac{7}{9}$ ● 0.77
29. 1.6 ● $\frac{8}{5}$	**30.** 0.875 ● $\frac{7}{8}$	**31.** $\frac{1}{6}$ ● 0.16	**32.** $\frac{5}{3}$ ● 1.67

Challenge!

Computers do not work with fractions. If you want to input a fraction, you use the form A / B (A divided by B) and your computer will work with it as a decimal.

Find the missing input or output.

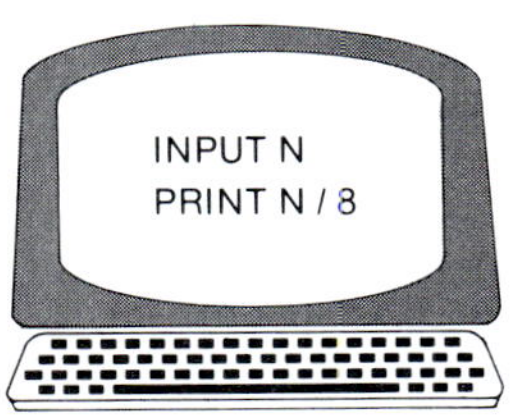

	Input	Output
	1	0.125
33.	3	?
34.	?	0.75
35.	?	0.875

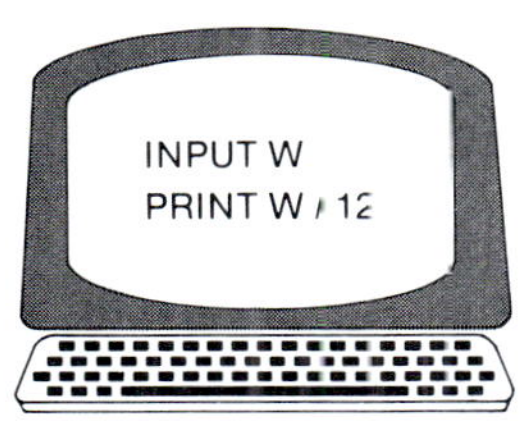

	Input	Output
	3	0.25
36.	6	?
37.	?	0.75
38.	?	1.0

Multiples and common multiples

Listed below are the first few multiples of 15 and 18.

multiples of 15: 15, 30, 45, 60, 75, 90, . . .

multiples of 18: 18, 36, 54, 72, 90, 108, . . .

90, 180, 270, . . . are multiples of both 15 and 18. They are called **common multiples** of 15 and 18. The **least common multiple (LCM)** of 15 and 18 is 90.

SHORTCUT

You can use the prime factorization of two numbers to find their least common multiple (LCM).

EXAMPLE 1.

$15 = 3 \times 5$

$18 = 2 \times 3 \times 3$

To find the LCM, multiply the GCF by the prime factors that are not common to both factorizations. Each common multiple of 15 and 18 contains the GCF, 3, and 2, 3, and 5 as factors. So the LCM of 15 and 18 is $3 \times 2 \times 3 \times 5$, or 90.

EXAMPLE 2.

$16 = 2 \times 2 \times 2 \times 2$

$24 = 2 \times 2 \times 2 \times 3$

The LCM of 16 and 24 is $2 \times 2 \times 2 \times 2 \times 3$, or 48.

EXERCISES
Give the LCM.

1. 12, 8	**2.** 6, 9	**3.** 15, 12	**4.** 10, 16	**5.** 16, 4
6. 18, 12	**7.** 16, 40	**8.** 14, 21	**9.** 27, 18	**10.** 18, 24
11. 9, 36	**12.** 18, 14	**13.** 12, 16	**14.** 24, 30	**15.** 30, 20
16. 50, 30	**17.** 36, 18	**18.** 33, 11	**19.** 25, 20	**20.** 21, 14

Give the LCM of the denominators.

21. $\frac{3}{2}, \frac{1}{3}$
22. $\frac{3}{4}, \frac{6}{5}$
23. $\frac{3}{4}, \frac{1}{2}$
24. $\frac{4}{5}, \frac{5}{7}$
25. $\frac{2}{3}, \frac{5}{4}$
26. $\frac{3}{4}, \frac{5}{6}$
27. $\frac{1}{2}, \frac{7}{5}$
28. $\frac{3}{5}, \frac{3}{8}$
29. $\frac{5}{6}, \frac{5}{3}$
30. $\frac{7}{4}, \frac{5}{9}$
31. $\frac{11}{6}, \frac{6}{7}$
32. $\frac{2}{3}, \frac{7}{5}$
33. $\frac{1}{6}, \frac{1}{9}$
34. $\frac{5}{12}, \frac{9}{16}$
35. $\frac{5}{9}, \frac{11}{12}$
36. $\frac{13}{16}, \frac{11}{24}$
37. $\frac{11}{20}, \frac{7}{30}$
38. $\frac{13}{18}, \frac{5}{24}$
39. $\frac{3}{16}, \frac{5}{32}$
40. $\frac{17}{18}, \frac{13}{27}$

True or false?

41. An even number is a multiple of 2.
42. An odd number is a multiple of 3.
43. Every whole number is a factor of 0.
44. Every whole number greater than 0 is a multiple of 1.
45. If a is a multiple of b, then b is a factor of a.

Challenge!

Guess our numbers.

46. The product of the LCM and the GCF is 7.
47. The product is 12. The LCM is 6.
48. The sum is 12. The GCF is 4.
49. The LCM is 12. The GCF is 2.
50. The LCM of three numbers is 48. Their GCF is 4. Their sum is 52.

Comparing fractions

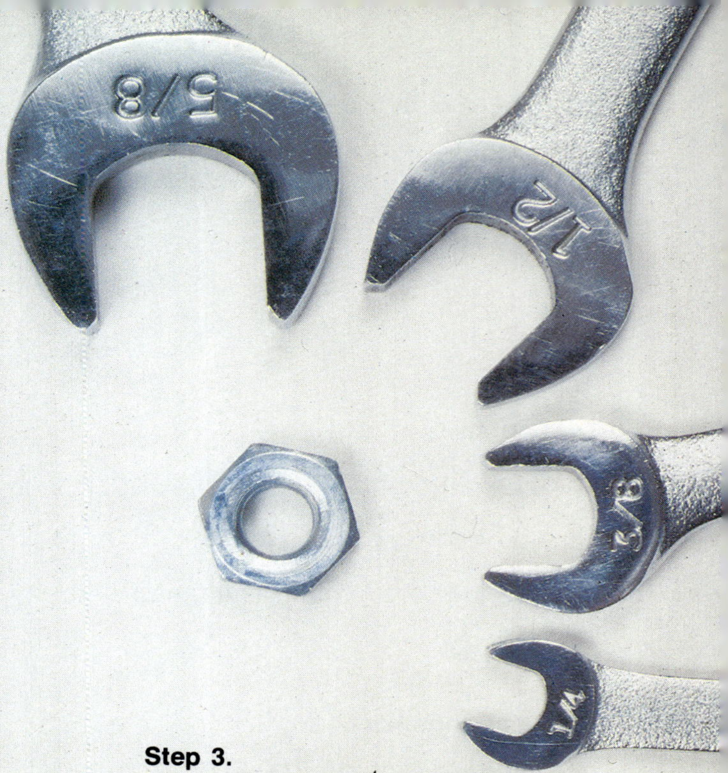

You can compare fractions that have the same denominator by comparing their numerators.

$$\frac{3}{8} < \frac{5}{8}$$

To compare fractions that have different denominators, find equivalent fractions with the same denominator **(common denominator).** A common multiple of the denominators of two fractions is a common denominator. The least common multiple of the denominators is the **least common denominator** (LCD).

EXAMPLE 1. Compare $\frac{2}{3}$ and $\frac{1}{2}$.

Step 1.
Find the least common denominator.

$$\frac{2}{3} \bullet \frac{1}{2}$$

$$\frac{}{6} \quad \frac{}{6}$$

Step 2.
Change to equivalent fractions.

$$\frac{2}{3} \bullet \frac{1}{2}$$

$$\frac{4}{6} \quad \frac{3}{6}$$

Step 3.
Compare. Since $\frac{4}{6}$ is greater than $\frac{3}{6}$, $\frac{2}{3}$ is greater than $\frac{1}{2}$.

$$\frac{2}{3} > \frac{1}{2}$$

$$\frac{4}{6} > \frac{3}{6}$$

Another way to compare two fractions is to change the fractions to equivalent decimals.

EXAMPLE 2. Compare $\frac{1}{4}$ and $\frac{3}{11}$.

Step 1.
Change to equivalent decimals.

$$\frac{1}{4} \bullet \frac{3}{11}$$

$$0.25 \quad 0.\overline{27}$$

Step 2.
Compare. Since 0.25 is less than $0.\overline{27}$, $\frac{1}{4}$ is less than $\frac{3}{11}$.

$$\frac{1}{4} < \frac{3}{11}$$

$$0.25 < 0.\overline{27}$$

EXERCISES

Give the least common denominator.

1. $\frac{1}{2}$ $\frac{3}{4}$ **2.** $\frac{5}{6}$ $\frac{1}{3}$ **3.** $\frac{3}{8}$ $\frac{1}{4}$

4. $\frac{1}{3}$ $\frac{1}{2}$ **5.** $\frac{5}{3}$ $\frac{1}{2}$ **6.** $\frac{1}{6}$ $\frac{1}{5}$

<, =, or >? Use the equivalent-fraction method.

7. $\frac{3}{8}$ ● $\frac{1}{2}$ **8.** $\frac{1}{2}$ ● $\frac{4}{7}$ **9.** $\frac{2}{3}$ ● $\frac{4}{6}$

10. $\frac{3}{2}$ ● $\frac{7}{4}$ **11.** $\frac{6}{7}$ ● $\frac{5}{6}$ **12.** $\frac{5}{8}$ ● $\frac{2}{3}$

13. $\frac{4}{12}$ ● $\frac{1}{3}$ **14.** $\frac{3}{2}$ ● $\frac{4}{3}$ **15.** $\frac{4}{7}$ ● $\frac{3}{5}$

< or >? Use the decimal method.

16. $\frac{3}{5}$ ● $\frac{1}{2}$ **17.** $\frac{4}{5}$ ● $\frac{9}{13}$ **18.** $\frac{5}{7}$ ● $\frac{8}{11}$

19. $\frac{4}{9}$ ● $\frac{3}{7}$ **20.** $\frac{5}{11}$ ● $\frac{1}{2}$ **21.** $\frac{4}{7}$ ● $\frac{7}{13}$

Round these fractions to 0, $\frac{1}{2}$, or 1.

22. $\frac{45}{95}$ **23.** $\frac{14}{15}$ **24.** $\frac{2}{27}$ **25.** $\frac{99}{104}$ **26.** $\frac{16}{35}$

Problem Solving

27. Will a bolt that is $\frac{7}{8}$ of an inch long go through a sheet of steel that is $\frac{9}{32}$ of an inch thick?

28. Will a bolt that is $\frac{3}{8}$ of an inch in diameter go through a $\frac{5}{16}$-inch hole?

29. Is a $\frac{5}{8}$-inch wrench too large or too small for a $\frac{3}{4}$-inch nut?

★ **30.** A certain set of wrenches contains the following sizes in inches:

$\frac{3}{4}$ $\frac{1}{2}$ $\frac{5}{16}$ $\frac{1}{4}$ $\frac{7}{16}$ $\frac{13}{16}$ $\frac{9}{16}$ $\frac{5}{8}$ $\frac{3}{8}$ $\frac{11}{16}$

Order the sizes from smallest to largest.

KEEPING SKILLS SHARP

<u>Estimate</u> to find the two incorrect answers.

1. a. $2.89 \div 0.96 = 30.1$

b. $35.8 \div 4.02 = 0.89$

c. $39.2 \div 1.4 = 28$

d. $355.3 \div 3.23 = 110$

Divide.

2. $0.3\overline{)57.3}$

3. $10\overline{)98.4}$

4. $15\overline{)351}$

5. $2.6\overline{)16.38}$

6. $0.08\overline{)0.1624}$

7. $5.7\overline{)548.91}$

8. $103\overline{)5356}$

9. $21.6\overline{)20.736}$

10. $12.4\overline{)47.12}$

11. $2.65\overline{)7.9765}$

12. $1.08\overline{)4.266}$

Adding fractions

To add fractions having the same denominator, add the numerators to get the numerator of the sum and use the common denominator for the denominator of the sum.

$$\frac{1}{8} + \frac{3}{8} = \frac{4}{8}$$

$$= \frac{1}{2} \leftarrow \textit{lowest terms}$$

When the fractions have different denominators, change to equivalent fractions with common denominators.

Step 1.
Find the least common denominator.

$$\frac{5}{12} + \frac{1}{4} = \frac{}{12} + \frac{}{12}$$

Step 2.
Change to equivalent fractions.

$$\frac{5}{12} + \frac{1}{4} = \frac{5}{12} + \frac{3}{12}$$

Step 3.

$$\frac{5}{12} + \frac{1}{4} = \frac{5}{12} + \frac{3}{12}$$

$$= \frac{8}{12}$$

$$= \frac{2}{3}$$

EXERCISES

Give each sum in lowest terms.

1. $\frac{1}{3} + \frac{1}{6}$	**2.** $\frac{1}{3} + \frac{1}{3}$	**3.** $\frac{1}{2} + \frac{1}{4}$	**4.** $\frac{1}{6} + \frac{1}{6}$
5. $\frac{1}{10} + \frac{3}{10}$	**6.** $\frac{3}{8} + \frac{1}{2}$	**7.** $\frac{2}{5} + \frac{1}{2}$	**8.** $\frac{1}{3} + \frac{1}{4}$
9. $\frac{1}{5} + \frac{1}{6}$	**10.** $\frac{1}{2} + \frac{1}{3}$	**11.** $\frac{5}{9} + \frac{2}{9}$	**12.** $\frac{1}{6} + \frac{1}{2}$
13. $\frac{3}{8} + \frac{3}{8}$	**14.** $\frac{3}{4} + \frac{1}{6}$	**15.** $\frac{1}{8} + \frac{2}{3}$	**16.** $\frac{3}{8} + \frac{1}{8}$
17. $\frac{1}{10} + \frac{3}{5}$	**18.** $\frac{1}{9} + \frac{2}{3}$	**19.** $\frac{1}{2} + \frac{0}{4}$	**20.** $\frac{1}{2} + \frac{1}{5}$

Add. Give each sum in lowest terms.

21. $\frac{1}{3} + \frac{3}{5}$
22. $\frac{1}{6} + \frac{3}{5}$
23. $\frac{1}{3} + \frac{5}{9}$
24. $\frac{1}{3} + \frac{3}{9}$
25. $\frac{1}{8} + \frac{3}{4}$
26. $\frac{1}{4} + \frac{2}{5}$

27. $\frac{1}{8} + \frac{1}{4}$
28. $\frac{2}{3} + \frac{1}{6}$
29. $\frac{2}{3} + \frac{1}{10}$
30. $\frac{3}{10} + \frac{3}{10}$
31. $\frac{1}{3} + \frac{5}{12}$
32. $\frac{1}{4} + \frac{5}{8}$

33. $\frac{5}{9} + \frac{1}{4}$
34. $\frac{3}{8} + \frac{2}{5}$
35. $\frac{2}{3} + \frac{1}{4}$
36. $\frac{1}{10} + \frac{1}{10}$
37. $\frac{1}{10} + \frac{3}{4}$
38. $\frac{1}{2} + \frac{1}{3}$

Problem Solving

39. Dolores bought $\frac{3}{4}$ of a pound of dried peaches and $\frac{1}{2}$ of a pound of dried apricots. How many pounds of dried fruit did she buy?

40. Alvin wrote $\frac{1}{3}$ of his English theme on Tuesday and $\frac{1}{2}$ of it on Wednesday. What fraction of his theme did he write during those two days?

41. Andrew read $\frac{1}{8}$ of a book on Monday, $\frac{1}{4}$ of the book on Tuesday, and $\frac{1}{2}$ of the book on Wednesday. What fraction of the book did he read during the three days?

42. A youth group held a car wash to raise money for a camp-out. They earned $\frac{3}{5}$ of the money needed. What fraction of the money did they still need to earn?

Challenge!

Which of these fractions would you add to get the following sums?

43. $\frac{5}{6}$
44. $\frac{5}{8}$
45. $\frac{7}{12}$
46. $\frac{7}{8}$
47. $\frac{31}{40}$
48. $\frac{13}{20}$

Subtracting fractions

To subtract fractions having the same denominator, subtract the numerators to get the numerator of the difference and use the common denominator for the denominator of the difference.

$$\frac{5}{8} - \frac{3}{8} = \frac{2}{8}$$
$$= \frac{1}{4} \leftarrow \textit{lowest terms}$$

When the fractions have different denominators, change to equivalent fractions with a common denominator.

Step 1.
Find the least common denominator.

$$\frac{5}{6} - \frac{5}{9} = \frac{\quad}{18} - \frac{\quad}{18}$$

Step 2.
Change to equivalent fractions.

$$\frac{5}{6} - \frac{5}{9} = \frac{15}{18} - \frac{10}{18}$$

Step 3.
Subtract.

$$\frac{5}{6} - \frac{5}{9} = \frac{15}{18} - \frac{10}{18}$$
$$= \frac{5}{18}$$

EXERCISES

Give each difference in lowest terms.

1. $\frac{2}{3} - \frac{1}{3}$ **2.** $\frac{3}{8} - \frac{1}{4}$ **3.** $\frac{2}{3} - \frac{1}{6}$ **4.** $\frac{5}{6} - \frac{1}{3}$

5. $\frac{2}{3} - \frac{1}{2}$ **6.** $\frac{3}{4} - \frac{2}{3}$ **7.** $\frac{1}{2} - \frac{1}{6}$ **8.** $\frac{5}{3} - \frac{4}{3}$

9. $\frac{3}{4} - \frac{5}{8}$ **10.** $\frac{5}{6} - \frac{2}{3}$ **11.** $\frac{5}{8} - \frac{1}{8}$ **12.** $\frac{1}{3} - \frac{2}{9}$

13. $\frac{3}{4} - \frac{1}{3}$ **14.** $\frac{3}{5} - \frac{1}{5}$ **15.** $\frac{7}{8} - \frac{3}{4}$ **16.** $\frac{7}{9} - \frac{2}{3}$

17. $\frac{5}{8} - \frac{1}{4}$ **18.** $\frac{9}{10} - \frac{1}{2}$ **19.** $\frac{7}{5} - \frac{3}{5}$ **20.** $\frac{5}{7} - \frac{5}{9}$

Subtract. Give each difference in lowest terms.

21. $\frac{3}{4} - \frac{1}{2}$ **22.** $\frac{2}{3} - \frac{1}{4}$ **23.** $\frac{3}{4} - \frac{5}{8}$ **24.** $\frac{5}{9} - \frac{1}{2}$ **25.** $\frac{5}{6} - \frac{3}{8}$ **26.** $\frac{8}{9} - \frac{3}{4}$

27. $\frac{5}{6} - \frac{0}{2}$ **28.** $\frac{5}{8} - \frac{0}{4}$ **29.** $\frac{5}{9} - \frac{2}{9}$ **30.** $\frac{2}{5} - \frac{3}{8}$ **31.** $\frac{3}{4} - \frac{3}{10}$ **32.** $\frac{8}{9} - \frac{2}{3}$

Round each fraction to the nearest half. Then estimate the sum or difference.

33. $\frac{10}{21} + \frac{49}{53}$ **34.** $\frac{13}{15} - \frac{3}{8}$ **35.** $\frac{1}{35} - \frac{2}{3}$ **36.** $\frac{9}{16} - \frac{23}{47}$

37. $\frac{17}{35} - \frac{9}{100}$ **38.** $\frac{15}{92} + \frac{67}{72}$ **39.** $\frac{11}{21} + \frac{4}{27}$ **40.** $\frac{13}{19} - \frac{2}{11}$

Problem Solving

41. Don is in $\frac{3}{8}$ of the scenes in the play. Val is in $\frac{1}{4}$ of the scenes. Who is in more scenes? What fraction more?

42. The school photographer took $\frac{1}{2}$ of a roll of film of the audience and $\frac{3}{8}$ of a roll of the performers. What fraction of a roll did he shoot?

43. $\frac{1}{2}$ of the tickets sold were student tickets and $\frac{1}{4}$ were adult tickets. The rest of the tickets sold were children's tickets. What fraction of the tickets were children's?

44. $\frac{1}{3}$ of the scenes had background music. $\frac{5}{8}$ of the scenes had vocals. What fraction of the scenes had neither background music nor vocals?

Challenge!

Which expression has the least value?

45.

r	s	t
1	2	3

a. $\frac{s}{t}$

b. $\frac{r}{s}$

c. $\frac{r}{t}$

d. $\frac{s}{t} - \frac{r}{s}$

46.

w	x	y
2	3	5

a. $\frac{w}{x}$

b. $\frac{x}{y}$

c. $\frac{w}{x} - \frac{x}{y}$

d. $\frac{w}{x} - \frac{w}{y}$

47.

f	g	h
3	4	8

a. $\frac{g}{h}$

b. $\frac{f}{g}$

c. $\frac{f}{g} - \frac{g}{h}$

d. $\frac{f}{h} + \frac{f}{h}$

Writing whole numbers and mixed numbers as fractions

Look at the pies. They show that there are 12 sixths in 2.

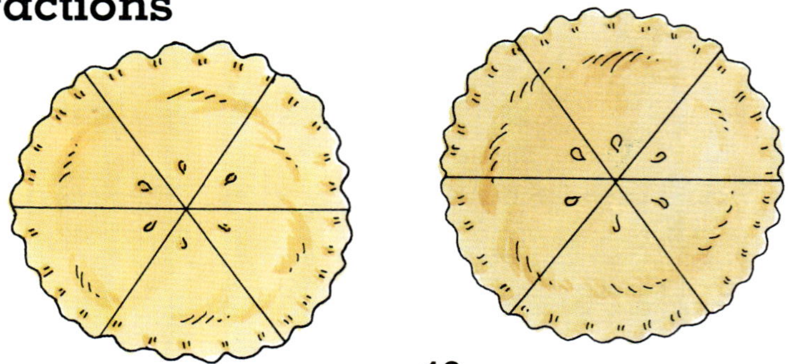

To write a whole number as a fraction, you can write the whole number over a denominator of 1. Then, to change to an equivalent fraction, you multiply both numerator and denominator by the same nonzero whole number.

$$2 = \frac{12}{6}$$

$$2 = \frac{2}{1}$$

$$\frac{2}{1} = \frac{12}{6} \quad (\times 6 \text{ numerator}, \times 6 \text{ denominator})$$

A **mixed number** has a whole-number part and a fraction part.

$2\frac{5}{6}$ is a mixed number.

It means $2 + \frac{5}{6}$.

To write a mixed number as a fraction, you can write the whole-number part as a fraction and then add.

$$\begin{aligned} 2\frac{5}{6} &= 2 + \frac{5}{6} \\ &= \frac{2}{1} + \frac{5}{6} \\ &= \frac{12}{6} + \frac{5}{6} \\ &= \frac{17}{6} \end{aligned}$$

$$2\frac{5}{6} = \frac{17}{6}$$

SHORTCUT

Multiply 6 by 2 to find how many sixths in 2. Then add 5 to find how many sixths in $2\frac{5}{6}$.

$(6 \times 2) + 5 = 17$

EXERCISES

Copy and complete.

1. $3 = \frac{?}{2}$
2. $2 = \frac{?}{8}$
3. $5 = \frac{?}{3}$
4. $4 = \frac{?}{5}$
5. $1 = \frac{?}{6}$
6. $8 = \frac{?}{9}$
7. $6 = \frac{?}{4}$
8. $4 = \frac{?}{7}$
9. $3 = \frac{?}{6}$
10. $5 = \frac{?}{10}$

Write each mixed number as a fraction.

11. $4\frac{1}{5}$
12. $2\frac{3}{8}$
13. $7\frac{2}{3}$
14. $5\frac{1}{2}$
15. $3\frac{1}{4}$
16. $2\frac{5}{6}$
17. $1\frac{7}{10}$
18. $5\frac{3}{4}$
19. $2\frac{7}{9}$
20. $6\frac{2}{5}$
21. $2\frac{1}{3}$
22. $9\frac{1}{2}$

Write each mixed number as a decimal.

$\frac{1}{4} = 0.25$

$3\frac{1}{4} = 3.25$

23. $1\frac{1}{2}$
24. $2\frac{3}{4}$
25. $4\frac{3}{8}$
26. $2\frac{7}{10}$
27. $3\frac{1}{8}$
28. $2\frac{3}{5}$
29. $1\frac{5}{8}$
30. $6\frac{1}{10}$

Complete.

	31.	32.	33.	34.	35.	36.	37.	38.	39.
Fraction	$\frac{5}{2}$	$\frac{7}{4}$	$\frac{11}{8}$	?	?	?	?	?	?
Mixed number	?	?	?	$2\frac{1}{4}$	$3\frac{2}{5}$	$1\frac{7}{8}$	?	?	?
Decimal	?	?	?	?	?	?	1.6	2.75	2.125

Complete.

	40.	41.	42.	43.	44.	45.
Fraction	$\frac{83}{3}$	$\frac{193}{6}$	$\frac{123}{9}$	?	?	?
Mixed number	?	?	?	$18\frac{1}{3}$	$12\frac{5}{16}$	$25\frac{4}{9}$
Decimal	?	?	?	?	?	?

Choose the better estimate.

46. $\frac{11}{12} - \frac{5}{8}$
   - **a.** greater than $\frac{1}{2}$
   - **b.** less than $\frac{1}{2}$

47. $\frac{9}{10} + \frac{24}{30}$
   - **a.** greater than 2
   - **b.** less than 2

48. $\frac{4}{7} + \frac{3}{5}$
   - **a.** greater than 1
   - **b.** less than 1

Writing fractions as whole numbers or as mixed numbers

When the numerator is greater than or equal to the denominator, the fraction is greater than or equal to 1. You can write the fraction as either a whole number or a mixed number by dividing the numerator by the denominator.

$$\frac{16}{8} = ?$$

number of coins ↓

number of eighths in a whole coin → $8\overline{)16}$ with quotient 2 ← *number of eighths*

So, $\frac{16}{8} = 2$

$$\frac{22}{8} = ?$$

$$\begin{array}{r} 2 \\ 8\overline{)22} \\ -16 \\ \hline 6 \end{array}$$

6 ← *number of eighths left over*

So, $\frac{22}{8} = 2\frac{6}{8}$

$= 2\frac{3}{4}$

Here is how to write your answers in **simplest form.**

Write a fraction that is less than 1 in lowest terms.

$$\frac{12}{20} = \frac{3}{5}$$

Write a mixed number so that its fraction part is less than 1 and is in lowest terms.

$$2\frac{10}{15} = 2\frac{2}{3}$$

Write a fraction that is greater than or equal to 1 as a whole number or as a mixed number in simplest form.

$$\frac{12}{4} = 3$$

$$\frac{14}{8} = 1\frac{6}{8} = 1\frac{3}{4}$$

Write in simplest form.

1. $\frac{7}{3}$ 2. $\frac{19}{5}$ 3. $\frac{12}{4}$ 4. $\frac{18}{8}$ 5. $\frac{8}{20}$ 6. $\frac{15}{2}$ 7. $\frac{21}{12}$ 8. $\frac{18}{3}$

9. $\frac{14}{20}$ 10. $1\frac{8}{12}$ 11. $\frac{15}{10}$ 12. $3\frac{2}{10}$ 13. $\frac{9}{4}$ 14. $\frac{30}{6}$ 15. $1\frac{10}{12}$ 16. $\frac{12}{18}$

Add or subtract. Give each answer in simplest form.

17. $\frac{1}{2}+\frac{5}{6}$ 18. $\frac{3}{2}-\frac{1}{8}$ 19. $\frac{1}{6}+\frac{5}{6}$ 20. $\frac{7}{3}-\frac{5}{6}$ 21. $\frac{11}{4}-\frac{3}{4}$

22. $\frac{3}{2}+\frac{2}{3}$ 23. $\frac{3}{4}+\frac{2}{5}$ 24. $\frac{5}{4}-\frac{1}{6}$ 25. $\frac{1}{10}+\frac{2}{5}$ 26. $\frac{5}{3}-\frac{2}{5}$

Round these mixed numbers to the nearest half.

27. $9\frac{3}{5}$ 28. $12\frac{1}{12}$ 29. $14\frac{7}{8}$ 30. $5\frac{7}{15}$ 31. $31\frac{11}{16}$ 32. $24\frac{14}{29}$

Problem Solving

33. Linda filled $\frac{3}{4}$ of a page in her coin album with Indian Head pennies and $\frac{7}{8}$ of a page with old Lincoln pennies. How many pages were filled with pennies?

34. One page was $\frac{1}{4}$-filled with French coins and $\frac{1}{8}$-filled with German coins. How much of the page was *not* filled with coins?

35. Buffalo nickels filled $\frac{3}{4}$ of a page, Jackson nickels filled $\frac{1}{2}$ of a page, and Liberty nickels filled $\frac{3}{8}$ of a page. How many pages did these nickels fill?

36. Carson City dollars filled $\frac{1}{2}$ of a page. Walking Liberty halves filled $\frac{5}{8}$ of a page. How much more of a page was filled with Walking Liberty halves?

Challenge!

Give the missing number.

37. 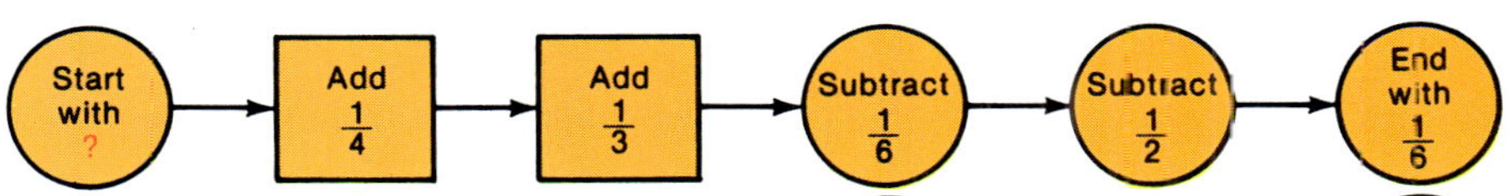

38. 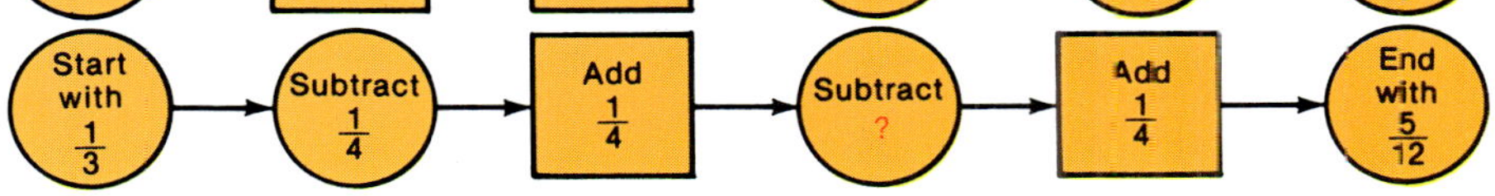

Adding mixed numbers

Here is how to add mixed numbers.

EXAMPLE 1.
Step 1. Change to fractions with a common denominator.
Step 2. Add fractions.
Step 3. Add whole numbers.

$$\begin{array}{rcr} 3\frac{3}{8} & = & 3\frac{3}{8} \\ +2\frac{1}{2} & = & +2\frac{4}{8} \\ \hline & & 5\frac{7}{8} \end{array}$$

EXAMPLE 2.
Step 1. Change to fractions with a common denominator.
Step 2. Add fractions.
Step 3. Add whole numbers.
Step 4. Regroup to write the sum as a mixed number in simplest form.

$$\begin{array}{rcr} 2\frac{3}{4} & = & 2\frac{9}{12} \\ +4\frac{1}{3} & = & +4\frac{4}{12} \\ \hline & & 6\frac{13}{12} = 7\frac{1}{12} \end{array}$$

SHORTCUT
MENTAL MATH

You can add and regroup in your head as you do with whole numbers.

$$\begin{array}{r} \overset{1}{2}\frac{3}{5} \\ +1\frac{4}{5} \\ \hline 4\frac{2}{5} \end{array} \qquad \frac{7}{5} = 1\frac{2}{5}$$

EXERCISES

Add. Write each sum in simplest form.

1. $3\frac{1}{2} + 2\frac{1}{4}$
2. $9\frac{3}{8} + 6\frac{1}{2}$
3. $11\frac{1}{6} + 3\frac{3}{4}$
4. $8\frac{3}{5} + 7\frac{1}{4}$
5. $5\frac{3}{8} + 18\frac{3}{4}$
6. $8\frac{2}{3} + 6\frac{1}{2}$

7. $2\frac{1}{2} + 3\frac{5}{6}$

8. $6\frac{2}{5} + 5\frac{3}{4}$

9. $7\frac{5}{8} + 6\frac{3}{4}$

10. $12\frac{3}{7} + 2\frac{1}{3}$

11. $9\frac{3}{4} + 1\frac{3}{8}$

12. $4\frac{1}{2} + \frac{7}{8}$

13. $6\frac{7}{8} + \frac{3}{4}$

14. $3\frac{3}{4} + 1\frac{1}{4}$

15. $6\frac{1}{2} + 7\frac{1}{2}$

16. $9\frac{2}{5} + 7\frac{3}{10}$

17. $5\frac{3}{8} + 6$

18. $2\frac{5}{8} + 10\frac{3}{8}$

19. $16\frac{4}{7} + 4\frac{1}{2}$

20. $15\frac{4}{5} + 2\frac{1}{5}$

21. $3\frac{7}{8} + \frac{1}{8}$

22. $\frac{6}{7} + 8\frac{3}{4}$

23. $\frac{2}{3} + 5\frac{4}{5}$

24. $6\frac{7}{8} + 2\frac{3}{5}$

Round each mixed number to the nearest half. Then estimate the sum.

25. $1\frac{5}{6} + 2\frac{1}{5}$
26. $11\frac{3}{7} + 9\frac{12}{13}$
27. $39\frac{13}{24} + 53\frac{5}{8}$
28. $41\frac{3}{25} + 8\frac{16}{19}$
29. $16\frac{2}{11} + 4\frac{3}{23}$

Problem Solving

30. Kim worked on the model airplane $3\frac{3}{4}$ hours on Saturday and $2\frac{1}{2}$ hours on Sunday. How many hours did she work in all?

★ 31. From the scale drawing, compute the total length of the model airplane.

★ 32. Compute the total wingspan.

★ 33. The finish color is a mixture of $1\frac{1}{4}$ ounces of white, $1\frac{1}{2}$ ounces of blue, and $\frac{3}{4}$ ounce of yellow paint. How many ounces of finish color did the plans call for?

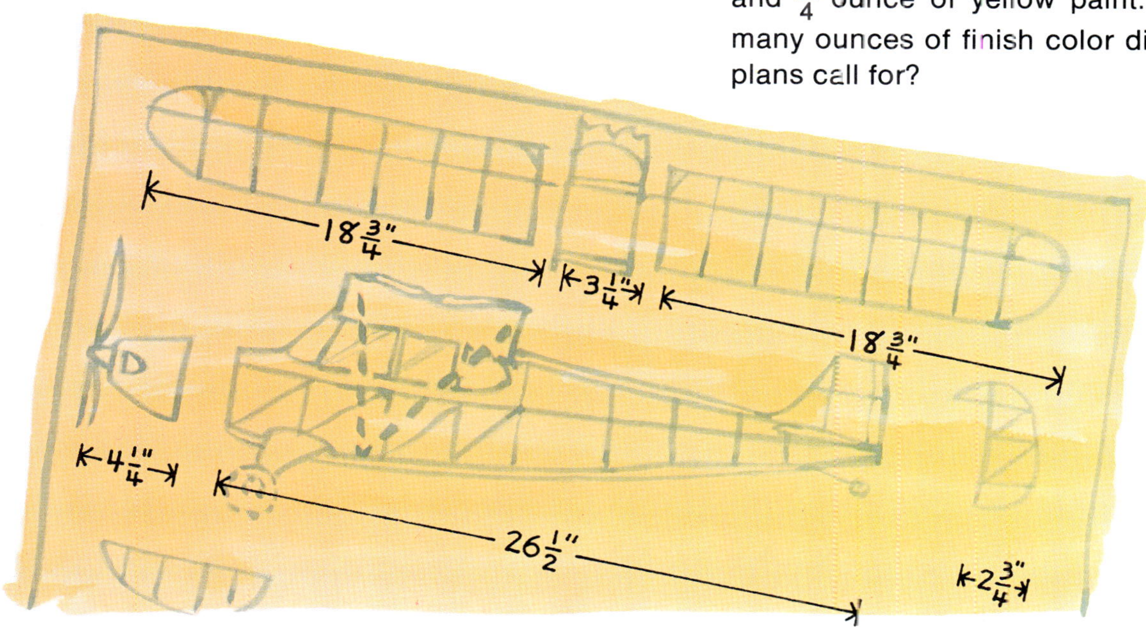

Subtracting mixed numbers

Here is how to subtract mixed numbers.

EXAMPLE 1. **Step 1.** Change to fractions with a common denominator.

Step 2. Subtract fractions.

Step 3. Subtract whole numbers.

$$\begin{array}{rcr} 5\frac{5}{8} & = & 5\frac{5}{8} \\ -2\frac{1}{4} & = & -2\frac{2}{8} \\ \hline & & 3\frac{3}{8} \end{array}$$

EXAMPLE 2. **Step 1.** Change to fractions with a common denominator.

Step 2. Regroup 1 for $\frac{20}{20}$.

Step 3. Subtract.

$$\begin{array}{rcrcr} 6\frac{3}{5} & = & 6\frac{12}{20} & = & 5\frac{32}{20} \\ -2\frac{3}{4} & = & -2\frac{15}{20} & = & -2\frac{15}{20} \\ \hline & & & & 3\frac{17}{20} \end{array}$$

SHORTCUT

Here is a short way to show the regrouping.

$$\begin{array}{rcr} 6\frac{3}{5} & = & 6\frac{12}{20} \\ -2\frac{3}{4} & = & -2\frac{15}{20} \\ \hline & & 3\frac{17}{20} \end{array}$$

EXERCISES

Subtract. Write each difference in simplest form.

1. $8\frac{1}{2} - 3\frac{1}{4}$
2. $9\frac{3}{5} - 2\frac{1}{2}$
3. $12\frac{3}{4} - 4\frac{2}{3}$
4. $16\frac{7}{8} - 9\frac{2}{3}$
5. $18\frac{1}{2} - 9\frac{3}{4}$
6. $15\frac{3}{8} - 7\frac{2}{3}$
7. $23\frac{1}{4} - 14\frac{1}{2}$
8. $18 - 12\frac{3}{4}$
9. $21\frac{1}{3} - 19\frac{2}{3}$
10. $26\frac{1}{3} - 14\frac{5}{6}$
11. $32\frac{9}{10} - 23\frac{2}{5}$
12. $35 - 26\frac{5}{8}$

13. $5\frac{1}{6} - 1\frac{1}{4}$

14. $10\frac{1}{3} - 6\frac{3}{4}$

15. $9 - 2\frac{2}{5}$

16. $6\frac{3}{8} - 5\frac{5}{6}$

17. $7 - \frac{3}{4}$

18. $50\frac{1}{3} - 16\frac{3}{8}$

19. $18\frac{5}{7} - 12\frac{1}{3}$

20. $34\frac{2}{9} - 18\frac{2}{3}$

High-Jump Results

Name	*Height in inches*
Bowers	$56\frac{7}{8}$
Davis	58
Faulin	$61\frac{5}{8}$
Jones	61
Morris	$62\frac{1}{4}$
Porter	$57\frac{3}{8}$
Richardson	$58\frac{1}{4}$
Wilson	$60\frac{1}{8}$

Problem Solving

21. How many jumped higher than $58\frac{1}{2}$ in.?

22. Who came in first? second? last?

23. How much higher did Faulin jump than Porter?

24. How much higher would Richardson have had to jump to tie Jones?

Round each mixed number to the nearest half. Then estimate the difference.

25. $6\frac{2}{15} - 5\frac{8}{9}$

26. $13\frac{15}{19} - 9\frac{3}{31}$

27. $22\frac{11}{21} - 15\frac{4}{9}$

Challenge!

Which two expressions have the same value?

28.

w	x	y
$9\frac{1}{2}$	$8\frac{1}{3}$	$5\frac{3}{4}$

a. $w + y$ b. $w - x$

c. $w - y$ d. $y + w$

29.

r	s	t
$4\frac{3}{4}$	$2\frac{1}{2}$	$1\frac{5}{8}$

a. $(r - s) - t$ b. $r - (s - t)$

c. $(r + s) + t$ d. $r + (s + t)$

KEEPING SKILLS SHARP

Estimate to find the two incorrect answers.

1. a. $8.5 \times 2.04 = 1.734$
   b. $12.4 \times 0.95 = 11.78$
   c. $1.8 \times 10.2 = 18.36$
   d. $0.11 \times 8.73 = 0.09603$

Multiply.

2. 5.8×2.9

3. 0.26×1.4

4. 7.4×0.35

5. 9.3×6.4

6. 0.52×18

7. 45×0.25

8. 3.81×16

9. 28.4×0.38

10. 2.97×0.58

Problem Solving Not enough information

How much did Juan and his 9-year-old sister Maria pay to get into Frontierland? *Missing fact:* Juan's age.

What is the missing fact not found in the picture or problem?
If you know the missing fact, solve the problem.

1. How many books of tickets would Juan and Maria have to buy so they could go on the 3 rides?
2. How many tickets did they have left after their first ride, which was the Raft to Fort Pike?
3. How long did they have to wait to see the Stunt Show?
4. How many hours was the intermission between Stunt Shows?
5. How many Badlands Railroad rides leave every hour?
6. How many people can ride on an 8-car Haunted Mine train?

Problem Solving Making a list

1. Study and understand.
2. Plan and do.
3. Answer and check.

Four students decided to sit in the first four seats of a roller coaster. How many different seating orders are there?

Step 1. Use letters to represent the students. List the seating orders with *A* in the first seat.

B second: ABCD ABDC
C second: ACBD ACDB
D second: ADBC ADCB

Step 2. List the seating orders with *B* in the first seat.

A second: BACD BADC
C second: BCAD BCDA
D second: BDAC BDCA

Step 3. Notice the pattern. There are 6 possible seating orders if *A* is first and 6 orders if *B* is first. There are also 6 seating orders if *C* is first and 6 orders if *D* is first.

Answer: There are 4×6, or 24, possible seating orders.

Solve by making a list.

1. Three people are going to ride a 3-seat water slide. How many different seating orders are possible?

2. Four people decided to buy tickets for the bumper cars. How many ways could they line up at the ticket office?

3. Suppose in the example given above that 5 students decided to sit in the first 5 seats of the roller coaster. How many different seating orders are possible?

★ 4. Three students want to sit together on the Ferris wheel, but each seat can hold only two people. How many ways can the seat be filled? [*Hint:* How many ways can they be seated if student *A* does not ride? if *B* does not ride?]

Chapter Checkup

Give the greatest common factor of the numerator and denominator of each fraction. [pages 130–131]

1. $\frac{8}{12}$ 2. $\frac{9}{16}$ 3. $\frac{35}{56}$ 4. $\frac{34}{51}$ 5. $\frac{18}{42}$ 6. $\frac{24}{40}$

Complete this table. Give each fraction in lowest terms. [pages 132–135]

	7.	8.	9.	10.	11.	12.
Fraction	$\frac{3}{10}$	$\frac{1}{8}$	$\frac{1}{15}$	?	?	?
Decimal	?	?	?	0.4	0.75	0.025

< **or** >**?** [pages 138–139]

13. $\frac{3}{7}$ ● $\frac{3}{5}$ 14. $\frac{2}{3}$ ● $\frac{3}{5}$ 15. $\frac{4}{7}$ ● $\frac{1}{2}$ 16. $\frac{5}{4}$ ● $\frac{3}{2}$

Give each sum or difference in lowest terms. [pages 140–143]

17. $\frac{3}{8} + \frac{1}{2}$ 18. $\frac{2}{9} + \frac{1}{3}$ 19. $\frac{1}{3} + \frac{2}{5}$ 20. $\frac{3}{10} + \frac{1}{4}$ 21. $\frac{1}{2} + \frac{1}{3}$

22. $\frac{7}{8} - \frac{1}{4}$ 23. $\frac{4}{5} - \frac{1}{2}$ 24. $\frac{3}{4} - \frac{2}{5}$ 25. $\frac{4}{9} - \frac{1}{6}$ 26. $\frac{3}{4} + \frac{5}{12}$

Complete this table. [pages 144–147]

	27.	28.	29.	30.	31.	32.
Fraction	$\frac{8}{3}$	$\frac{12}{5}$	$\frac{34}{7}$	?	?	?
Mixed number	?	?	?	$3\frac{1}{2}$	$5\frac{3}{4}$	$9\frac{5}{8}$

Add or subtract. Give answers in simplest form. [pages 148–151]

33. $3\frac{1}{2} + 2\frac{1}{4}$ 34. $4\frac{3}{5} + 2\frac{4}{5}$ 35. $1\frac{2}{3} + \frac{5}{6}$ 36. $4\frac{7}{8} - 1\frac{1}{4}$ 37. $3\frac{2}{7} - 1\frac{5}{7}$ 38. $6\frac{1}{2} - 3\frac{5}{8}$

Solve. [pages 152, 153]

39. The school long-jump record was $15\frac{1}{2}$ feet. Andrew jumped $13\frac{3}{4}$ feet. By how much did he miss the record?

40. Adele bought $8\frac{1}{4}$ pounds of fruit. She ate $\frac{3}{8}$ pound of the fruit. How much fruit was there then?

Chapter Project

ADDING AND SUBTRACTING MIXED NUMBERS

SLIDE RULERS AND ADDITION

You will need two rulers marked in inches and fractions of inches. Follow these steps to see how to use your rulers to add $1\frac{3}{4}$ and $2\frac{1}{8}$.

Step 1. Slide ruler A so that its 0 mark is directly under the mark for $1\frac{3}{4}$ inches on ruler B.

Step 2. Locate the mark for $2\frac{1}{8}$ inches on ruler A. Read the answer on ruler B directly above the mark for $2\frac{1}{8}$ inches on ruler A.

$$1\frac{3}{4} + 2\frac{1}{8} = 3\frac{7}{8}$$

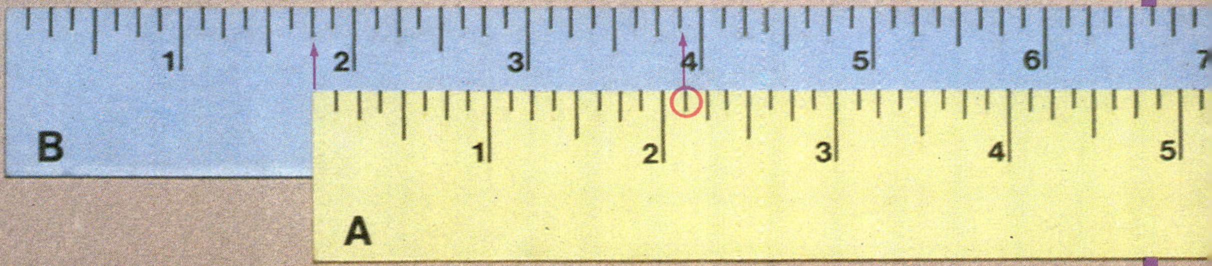

Use your slide rulers to find each sum.

1. $2\frac{1}{2} + 1\frac{3}{4}$
2. $4\frac{1}{4} + 2\frac{1}{2}$
3. $1\frac{1}{8} + 2\frac{3}{4}$
4. $1\frac{3}{8} + 2\frac{1}{4}$
5. $2\frac{5}{8} + 3\frac{1}{2}$
6. $2\frac{7}{8} + 3\frac{1}{4}$
7. $2\frac{5}{8} + 3\frac{5}{8}$
8. $3\frac{3}{8} + 1\frac{7}{8}$
9. $4\frac{1}{2} + 3\frac{7}{8}$

Find a way to use your slide rulers to give each difference.

10. $3\frac{1}{2} - 1\frac{1}{2}$
11. $5\frac{3}{4} - 2\frac{1}{2}$
12. $6\frac{3}{8} - 4\frac{3}{4}$

Chapter Review

Divide both terms by their greatest common factor.

$$\frac{6}{15} = \frac{2}{5} \quad (\div 3)$$

Write in lowest terms.

1. $\frac{8}{12}$ 2. $\frac{15}{10}$ 3. $\frac{8}{18}$ 4. $\frac{12}{30}$

$$\frac{3}{4} = ? \qquad 4\overline{)3.00} = 0.75$$

$$\frac{3}{4} = 0.75$$

Write as a decimal.

5. $\frac{3}{5}$ 6. $\frac{5}{4}$ 7. $\frac{7}{9}$ 8. $\frac{3}{8}$

$$\frac{1}{5} + \frac{2}{3} = \frac{3}{15} + \frac{10}{15} = \frac{13}{15}$$

Add. Give each sum in lowest terms.

9. $\frac{1}{4} + \frac{2}{3}$ 10. $\frac{3}{8} + \frac{1}{6}$ 11. $\frac{1}{10} + \frac{2}{5}$ 12. $\frac{2}{5} + \frac{1}{4}$

$$\frac{3}{4} - \frac{1}{3} = \frac{9}{12} - \frac{4}{12} = \frac{5}{12}$$

Subtract. Give each difference in lowest terms.

13. $\frac{8}{9} - \frac{2}{3}$ 14. $\frac{4}{5} - \frac{1}{2}$ 15. $\frac{1}{2} - \frac{1}{6}$ 16. $\frac{9}{10} - \frac{3}{4}$

$$2\frac{3}{5} + 1\frac{2}{3} = 2\frac{9}{15} + 1\frac{10}{15} = 4\frac{4}{15}$$

Add. Give each sum in simplest form.

17. $2\frac{3}{7} + 3\frac{1}{7}$ 18. $2\frac{5}{8} + 4\frac{3}{8}$ 19. $4\frac{1}{4} + 3\frac{5}{6}$ 20. $2\frac{3}{4} + \frac{7}{8}$

$$3\frac{1}{5} - 1\frac{4}{5} = 2\frac{6}{5} - 1\frac{4}{5} = 1\frac{2}{5}$$

Subtract. Give each difference in simplest form.

21. $4\frac{3}{8} - 3\frac{1}{8}$ 22. $4\frac{3}{10} - 1\frac{4}{5}$ 23. $5\frac{1}{3} - 3\frac{1}{2}$ 24. $8 - 1\frac{5}{6}$

Chapter Challenge

DEFECTIVE, PERFECT, AND ABUNDANT NUMBERS

The ancient Greeks classified whole numbers in many ways. One way was by the sum of the factors smaller than the given number.

If the sum is less than the number, the number is **defective;** if the sum is equal to the number, the number is **perfect;** and if the sum is greater than the number, the number is **abundant.**

Number	*Factors less than number*	*Sum of factors less than number*	*Kind of number*
4	1, 2	1 + 2 = 3	defective
6	1, 2, 3	1 + 2 + 3 = 6	perfect
8	1, 2, 4	1 + 2 + 4 = 7	defective
12	1, 2, 3, 4, 6	1 + 2 + 3 + 4 + 6 = 16	abundant

1. Classify the whole numbers 1 through 30 as defective, perfect, or abundant.
2. Which two numbers less than 30 are perfect numbers?

Euclid (an ancient Greek) discovered the following fact about prime numbers and perfect numbers:

If $2^n - 1$ is a prime number, then $(2^n - 1)2^{n-1}$ is a perfect number.

EXAMPLE. Let $n = 5$.

Since $2^5 - 1$ is prime, we know that $(2^5 - 1)2^{5-1}$ is perfect.

3. For $n = 5$, what perfect number do you get?
4. What is the next perfect number that can be found this way?

Cumulative Checkup

STANDARDIZED FORMAT

Choose the correct letter.

1. The standard numeral for two and three hundredths is

a. 2.3
b. 0.23
c. 2.03
d. none of these

2. Subtract.

$4.3 - 0.29$

a. 4.01
b. 3.01
c. 0.14
d. none of these

3. Multiply.

$$\begin{array}{r} 3.2 \\ \times 0.06 \\ \hline \end{array}$$

a. 19.2
b. 1.92
c. 0.192
d. none of these

4. Divide.

$0.32\overline{)1.984}$

a. 0.62
b. 6.2
c. 62
d. none of these

5. Solve.

$4y - 6 = 34$

a. 7
b. 10
c. 40
d. none of these

6. 4.7 cm = __?__ m

a. 0.047 m
b. 470 m
c. 4700 m
d. none of these

7. What is the circumference? Use 3.14 for π.

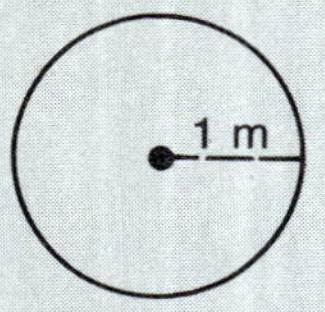

a. 6.28 m
b. 3.14 m
c. 2 m
d. none of these

8. The prime factorization of 48 is

a. $2^4 \times 3$
b. $4^2 \times 3$
c. $2^8 \times 3$
d. none of these

9. Add.

$$\begin{array}{r} 3\frac{2}{3} \\ +2\frac{3}{4} \\ \hline \end{array}$$

a. $6\frac{5}{12}$
b. $5\frac{1}{12}$
c. $5\frac{5}{12}$
d. none of these

10. Subtract.

$$\begin{array}{r} 4\frac{1}{3} \\ -1\frac{1}{2} \\ \hline \end{array}$$

a. $3\frac{1}{6}$
b. $3\frac{5}{6}$
c. $2\frac{1}{6}$
d. none of these

11. The TV program was $\frac{1}{2}$ hour long. 6 minutes were used for ads. How many minutes were not used for ads?

a. 24 **b.** 44
c. 54 **d.** none of these

12. Sandra bought 5 12¢ stamps and 15 18¢ stamps. What was the total cost of the stamps?

a. \$6 **b.** \$3.30
c. \$2.10 **d.** none of these

5

Fractions—Multiplication and Division

The white fabric required for this quilt was $\frac{1}{8}$ yard for each circle and an additional $\frac{3}{4}$ yard for the rest of the design. How much white fabric was needed for a quilt with 16 circles?

Multiplying fractions

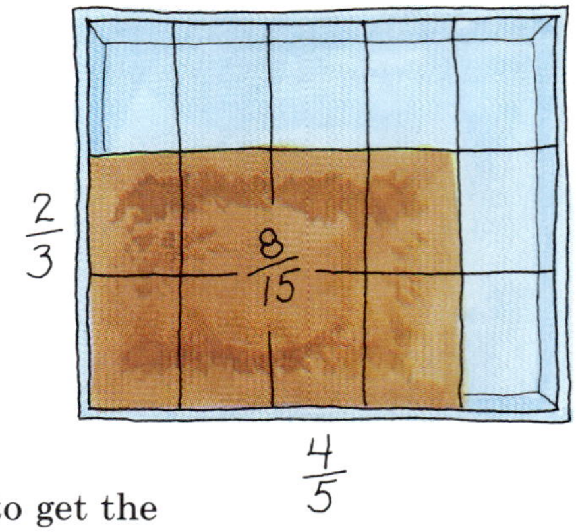

We can multiply to find the fraction of the corn bread that is left in the pan.

The length that remains is $\frac{4}{5}$ of the pan's length.

The width that remains is $\frac{2}{3}$ of the pan's width.

The area that remains is $\frac{8}{15}$ of the pan's area.

To multiply two fractions, multiply the numerators to get the numerator of the product, and multiply the denominators to get the denominator of the product.

$$\frac{4}{5} \times \frac{2}{3} = \frac{8}{15}$$

Here is a shortcut often used when multiplying fractions. The shortcut, called **canceling,** is to divide both a numerator and a denominator by a common factor *before* multiplying.

EXAMPLE 1.

Long method

Multiply, and then divide by the common factor 3.

$$\frac{3}{8} \times \frac{5}{24} = \frac{15}{192} = \frac{5}{64}$$

Canceling shortcut

Divide by the common factor 3 and then multiply.

$$\frac{\overset{1}{\cancel{3}}}{8} \times \frac{5}{\underset{8}{\cancel{24}}} = \frac{5}{64}$$

When you use the canceling shortcut, you multiply smaller numbers and the product is in lowest terms.

EXAMPLE 2.

Long method

Multiply, and then divide by the common factor 15.

$$\frac{5}{9} \times \frac{21}{10} = \frac{105}{90} = \frac{7}{6} = 1\frac{1}{6}$$

Canceling shortcut

Divide by the common factors 3 and 5 and then multiply.

$$\frac{\overset{1}{\cancel{5}}}{\underset{3}{\cancel{9}}} \times \frac{\overset{7}{\cancel{21}}}{\underset{2}{\cancel{10}}} = \frac{7}{6} = 1\frac{1}{6}$$

EXERCISES

Give each product in simplest form.

1. $\frac{1}{2} \times \frac{1}{3}$
2. $\frac{2}{3} \times \frac{4}{5}$
3. $\frac{3}{8} \times \frac{4}{5}$
4. $\frac{1}{6} \times \frac{2}{9}$
5. $\frac{3}{7} \times \frac{5}{3}$
6. $\frac{6}{5} \times \frac{51}{9}$
7. $\frac{12}{4} \times \frac{4}{3}$
8. $8 \times \frac{1}{2}$
9. $\frac{3}{5} \times \frac{0}{2}$
10. $6 \times \frac{2}{3}$
11. $\frac{2}{3} \times \frac{0}{5}$
12. $\frac{4}{3} \times 5$
13. $\frac{2}{3} \times \frac{3}{2}$
14. $\frac{5}{4} \times \frac{4}{5}$
15. $\frac{6}{7} \times \frac{6}{6}$

Use canceling to give each product in simplest form.

16. $\frac{3}{4} \times \frac{2}{9}$
17. $\frac{5}{8} \times \frac{3}{20}$
18. $\frac{1}{2} \times \frac{8}{5}$
19. $6 \times \frac{3}{4}$
20. $\frac{2}{9} \times 3$
21. $\frac{2}{5} \times \frac{5}{4}$
22. $\frac{9}{2} \times \frac{8}{21}$
23. $\frac{7}{2} \times \frac{8}{12}$
24. $\frac{4}{14} \times \frac{21}{6}$
25. $\frac{18}{3} \times \frac{15}{27}$
26. $\frac{2}{5} \times \frac{5}{2}$
27. $\frac{12}{16} \times \frac{16}{12}$
28. $\frac{7}{12} \times \frac{12}{7}$
29. $\frac{15}{36} \times \frac{36}{15}$
30. $\frac{38}{45} \times \frac{45}{38}$

Round each fraction to $\frac{1}{2}$, 1, or 2. Then estimate the product.

31. $\frac{35}{39} \times \frac{7}{8}$
32. $\frac{49}{23} \times \frac{9}{20}$
33. $\frac{25}{11} \times \frac{32}{17}$
34. $\frac{22}{51} \times \frac{17}{35}$
35. $\frac{8}{15} \times \frac{13}{14}$

Problem Solving

36. A muffin recipe calls for $\frac{3}{4}$ of a cup of flour. How much flour is needed to make $\frac{2}{3}$ of a recipe?

37. A roll recipe calls for $\frac{2}{3}$ of a cup of whole-wheat flour. How much flour is needed to make $\frac{3}{2}$ of a recipe?

Challenge!

38. The total number of muffins is between 55 and 65. How many does each person have?

I have a better idea. Give me $\frac{1}{3}$ of your muffins and I'll have four times as many as you do!

Give me $\frac{2}{7}$ of your muffins and I'll have as many as you do.

Reciprocals

Work in groups. Get squared paper and a calculator.

PART A

1. Fill in the blanks so that each pair of fractions has a product of 1.
   a. $\frac{2}{3} \times \frac{?}{2} = 1$ **b.** $\frac{4}{7} \times \frac{7}{?} = 1$ **c.** $\frac{2}{5} \times \frac{?}{?} = 1$
2. Make up a rule to find the reciprocal of a number. Then give examples.
3. Use your rule to find the reciprocal of each of these numbers.
   a. $\frac{9}{7}$ **b.** 7 **c.** $2\frac{1}{3}$ **d.** 0.4 **e.** 1.5

   Does your rule work for each number? How do you know?
4. What does your rule give for the reciprocal of 0? Does that make sense? Why or why not?
5. Make up a step-by-step method for finding the reciprocal of
   a. a whole number.
   b. a mixed number.
   c. a decimal number.
6. Compare your reciprocal rule and step-by-step methods with those of other groups.

PART B

Draw a rectangle 6 squares long and 4 squares wide on your squared paper.

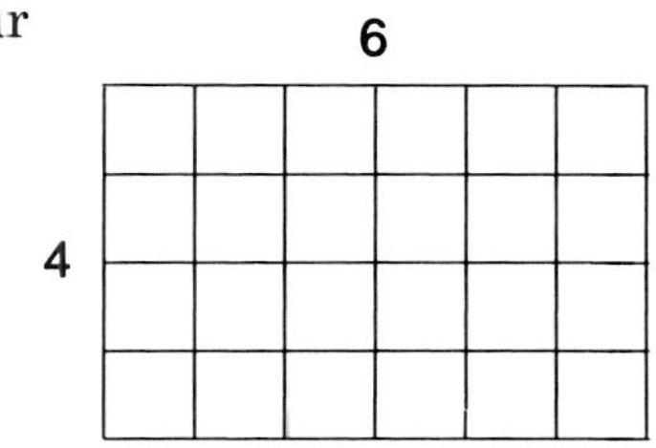

7. Decide on a way of dividing the rectangle into pieces $\frac{2}{3}$ of its size and $\frac{1}{3}$ of its size.
   a. Cut off $\frac{1}{3}$ of the rectangle so $\frac{2}{3}$ is left.
   b. Take the piece that is left over. Draw a rectangle $1\frac{1}{2}$ times larger.
   c. What is the area of the new rectangle? What was the area of the original 6-by-4 rectangle? Compare these two areas. Are your answers reasonable? How does your rule from Part A explain this?

8. Draw another 6-by-4 rectangle on squared paper.
   a. Find other ways of taking $\frac{2}{3}$ of this rectangle. How many new ways can you find?
   b. Use one of these ways to cut out $\frac{2}{3}$ of the rectangle. What would happen if you made this piece $\frac{3}{2}$ times larger?

9. Think of a number that is divisible by both 2 and 3.
   a. Using squared paper, cut out a rectangle that has this number of squares.
   b. Suppose you cut out a piece that was $\frac{2}{3}$ the size of this rectangle. Then you increased this piece to $1\frac{1}{2}$ times its size. What would the final area be? How can you check?

10. Use squared paper to test your rule for fractions other than $\frac{2}{3}$. Discuss your results.

11. Did you use estimation or a method of computation (mental math, pencil and paper, calculator) for this lesson? Why? Discuss your methods with others in your group and decide which were best.

12. Discuss within your group and with another group what you have discovered about reciprocals.

Dividing by fractions

Work in groups. Get a calculator. As you do these problems, decide whether to estimate or calculate.

PART A

Use squared paper, objects, pictures, a ruler, or another method of your choice to solve each problem.

1. Mary has a wire 6 feet long.
   a. How many $\frac{1}{4}$-foot pieces can she cut from her wire?
   b. How many $\frac{1}{3}$-foot pieces can she cut from her 6-foot-long wire?
2. If Mary's wire was 7 feet long, how many $\frac{1}{4}$-foot pieces could she cut?
3. Compare your strategies and solutions for the problems above.
4. a. Make up a rule for dividing a number by a fraction with a numerator of 1.
   b. Compare your rule and strategies with another group's.

PART B

5. Mark has 12 yards of material to make small pillows. He needs $\frac{3}{4}$ of a yard of material to make each pillow.
   a. Use your rule from problem 4a to find how many $\frac{1}{4}$-yard pieces there are in 12 yards.
   b. Now that you know how many $\frac{1}{4}$-yard pieces there are in 12 yards, use this to find how many $\frac{3}{4}$-yard pieces there are in 12 yards of material.
   c. How can you check your answer?

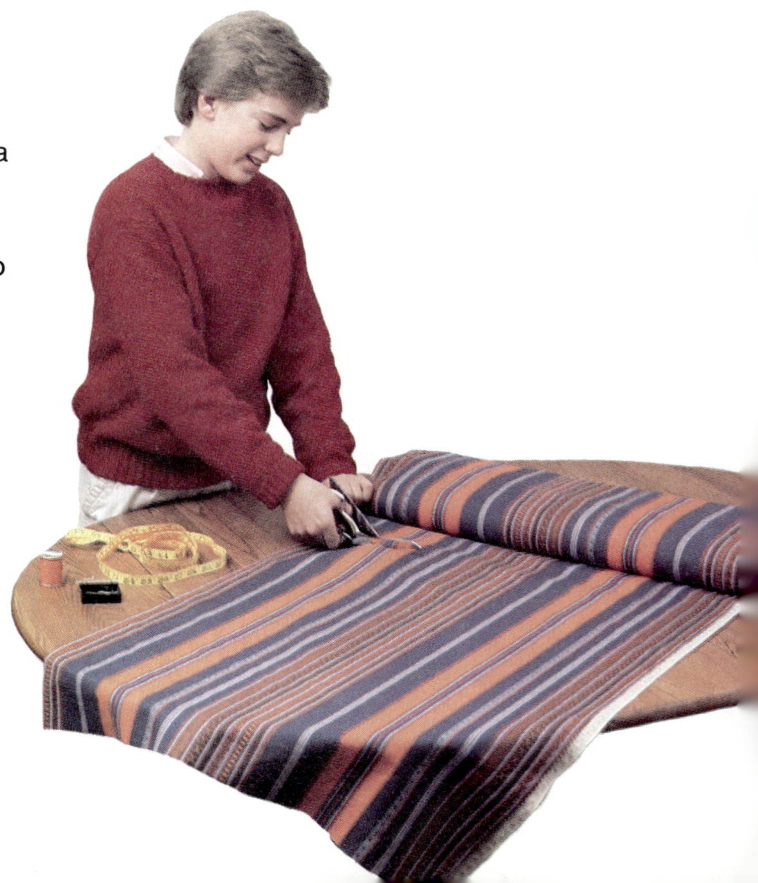

6. a. What two operations (or steps) did you use in 5a and 5b?
   b. Multiplying by what single fraction accomplishes both of those steps?
   c. How is that fraction related to the original fraction, $\frac{3}{4}$?
   d. Make up a rule for using multiplication to divide by any fraction.

7. Test your rule on the following problem: Sam is making cookies for a bake sale. Each batch of cookies uses $\frac{3}{8}$ of a pound of butter. If Sam has 6 pounds of butter, how many batches of cookies can he make?

8. Check your answer from question 8 in each of the following ways:
   a. Use squared paper, pictures, or objects.
   b. Calculate.

PART C

If your rule for dividing by any fraction worked, try the following problems. If not, first check with other groups or with your teacher.

9. A jeweler wished to separate $\frac{7}{8}$ of an ounce of gold into pieces each weighing $\frac{3}{16}$ of an ounce.
   a. How many $\frac{3}{16}$-ounce pieces can he get from the $\frac{7}{8}$ of an ounce of gold?
   b. Is your answer reasonable? How can you tell?

10. On her small farm, Karen wants to plant a different vegetable on every $1\frac{1}{2}$ acres of land. How many different vegetables can Karen plant if she has $13\frac{1}{2}$ acres of land? Decide on the best way to check your answer.

11. a. As a group, write down a step-by-step method for solving division problems involving fractions and whole numbers.
    b. Write down a step-by-step method for solving division problems involving fractions and mixed numbers.
    c. Test your two methods by making up and exchanging problems with others in your group.
    d. Compare your step-by-step methods and the strategies you used for checking your answers with those of another group.

Practice

Give each sum in simplest form.

1. $\frac{1}{2} + \frac{1}{4}$
2. $\frac{1}{3} + \frac{1}{2}$
3. $\frac{1}{3} + \frac{1}{4}$
4. $\frac{2}{3} + \frac{1}{4}$
5. $\frac{3}{8} + \frac{1}{4}$
6. $\frac{5}{8} + \frac{1}{6}$
7. $\frac{3}{5} + \frac{1}{4}$
8. $\frac{3}{4} + \frac{1}{6}$
9. $\frac{5}{8} + \frac{1}{7}$
10. $\frac{2}{3} + \frac{1}{6}$
11. $2 + \frac{4}{5}$
12. $\frac{3}{10} + \frac{1}{6}$
13. $\frac{5}{9} + \frac{3}{8}$
14. $\frac{4}{7} + \frac{2}{5}$
15. $\frac{5}{7} + \frac{3}{8}$
16. $1 + \frac{1}{10}$

Give each difference in simplest form.

17. $\frac{3}{4} - \frac{1}{2}$
18. $\frac{1}{2} - \frac{1}{3}$
19. $\frac{2}{3} - \frac{1}{2}$
20. $\frac{1}{3} - \frac{1}{4}$
21. $\frac{3}{4} - \frac{1}{6}$
22. $\frac{3}{8} - \frac{1}{6}$
23. $\frac{8}{7} - 1$
24. $\frac{2}{5} - \frac{1}{10}$
25. $\frac{4}{5} - \frac{1}{4}$
26. $3 - \frac{4}{5}$
27. $\frac{4}{3} - \frac{3}{8}$
28. $\frac{3}{7} - \frac{1}{5}$
29. $\frac{7}{10} - \frac{2}{3}$
30. $\frac{5}{6} - \frac{1}{9}$
31. $\frac{9}{10} - \frac{3}{8}$
32. $\frac{4}{9} - \frac{1}{10}$

Give each product in simplest form.

33. $\frac{2}{3} \times \frac{1}{2}$
34. $\frac{3}{4} \times \frac{2}{3}$
35. $\frac{4}{5} \times \frac{3}{7}$
36. $\frac{2}{5} \times \frac{3}{2}$
37. $\frac{5}{6} \times \frac{2}{3}$
38. $\frac{2}{3} \times 7$
39. $\frac{7}{5} \times \frac{3}{8}$
40. $\frac{4}{5} \times \frac{2}{3}$
41. $\frac{7}{8} \times \frac{4}{5}$
42. $\frac{2}{9} \times \frac{3}{4}$
43. $4 \times \frac{3}{8}$
44. $\frac{4}{5} \times \frac{3}{8}$
45. $\frac{5}{12} \times \frac{4}{3}$
46. $\frac{3}{2} \times 6$
47. $\frac{3}{4} \times \frac{4}{3}$
48. $\frac{2}{5} \times \frac{5}{2}$

Give each quotient in simplest form.

49. $\frac{2}{3} \div 2$
50. $\frac{2}{3} \div \frac{1}{2}$
51. $8 \div \frac{1}{2}$
52. $5 \div 2$
53. $\frac{3}{4} \div \frac{2}{3}$
54. $\frac{3}{4} \div \frac{3}{2}$
55. $\frac{5}{6} \div 3$
56. $\frac{4}{5} \div \frac{1}{5}$
57. $\frac{3}{8} \div 3$
58. $\frac{7}{8} \div \frac{1}{4}$
59. $\frac{3}{7} \div \frac{2}{5}$
60. $\frac{6}{7} \div \frac{2}{3}$
61. $4 \div \frac{1}{2}$
62. $\frac{7}{8} \div \frac{3}{4}$
63. $\frac{5}{12} \div \frac{1}{3}$
64. $\frac{9}{10} \div 3$

Problem Solving Making a table

1. Study and understand.
2. Plan and do.
3. Answer and check.

The painter can paint $\frac{1}{5}$ of the house in 1 day. The assistant can paint $\frac{1}{8}$ of the house in 1 day. Together, they can paint $\frac{1}{5} + \frac{1}{8}$ of the house in 1 day.

A painter can paint a house in 5 days and the painter's assistant can paint the same house in 8 days. About how long would it take them to paint the house working together?

	Day 1	*Day 2*	*Day 3*
Painter	$\frac{1}{5}$	$\frac{1}{5}$	$\frac{1}{5}$
Assistant	$\frac{1}{8}$	$\frac{1}{8}$	$\frac{1}{8}$
Together	$\frac{13}{40}$	$\frac{13}{40}$	$\frac{13}{40}$
Total	$\frac{13}{40}$	$\frac{26}{40}$	$\frac{39}{40}$

After 3 days, $\frac{39}{40}$ of the house would be painted.

Answer: It would take them a little more than 3 days to paint the house.

Solve by making a table.

1. One painter can paint a house in 6 days, and another can paint the same house in 9 days. About how long would it take them to paint the house together?

2. One typist can type 7 pages per hour, and another can type 6 pages per hour. About how long would it take the two typists to type a 100-page report?

3. One printing press can print 600 posters an hour. Another press can print 800 of the posters in an hour. If both presses are used on the same job, how long would it take to print 7000 posters?

★ 4. One hose can fill a swimming pool in 6 days, and another bigger hose can fill the pool in 5 days. If both hoses were used, about how long would it take to fill the pool?

Solving fraction equations

The regular price of a Star Wars poster is \$2.40. It is on sale for $\frac{2}{3}$ of its regular price. What is its sale price?

To find a fraction of a number, we can multiply the number by the fraction. To find the sale price, we can solve this equation:

Fraction of regular price → $\frac{2}{3}$; Regular price → \$2.40; Sale price → s

$$\frac{2}{3} \times \$2.40 = s$$

$$\frac{\$4.80}{3} = s$$

$$\$1.60 = s$$

The sale price is \$1.60.

The Superman poster is on sale for $\frac{3}{4}$ of its regular price. The sale price is \$1.80. What is its regular price?

To find the regular price, we can solve this equation:

Fraction of regular price → $\frac{3}{4}$; Regular price → r; Sale price → \$1.80

$$\frac{3}{4}r = \$1.80$$

Multiply both sides by the reciprocal of $\frac{3}{4}$:

$$\frac{4}{3} \times \frac{3}{4}r = \frac{4}{3} \times \$1.80$$

$$r = \frac{\$7.20}{3}$$

$$r = \$2.40$$

The regular price is \$2.40.

EXERCISES

Solve.

1. $\frac{1}{3} \times 18 = n$
2. $n = \frac{1}{4} \times 21$
3. $n = \frac{3}{2} \times 32$
4. $\frac{3}{4} \times 48 = n$
5. $n = \frac{3}{5} \times 22$
6. $\frac{5}{2} \times 47 = n$
7. $\frac{3}{8} \times \$16 = n$
8. $n = \frac{5}{9} \times \$36$
9. $\frac{1}{4} \times \$8.60 = n$
10. $\frac{2}{3} \times \$9.30 = n$

Solve.

11. $\frac{1}{4}n = 6$
12. $\frac{1}{5}n = 8$
13. $16 = \frac{2}{3}n$
14. $27 = \frac{3}{8}n$
15. $\frac{3}{2}n = 36$
16. $\frac{4}{3}n = 48$
17. $\frac{1}{3}n = \$9$
18. $\$18 = \frac{2}{3}n$
19. $\$6.30 = \frac{3}{8}n$
20. $\frac{4}{5}n = \$9.12$

Problem Solving

Write and solve an equation.

21. There are 20 large science-fiction posters on sale for $\frac{4}{5}$ of the regular price. What is the sale price of a poster that usually sells for $2.20?
22. All 11″ by 17″ posters of rock groups are on sale for $\frac{3}{4}$ of the regular price. What is the sale price of a poster that usually sells for $1.90?
23. The Wolf Man poster is on sale for $\frac{2}{3}$ of its regular price. The sale price is $1.72. What is the regular price?
24. There are 15 old posters of movie stars on sale for $\frac{3}{5}$ of the regular price. What is the regular price of a poster that costs $1.44?

KEEPING SKILLS SHARP

Divide. Round each quotient to the nearest hundredth.

1. $0.3\overline{)1.4}$
2. $0.08\overline{)3.57}$
3. $0.6\overline{)0.25}$
4. $0.12\overline{)2.13}$
5. $3.4\overline{)2.87}$
6. $5.1\overline{)5.94}$
7. $7.4\overline{)0.382}$
8. $12.5\overline{)6.468}$
9. $3.94\overline{)87.5}$
10. $0.514\overline{)0.6381}$
11. $0.821\overline{)52.60}$
12. $6.35\overline{)9.758}$
13. $92.4\overline{)41.092}$
14. $83.7\overline{)26.751}$
15. $1.55\overline{)9.368}$

Multiplying mixed numbers

Here are three ways to multiply mixed numbers.

METHOD 1.

Step 1.
Change to fractions.

$$3\frac{1}{4} \times 1\frac{2}{3} = \frac{13}{4} \times \frac{5}{3}$$

Step 2.
Multiply.

$$3\frac{1}{4} \times 1\frac{2}{3} = \frac{13}{4} \times \frac{5}{3} = \frac{65}{12}$$

Step 3.
Change the product to a mixed number.

$$3\frac{1}{4} \times 1\frac{2}{3} = \frac{13}{4} \times \frac{5}{3} = \frac{65}{12} = 5\frac{5}{12}$$

Estimate the product by rounding each factor to the nearest whole number. Is the answer $5\frac{5}{12}$ reasonable?

METHOD 2. You can use this method when you are multiplying by a whole number.

Step 1.
Multiply the fraction.

$$\begin{array}{r} 8\frac{1}{2} \\ \times 5 \\ \hline 2\frac{1}{2} \end{array}$$

Step 2.
Multiply the whole number.

$$\begin{array}{r} 8\frac{1}{2} \\ \times 5 \\ \hline 2\frac{1}{2} \\ 40 \end{array}$$

Step 3.
Add.

$$\begin{array}{r} 8\frac{1}{2} \\ \times 5 \\ \hline 2\frac{1}{2} \\ 40 \\ \hline 42\frac{1}{2} \end{array}$$

METHOD 3. You can use this method when the fractions can easily be changed to decimals.

Step 1.
Change each number to a decimal.

$$2\frac{4}{5} \times 3\frac{1}{2} = 2.8 \times 3.5$$

Step 2.
Multiply the decimals.

$$\begin{array}{r} 2.8 \\ \times 3.5 \\ \hline 140 \\ 84 \\ \hline 9.80 \end{array}$$

Step 3.
Change the product to a fraction.

$$9.80 = 9\frac{4}{5}$$

So, $2\frac{4}{5} \times 3\frac{1}{2} = 9\frac{4}{5}$

Before multiplying, you should think about the numbers involved and pick the method you think is easiest.

EXERCISES

First give the product in simplest form. Then estimate the product to see if your answer is reasonable.

1. $1\frac{1}{2} \times 4$
2. $2\frac{1}{2} \times 3$
3. $3\frac{1}{3} \times 4$
4. $2\frac{3}{4} \times 3$
5. $4\frac{1}{3} \times 2$
6. $6 \times 2\frac{2}{3}$
7. $4 \times 3\frac{3}{8}$
8. $6 \times 12\frac{3}{5}$
9. $8 \times 15\frac{1}{4}$
10. $10 \times 16\frac{1}{3}$
11. $1\frac{1}{2} \times 1\frac{1}{2}$
12. $2\frac{1}{4} \times 1\frac{1}{2}$
13. $3\frac{1}{8} \times 4\frac{1}{3}$
14. $5\frac{3}{5} \times 2\frac{1}{2}$
15. $6\frac{3}{5} \times 5\frac{1}{4}$
16. $4\frac{2}{3} \times 2\frac{3}{8}$
17. $2\frac{1}{2} \times 5\frac{1}{4}$
18. $8 \times 7\frac{1}{2}$
19. $1\frac{3}{8} \times 2\frac{3}{8}$
20. $3\frac{1}{4} \times 3$
21. $9\frac{1}{2} \times 6\frac{1}{4}$
22. $7\frac{1}{3} \times 8\frac{1}{5}$
23. $2\frac{1}{4} \times 2\frac{3}{8}$
24. $3\frac{1}{3} \times 2\frac{5}{8}$
25. $1\frac{2}{3} \times 5\frac{3}{4}$

Problem Solving

This time sheet shows the number of hours that the part-time help worked in the computer room at the library during one week.

Hours Worked						
Employee	*Mon*	*Tues*	*Wed*	*Thur*	*Fri*	*Sat*
Allan	$1\frac{3}{4}$	$2\frac{1}{2}$	$3\frac{1}{4}$	$1\frac{3}{4}$	$2\frac{3}{4}$	X
Becker	$2\frac{3}{4}$	$3\frac{1}{2}$	X	$2\frac{3}{4}$	X	$6\frac{1}{2}$
Davis	$2\frac{3}{4}$	X	3	$2\frac{1}{4}$	$1\frac{3}{4}$	5
Folder	X	$3\frac{1}{4}$	$2\frac{1}{4}$	X	$2\frac{1}{2}$	$7\frac{1}{2}$
Logan	$2\frac{1}{4}$	2	3	$2\frac{3}{4}$	$3\frac{1}{4}$	X
Monroe	$1\frac{3}{4}$	$3\frac{1}{4}$	X	$2\frac{3}{4}$	X	8
Sanchez	$2\frac{1}{2}$	X	$3\frac{1}{2}$	4	$5\frac{1}{2}$	$6\frac{1}{4}$

26. Who worked the most hours on Saturday?
27. On which day did Allan work the most hours?
28. How many hours did Sanchez work during the week?
29. Who worked more hours, Folder or Monroe? How many more?
30. How many hours of part-time help were used on Thursday?
31. Were more hours of part-time help used on Wednesday or Thursday? How many more?
32. Allan is paid \$3.55 an hour. How much did Allan earn for the week?
33. During the week, Sanchez earned \$78.30. How much is this an hour?
34. Davis earns \$3.60 an hour and Logan \$3.80. Who earned more during this particular week? How much more?
35. Becker is paid \$3.50 an hour. What were Becker's earnings for the week?

Dividing mixed numbers

To divide mixed numbers, change each mixed number to an equivalent fraction and divide.

$$10\frac{2}{7} \div 3\frac{5}{7} = \frac{72}{7} \div \frac{26}{7}$$
$$= \frac{\overset{36}{\cancel{72}}}{\underset{1}{\cancel{7}}} \times \frac{\overset{1}{\cancel{7}}}{\underset{13}{\cancel{26}}}$$
$$= \frac{36}{13}$$
$$= 2\frac{10}{13}$$

<u>Estimate</u> the quotient by rounding each mixed number to the nearest whole number. Is the answer $2\frac{10}{13}$ reasonable?

We can find a quotient by changing each number to a decimal and dividing:

$9\frac{3}{4} \div 1\frac{1}{2} = ?$

```
      65
1.5)9.75
   -90
     75
    -75
      0
```

$6.5 = 6\frac{1}{2}$

So, $9\frac{3}{4} \div 1\frac{1}{2} = 6\frac{1}{2}$

Use whichever method is easier.

EXERCISES

First give the quotient in simplest form. Then <u>estimate</u> the quotient to see if your answer is reasonable.

1. $6 \div 2\frac{1}{2}$

2. $4 \div 1\frac{1}{4}$

3. $4\frac{3}{4} \div 2$

4. $4 \div 1\frac{1}{2}$

5. $4\frac{1}{2} \div 1\frac{1}{2}$

6. $8\frac{3}{4} \div 2\frac{1}{3}$

7. $6\frac{3}{8} \div 3$

8. $9\frac{2}{3} \div 3\frac{3}{4}$

9. $8\frac{1}{2} \div 3\frac{2}{5}$

10. $9\frac{1}{4} \div 1\frac{1}{3}$

11. $8\frac{1}{2} \div 2\frac{1}{4}$

12. $4\frac{2}{3} \div 1\frac{3}{4}$

13. $16 \div 2\frac{1}{2}$

14. $12\frac{1}{2} \div 8\frac{3}{4}$

15. $6\frac{7}{8} \div 1\frac{3}{4}$

16. $8\frac{2}{3} \div 8\frac{2}{3}$

17. $5\frac{2}{3} \div 2\frac{1}{4}$

18. $3\frac{3}{4} \div 4\frac{2}{3}$

Problem Solving

19. To advertise the car wash, the students ordered 1200 handouts from a local printer. The printer charged $1\frac{1}{4}$¢ each for the first 1000 and $\frac{3}{4}$¢ each for the remaining handouts. What was the total cost?

20. Nine students signed up to work $3\frac{1}{2}$ hours in the morning and 12 students signed up to work $4\frac{1}{2}$ hours in the afternoon. How many hours were volunteered?

21. A merchant sold them 4 boxes of detergent for $\frac{2}{3}$ of the regular price. The regular price was $1.80 a box. What was the total price of the detergent?

22. One team washed 6 cars in $1\frac{1}{4}$ hours. What was their average time in hours for washing a car? What was the average time in minutes?

★ **23.** During the $3\frac{1}{2}$-hour morning session, 9 students washed an average of one car every $5\frac{1}{4}$ minutes. How many cars did they wash during the morning?

★ **24.** During the day they washed 76 cars for $1.75 each and vacuumed 37 cars for $.50 each. Their total expenses were $28.42. How much profit did they make?

Problem Solving

Three-step method

How big is a million dollars?

A million \$1 bills laid end to end would be about $96\frac{9}{10}$ miles long. A million dollars in \$10 bills would be $\frac{1}{10}$ as long, or about $9\frac{7}{10}$ miles long.

Solve these million-dollar problems. Remember to follow these steps:

1. Study and understand.
2. Plan and do.
3. Answer and check.

EXERCISES

Solve.

1. A million pennies laid end to end would be about $11\frac{5}{6}$ miles long.

   a. The diameter of a quarter is about $1\frac{1}{4}$ times the diameter of a penny. About how many miles long would a million quarters laid end to end be?

   b. About how many miles long would a million dollars' worth of quarters be?

2. The Sears Tower in Chicago is 1454 feet high.

   a. A million pennies placed one on top of another would make a stack about $3\frac{4}{9}$ times as high as the Sears Tower. About how many feet high would the stack be?

   b. About how many feet high would a stack of pennies worth a million dollars be?

3. A million dollars in new $5 bills weighs about 408.16 pounds and has a volume of about 7.96 cubic feet.

   a. A $5 bill and a $1 bill weigh the same. About how many pounds would a million dollars in $1 bills weigh?

   b. A $5 bill and a $1 bill have the same dimensions. About how many cubic feet would a million dollars in $1 bills be?

4. A million nickels weigh about 10,047 pounds.

   a. A penny weighs about $\frac{5}{8}$ as much as a nickel. About how many pounds would a million pennies weigh?

   b. About how many pounds would a million dollars in pennies weigh?

5. You can earn a million dollars if you get a job that pays $3 an hour and work 8 hours a day, 250 days a year, for $166\frac{2}{3}$ years.

   a. Suppose you earned $15 an hour and worked the same number of hours a year. How many years would it take you to earn a million dollars?

   b. Suppose you earned $24 an hour and worked the same number of hours a year. How many years would it take you to earn a million dollars?

Chapter Checkup

Multiply. Give each product in simplest form. [pages 160–161, 163, 166]

1. $\frac{3}{8} \times \frac{1}{2}$
2. $\frac{4}{5} \times 6$
3. $\frac{3}{4} \times \frac{4}{3}$
4. $\frac{7}{8} \times \frac{3}{4}$
5. $\frac{8}{15} \times \frac{3}{4}$
6. $\frac{1}{3} \times 2$
7. $\frac{4}{5} \times \frac{3}{8}$
8. $\frac{5}{7} \times \frac{2}{3}$

Divide. Give each quotient in simplest form. [pages 164–166]

9. $\frac{3}{5} \div \frac{2}{3}$
10. $\frac{5}{8} \div \frac{1}{2}$
11. $6 \div \frac{2}{3}$
12. $\frac{3}{4} \div 2$
13. $\frac{4}{5} \div 2$
14. $\frac{3}{8} \div \frac{3}{8}$
15. $\frac{3}{5} \div \frac{5}{6}$
16. $\frac{4}{5} \div \frac{2}{3}$

Multiply. Give each product in simplest form. [pages 170–171]

17. $2\frac{1}{2} \times 6$
18. $3\frac{1}{2} \times 1\frac{1}{4}$
19. $3\frac{3}{4} \times 2\frac{1}{2}$
20. $4\frac{2}{3} \times 3\frac{1}{4}$
21. $5\frac{1}{3} \times 2\frac{1}{4}$
22. $8 \times 1\frac{3}{10}$
23. $2\frac{2}{3} \times \frac{3}{8}$
24. $1\frac{3}{5} \times 2\frac{5}{8}$

Divide. Give each quotient in simplest form. [pages 172–173]

25. $5 \div 2\frac{1}{2}$
26. $3\frac{3}{4} \div 3$
27. $3\frac{1}{2} \div 2\frac{2}{3}$
28. $1\frac{3}{4} \div 4\frac{1}{2}$
29. $2\frac{2}{5} \div 1\frac{7}{8}$
30. $6 \div 1\frac{4}{5}$
31. $5\frac{1}{3} \div 4$
32. $3\frac{3}{5} \div 7\frac{1}{2}$

Solve. [pages 167–169, 174–175]

33. John had \$6.48. He spent $\frac{2}{3}$ of this amount to buy a book. How much did the book cost?

34. When Mia works on Saturday, she is paid \$5.70 an hour. This amount is $1\frac{1}{2}$ times as much as she is paid when she works on a weekday. How much is she paid an hour on a weekday?

Chapter Project

STOCK QUOTATIONS

The facts below are about the common stock of the Raytheon Co.

Today's high was $\$48\frac{1}{4}$.

Today's low was $\$47\frac{5}{8}$.

When the market closed, the stock was selling at $\$48\frac{1}{8}$ a share.

Ralston	.84	21	534	25	$24\frac{1}{2}$	$24\frac{3}{4}$	$-\frac{3}{8}$
Rmada Inn		...	945	$10\frac{5}{8}$	10	$10\frac{3}{8}$	$-\frac{1}{4}$
Rmpc	1.35b	16	34	$26\frac{3}{4}$	$26\frac{1}{2}$	$26\frac{3}{4}$	$-\frac{1}{4}$
Ranco	.84	20	56	18	$17\frac{3}{4}$	$17\frac{3}{4}$	$-\frac{1}{4}$
Range Oil		15	283	$11\frac{1}{4}$	11	$11\frac{1}{8}$	$-\frac{1}{8}$
Raymark		...	19	$13\frac{3}{4}$	$13\frac{3}{8}$	$13\frac{1}{2}$	$-\frac{1}{4}$
Raychm	.44	24	99	79	$78\frac{5}{8}$	79	$+\frac{3}{4}$
Rayml	.50h	12	29	$27\frac{1}{8}$	$27\frac{1}{8}$	$27\frac{1}{8}$	...
Raythn	1.40	13	3244	$48\frac{1}{4}$	$47\frac{5}{8}$	$48\frac{1}{8}$	$-\frac{1}{2}$
RB	.16	242	1039	$26\frac{7}{8}$	$26\frac{1}{4}$	$26\frac{3}{8}$	$+\frac{1}{4}$
R C A	.90	20	16982	$34\frac{1}{8}$	$32\frac{5}{8}$	33	−1
RCA cv pf	4	...	54	r78	77	$77\frac{1}{4}$	$+\frac{1}{4}$
RCA pf	$3\frac{1}{2}$	...	z530	r35	$33\frac{3}{4}$	34	+1
RCA pf	2.12	...	1052	$28\frac{1}{8}$	$27\frac{1}{4}$	$27\frac{3}{4}$	$-\frac{7}{8}$

1. Find the stock quotations in a newspaper. Choose one of the stock listings. How many shares could you buy for $1000?

EXAMPLE. Suppose you choose Raytheon at the closing price.

$$\begin{aligned} \$1000 \div 48\frac{1}{8} &= 1000 \div \frac{385}{8} \\ &= 1000 \times \frac{8}{385} \\ &= 1000 \times \frac{8}{385} \\ &= \frac{64}{3} \\ &= 21\frac{1}{3} \end{aligned}$$

So, you could buy 21 shares with $1000.

2. Every day for a week, record the closing price of the stock you chose. For each day, compute the value of the stock you "bought." Graph the daily value of your shares of stock.

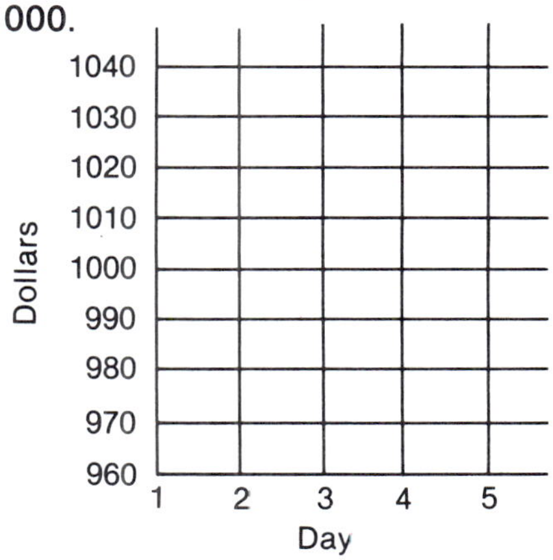

Chapter Review

Multiply numerators.
Multiply denominators.

$$\frac{3}{8} \times \frac{6}{5} = \frac{9}{20}$$

Multiply. Give each product in simplest form.

1. $\frac{4}{5} \times \frac{1}{3}$
2. $\frac{5}{6} \times \frac{3}{4}$
3. $\frac{2}{3} \times \frac{1}{6}$
4. $\frac{5}{8} \times \frac{3}{4}$
5. $\frac{2}{5} \times 10$
6. $8 \times \frac{5}{4}$
7. $\frac{2}{3} \times \frac{3}{4}$
8. $\frac{5}{6} \times \frac{2}{5}$
9. $\frac{3}{8} \times \frac{9}{10}$

To divide by a fraction, multiply by its reciprocal.

$$\frac{3}{8} \div \frac{6}{5} = \frac{3}{8} \times \frac{5}{6}$$

$$= \frac{5}{16}$$

Divide. Give each quotient in simplest form.

10. $\frac{3}{4} \div \frac{1}{4}$
11. $\frac{5}{8} \div \frac{3}{8}$
12. $9 \div \frac{2}{3}$
13. $\frac{3}{5} \div \frac{2}{3}$
14. $\frac{3}{5} \div \frac{3}{2}$
15. $\frac{5}{6} \div \frac{10}{3}$
16. $\frac{2}{3} \div \frac{2}{3}$
17. $\frac{1}{4} \div \frac{2}{5}$
18. $\frac{3}{8} \div \frac{9}{10}$

$$2\frac{3}{4} \times 3\frac{1}{2} = \frac{11}{4} \times \frac{7}{2}$$

$$= \frac{77}{8}$$

$$= 9\frac{5}{8}$$

Multiply. Give each product in simplest form.

19. $1\frac{3}{4} \times 1\frac{1}{3}$
20. $3\frac{1}{2} \times 2$
21. $6 \times 1\frac{3}{4}$
22. $2\frac{1}{4} \times \frac{1}{2}$
23. $3\frac{3}{8} \times 1\frac{3}{5}$
24. $4\frac{2}{3} \times 3\frac{5}{6}$

$$2\frac{3}{4} \div 3\frac{1}{2} = \frac{11}{4} \div \frac{7}{2}$$

$$= \frac{11}{4} \times \frac{2}{7}$$

$$= \frac{11}{14}$$

Divide. Give each quotient in simplest form.

25. $2\frac{1}{4} \div 1\frac{3}{4}$
26. $3\frac{1}{2} \div 2$
27. $6 \div 1\frac{1}{2}$
28. $2\frac{3}{4} \div \frac{1}{2}$
29. $2\frac{5}{8} \div 1\frac{4}{5}$
30. $1\frac{4}{5} \div 2\frac{5}{8}$

Chapter Challenge

REPEATING DECIMALS

The examples below show how to change a repeating decimal to an equivalent fraction.

EXAMPLE 1. Change $0.\overline{3}$ to a fraction.

Let n be the fraction.

$n = 0.\overline{3}$

$10n = 3.\overline{3}$

When one digit repeats, multiply each side of the equation by 10.

$10n - n = 3.\overline{3} - 0.\overline{3}$

Subtract n from each side. Since $n = 0.\overline{3}$, we subtract $0.\overline{3}$ from the right side of the equation.

$9n = 3$

$n = \frac{3}{9}$

$= \frac{1}{3}$

EXAMPLE 2. Change $0.\overline{36}$ to a fraction.

$n = 0.\overline{36}$

$100n = 36.\overline{36}$

When two digits repeat, multiply each side of the equation by 100.

$100n - n = 36.\overline{36} - 0.\overline{36}$

$99n = 36$

$n = \frac{36}{99}$

$= \frac{4}{11}$

Change each repeating decimal to its equivalent fraction in lowest terms.

1. $0.\overline{6}$ **2.** $0.\overline{4}$ **3.** $0.\overline{7}$ **4.** $0.\overline{8}$ **5.** $0.\overline{16}$ **6.** $0.\overline{45}$

7. $0.\overline{72}$ **8.** $0.\overline{09}$ **9.** $0.\overline{39}$ **10.** $0.\overline{132}$ **11.** $0.\overline{037}$ **12.** $0.\overline{123}$

Cumulative Checkup

STANDARDIZED FORMAT

Choose the correct letter.

1. 39.624 rounded to the nearest whole number is

a. 40
b. 39
c. 39.6
d. none of these

2. Add.

35.7 + 6.84

a. 10.41
b. 104.1
c. 41.54
d. none of these

3. Multiply.

$$\begin{array}{r} 4.16 \\ \times 20.3 \\ \hline \end{array}$$

a. 84.448
b. 9.568
c. 80.448
d. none of these

4. Divide.

14.496 ÷ 4.8

a. 3.2
b. 3.02
c. 30.2
d. none of these

5. Solve.

$2x - 5 = {}^{-}7$

a. $^{-}1$
b. 1
c. 6
d. none of these

6. 76.1 mm = __?__ m

a. 76,100
b. 0.761
c. 0.0761
d. none of these

7. Subtract.

$\frac{7}{10} - \frac{1}{2}$

a. $\frac{1}{5}$
b. $1\frac{2}{5}$
c. $\frac{3}{4}$
d. none of these

8. Add.

$3\frac{5}{8} + 2\frac{3}{4}$

a. $5\frac{3}{8}$
b. $6\frac{3}{8}$
c. $5\frac{2}{3}$
d. none of these

9. Multiply.

$\frac{5}{8} \times \frac{4}{5}$

a. $\frac{1}{2}$
b. $\frac{25}{32}$
c. 2
d. none of these

10. Divide.

$1\frac{1}{3} \div \frac{3}{4}$

a. $1\frac{7}{9}$
b. $\frac{9}{16}$
c. 1
d. none of these

11. José watched TV for $\frac{2}{5}$ of an hour. How many minutes was that?

a. 40 **b.** 24
c. 12 **d.** none of these

12. Betty had $1. She bought 4 pears for 17¢ each. How much money did she have left?

a. 32¢ **b.** 83¢
c. 68¢ **d.** none of these

6

Measurement—Area and Volume

The volume of this large storage tank is about 98,125 cubic meters. If it was full of water, how many Olympic-size swimming pools (50 meters by 20 meters by 4 meters) could it fill?

Area—rectangles and squares

The **area** of a region is the number of square units needed to cover the region.

RECTANGLE
The area of a rectangle is its length times its width.

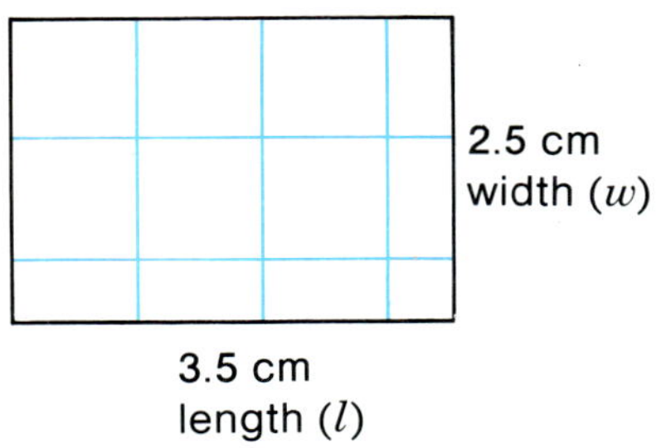

Formula $A = lw$

$= (3.5 \text{ cm})(2.5 \text{ cm})$

$= 8.75 \text{ cm}^2$

(cm^2 stands for square centimeters)

SQUARE
The area of a square is the square of the length of a side.

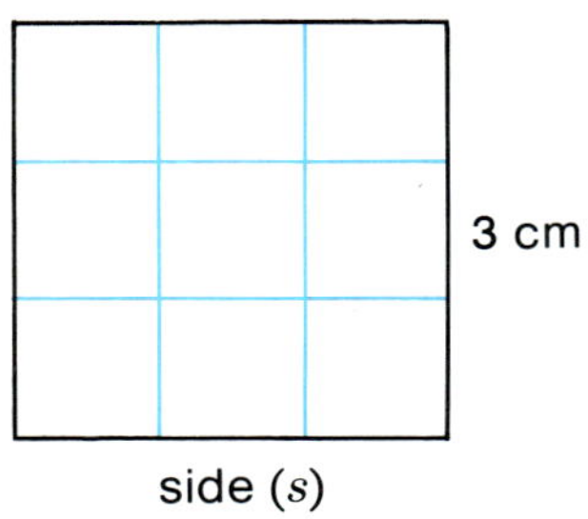

Formula $A = s^2$

$= (3 \text{ cm})(3 \text{ cm})$

$= 9 \text{ cm}^2$

You can count the squares to check each answer.

EXERCISES

Tell whether the question involves perimeter or area. (*Remember:* The perimeter of a figure is the distance around it.)

1. How much fence is needed to fence in a yard?
2. How much tile is needed for your classroom floor?
3. How much grass seed is needed to seed a lawn?
4. How much ceiling molding is needed to go around a room?

Find the area of each region.

5. 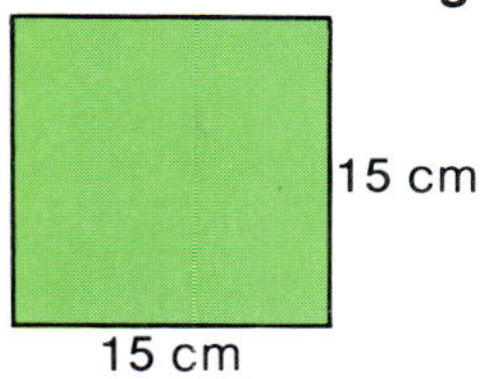

6. 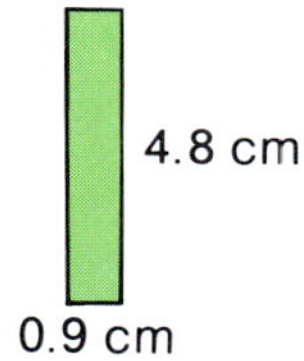

7. 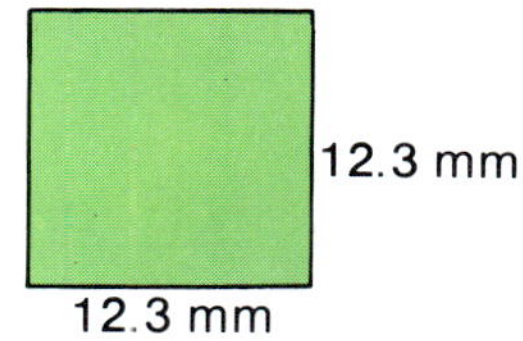

8. 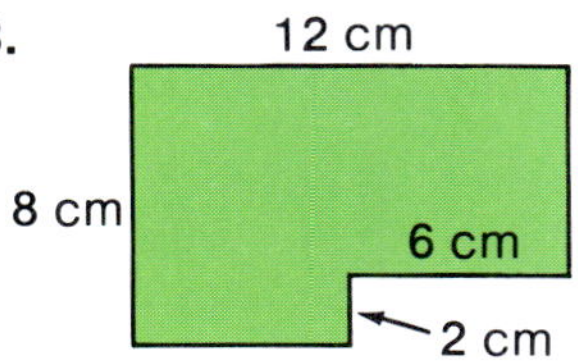

9. 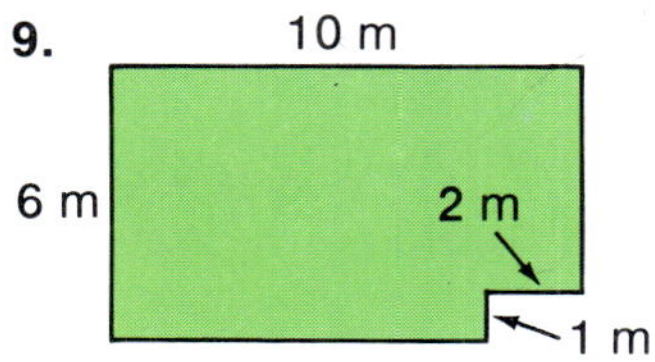

10. 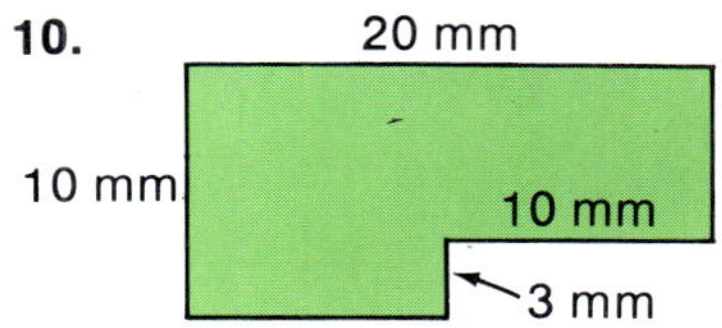

Find the perimeter and area of each region.

11. 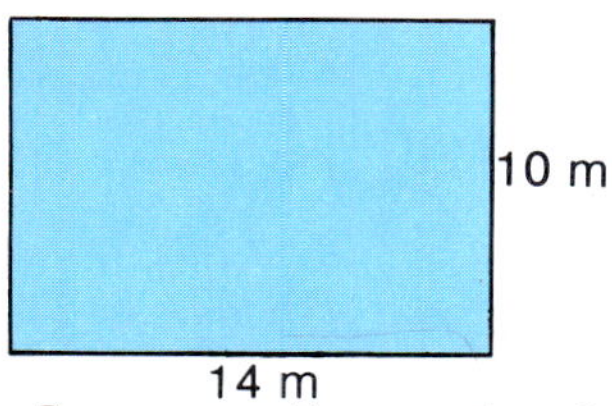

12. 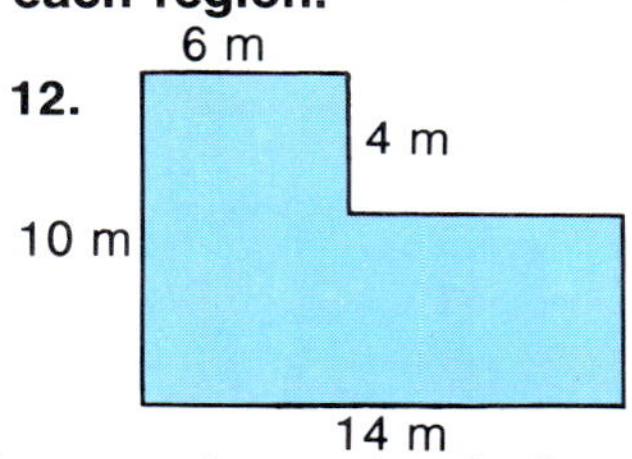

13. 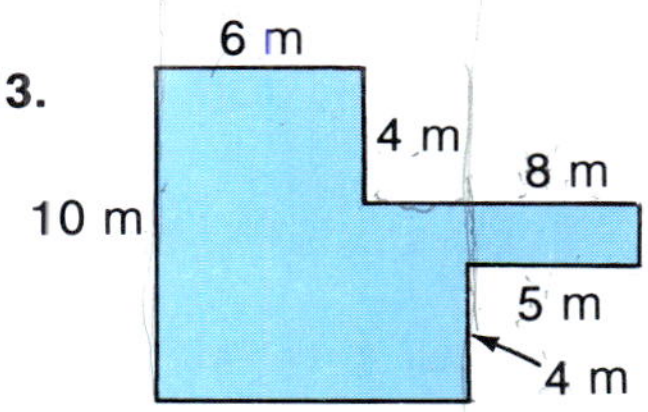

14. Compare the perimeters in exercises 11–13. Compare the areas in exercises 11–13.

Problem Solving

[*Hint:* You may want to draw a picture.]

15. A rectangular floor 6.2 m by 4.5 m is to be covered with carpet costing \$14.75 a square meter. How much will the carpet cost? Round your answer to the nearest dollar.

16. A solid wooden fence is 1.2 m high and 84 m long. How many cans of paint are needed to paint both sides if 1 can will cover 30.2 m^2? Round your answer up to the nearest whole number.

17. **a.** How many square centimeters are in a square meter?
    b. What is the area of the floor in exercise 15 in square centimeters?

★ 18. A ballroom is being changed into a fitness center. The room measures 30 m by 40 m. Its floor, except for a rectangular region 15 m by 19 m, will be tiled. Each square tile is 30 cm on a side. How many tiles will be needed?

Area—parallelograms

A **parallelogram** is a four-sided figure whose opposite sides are parallel.

The base of this parallelogram is 3 cm and the height is 2 cm.

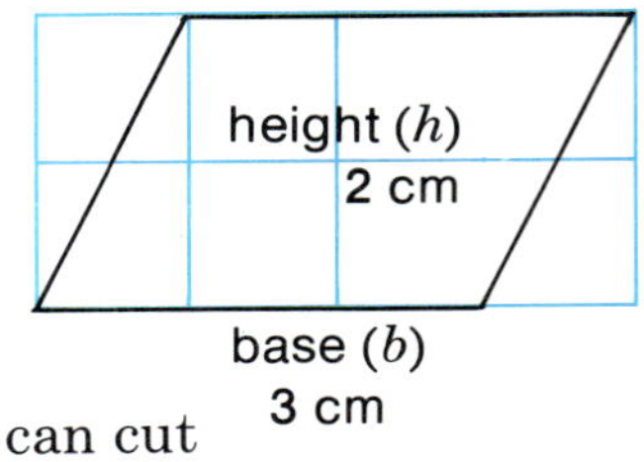

To find a formula for the area of the parallelogram, we can cut the parallelogram into two pieces and arrange the pieces to make a rectangle:

Step 1.

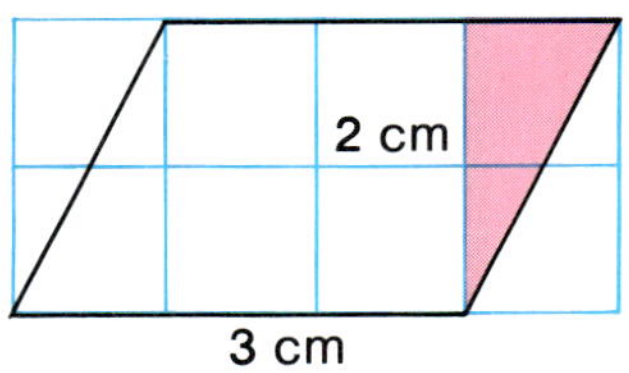

Step 2.

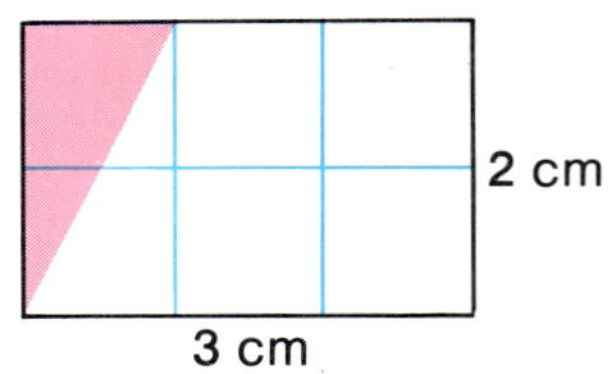

The parallelogram and the rectangle have the same base and the same height. They have the same area of 6 cm^2.

We can find the area of a parallelogram by multiplying base and height.

Formula $A = bh$

$= (3\text{ cm})(2\text{ cm})$

$= 6\text{ cm}^2$

EXAMPLE.

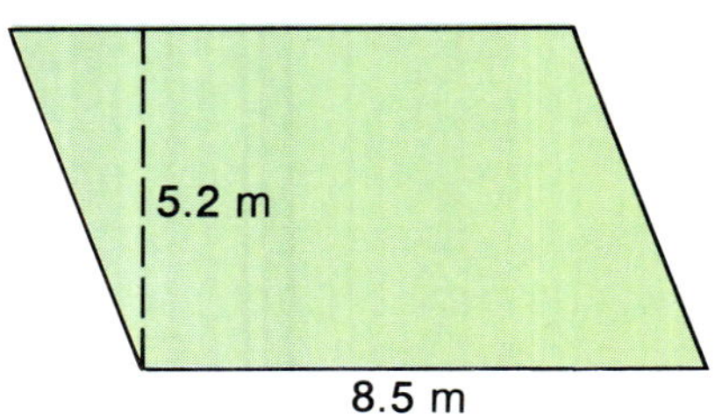

$A = bh$

$= (8.5\text{ m})(5.2\text{ m})$

$= 44.20\text{ m}^2$

EXERCISES

Find the area of each shaded region.

1.

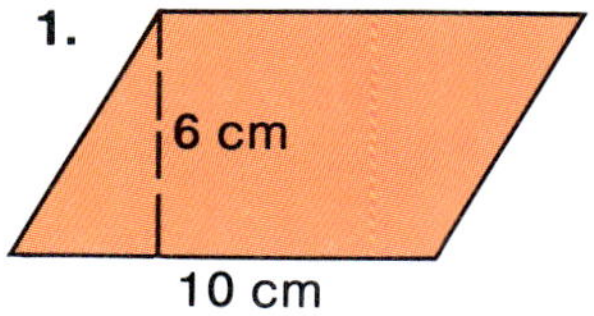

2.

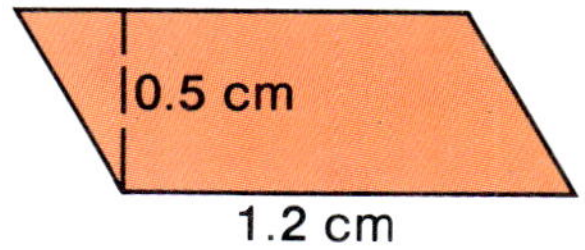

3.

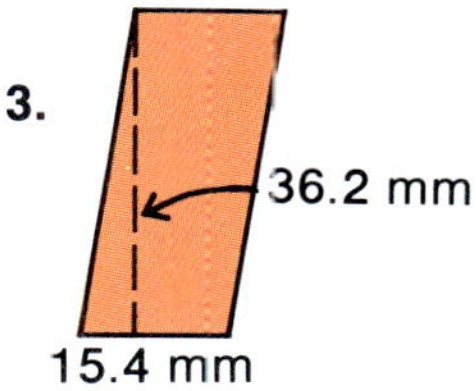

4.

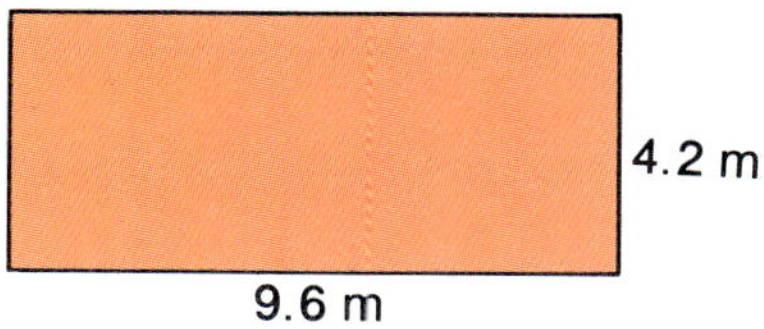

5.

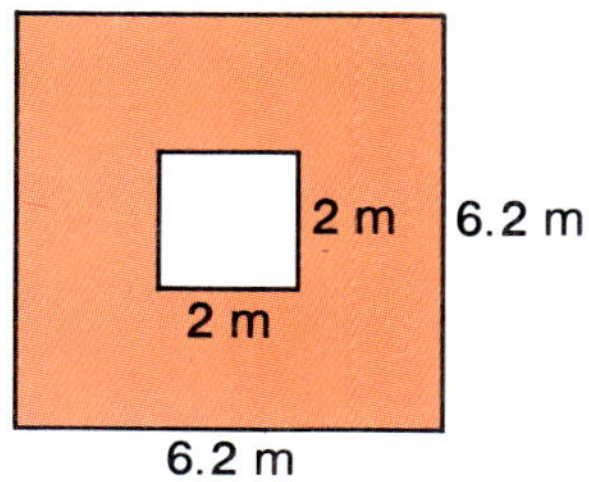

6.

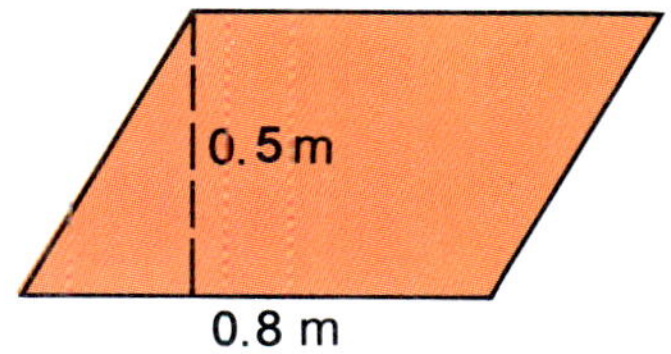

Find the perimeter and area of each region.

7.

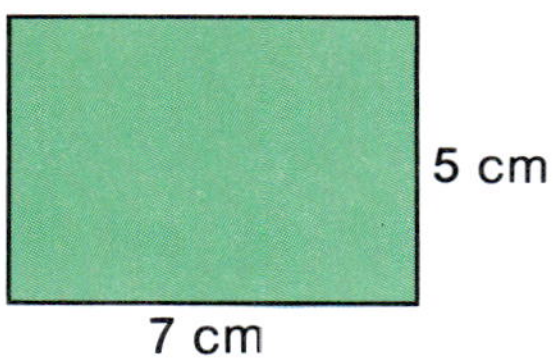

8.

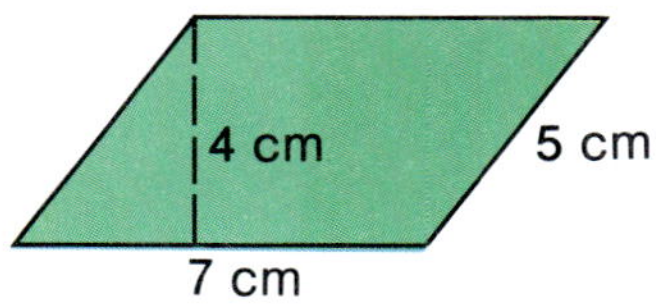

9.

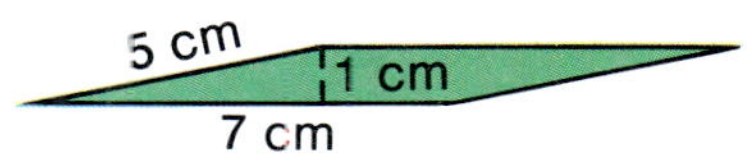

10. Compare the perimeters in exercises 7–9. Compare the areas in exercises 7–9.

★ **11.** A rectangular piece of stained glass was 10 cm long and 5 cm wide. It was cut to get a rectangular piece 9 cm long and 4 cm wide. What fraction of the original piece was used?

Challenge!

Solve.

12. How many squares do you see? Draw a picture. Show how to move 2 toothpicks to form 7 squares. [*Hint:* Not all the squares will have the same area.]

Area—triangles and trapezoids

The two triangles have the same size and shape. We can put them together to make a parallelogram.

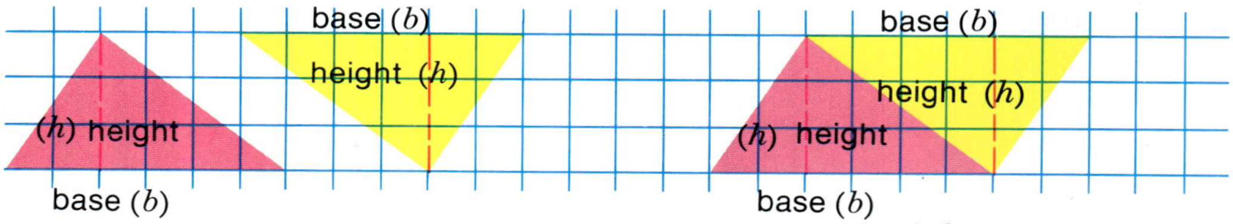

Since the area of the red triangle is one half of the area of the parallelogram, we get the following formula for the area of a triangle:

Formula $A = \frac{1}{2}bh$

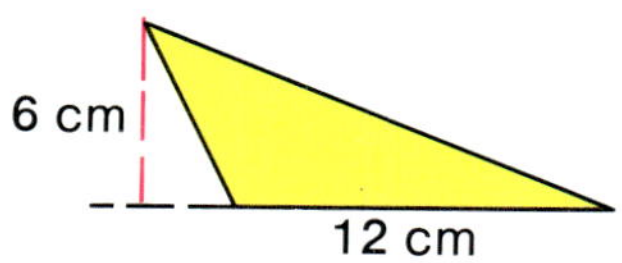

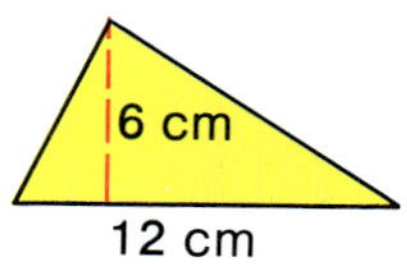

$$
\begin{aligned}
A &= \frac{1}{2}bh \\
&= \frac{1}{2}(12 \text{ cm})(6 \text{ cm}) \\
&= 36 \text{ cm}^2
\end{aligned}
$$

A **trapezoid** is a four-sided figure with two opposite sides parallel.

The two trapezoids are the same size and shape. If we put them together, we get a parallelogram.

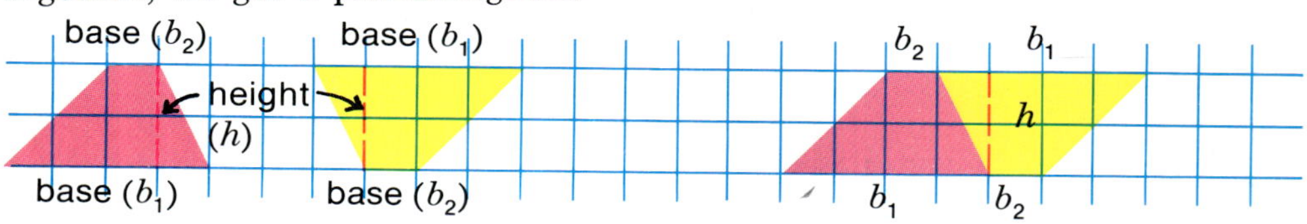

Since the area of the red trapezoid is one half of the area of the parallelogram, we get the following formula for the area of a trapezoid:

Formula $A = \frac{1}{2}(b_1 + b_2)h$

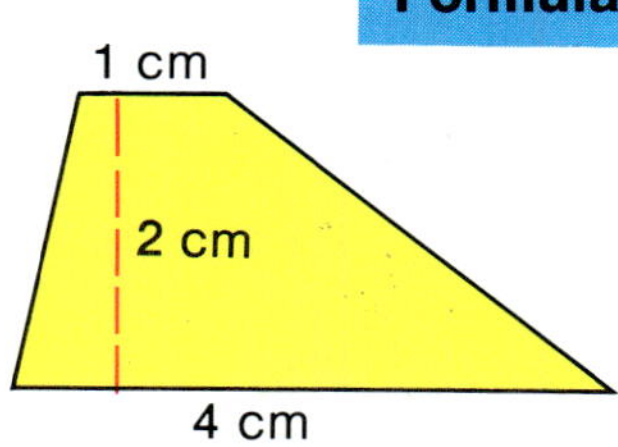

$$
\begin{aligned}
A &= \frac{1}{2}(b_1 + b_2)h \\
&= \frac{1}{2}(4 \text{ cm} + 1 \text{ cm})(2 \text{ cm}) \\
&= \frac{1}{2}(10 \text{ cm}^2) \\
&= 5 \text{ cm}^2
\end{aligned}
$$

EXERCISES

Compute each area.

1. 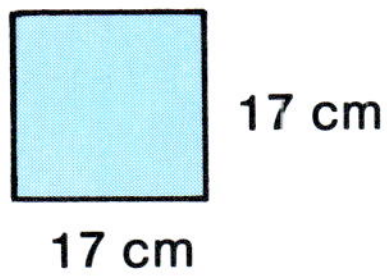

2. 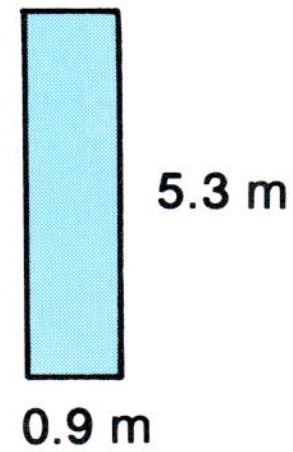

3. 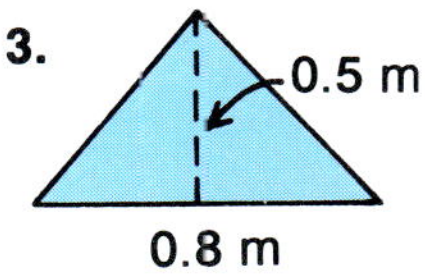

4. 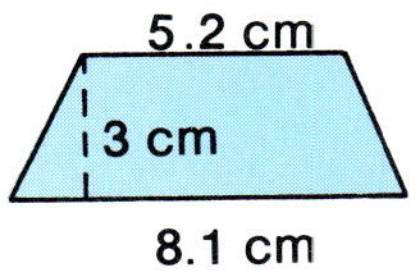

5. 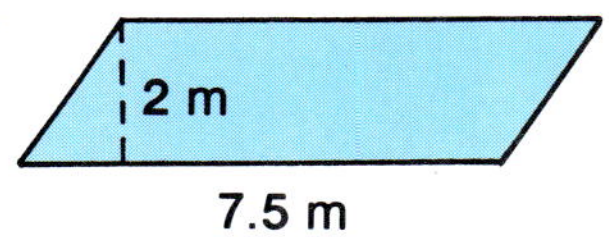

6. 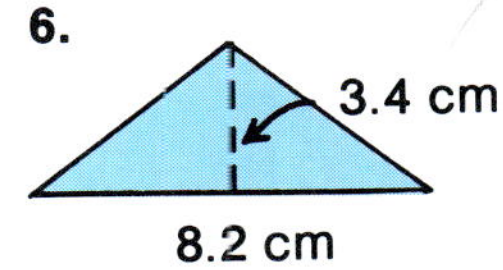

First make the necessary measurements to the nearest millimeter. Then compute the area.

7. 

8. 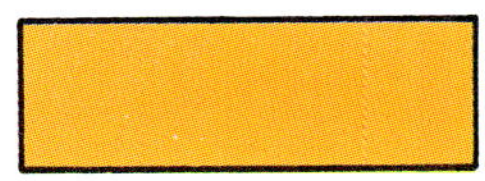

9. 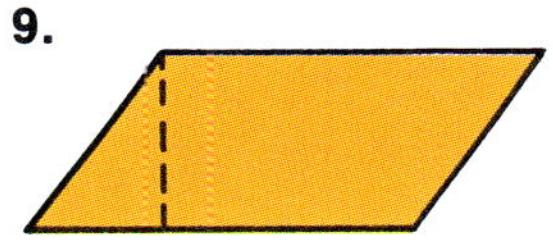

10. 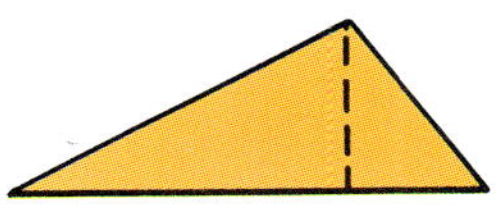

11. 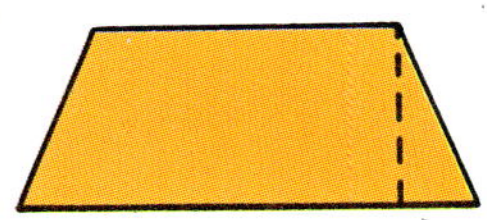

12. 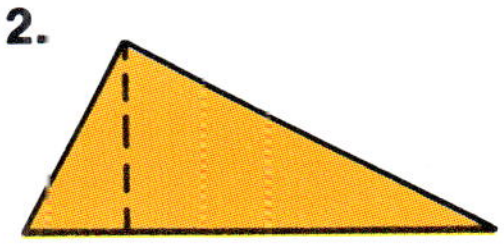

Find the area of the shaded region.

13. 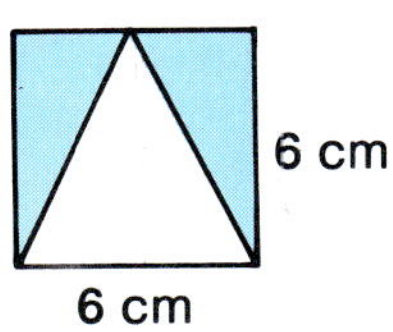

14. 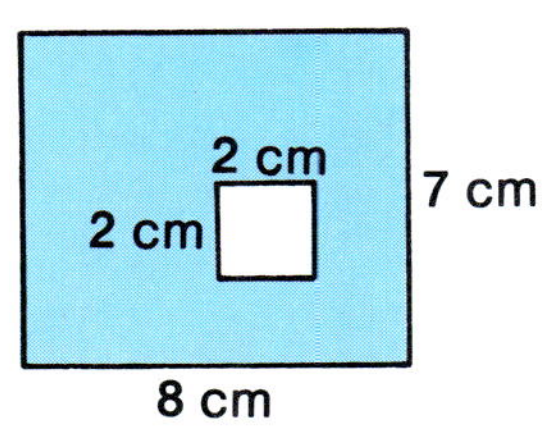

15. 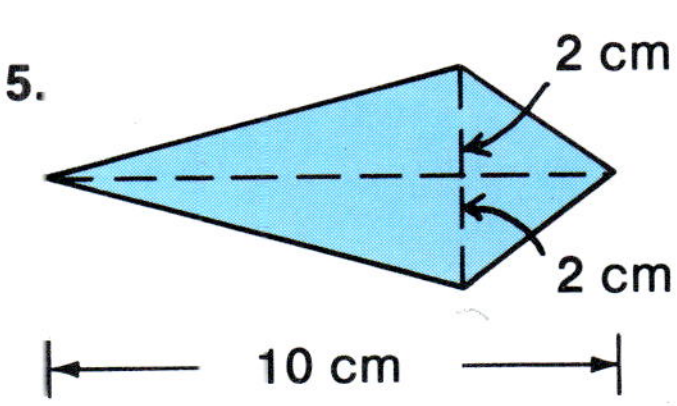

Problem Solving

16. A bumper sticker is shaped like a trapezoid. Its bases are 40 cm and 35 cm, and its height is 7 cm. What is the area of the bumper sticker?

17. A baseball pennant is shaped like a triangle. Its base is 30 cm and its area is 1440 cm^2. What is its height?

Area—circles

These pictures will help you to see that the formula for the area of a circle is reasonable.

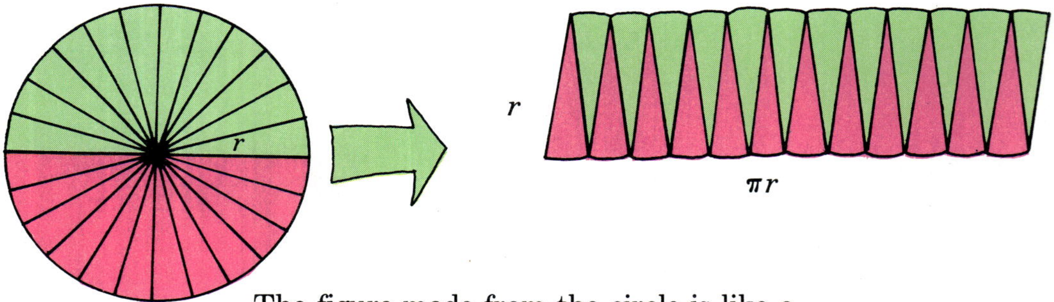

The figure made from the circle is like a parallelogram with height r (radius of the circle) and base πr (half the circumference of the circle).

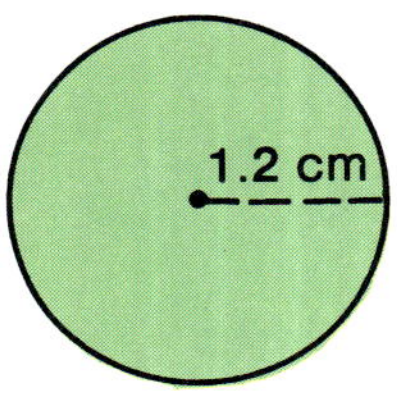

Formula $A = \pi r^2$

$A \approx 3.14(1.2\text{ cm})^2$

$\approx 3.14(1.44\text{ cm}^2)$

$\approx 4.52\text{ cm}^2$

Remember:
$\pi r^2 = \pi \times r \times r$

Read $\approx$ as "is approximately equal to."

EXERCISES

Compute the area to the nearest hundredth. Use 3.14 as an approximation for π.

1. 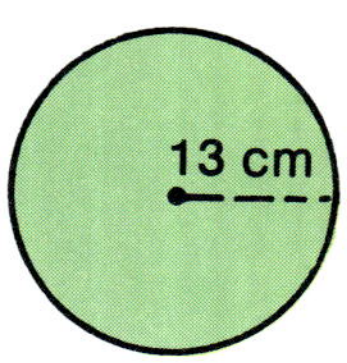

2. 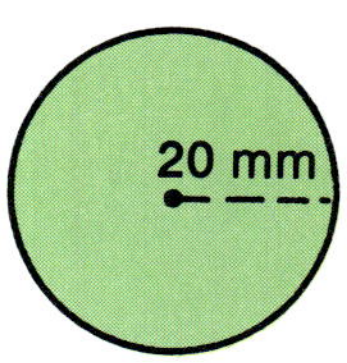

3. 

4. 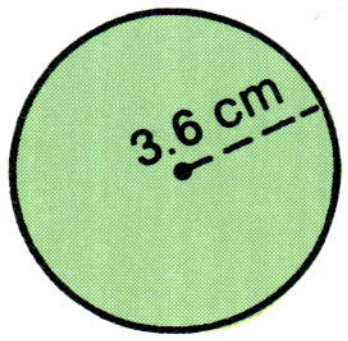

5. 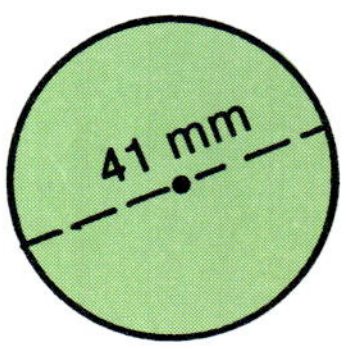

6. 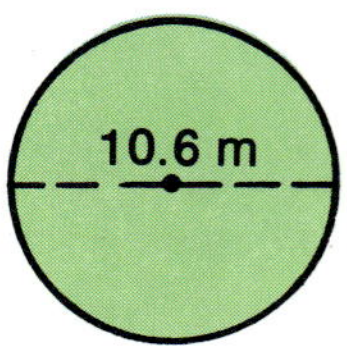

First measure the diameter to the nearest millimeter. Then compute the area. Use 3.14 as an approximation for π.

Compute the area to the nearest 0.01 m². Use 3.14 as an approximation for π.

13.

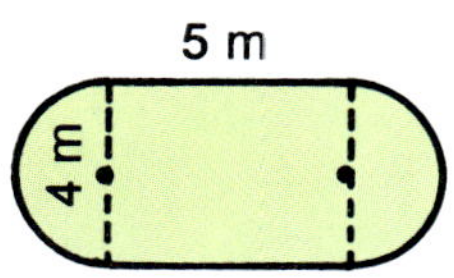

14.

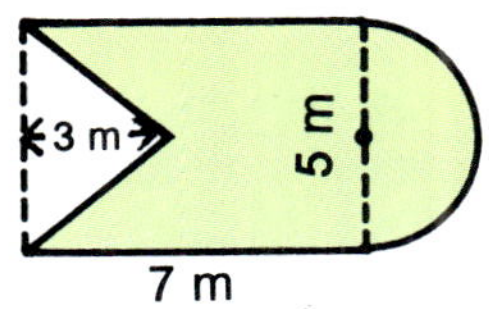

Problem Solving

Use 3.14 as an approximation for π. [*Hint:* First draw a picture.]

15. What is the area of the largest circle that can be cut from a square measuring 6.8 cm on a side?

16. A horse is tied to a stake in the middle of a large pasture. The farthest the horse can move from the stake is 15 meters. Over how much area can the horse graze?

KEEPING SKILLS SHARP

Give the product in simplest form.

1. $2 \times 3\frac{1}{3}$
2. $3 \times 2\frac{1}{4}$
3. $4\frac{1}{5} \times 4$
4. $2\frac{2}{3} \times 5$
5. $7\frac{3}{4} \times 3$
6. $9 \times 3\frac{2}{5}$
7. $3\frac{3}{4} \times 4\frac{1}{4}$
8. $1\frac{1}{5} \times 4\frac{3}{8}$
9. $5\frac{1}{6} \times 1\frac{3}{4}$
10. $4\frac{1}{3} \times 6\frac{1}{5}$
11. $2\frac{1}{9} \times 8\frac{3}{4}$
12. $7\frac{2}{5} \times 9\frac{2}{5}$
13. $2\frac{1}{2} \times 1\frac{1}{4}$
14. $3\frac{2}{3} \times 2\frac{1}{3}$
15. $2\frac{1}{2} \times 3\frac{2}{5}$
16. $4 \times 2\frac{5}{6}$

Surface area—prisms and cylinders

A **prism** has two **bases** that are the same size and shape and are in parallel planes. The other faces are all rectangles.

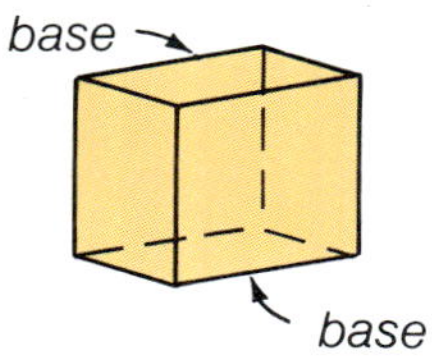

2 bases
4 other faces

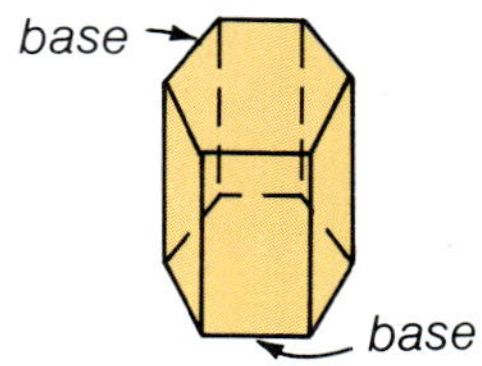

2 bases
6 other faces

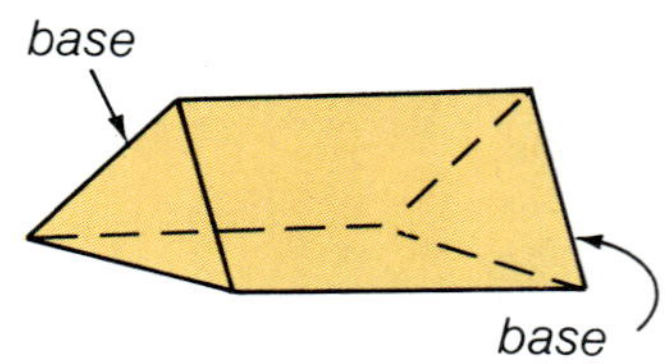

2 bases
3 other faces

The **surface area** of a prism is the sum of the areas of the faces.

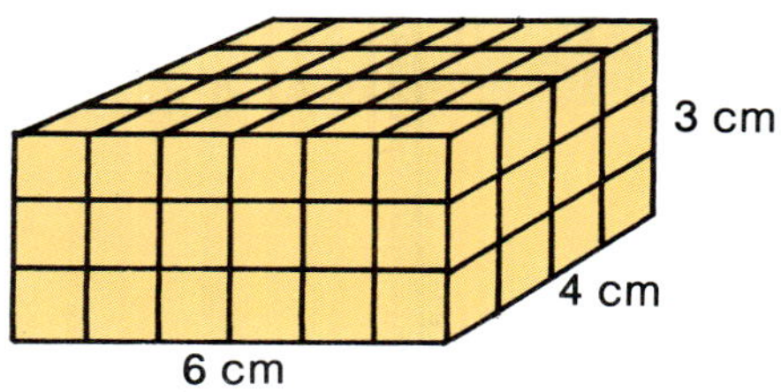

The areas of the faces are:

top	24 cm^2
bottom	24 cm^2
right	12 cm^2
left	12 cm^2
front	18 cm^2
back	18 cm^2
The surface area is	108 cm^2

A **cylinder** has two circular bases that are the same size and are in parallel planes. It has one curved face.

To find the surface area of a cylinder, you can think about cutting the cylinder apart like this:

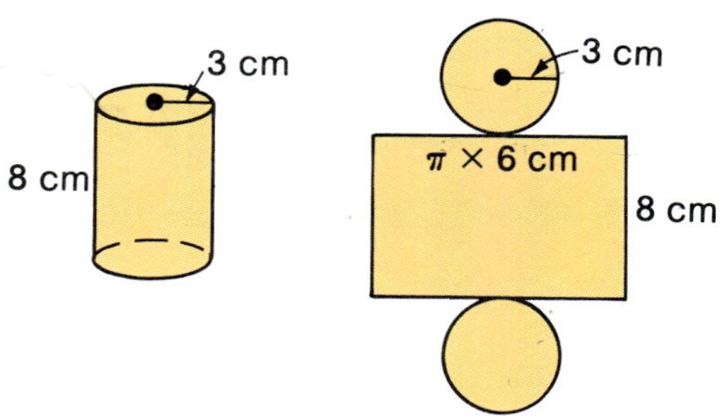

Using 3.14 as an approximation of π, the areas of the faces are:

base	$3.14 \times 3^2 =$	28.26 cm^2
base	$3.14 \times 3^2 =$	28.26 cm^2
curved face	$8 \times 6 \times 3.14 =$	150.72 cm^2
	The surface area is about	207.24 cm^2

Notice that the length of the rectangle is the circumference of each base.

EXERCISES

Find each surface area. Use 3.14 as an approximation for π.

1.
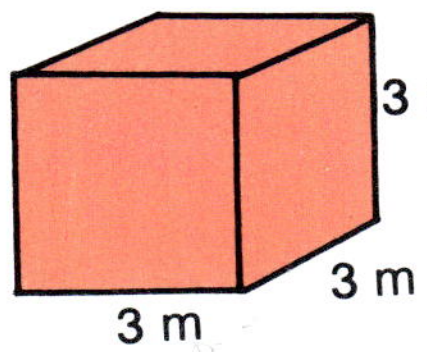

2.
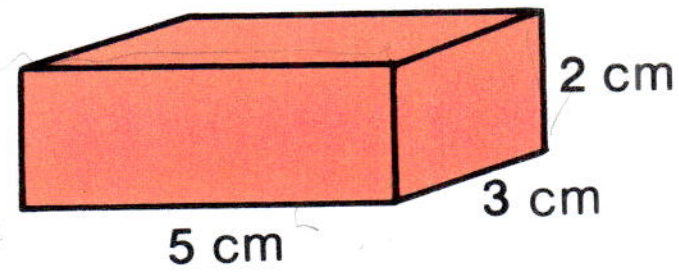

3.
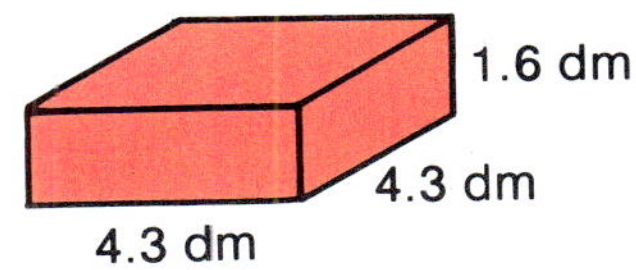

4.
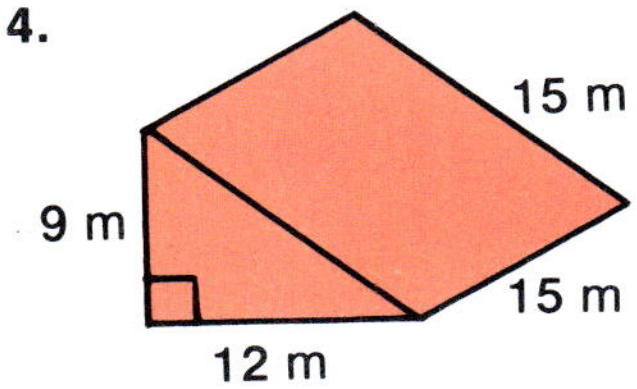

5.
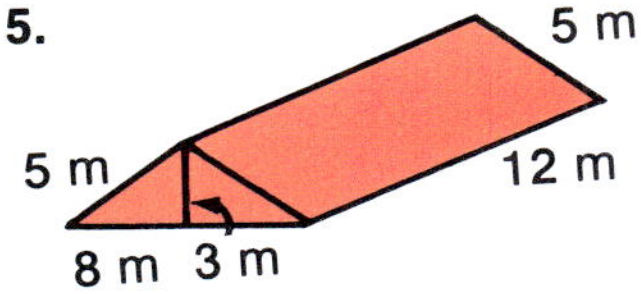

6.
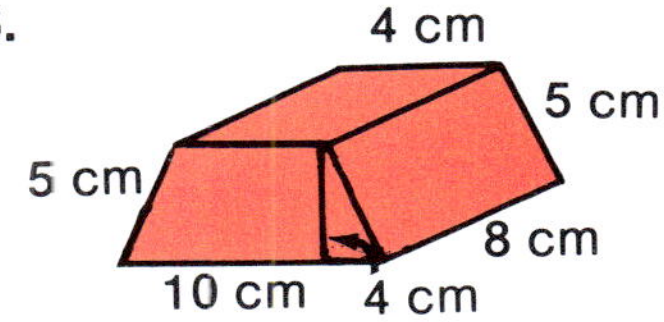

7.
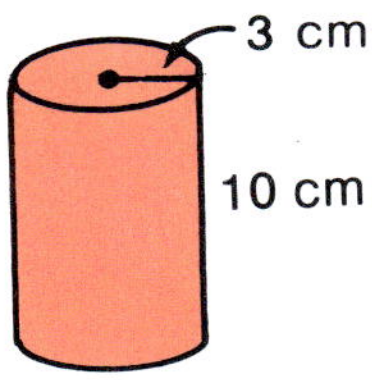

8.
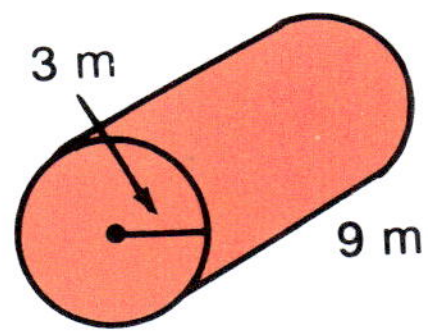

9.
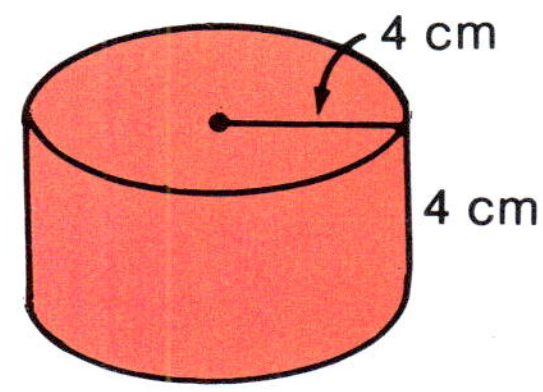

Problem Solving

Use 3.14 as an approximation for π. [*Hint:* Draw a picture.]

10. A cylindrical hatbox has a diameter of 38 cm and a height of 26 cm. How much paper is needed to cover the hatbox?

11. What is the surface area of a box that has a square base 32.5 cm on a side and is 40 cm high?

12. A cylindrical silo has a height of 15.5 meters and a diameter of 4.2 meters. If 1 can of paint covers 10 square meters, how many cans must be bought to paint the curved face with 1 coat?

13. An open cylinder is made from a square piece of sheet metal measuring 5.25 m on a side. What will be the radius of the cylinder? Round your answer to the nearest 0.01 m.

Challenge!

★ **14.** Determine how 64 blocks can be stacked to get a rectangular prism having the least surface area.

Volume

To find the **volume** of this box, we can pick a unit cube, for example, a cubic centimeter (cm^3),

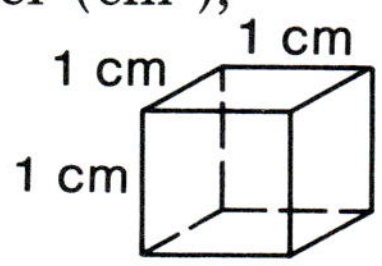

and then count the number of these cubic units it takes to fill the box.

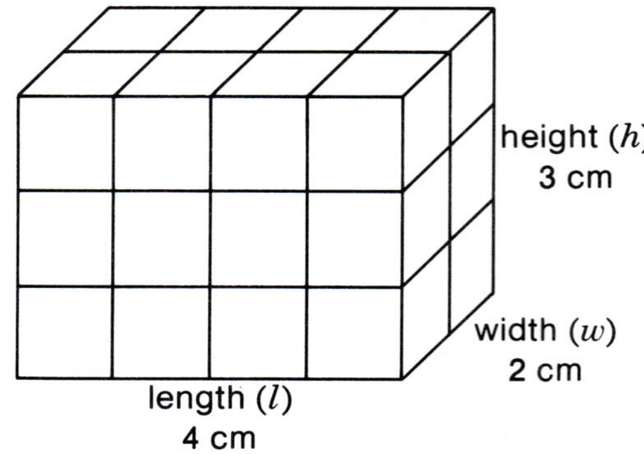

We could also find the volume of the box by first multiplying the length by the width to find the number of cubes in one layer and then multiplying that number by the height to find the total number of cubes.

Formula $V = lwh$

$= (4 \text{ cm})(2 \text{ cm})(3 \text{ cm})$

$= 24 \text{ cm}^3$

EXERCISES

Find each volume.

1.

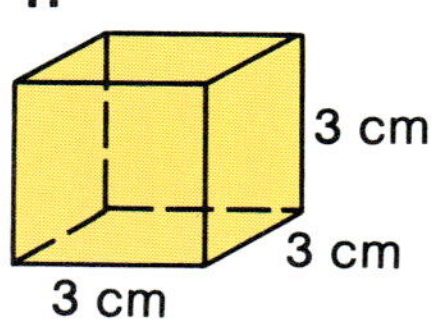

2.

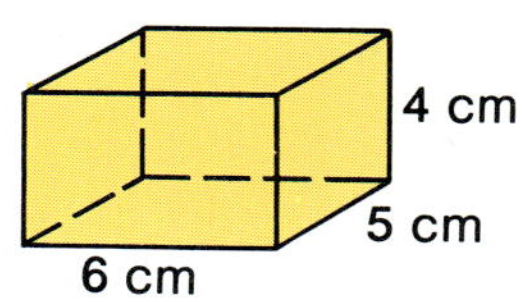

3.

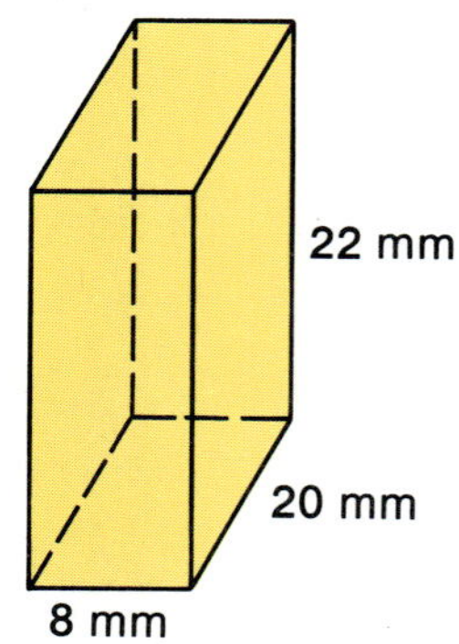

4.

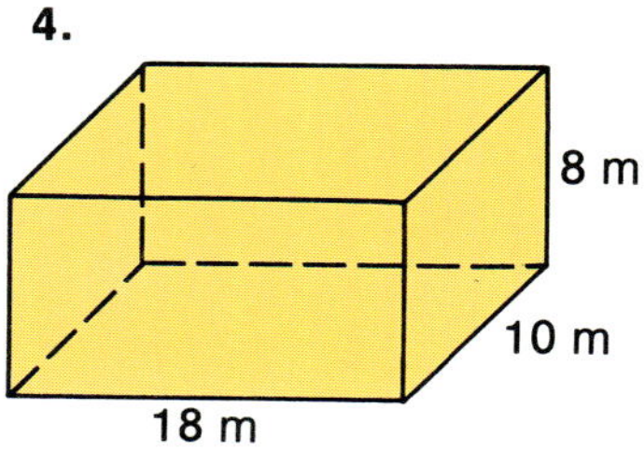

5.

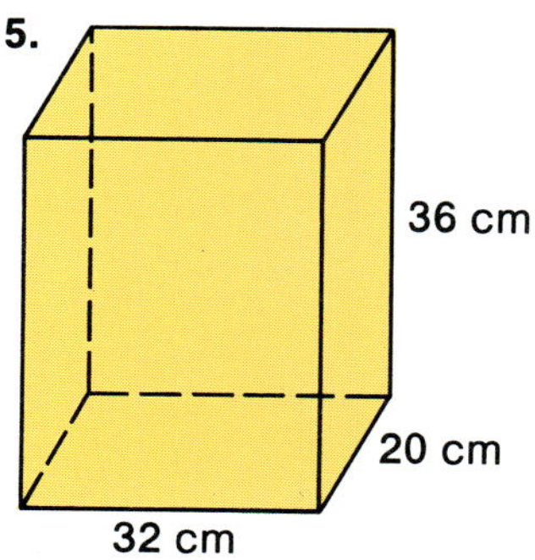

6.

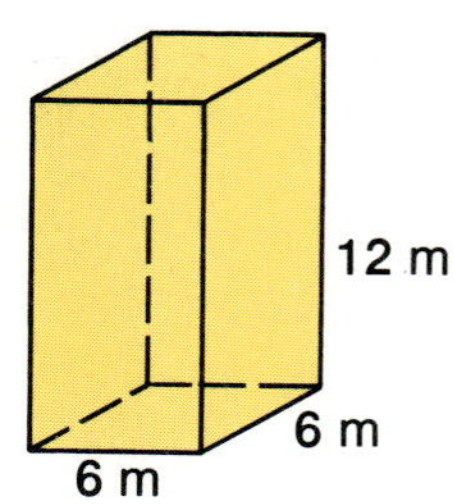

7.

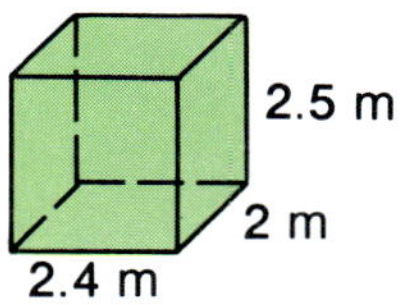

8.

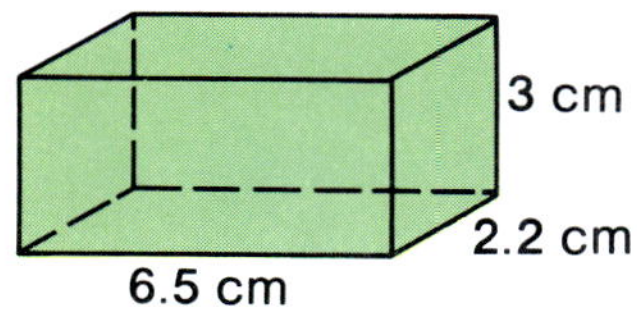

9.

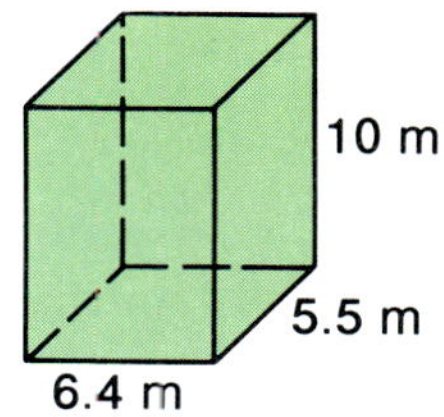

★ 10.

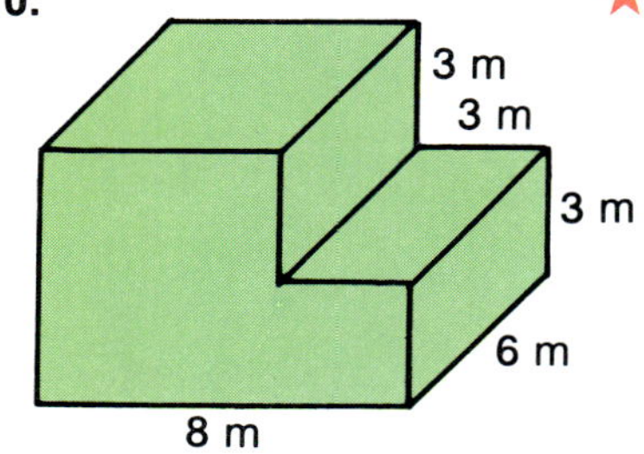

★ 11.

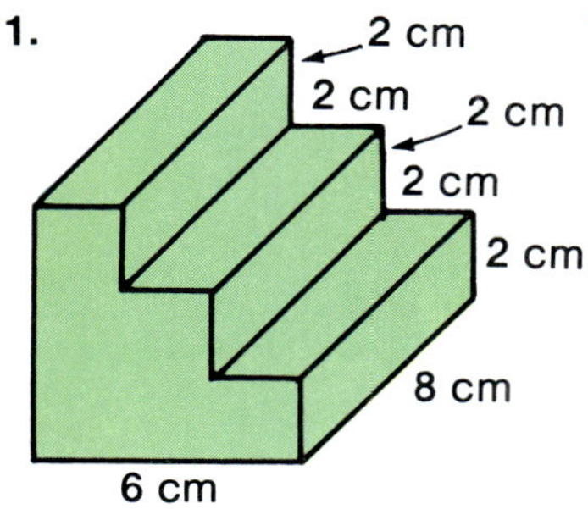

★ 12.

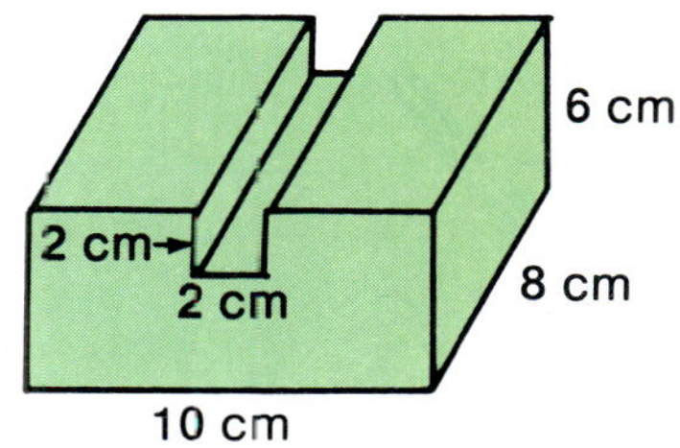

Tell whether the question involves perimeter, area, or volume.

13. How much sand is needed to fill a trench?

14. How many shrubs are needed to border a yard?

15. How much sod is needed to cover a lawn?

16. How much concrete is needed for a wall?

17. How much fence is needed to fence in a patio?

Problem Solving

A rectangular swimming pool is 12 m long, 6 m wide, and 2 m deep.

18. How many cans of waterproof paint are needed to paint the inside of the pool if 1 can covers 20 m^2?

19. How many m^3 of water does it take to fill the pool to 1 dm from the top?

20. What are the dimensions of a plastic cover that covers the top of the pool and extends 3 dm past each edge?

21. How many cubic decimeters are in a cubic meter?

22. What is the volume of the water in the pool in cubic decimeters? (See exercise 19.)

★ 23. What is the volume of the water in the pool in cubic kilometers? (See exercise 19.)

Volume—prisms and cylinders

To find the volume of a prism or cylinder, you can find the area of a base (B), which is the number of unit cubes in one layer, and multiply that number by the height (h), which is the number of layers.

Recall that a prism has two parallel bases that are the same size and shape. The remaining faces are rectangles.

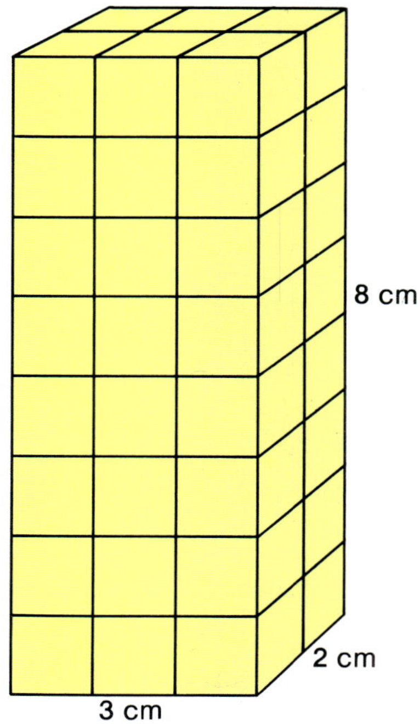

Formula $V = Bh$

$= (3 \text{ cm} \times 2 \text{ cm})(8 \text{ cm})$

$= (6 \text{ cm}^2)(8 \text{ cm})$

$= 48 \text{ cm}^3$

EXAMPLE 1.

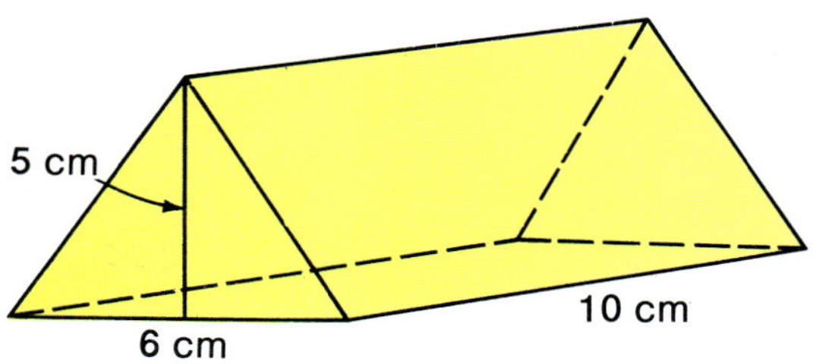

$V = Bh$

$= \left(\frac{1}{2} \times 6 \text{ cm} \times 5 \text{ cm}\right)(10 \text{ cm})$

$= (15 \text{ cm}^2)(10 \text{ cm})$

$= 150 \text{ cm}^3$

EXAMPLE 2.

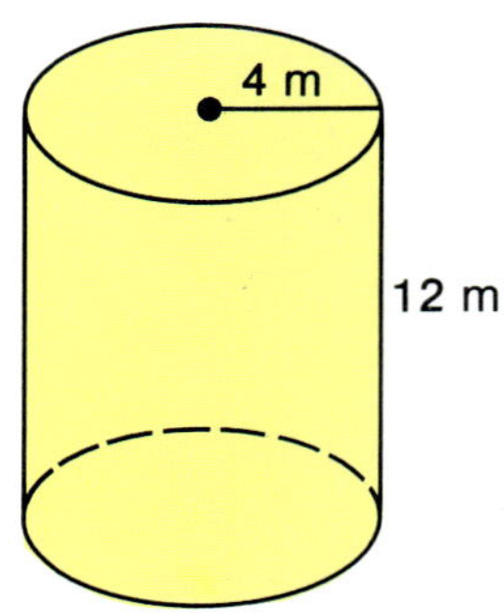

$V = Bh$

$= (\pi \times (4 \text{ m})^2)(12 \text{ m})$

$\approx (3.14 \times 16 \text{ m}^2)(12 \text{ m})$

$\approx (50.24 \text{ m}^2)(12 \text{ m})$

$\approx 602.88 \text{ m}^3$

EXERCISES

Compute each volume to the nearest hundredth. Use 3.14 as an approximation for π.

1.

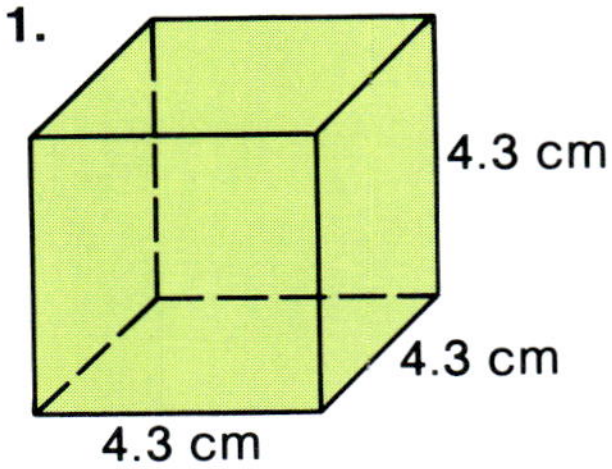

2.

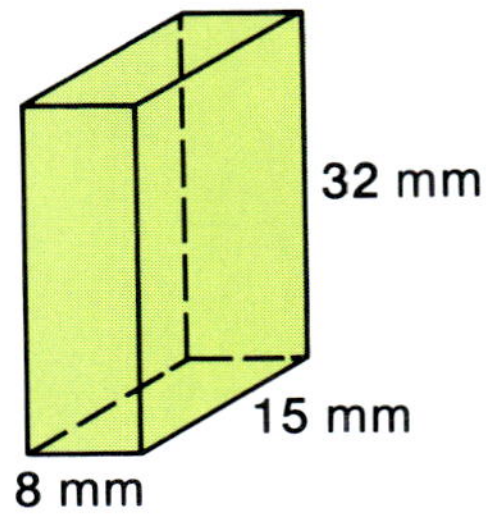

3.

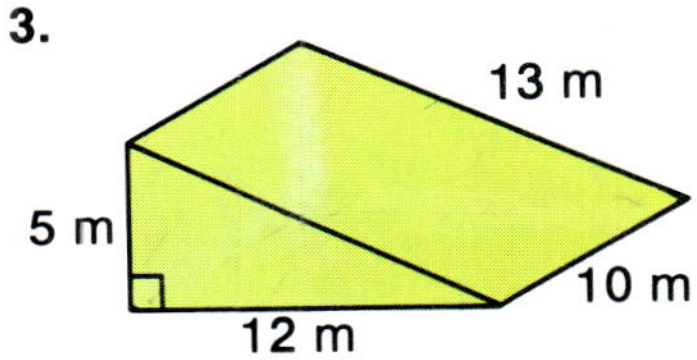

4.

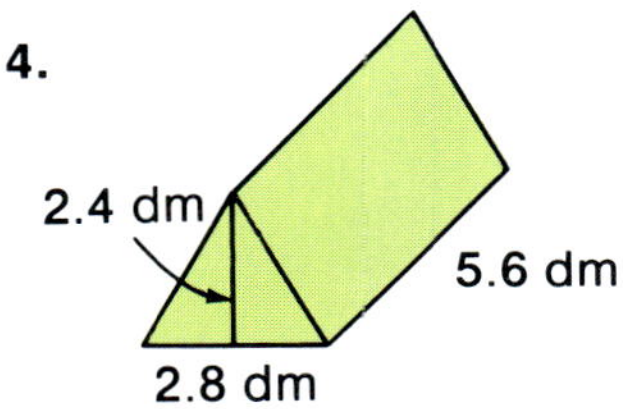

5.

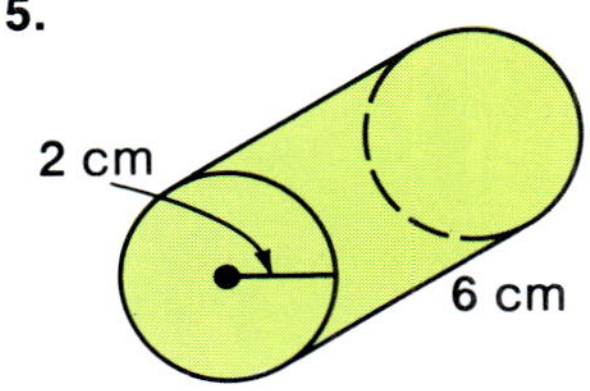

6.

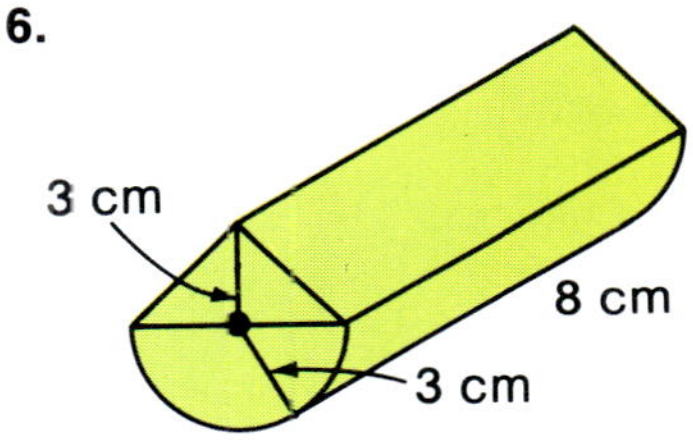

Problem Solving

Use 3.14 as an approximation for π.

7. What is the volume of a cube whose sides are 3 cm long?

8. What is the volume of a cube with a side 2 times as long as the one in exercise 7?

9. Look at your answers to exercises 7 and 8. What happens to the volume of a cube when the length of its edge is doubled?

10. A metal cylinder is 30 cm in height and has a radius of 6 cm. A 2-cm hole is drilled from the center of one base to the center of the other base. What is the volume of the metal that remains?

Challenge!

First estimate. Then compute.

11. Are there more centimeters in 2 kilometers than there are seconds in 2 weeks?

12. Are there more cubic centimeters in a cubic meter than there are minutes in a year?

Volume—pyramids, cones, and spheres

PYRAMID

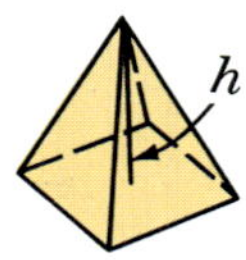

The volume of a pyramid is $\frac{1}{3}$ the volume of a prism having the same base and height.

Formula $V = \frac{1}{3}Bh$

EXAMPLE 1.

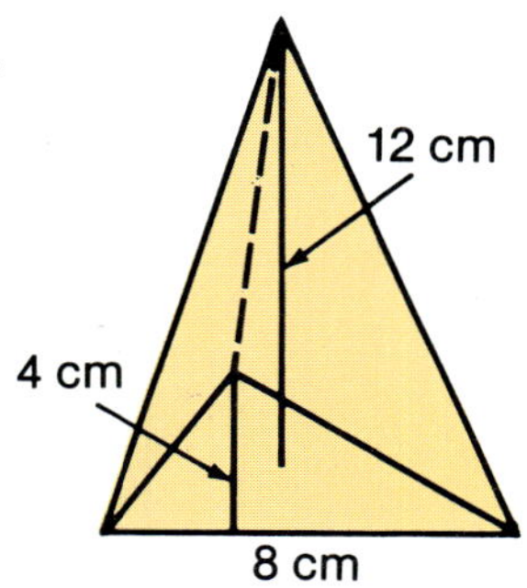

$$V = \frac{1}{3}Bh$$
$$= \frac{1}{3}\left(\frac{1}{2} \times 8 \text{ cm} \times 4 \text{ cm}\right)(12 \text{ cm})$$
$$= \frac{1}{3}(192 \text{ cm}^3)$$
$$= 64 \text{ cm}^3$$

CONE

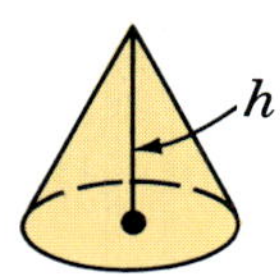

The volume of a cone is $\frac{1}{3}$ the volume of a cylinder having the same base and height.

Formula $V = \frac{1}{3}Bh$

EXAMPLE 2.

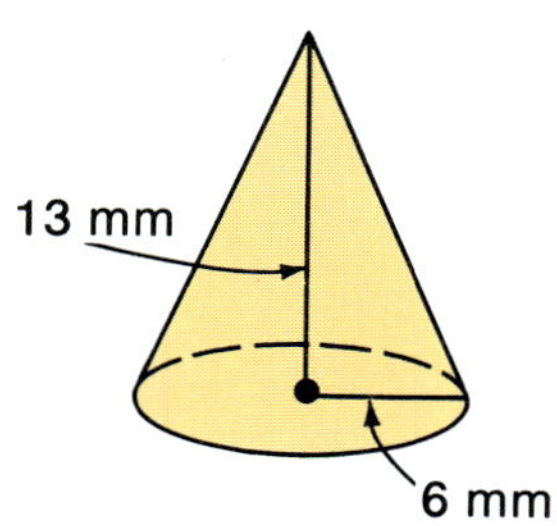

$$V = \frac{1}{3}Bh$$
$$= \frac{1}{3}(\pi \times 6 \text{ mm} \times 6 \text{ mm})(13 \text{ mm})$$
$$\approx \frac{1}{3}(3.14 \times 6 \text{ mm} \times 6 \text{ mm})(13 \text{ mm})$$
$$\approx \frac{1}{3}(1469.52 \text{ mm}^3)$$
$$\approx 489.84 \text{ mm}^3$$

The volume of a sphere is $\frac{4}{3}\pi$ times the cube of the radius.

Formula $V = \frac{4}{3}\pi r^3$

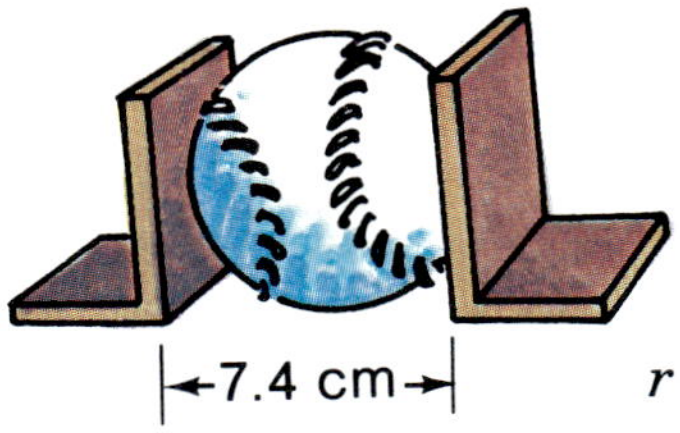

$r = 3.7$ cm

$\approx \frac{4}{3}(3.14)(3.7 \text{ cm})^3$

$\approx \frac{4}{3}(3.14)(50.653 \text{ cm}^3)$

$\approx 212.067 \text{ cm}^3$

EXERCISES

Give each volume to the nearest hundredth. Use 3.14 as an approximation for π.

1. 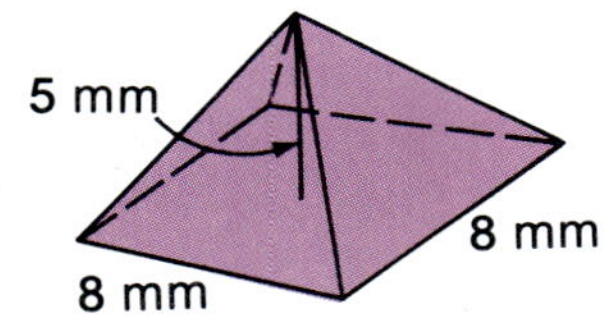

2. 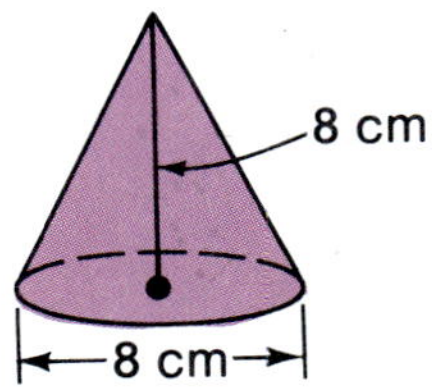

3. 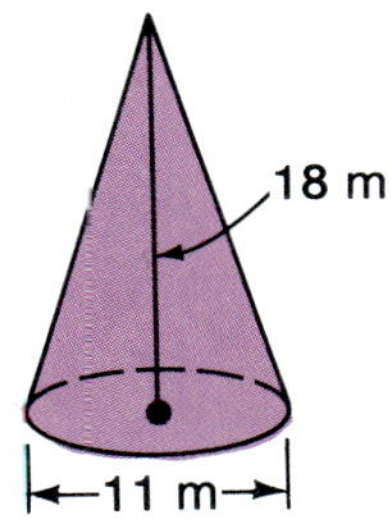

4. 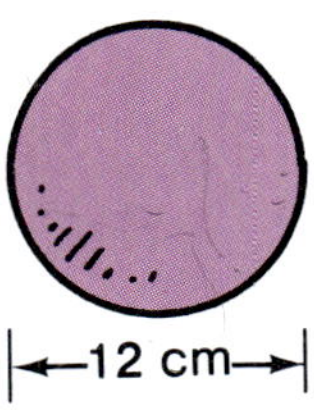

5. 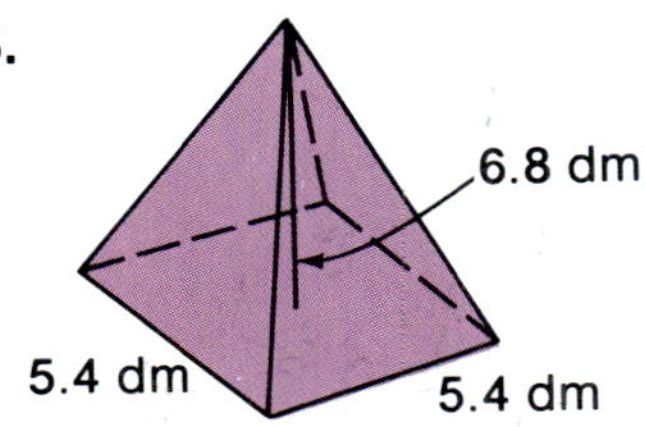

6. 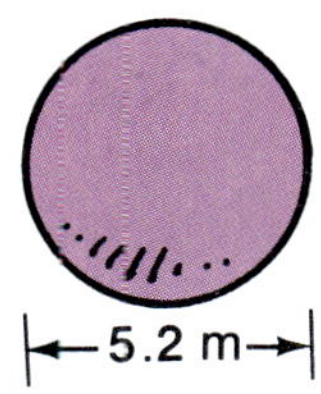

Problem Solving

Give each answer to the nearest hundredth. Use 3.14 as an approximation for π.

7. What is the volume of a spherical water tank that has a diameter of 16 m?

8. A drinking cup shaped like a cone has a radius of 8.4 cm and a height of 11.6 cm. What is its volume?

9. A certain railroad tank car in the shape of a cylinder has a length of 19.8 m and a diameter of 2.4 m. What is its volume?

10. The Great Pyramid of Egypt had a square base about 234.1 m on a side. Its height was about 146.9 m. What was its approximate volume?

Problem Solving Exploring area and perimeter

Work in groups. Get some squared paper and a calculator.

PART A

Farmer Jill has 24 m of fence to make a pigpen in the corner of her barn. She wants to make a rectangular pen with the largest possible area.

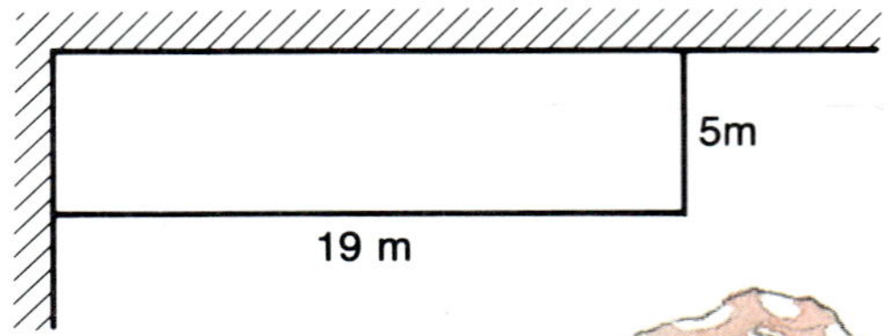

1. At first, Jill plans to use her 24 m of fence to make a pen 5 m wide and 19 m long. What would the area of this pen be?
2. Can you think of a better solution? Make a scale drawing of the sides of the barn. Then draw a pen that uses 24 m of fence but has a larger area than Jill's.
3. Find the best solution. How can you be sure your pen has the largest area? Compare and discuss your results with your group. What methods did you use?
4. Suppose Jill had 41 m of fence. Predict the length and width of the pen with the largest area. Check your prediction.

PART B

Jill wants to build another rectangular pen. But this one will be along a wall instead of in a corner. She still wishes to use 24 m of fence and wants the pen with the largest area.

5. Discuss how this problem is different from the problem in Part A.

6. Predict the shape of the pen with the largest area. What would the length and width of this pen be? What would its area be?

7. Find the pen with the largest area. Check by finding the areas of pens with smaller and larger widths.

8. Was your prediction of the shape correct? Discuss what happened.

9. Which pen can have more area using 24 m of fence, one built in a corner or one built along a wall? Discuss why this is so.

PART C

Suppose Jill wished to build a rectangular pen in the middle of the barn with 24 m of fence.

10. How would the area of this pen compare with the area of one built in a corner and one built along a wall? Discuss the reasons for your answers.

11. For a pen built in the middle of the barn, predict the shape that will have the largest area. Check your answer.

12. With another group, discuss the strategies used to solve the problems in this lesson and the results you got.

13. Did you use a method of computation (mental math, pencil and paper, calculator) or estimation to solve the problems in this lesson? Why? Discuss.

Conversion fractions

Here are some **conversion fractions.** The numerator and denominator of each fraction are the same length.

$\frac{1\text{ ft}}{12\text{ in.}}$ $\frac{12\text{ in.}}{1\text{ ft}}$ $\frac{1\text{ yd}}{3\text{ ft}}$ $\frac{3\text{ ft}}{1\text{ yd}}$

SHORTCUT

Change from one unit to another by using a conversion fraction.

EXAMPLE 1. Change 4 feet to inches.

$$4\text{ ft} = 4\text{ ft} \times \frac{12\text{ in.}}{1\text{ ft}}$$
$$= 4\,\cancel{\text{ft}} \times \frac{12\text{ in.}}{1\,\cancel{\text{ft}}}$$
$$= 48\text{ in.}$$

Cancel units as you would cancel numbers.

EXAMPLE 2. Change 28 inches to feet.

$$28\text{ in.} = 28\text{ in.} \times \frac{1\text{ ft}}{12\text{ in.}}$$
$$= 28\,\cancel{\text{in.}} \times \frac{1\text{ ft}}{12\,\cancel{\text{in.}}}$$
$$= \frac{28}{12}\text{ ft}$$
$$= 2\tfrac{1}{3}\text{ ft}$$

Notice that the numerator of the conversion fraction contains the unit you are changing to and the denominator contains the unit you are changing from.

EXERCISES

Use a conversion fraction to change the following.

1. 28 yd to ft

2. 28 ft to yd

3. 208 in. to ft

4. 108 in. to yd

5. 16 ft to in.

6. 10 yd to in.

7. 2640 yd to mi

8. 2640 ft to mi

Solve by using a conversion fraction.

9. A height record for walking stilts was 21 feet. How many inches is that?

10. A distance record for tightrope walking was 3790 yards. How many miles is that?

We can also use conversion fractions for area and volume.

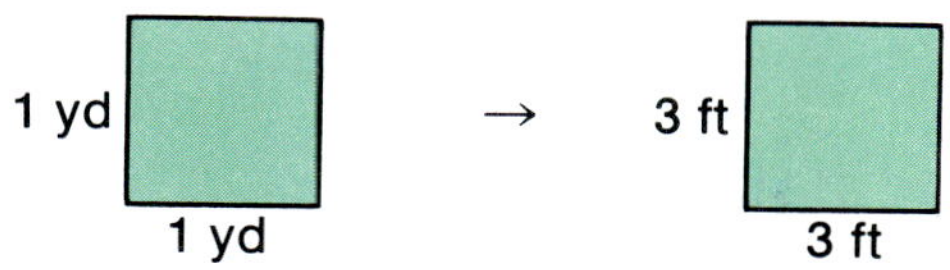

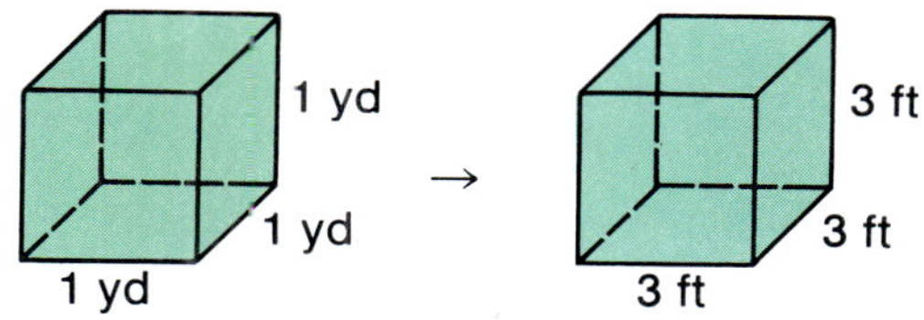

$1 \text{ yd}^2 = 9 \text{ ft}^2$

$1 \text{ yd}^3 = 27 \text{ ft}^3$

Conversion fractions: $\frac{1 \text{ yd}^2}{9 \text{ ft}^2}$ $\frac{9 \text{ ft}^2}{1 \text{ yd}^2}$ $\frac{1 \text{ yd}^3}{27 \text{ ft}^3}$ $\frac{27 \text{ ft}^3}{1 \text{ yd}^3}$

EXAMPLE 3. Change $2\frac{1}{3}$ yd² to ft².

$$2\frac{1}{3} \text{ yd}^2 = 2\frac{1}{3} \text{ yd}^2 \times \frac{9 \text{ ft}^2}{1 \text{ yd}^2}$$
$$= 2\frac{1}{3} \cancel{\text{yd}^2} \times \frac{9 \text{ ft}^2}{1 \cancel{\text{yd}^2}}$$
$$= 21 \text{ ft}^2$$

EXAMPLE 4. Change 60 ft³ to yd³.

$$60 \text{ ft}^3 = 60 \text{ ft}^3 \times \frac{1 \text{ yd}^3}{27 \text{ ft}^3}$$
$$= 60 \cancel{\text{ft}^3} \times \frac{1 \text{ yd}^3}{27 \cancel{\text{ft}^3}}$$
$$= \frac{60}{27} \text{ yd}^3$$
$$= 2\frac{2}{9} \text{ yd}^3$$

Copy and complete each conversion fraction.

11. $\frac{1 \text{ ft}^2}{? \text{ in.}^2}$ **12.** $\frac{? \text{ in.}^2}{1 \text{ ft}^2}$ **13.** $\frac{1 \text{ ft}^3}{? \text{ in.}^3}$ **14.** $\frac{? \text{ in.}^3}{1 \text{ ft}^3}$

Use a conversion fraction to change the following.

15. 5 ft² to in.² **16.** 6 ft³ to in.³ **17.** $1\frac{2}{3}$ yd² to ft²

18. 50 ft³ to yd³ **19.** 300 in.² to ft² **20.** 72 in.² to ft²

Problem Solving

21. What is the area in ft² of a rectangle 2 yd long and $1\frac{1}{3}$ yd wide?

22. What is the volume in yd³ of a box 6 ft long, 5 ft wide, and 4 ft high?

Area—customary units

When measurements are given in more than one unit, change to a single unit before multiplying. Here are two ways to find the area of a rectangle 3 ft 9 in. long and 1 ft 4 in. wide.

Change to inches.

$A = lw$

$= (3\text{ ft } 9\text{ in.})(1\text{ ft } 4\text{ in.})$

$= (45\text{ in.})(16\text{ in.})$

$= 720\text{ in.}^2$

Change to feet.

$A = lw$

$= (3\text{ ft } 9\text{ in.})(1\text{ ft } 4\text{ in.})$

$= \left(3\frac{3}{4}\text{ ft}\right)\left(1\frac{1}{3}\text{ ft}\right)$

$= \left(\frac{\overset{5}{\cancel{15}}}{\underset{1}{\cancel{4}}}\text{ ft}\right)\left(\frac{\overset{1}{\cancel{4}}}{\underset{1}{\cancel{3}}}\text{ ft}\right)$

$= 5\text{ ft}^2$

We can compare these answers by changing 5 ft^2 to square inches. Using a conversion fraction, we have

$$5\text{ ft}^2 \times \frac{144\text{ in.}^2}{1\text{ ft}^2} = 720\text{ in.}^2$$

EXERCISES

Give the number of

1. square feet in a square yard.
2. square inches in a square foot.
3. square inches in a square yard.
4. square inches in $2\frac{1}{2}$ square feet.
5. square yards in 36 square feet.
6. square feet in 436 square inches.

Give each area. Use $\frac{22}{7}$ as an approximation for π.

7.
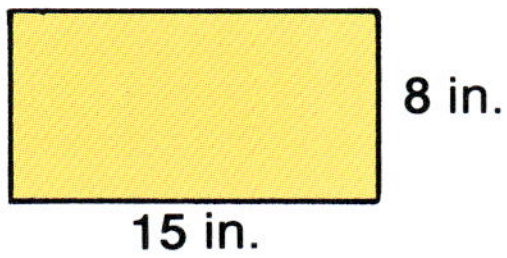

8.
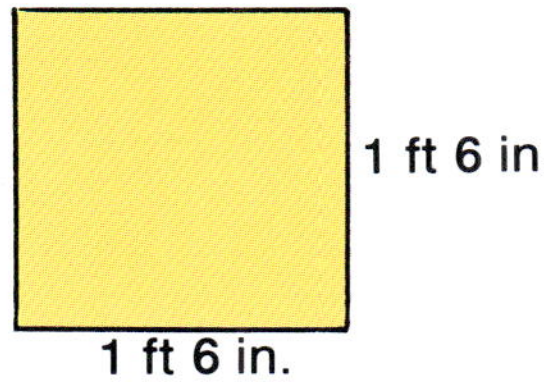

9.
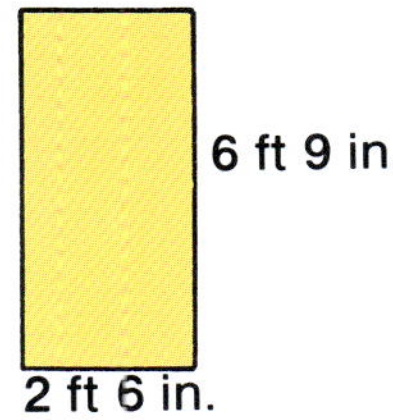

10.
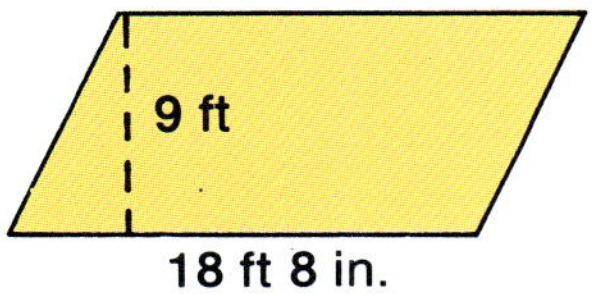

11.
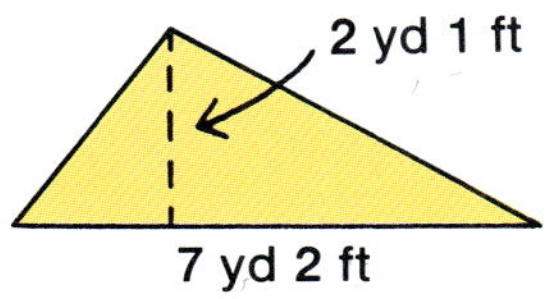

12.
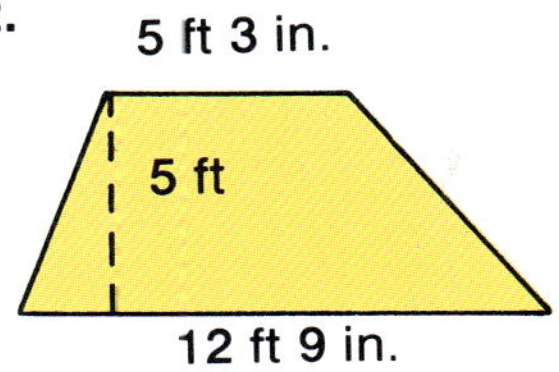

13.
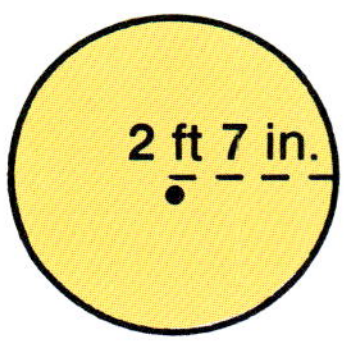

14.
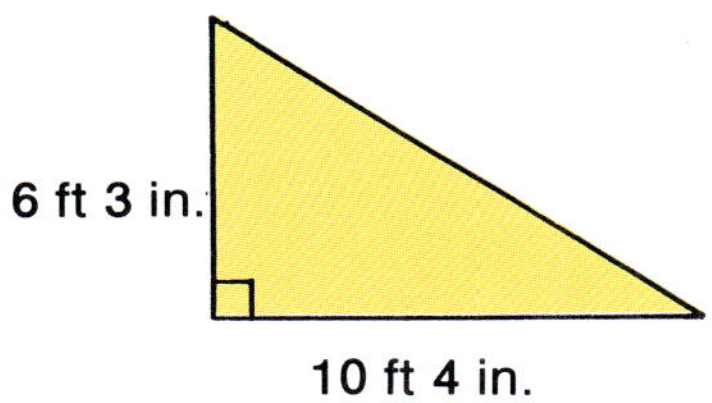

15.
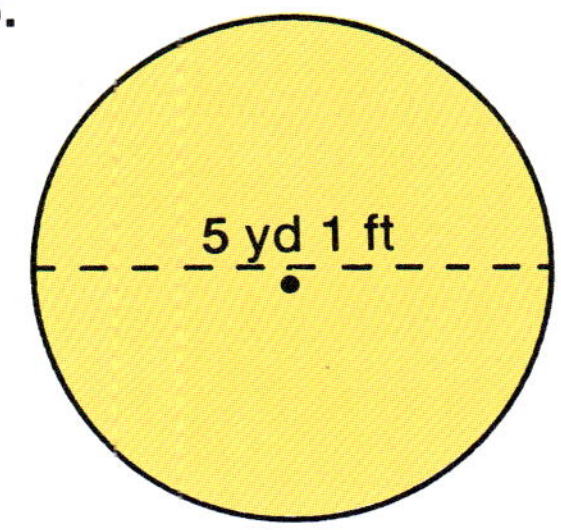

Problem Solving

Use $\frac{22}{7}$ as an approximation for π.

16. A square flower bed measures 10 ft 6 in. on a side. The width of a rectangular garden having the same area is 5 ft 3 in. What is its length?

17. A circular garden with a radius of 9 yd 1 ft has a brick walk around it that is 2 yd 1 ft wide. What is the area of the walk?

Challenge!

Solve.

18. The acre is a customary unit for measuring land area. There are 640 acres in a square mile. How many square feet are in an acre?

19. A football field, including the end zones, is 360 feet long by 160 feet wide. What is the area of a football field in acres?

Volume—customary units

When measurements are given in more than one unit, change to a single unit before multiplying.

Study these examples.

EXAMPLE 1.

$V = Bh$

$= (1 \text{ ft } 9 \text{ in.} \times 10 \text{ in.})(1 \text{ ft } 3 \text{ in.})$

$= (21 \text{ in.} \times 10 \text{ in.})(15 \text{ in.})$

$= (210 \text{ in.}^2)(15 \text{ in.})$

$= 3150 \text{ in.}^3$

EXAMPLE 2.

$V = Bh$

$= \pi(7 \text{ in.} \times 7 \text{ in.})(15 \text{ in.})$

$= \pi(49 \text{ in.}^2)(15 \text{ in.})$

$= \pi(735 \text{ in.}^3)$

$\approx \left(\frac{22}{7}\right)(735 \text{ in.}^3)$

$\approx 2310 \text{ in.}^3$

EXERCISES
Give the number of

1. cubic feet in a cubic yard.
2. cubic inches in a cubic foot.
3. cubic inches in a cubic yard.
4. cubic feet in 5 cubic yards.
5. cubic feet in $3\frac{1}{3}$ cubic yards.
6. cubic yards in 84 cubic feet.

Give each volume. Use $\frac{22}{7}$ as an approximation for π.

7. 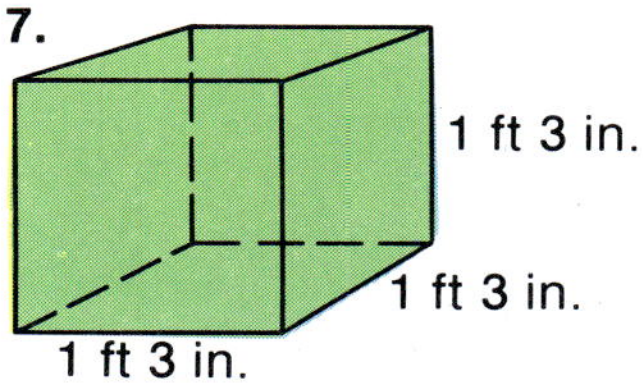

8. 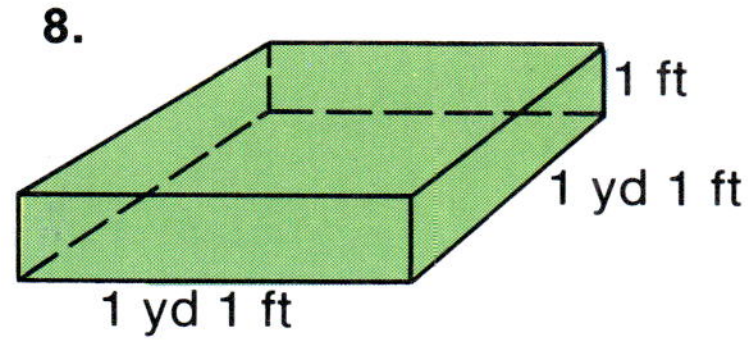

9. 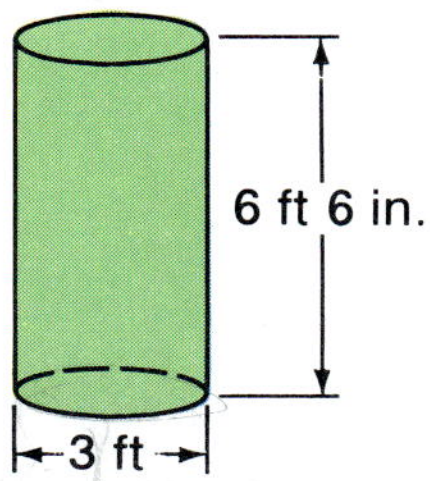

10. 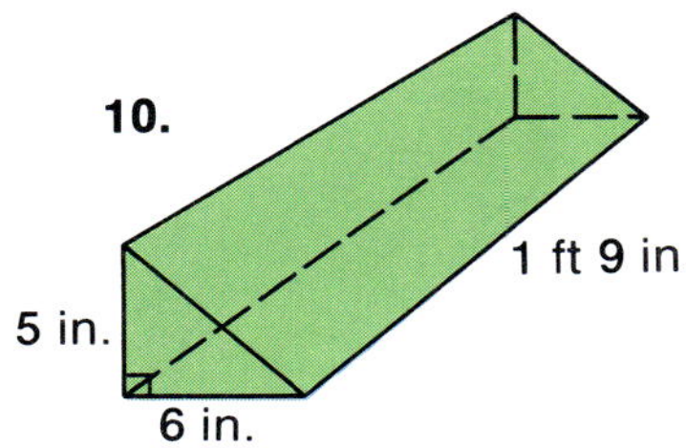

11. 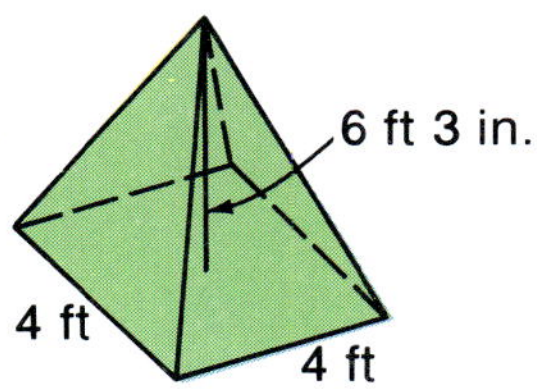

12. 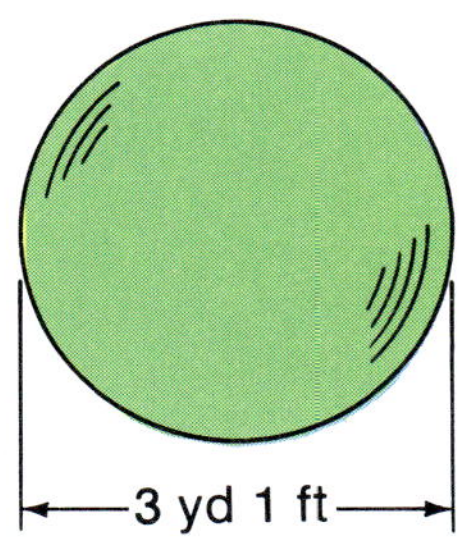

Problem Solving

Use $\frac{22}{7}$ as an approximation for π.

13. A rectangular box is 2 ft by 3 ft by 6 ft. What is its volume in cubic feet? In cubic yards?

14. A cylindrical tank has a diameter of 1 yard and a height of 2 yd 1 ft. What is its volume?

15. How many cubic yards of concrete are needed to pour a garage floor that is 25 ft by 25 ft and 4 inches thick?

16. A swimming pool is 16 ft wide and 36 ft long. When it is full, the average water depth is 4 ft 6 in. If the water was pumped into the pool at a rate of 2.5 ft^3 per minute, how many hours did it take to fill the pool?

KEEPING SKILLS SHARP

Divide. Give the quotient in simplest form.

1. $5 \div 1\frac{1}{2}$
2. $7\frac{2}{3} \div 4$
3. $4\frac{5}{9} \div 1$
4. $3\frac{3}{4} \div 3$
5. $4 \div 1\frac{1}{9}$
6. $5\frac{1}{2} \div 1\frac{1}{3}$
7. $3\frac{1}{4} \div 2$
8. $5 \div 2\frac{1}{3}$
9. $4\frac{1}{2} \div 2$
10. $4\frac{2}{3} \div 1\frac{1}{4}$
11. $6\frac{1}{2} \div 2\frac{2}{3}$
12. $8\frac{3}{4} \div 3$
13. $5 \div 3\frac{1}{4}$
14. $6\frac{1}{2} \div 3$
15. $5\frac{2}{3} \div 1\frac{1}{4}$
16. $4\frac{3}{8} \div 2\frac{1}{4}$

Chapter Checkup

Find the area. [pages 182–189] **Use 3.14 as an approximation for π.**

1.

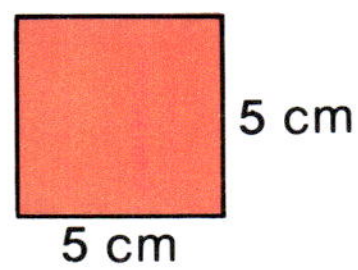

2.

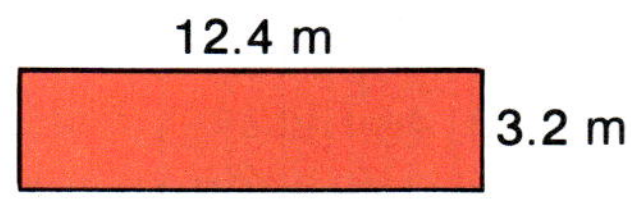

3.

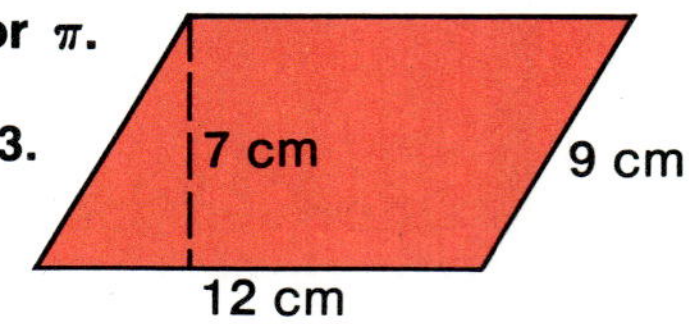

4.

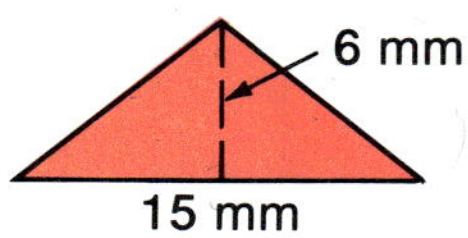

5.

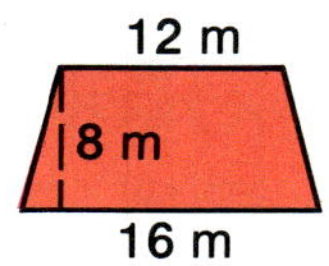

6.

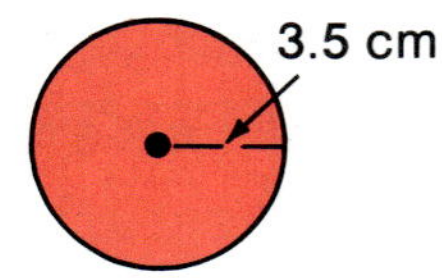

Find the surface area. [pages 190–191] **Use 3.14 as an approximation for π.**

7.

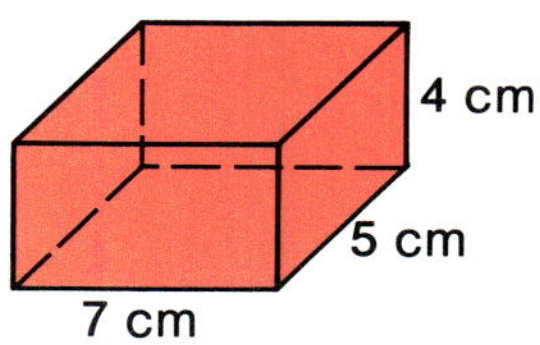

8.

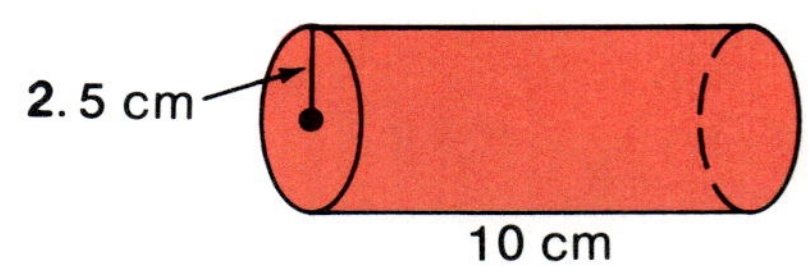

Find the volume. [pages 192–198] **Use 3.14 as an approximation for π.**

9.

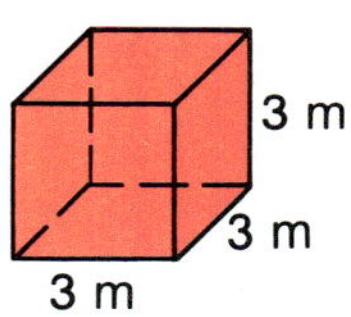

10.

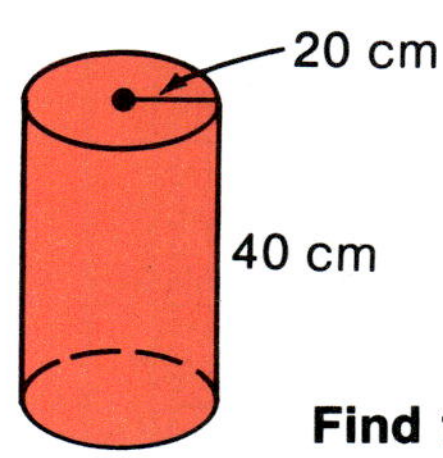

11.

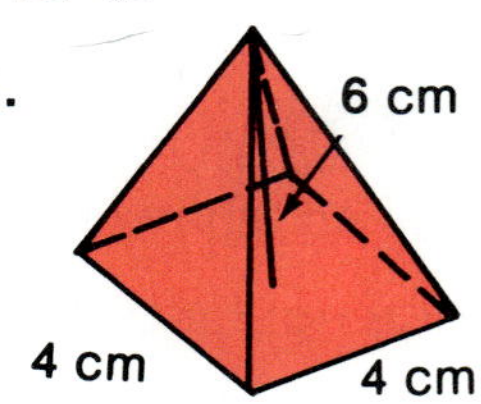

Find the area. [pages 202–203]

12.

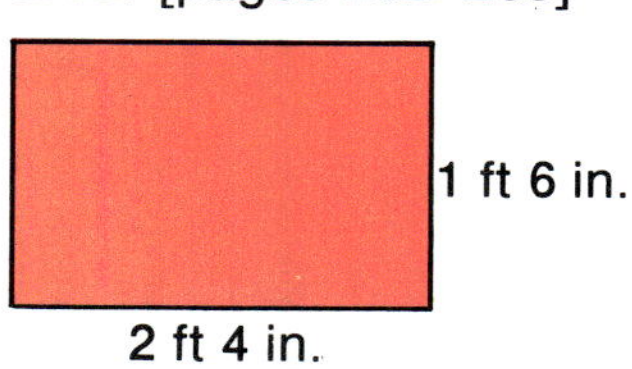

Find the volume. [pages 204–205]

13.

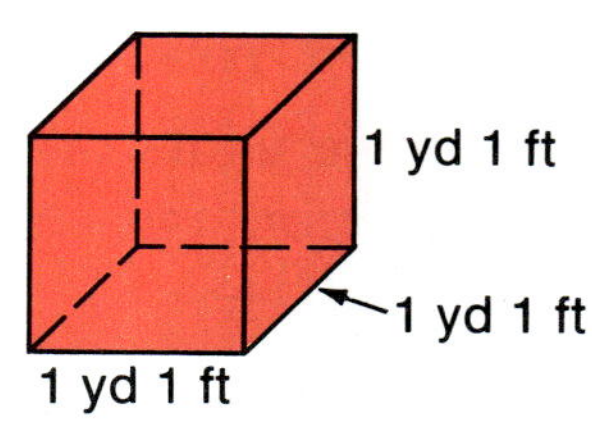

Solve. [page 199]

14. A sheep is in a 23-m by 32-m pen. How much area does the sheep have for grazing?

15. How many cubic meters of concrete are needed to fill a form that is 6 m wide, 10 m long, and 0.1 m thick?

Chapter Project

DENSITY

You can find the volume of an irregular-shaped solid by following these steps:

Step 1. Set a container inside a rectangular pan. Fill the container right to the top with water.

Step 2. Completely submerge the solid object.

Step 3. Determine the volume of the displaced water in the pan.

The volume of the displaced water is equal to the volume of the solid object. *Remember:* 1 mL = 1 cm^3.

1. Get several irregular solids and find the volume of each.
2. The **density** of a substance is the ratio of its mass to its volume. In the metric system, density can be expressed in grams per cubic centimeter. Find the mass of each of the objects and compute the density by dividing its mass in grams by its volume in cm^3. Round each density to the nearest tenth.

Chapter Review

Area of a parallelogram

$$A = bh$$

Area of a triangle

$$A = \frac{1}{2}bh$$

Area of a trapezoid

$$A = \frac{1}{2}(b_1 + b_2)h$$

Area of a circle

$$A = \pi r^2$$

The surface area of a prism or cylinder is the sum of the areas of the faces.

Volume of a prism or cylinder

$$V = Bh$$

(B is the area of the base.)

Volume of a cone or pyramid

$$V = \frac{1}{3}Bh$$

Give each area. Use 3.14 as an approximation for π.

1.

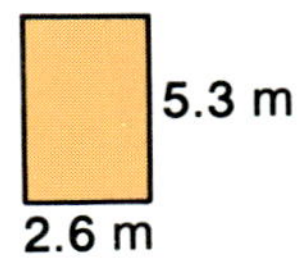

2.

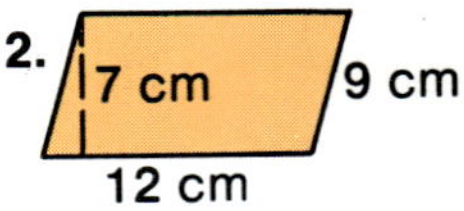

3.

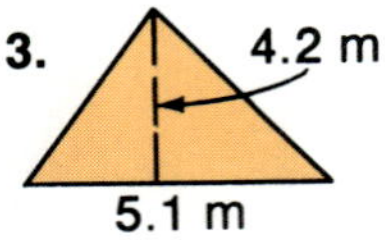

4.

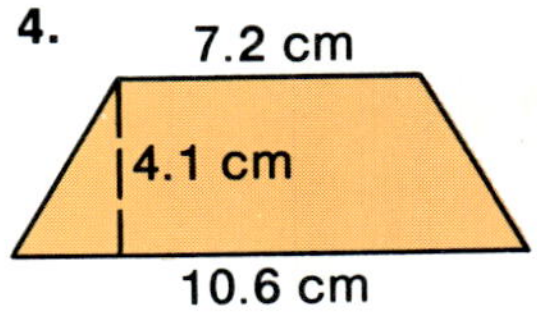

5.

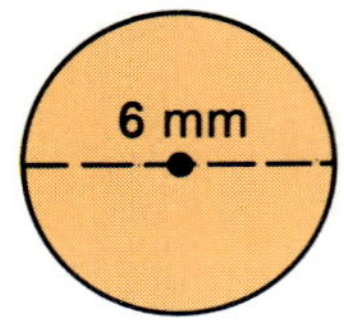

Give the surface area. Use 3.14 as an approximation for π.

6.

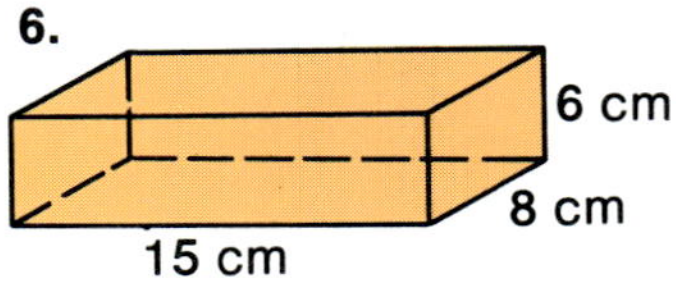

7.

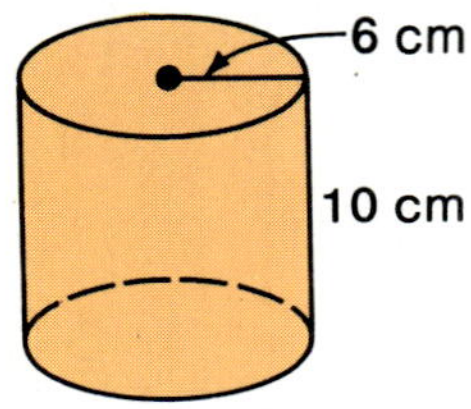

Give each volume. Use 3.14 as an approximation for π.

8.

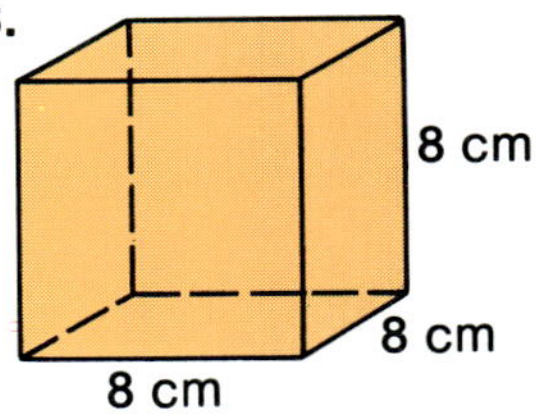

9.

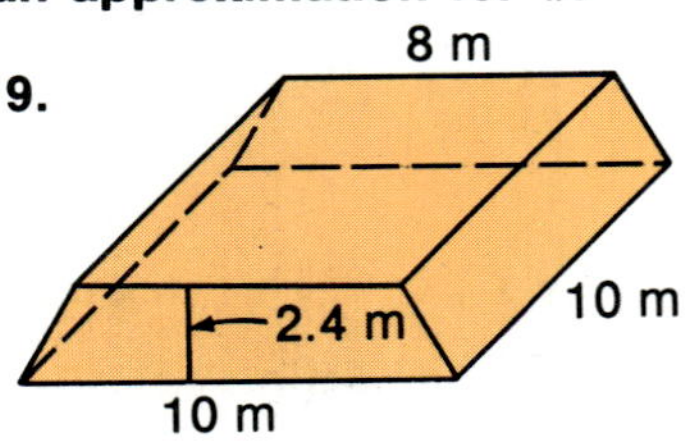

10.

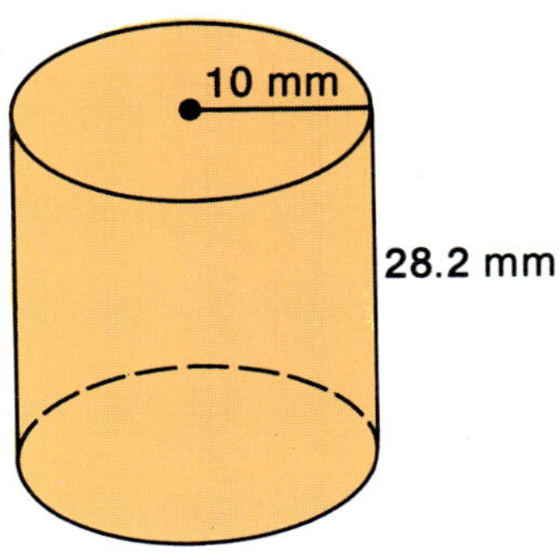

11.

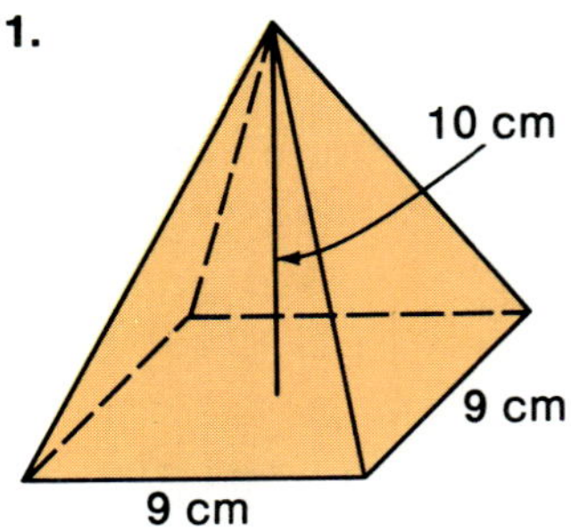

Chapter Challenge

ESTIMATION OF QUANTITY

1. Suppose that a circular pizza with a 9-inch diameter is just enough for 1 person. How many circular pizzas with a 15-inch diameter would you have to buy to feed 20 people?
2. What would your answer be if the pizzas were square instead of round?
3. A cheese ball with a 3-inch diameter is just enough for 3 people. How many cheese balls with a 6-inch diameter are needed to feed 30 people?

Cumulative Checkup

STANDARDIZED FORMAT

Choose the correct letter.

1. Which number is less than 382.529?

a. 382.6
b. 382.592
c. 382.06
d. none of these

2. Give the quotient.

$0.13\overline{)7.8}$

a. 0.6
b. 6
c. 60
d. none of these

3. $8.25 \times 10^4 = \underline{?}$

a. 82,500
b. 825,000
c. 8,250,000
d. none of these

4. The prime factorization of 54 is

a. 2×3^3
b. $3^3 \times 6$
c. $2 \times 3 \times 9$
d. none of these

5. 12.5 cm = $\underline{?}$

a. 0.125 km
b. 12,500 mm
c. 125 m
d. none of these

6. $<$, $=$, or $>$?

$\frac{7}{10}$ ⓘ? $\frac{3}{4}$

a. $<$
b. $=$
c. $>$
d. none of these

7. Subtract.

$8\frac{1}{4} - 3\frac{1}{2}$

a. $4\frac{3}{4}$
b. $5\frac{3}{4}$
c. $5\frac{1}{4}$
d. none of these

8. Multiply.

$\frac{3}{4} \times \frac{6}{7}$

a. $\frac{7}{8}$
b. $\frac{9}{14}$
c. $\frac{3}{14}$
d. none of these

9. Find the area.

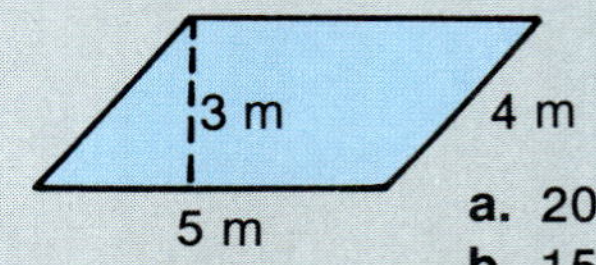

a. 20 m²
b. 15 m²
c. 40 m²
d. none of these

10. Find the surface area.

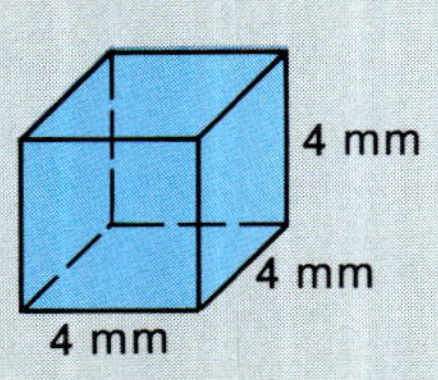

a. 24 mm²
b. 64 mm²
c. 96 mm²
d. none of these

11. A 6-m by 4-m floor was covered with 0.1-m square tiles. How many tiles were there?

a. 24 **b.** 240
c. 2400 **d.** none of these

12. Roger had $\frac{1}{4}$ the number of tapes that David had. Roger had 8 tapes. How many tapes did David have?

a. 32 **b.** 12
c. 2 **d.** none of these

7

Ratio and Proportion

To construct this model, the architect lets 3 centimeters on the model represent 150 centimeters on the full-size building. What is the scale of the model?

READY OR NOT?

Are the two fractions equivalent?

1. $\frac{1}{2}$ $\frac{2}{4}$

2. $\frac{2}{3}$ $\frac{6}{9}$

3. $\frac{4}{1}$ $\frac{8}{2}$

4. $\frac{5}{10}$ $\frac{1}{2}$

5. $\frac{3}{4}$ $\frac{8}{12}$

6. $\frac{5}{6}$ $\frac{15}{18}$

7. $\frac{3}{5}$ $\frac{12}{15}$

8. $\frac{16}{20}$ $\frac{8}{10}$

9. $\frac{8}{5}$ $\frac{7}{4}$

10. $\frac{5}{6}$ $\frac{6}{7}$

Ratios

An artist made a special shade of green by mixing 2 parts of canary yellow with 4 parts of sky blue. If the artist wished to tell someone how to mix the same color, which of these ways would be correct?

a. "Use 2 parts of yellow for each 4 parts of blue."
b. "Use 4 parts of yellow for each 2 parts of blue."
c. "Use 4 parts of yellow for each 8 parts of blue."
d. "Use 2 parts more of blue than yellow."

Statements **b** and **d** are incorrect. Think of a way to follow the directions in **d** and get the wrong color.

A **ratio** compares two numbers. The ratio of parts of yellow to parts of blue is 2 to 4. Here are three ways to write the ratio:

2 to 4 $\qquad$ 2 : 4 $\qquad$ $\frac{2}{4}$

You can read each ratio as "2 to 4."

Remember that order is important when working with ratios. The ratio of yellow to blue is $\frac{2}{4}$ and the ratio of blue to yellow is $\frac{4}{2}$.

These ratios are equal:

$$\frac{1}{2} = \frac{2}{4} = \frac{3}{6}$$

You can find equal ratios by thinking about equivalent fractions.

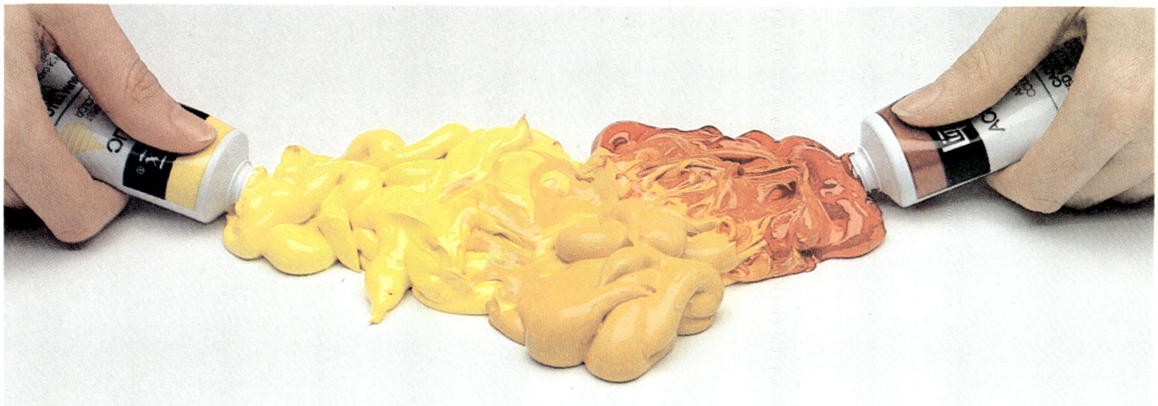

EXERCISES

The artist made some orange paint by mixing 6 parts of yellow with 2 parts of red.

1. Did the artist use more yellow paint than red paint?
2. What is the ratio of parts of yellow to parts of red?
3. What is the ratio of parts of red to parts of yellow?
4. Suppose that you wanted to mix 16 grams of the orange paint. How many grams of each color would you use?
5. Suppose that you had only 3 g of yellow paint and 3 g of red. How many grams of orange could you make?

Write each ratio as a fraction in lowest terms.

6. 9 : 6
7. $\frac{8}{6}$
8. 21 to 14
9. $\frac{14}{21}$
10. 36 to 24
11. 18 meters to 12 meters
12. 12 grams to 20 grams
13. 12 hours to 16 hours
14. 32 kg to 18 kg
15. 24 min to 18 min
16. 14 liters to 8 liters

Complete the equal ratios.

17. $\frac{2}{3} = \frac{?}{6}$
18. $\frac{1}{4} = \frac{?}{12}$
19. $\frac{10}{4} = \frac{5}{?}$
20. $\frac{18}{12} = \frac{9}{?}$
21. $\frac{3}{4} = \frac{?}{24}$
22. $\frac{3}{?} = \frac{15}{20}$
23. $\frac{8}{3} = \frac{?}{21}$
24. $\frac{?}{16} = \frac{5}{4}$
25. $\frac{?}{3} = \frac{10}{30}$
26. $\frac{1}{3} = \frac{?}{12}$
27. $\frac{4}{16} = \frac{?}{4}$
28. $\frac{12}{?} = \frac{24}{22}$
29. $\frac{10}{15} = \frac{?}{3}$
30. $\frac{15}{30} = \frac{?}{2}$
31. $\frac{7}{9} = \frac{?}{36}$
32. 27 : ? = 9 : 5
33. 24 : 36 = ? : 6
34. 20 : 32 = 5 : ?

Proportions

An equation stating that two ratios are equal is called a **proportion.** Every proportion has a related multiplication equation.

Proportion $\frac{3}{5} = \frac{6}{10}$

Related multiplication equation $3 \times 10 = 5 \times 6$

The two products 3×10 and 5×6 are called **cross products.**

If the ratios are equal, then the cross products are equal. Also, if the cross products are equal, then the ratios are equal. This gives us an easy way to decide whether the two ratios are equal.

EXAMPLE 1.

The cross products are equal, so the ratios are equal:

$$\frac{4}{6} = \frac{6}{9}$$

EXAMPLE 2.

The cross products are not equal, so the ratios are not equal:

$$\frac{3}{5} \neq \frac{6}{11}$$

Read $\neq$ as "is not equal to."

EXAMPLE 3.

Jan made 3 free throws out of 5 attempts. At that ratio, how many free throws would she make out of 35 attempts?

We can solve the problem by solving a proportion.

$\frac{3}{5} = \frac{n}{35}$ ← *free throws made* (n) ← *free throws attempted* (35)

$5n = 3 \times 35$

$5n = 105$

$n = 21$

To solve a proportion, solve its multiplication equation.

She would make 21 free throws.

EXERCISES

$=$ or $\neq$?

1. $\frac{3}{4}$ ● $\frac{7}{9}$
2. $\frac{7}{2}$ ● $\frac{11}{3}$
3. $\frac{2}{3}$ ● $\frac{3}{4\frac{1}{2}}$
4. $\frac{2\frac{1}{4}}{8}$ ● $\frac{5}{16}$
5. $\frac{3.5}{2}$ ● $\frac{10.5}{6}$
6. $\frac{4}{5}$ ● $\frac{12.8}{16}$

Solve each proportion.

7. $\frac{3}{8} = \frac{n}{32}$
8. $\frac{3}{4} = \frac{n}{20}$
9. $\frac{5}{7} = \frac{35}{n}$
10. $\frac{n}{18} = \frac{2}{3}$
11. $\frac{2}{9} = \frac{12}{n}$
12. $\frac{5}{n} = \frac{20}{40}$
13. $\frac{2.7}{8.1} = \frac{n}{18}$
14. $\frac{4.2}{1.4} = \frac{15}{n}$
15. $\frac{0.13}{0.52} = \frac{3.2}{n}$
16. $\frac{n}{7} = \frac{2}{3}$
17. $\frac{3}{4} = \frac{11}{n}$
18. $\frac{5}{8} = \frac{8}{n}$

Problem Solving

GAME RECORD		
Player	*Free Throws Made*	*Free Throws Attempted*
Sarah D.	5	12
Jill G.	6	9
Maria M.	5	10
Ingrid P.	4	6
Paige R.	9	14
Sue V.	10	17

19. What was Jill's ratio of free throws made to free throws attempted?
20. What was Paige's ratio of free throws missed to free throws made?
21. By the ratio shown in the table, how many free throws would Sue make if she shot 50 times? Round your answer to the nearest whole number.

KEEPING SKILLS SHARP

Write as decimals.

1. $\frac{1}{2}$
2. $\frac{1}{3}$
3. $\frac{2}{3}$
4. $\frac{5}{8}$
5. $\frac{7}{4}$
6. $\frac{9}{2}$
7. $\frac{16}{5}$
8. $\frac{16}{10}$
9. $\frac{24}{5}$
10. $\frac{32}{3}$
11. $\frac{37}{2}$
12. $\frac{43}{8}$
13. $\frac{22}{5}$
14. $\frac{34}{4}$
15. $\frac{27}{5}$
16. $\frac{19}{8}$
17. $\frac{23}{4}$
18. $\frac{35}{8}$
19. $\frac{37}{3}$
20. $\frac{43}{6}$

Proportions—estimating the solution

Earlier you learned that you can find an equivalent fraction by multiplying the numerator and denominator of a fraction by the same nonzero number. You can use this fact to estimate the solution of a proportion. Study these examples.

EXAMPLE 1.

$\frac{5}{9} = \frac{n}{19}$

19 is about 9 × 2. So n is about 5 × 2, or 10.

$9n = 5 \times 19$

$9n = 95$

$n = 10\frac{5}{9}$

Is the solution close to the estimate?

EXAMPLE 2.

$n:17 = 3:2$

$\frac{n}{17} = \frac{3}{2}$

17 is about 8 × 2. So n is about 8 × 3, or 24.

$2n = 3 \times 17$

$2n = 51$

$n = 25\frac{1}{2}$

EXERCISES

First estimate the solution. Then solve the proportion.

1. $\frac{n}{5} = \frac{8}{11}$	**2.** $\frac{6}{n} = \frac{23}{16}$	**3.** $\frac{5}{9} = \frac{n}{29}$	**4.** $\frac{22}{15} = \frac{7}{n}$	**5.** $\frac{n}{12} = \frac{9}{35}$
6. $\frac{9}{n} = \frac{14}{8}$	**7.** $\frac{16}{9} = \frac{2}{n}$	**8.** $\frac{n}{8} = \frac{18}{23}$	**9.** $\frac{5}{12} = \frac{1}{n}$	**10.** $\frac{14}{29} = \frac{n}{9}$
11. $\frac{8}{5} = \frac{n}{4}$	**12.** $\frac{15}{n} = \frac{7}{2}$	**13.** $\frac{11}{5} = \frac{2}{n}$	**14.** $\frac{25}{16} = \frac{n}{3}$	**15.** $\frac{n}{42} = \frac{15}{20}$
16. $\frac{n}{9} = \frac{8}{28}$	**17.** $\frac{20}{11} = \frac{4}{n}$	**18.** $\frac{15}{23} = \frac{n}{6}$	**19.** $\frac{27}{n} = \frac{13}{7}$	**20.** $\frac{43}{18} = \frac{11}{n}$

21. $3:8 = n:32$	**22.** $7:n = 15:26$	**23.** $9:2 = n:9$	**24.** $13:n = 6:4$
25. $n:11 = 4:6$	**26.** $36:15 = 9:n$	**27.** $40:n = 19:32$	**28.** $35:17 = n:9$
29. $36:n = 15:7$	**30.** $n:20 = 32:60$	**31.** $15:47 = n:27$	**32.** $50:74 = 21:n$

Problem Solving Finding a pattern

1. Study and understand.
2. Plan and do.
3. Answer and check.

Attendance at the roller-skating rink increased steadily each weekend.

Weekend	1	2	3	4	5
Attendance	500	550	605	665	?

If the increase continues at the same rate, what attendance can be expected for the 5th weekend?

Step 1. First, divide 550 by 500.
Next, divide 605 by 550.
Finally, divide 665 by 605.

Weekend	1	2	3	4	5
Attendance	500	550	605	665	?

Step 2. Multiply 665 by 1.1.

Weekend	1	2	3	4	5
Attendance	500	550	605	665	731

Answer: The expected attendance for the 5th weekend is about 731.

Solve by finding a pattern.

1. The number of skaters in the special events increased.

Event	1	2	3	4
Entries	16	24	36	54

If the increase continues at the same rate, how many skaters will enter the next event?

2. As the weather turned warmer in the spring, the attendance decreased.

Weekend	12	13	14	15
Attendance	800	400	200	100

What attendance can be expected for the 16th weekend?

Problem Solving Using proportions

A popular model-railroad scale is the HO scale. This scale is 1 : 87. This means that the ratio of the length of any model part to the real part is 1 to 87.

You can use proportions to solve problems about HO-scale models.

EXAMPLE. A baggage car is 18.3 meters long. How long would the HO-scale model be?

length of model → (numerator of $\frac{1}{87}$) ; *length of real car* → (denominator of $\frac{1}{87}$)

$$\frac{1}{87} = \frac{n}{18.3}$$

← *length of model* (numerator of $\frac{n}{18.3}$) ; ← *length of real car* (denominator of $\frac{n}{18.3}$)

$$87n = 18.3$$

$$n \approx 0.21$$

The model is about 0.21 meter, or 21 centimeters, long.

EXERCISES

Solve.

1. How long is an HO-scale model of a chemical tank car that is 10.7 meters in length?

2. How long is a Pullman car if an HO-scale model of it is 24.54 cm long?

3. Model railroaders build scenery to scale for their layouts. An HO-scale model bridge is 44.54 cm long. How long would the real bridge be?

4. An HO-scale model water tower is 15.8 cm tall. How tall would the real tower be?

Another popular scale in model railroading is the N scale, in which the ratio is 1 : 160.

5. How long would an N-scale model of an 18-meter passenger car be?

6. How many meters long is a hopper car if its N-scale model is 6.9 cm long?

7. Both an N-scale model and an HO-scale model were made of an old steam engine that was 17 meters long. Which model was longer? How much longer?

8. Standard railroad tracks are spaced 1.439 meters apart. How far apart would the tracks be in an N-scale model?

9. A model-railroad club constructed an N-scale model layout that had 105 meters of main-line track. How many kilometers of real track did their layout represent?

10. An HO-scale model of a tank car is 12.25 cm long. How long would an N-scale model of the same tank car be?

11. Another scale used in model railroading is the O scale. The O scale is 1 : 48. Which are longer, O-scale models or HO-scale models?

Problem Solving Using rates

A **rate** is a ratio of two quantities. Here are some examples.

Jogged 2 km in 13 minutes.

Rate: $\frac{2 \text{ km}}{13 \text{ min}}$

Rate: $\frac{3 \text{ apples}}{35¢}$

Rate: $\frac{\$49.50}{1 \text{ dozen}}$

Earned \$5.50 in $2\frac{1}{2}$ hours.

Rate: $\frac{\$5.50}{2\frac{1}{2} \text{ hours}}$

A proportion may be used to solve a rate problem.

How much paint is needed to paint 170 square meters?

$$\begin{matrix} \textit{liters} \rightarrow \\ \textit{area} \rightarrow \end{matrix} \frac{4}{35} = \frac{n}{170} \begin{matrix} \leftarrow \textit{liters} \\ \leftarrow \textit{area} \end{matrix}$$

$$35n = 680$$

$$n \approx 19.4$$

So, 19.4 L of paint are needed. How many cans of paint would you have to buy?

EXERCISES

Solve by using a proportion.

1. Small green peppers are on sale for 84¢ a dozen. What is the price of 20 peppers?
2. Oranges are on sale for 6 for 79¢. Find the price of 11 oranges. Round the answer to the nearest cent.
3. If 12 pencils cost 43¢, how much would 9 pencils cost? Round the answer to the nearest cent.
4. If peanuts are priced at $2.40 a kg, how many kg can be bought for $3.50? Round the answer to the nearest tenth of a kilogram.
5. George hiked 18.5 km in 4 hours. At that rate, how many kilometers could he hike in $6\frac{1}{2}$ hours? Round the answer to the nearest tenth km.
6. Mrs. Allan drove 252 km in 3 hours. At that rate, how long would it take her to drive 386 km? Round the answer to the nearest tenth of an hour.
7. John hit 5 home runs in 84 times at bat. At that rate, how many times at bat would he need to hit 7 home runs? Round the answer to the nearest whole number.
8. A computer printer can print 120 characters (letters or symbols) in a second. At that rate, how many minutes would it take to print 8000 characters? Round the answer to the nearest minute.
9. Donna earned $8.72 for 4 hours of baby-sitting. At that rate, what would she earn for 3 hours of baby-sitting?
10. A certain recipe calls for $\frac{1}{2}$ teaspoon of vanilla and 5 tablespoons of butter. How much vanilla is needed for 15 tablespoons of butter?
11. Mr. Jackson's automobile used 48 liters of gasoline in 280 km. At that rate, how much gasoline would be needed for a 456-km trip? Round the answer to the nearest tenth liter.

KEEPING SKILLS SHARP

Add or subtract. Give answers in simplest form.

1. $\frac{3}{5}+\frac{1}{5}$
2. $\frac{3}{8}+\frac{1}{2}$
3. $\frac{4}{5}+\frac{1}{3}$
4. $\frac{1}{4}+\frac{2}{3}$
5. $\frac{3}{7}+\frac{1}{8}$
6. $\frac{5}{3}+\frac{3}{5}$
7. $\frac{7}{8}+\frac{5}{6}$
8. $\frac{3}{10}+\frac{1}{4}$
9. $\frac{5}{7}-\frac{1}{7}$
10. $\frac{5}{6}-\frac{1}{2}$
11. $\frac{7}{8}-\frac{2}{3}$
12. $\frac{4}{5}-\frac{2}{7}$
13. $\frac{11}{12}-\frac{2}{9}$
14. $\frac{5}{3}-\frac{3}{5}$
15. $\frac{7}{8}-\frac{5}{6}$
16. $\frac{9}{10}-\frac{3}{4}$

Using scale

Work in groups. Get squared paper, a meterstick, and a calculator.

Maribel works in a factory that makes rose rugs. The factory makes just one kind of rug, but in many different sizes. Every rose rug is a square with a single rose. The standard rose rug has sides that are 5 m long and a print of a rose that is 4 m long. Maribel is planning a rose rug with sides that are 15 m long.

Standard rose rug

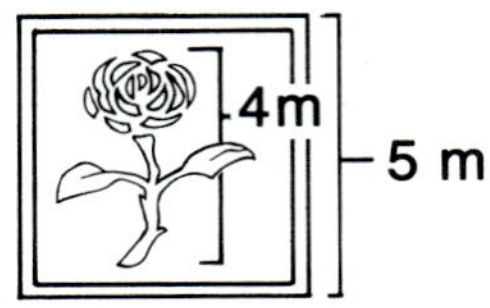

Maribel's rose rug

1. a. Compare one side of Maribel's rug with one side of the standard rug. How many times longer will it be?
   b. How long will the rose on Maribel's rug be? How many times longer is that than the rose on the standard rug?
   c. Find the area that each rug will cover. How many times more area will Maribel's rug cover than the standard rug?
   d. Compare the changes in length with the changes in area. Are your answers reasonable? How can you tell?
2. Some rug dye spilled on Maribel's chart for different rug sizes.
   a. Copy the chart and fill in the missing information.
   b. Discuss the strategies you used to find the missing information.

Rug Length	Rose Length	Rug Area
1 m		
2 m		
		9 m²
5 m	4 m	
	8 m	
15 m		
	14.4 m	
		400 m²

3. Imagine your classroom is empty. You wish to install the largest rose rug that will fit. The side of a rose rug must be a whole number of meters long.
   a. For your classroom's rug, estimate the side of the rug, the rose's length, and the rug area.
   b. Now measure to find the largest rose rug for the classroom. Which measurement(s) do you need to make? Why?
   c. Calculate the rose length and the area for the rug you need. Were your estimates close?

4. Maribel wanted to cover completely the floor of a 13 m x 7 m rectangular room with rose rugs. Remember, the sides must be a whole number of meters long.
   a. Design a way of covering this floor with rose rugs.
   b. Find another way to cover the floor using a different number of rugs.
   c. Compare your answers with others in your group and with those of other groups.

5. A rose rug costs \$20 plus \$10 per square meter of area.
   a. Which of your ways of covering the floor is more expensive? Why is this so?
   b. Find the least expensive way of covering the floor.

6. Suppose the 1-meter-long rug weighed 3.2 kg.
   a. How heavy would the 5-meter-long rug be?
   b. How many times heavier would it be than the 1-meter-long rug?

7. If you double each side of a rose rug, what happens to the rose's length, the rug's area, and the rug's weight? What happens to each if you triple each side? Try to make up a rule to describe these changes.

8. Discuss your strategies and conclusions with another group.

9. For which computations was the calculator useful? When was mental math or paper-and-pencil computation best? Why? Discuss.

Scale drawings and models

Work in groups. Get squared paper, string, and a calculator.

PART A

This diagram of a two-room cabin is not drawn to scale.

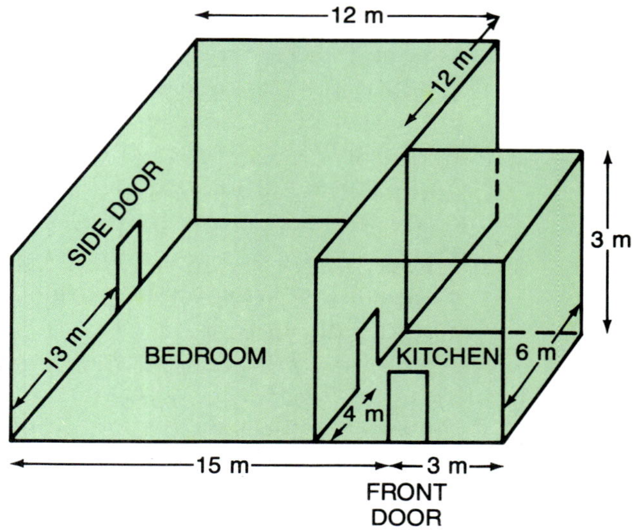

1. On squared paper, make a scale drawing of the cabin's floor plan. You can use a scale of 1 cm : 1 m, or any scale that will fit on your paper.

2. Use your scale drawing to find the length of the shortest path from the side door through the cabin to the front door.

3. **a.** Find the actual dimensions of the bedroom and the kitchen.
   b. Find the actual areas of the bedroom and the kitchen.
   c. Find each of these ratios and write in lowest terms:

   $\frac{\text{Bedroom length}}{\text{Kitchen length}}$ $\frac{\text{Bedroom width}}{\text{Kitchen width}}$ $\frac{\text{Bedroom area}}{\text{Kitchen area}}$

   d. Find the product of the first and second ratios. Compare this product with the third ratio. What do you notice? Explain why this happens.

4. For each pair, find the ratio of the actual dimensions to the scale drawing dimensions. Then find how many times longer the first item is than the second item. *Remember:* 1 m = 100 cm

   a. $\frac{\text{Actual length of kitchen in cm}}{\text{Length of kitchen in scale drawing}}$ **b.** $\frac{\text{Actual length of shortest path through cabin in cm}}{\text{Length of shortest path through scale drawing of cabin}}$

5. Find the ratio of the actual area of the kitchen in cm^2 to the area of the scale drawing of the kitchen. How many times larger is the actual area?

6. Compare your answers from problems 4 and 5. Discuss the similarities and differences that you find.

7. Use the ratios you have already found to predict this ratio:

$$\frac{\text{Actual area of bedroom}}{\text{Area of scale drawing of bedroom}}$$

Was your prediction correct?

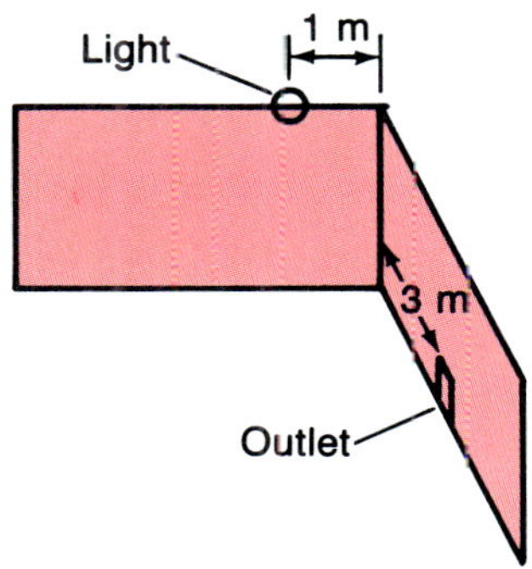

PART B

8. The kitchen of the two-room cabin is 3 m high. Make a partial three-dimensional scale model of the kitchen using these two steps:
   a. Make a scale drawing of the two kitchen walls that have no doors.
   b. Fold your drawing along the line where the walls meet.

9. Suppose there is a light at the top of the back wall as shown. You need a wire to go along the walls to the electrical outlet at the bottom of the side wall.
   a. Use your scale model to find the shortest length of wire needed.
   b. Check your answer by finding the lengths of other possible paths for the wire and then comparing lengths.

10. a. Find the volume of the kitchen in cubic meters.
   b. Find the length, width, and height of the kitchen in centimeters. Use these dimensions to find the volume of the kitchen in cubic centimeters.
   c. What would be the volume of your scale model of the kitchen if it were complete?
   d. Use your answers to b and c to find this ratio (use cubic centimeters for both volumes):

$$\frac{\text{Volume of actual kitchen}}{\text{Volume of your scale model of kitchen}}$$

   e. Compare this ratio with the ratios you found in problems 4a and 5. Discuss your comparisons.

11. For which parts of the lesson was the calculator helpful?

Problem Solving

Using a chart

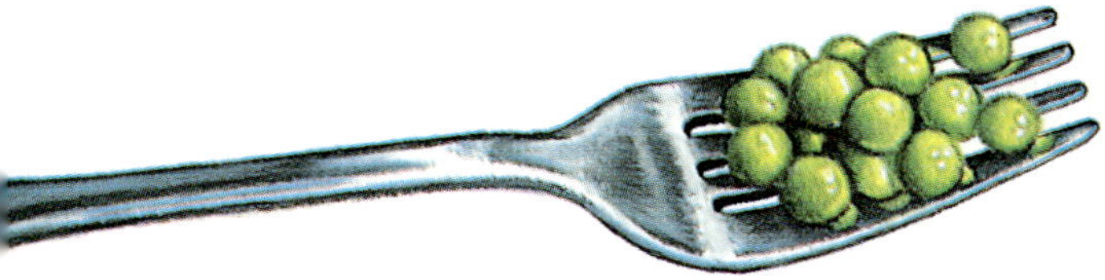

A balanced diet includes food from these categories:

- cereals and bread
- milk
- meat
- citrus fruit
- potatoes
- leafy green and yellow vegetables
- other fruit and vegetables
- eggs
- dried beans, peas, and nuts

The table shows the number of calories (the amount of energy) that some foods provide.

Food	*Calories*
apple	85
bacon, 2 slices	96
beans, green, $\frac{1}{2}$ cup	15
bread and butter, 1 slice	92
carrot, raw, 1 large	42
cereal, dry, 1 cup, with milk and sugar	205
cheese, American, 1 slice	108
cola, 8-oz glass	105
cookie, chocolate chip	50
egg, fried, 1	108
halibut, broiled, 4 oz	206
ham, 2 slices	248
ice cream, $\frac{2}{3}$ cup	186
milk, whole, 8 oz	160
orange juice, 4 oz	56
pancake, 1, 2 tablespoons of syrup	210
peas, $\frac{1}{2}$ cup	58
pie, fruit, $\frac{1}{6}$ of 9″ pie	400
pork chop, 6 oz	308
spaghetti, 1 cup; meat sauce, 5 oz	396

EXERCISES

Find the number of calories in

1. 9 oz of orange juice.

2. 7 oz of whole milk.

3. 10 oz of cola.

4. $1\frac{1}{2}$ cups of ice cream.

5. How many cups of ice cream make 480 calories? Round your answer to the nearest tenth of a cup.

6. How many calories are in this breakfast?
1 fried egg
2 slices of bacon
1 slice of bread and butter
6 oz of whole milk
6 oz of orange juice

Along with a proper diet, you need to exercise. By exercising, you can burn off calories that would otherwise go to a sometimes-unwanted weight gain. The table shows the approximate number of calories burned off each minute by different kinds of exercise. (It is based on a 100-pound person.)

Type of Exercise	*Calories per Minute*
Fast walking	3.5
Bicycling	5.4
Jogging-and-walking	6.3
Jogging	7.1
Swimming	8.2

Round each answer to the nearest tenth.

7. How many calories are used bicycling for 45 minutes?

8. How many calories are used in jogging-and-walking for $1\frac{1}{4}$ hours?

9. How many minutes of swimming are needed to burn up 500 calories?

10. Suppose that you ate 3 pancakes with syrup. How many minutes would you have to jog to burn up the calories?

11. Suppose that for lunch you had
1 apple
1 cheese sandwich
(2 slices of bread and butter and 2 slices of American cheese)
10 oz of whole milk
How many minutes would you have to walk to burn off the calories? If you averaged 3.5 miles per hour, how far would the lunch take you?

★ **12.** Suppose that you ate $\frac{1}{4}$ of an 8″ fruit pie. How many minutes would you have to jog to burn off the calories?

Chapter Checkup

Look at the cards and give the ratio. [pages 212–213]

1. odd numbers to even numbers
2. even numbers to odd numbers
3. prime numbers to composite numbers
4. factors of 12 to multiples of 5

Write each ratio as a fraction in lowest terms. [pages 212–213]

5. $\frac{6}{8}$ 6. 6:9 7. 5 to 10 8. 15:12 9. 12 to 15

Solve each proportion. [pages 214–216]

10. $\frac{3}{4} = \frac{9}{n}$ 11. $\frac{5}{7} = \frac{n}{35}$ 12. $\frac{3}{8} = \frac{n}{24}$ 13. $\frac{16}{15} = \frac{32}{n}$

14. $\frac{n}{8} = \frac{7}{4}$ 15. $\frac{6}{n} = \frac{2}{9}$ 16. $\frac{n}{12} = \frac{7}{5}$ 17. $\frac{19}{n} = \frac{4}{17}$

Round each solution to the nearest hundredth. [page 222]

18. $\frac{n}{13} = \frac{9}{3}$ 19. $\frac{23}{n} = \frac{6}{11}$ 20. $\frac{13}{7} = \frac{n}{5}$ 21. $\frac{9}{22} = \frac{7}{n}$

Solve by using a proportion. [pages 217–219, 220–227]

22. A model of a train is built to a scale of 1:80. The model engine is 20 cm long. How long is the actual engine?

23. The scale on a map is 2 cm : 7.5 km. How far apart are two cities that are 16.5 cm apart on the map?

24. A junior-high class collected 2450 aluminum cans in 4 days. At that rate, how many cans would the class collect in 10 days?

25. A bicyclist rode 35 km in 2.4 hours. At that rate, how many km would he ride in 3.5 hours? Round your answer to the nearest km.

Chapter Project

USING PROPORTIONS TO ESTIMATE

How much does a thumbtack weigh?

1. Notice that 14 thumbtacks balance 5 grams.
   To answer the question, solve this proportion:

$$\begin{matrix}\textit{number of tacks} \rightarrow \\ \textit{number of grams} \rightarrow\end{matrix} \frac{14}{5} = \frac{1}{n} \begin{matrix}\leftarrow \textit{number of tacks} \\ \leftarrow \textit{number of grams}\end{matrix}$$

   Round your answer to the nearest hundredth of a gram.

2. Use the method shown above to find the weight of one

   a. small paper clip.
   b. large paper clip.
   c. straight pin.
   d. paper staple.

3. Use the the method to determine the weight of some other small objects.

Chapter Review

The ratio of blue cards to red cards is

$\frac{2}{3}$ ← *blue* / ← *red*

$$\frac{5}{8} = \frac{3}{n}$$
$$5n = 8 \times 3$$
$$5n = 24$$
$$n = \frac{24}{5}$$
$$n = 4\frac{4}{5}$$

A jogger ran 15 km in 2 hours. At that rate, how far could she run in 3.5 hours?

distance → / *time* → $\frac{15}{2} = \frac{n}{3.5}$ ← *distance* / ← *time*

$$2n = 15 \times 3.5$$
$$2n = 52.5$$
$$n = 26.25$$

Answer: 26.25 km

Give each ratio.

1. yellow cards to green cards
2. green cards to yellow cards
3. vowels to consonants
4. consonants to vowels

Solve each proportion.

5. $\frac{4}{5} = \frac{12}{n}$
6. $\frac{5}{4} = \frac{n}{10}$
7. $\frac{35}{n} = \frac{5}{100}$
8. $\frac{n}{50} = \frac{3}{100}$
9. $\frac{3}{7} = \frac{n}{50}$
10. $\frac{4}{5} = \frac{25}{n}$

Solve.

11. An airplane flew 1820 km in 2.5 hours. At that rate, how far did it fly in 1 hour?
12. A racing car traveled 532 km in 2 hours. At that rate, how long would it take to travel 800 km?
13. The ratio of boys to girls in a certain school is $\frac{5}{6}$. If there are 420 girls, how many boys are there?
14. A recipe for Orange Fizz calls for 1.5 liters of water to 3 liters of orange juice. How much soda water is needed for 2.5 liters of orange juice?

Chapter Challenge

GOLDEN RECTANGLES

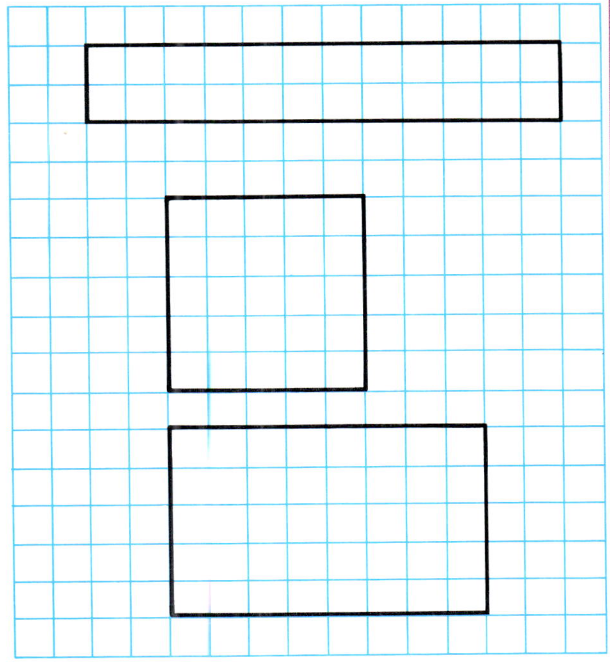

Ratios can describe the shapes of rectangles. In this "thin" rectangle, the ratio of width to length is 1 : 6.

In the "fattest" rectangle (which is a square) the ratio of width to length is 1 : 1.

The ancient Greeks considered a rectangle shaped like this one as the most pleasing to the eye. It is called a *golden rectangle,* and the ratio of its width to its length is about 1 : 1.618.

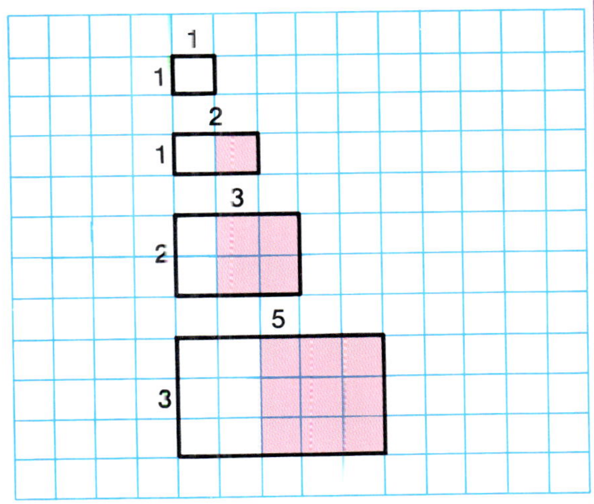

1. Study this pattern of rectangles. The second rectangle was made by adding a square to the first square. The third rectangle was made by adding a square to the second rectangle. And so on.
2. Get a piece of graph paper. Copy the rectangles and continue the pattern.
3. Compute the ratio of width to length for each rectangle. Are any of your ratios close to the golden ratio of 1 : 1.618?

Cumulative Checkup

STANDARDIZED FORMAT

Choose the correct letter.

1. 9.9982 rounded to the nearest tenth is

a. 9.9
b. 9.10
c. 10.0
d. none of these

2. Multiply.

$$\begin{array}{r} 4.02 \\ \times 30.8 \\ \hline \end{array}$$

a. 12.3816
b. 1.5816
c. 1.5276
d. none of these

3. Solve.

$4a + 6 = 14$

a. 5
b. 20
c. 2
d. none of these

4. 5.2 L = ? mL

a. 0.052
b. 5200
c. 520
d. none of these

5. Give the difference.

$\frac{5}{3} - \frac{1}{2}$

a. 4
b. $1\frac{1}{6}$
c. $\frac{2}{3}$
d. none of these

6. Give the product.

$1\frac{3}{4} \times 2\frac{4}{5}$

a. $4\frac{9}{10}$
b. $4\frac{11}{20}$
c. $\frac{5}{8}$
d. none of these

7. Find the area. Use 3.14 as an approximation for π.

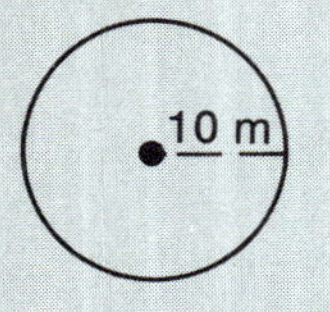

a. 31.4 m^2
b. 314 m^2
c. 62.8 m^2
d. none of these

8. Find the volume.

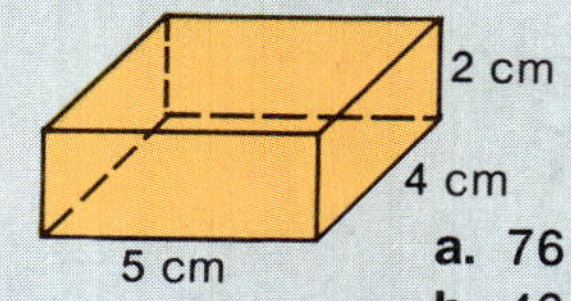

a. 76 cm^3
b. 40 cm^3
c. 11 cm^3
d. none of these

9. The ratio of consonants to vowels in the word NUMBER is

a. 2 : 1
b. 2 : 4
c. 4 : 6
d. none of these

10. Solve.

$\frac{3}{8} = \frac{x}{100}$

a. 37.5
b. $266\frac{2}{3}$
c. 0.24
d. none of these

11. There are 120 people. Of these, $\frac{2}{3}$ like baseball. How many do not like baseball?

a. 80 **b.** 90
c. 40 **d.** none of these

12. 1 cm on a map stands for 0.5 km. What length does 8 cm on the map stand for?

a. 4 km **b.** 16 km
c. 40 km **d.** none of these

8

Percent

The vest that Chad is looking at has a regular price of $39. What is the sale price?

READY OR NOT?

Solve each proportion.

1. $\frac{3}{5} = \frac{x}{100}$
2. $\frac{x}{45} = \frac{40}{100}$
3. $\frac{40}{x} = \frac{125}{100}$
4. $\frac{18}{n} = \frac{37\frac{1}{2}}{100}$
5. $\frac{2}{3} = \frac{x}{100}$
6. $\frac{x}{36} = \frac{25}{100}$
7. $\frac{x}{24} = \frac{75}{100}$
8. $\frac{5}{6} = \frac{x}{100}$
9. $\frac{12}{x} = \frac{33\frac{1}{3}}{100}$
10. $\frac{x}{72} = \frac{12\frac{1}{2}}{100}$
11. $\frac{x}{100} = \frac{150}{50}$
12. $\frac{150}{75} = \frac{x}{100}$

Writing a percent as a fraction

The Emerson School Pep Club has a 100-member card section. They first make each design on graph paper.

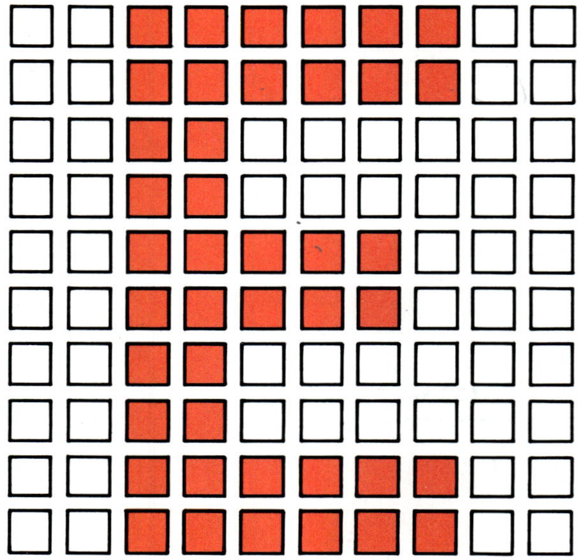

To make this E, 42 out of the 100 cards need to be red. That is 42% (42 percent) of the cards. **Percent** means "hundredths."

$$42\% = \frac{42}{100}$$

To write a percent as a fraction, mixed number, or whole number, you can first write the percent as a fraction with a denominator of 100. Then you can write the fraction in simplest form.

$$10\% = \frac{10}{100} = \frac{1}{10}$$

$$42\% = \frac{42}{100} = \frac{21}{50}$$

$$150\% = \frac{150}{100} = 1\frac{1}{2}$$

$$100\% = \frac{100}{100} = 1$$

EXERCISES

Write as a fraction, mixed number, or whole number.

1. 20%	**2.** 300%	**3.** 10%	**4.** 60%	**5.** 96%
6. 120%	**7.** 65%	**8.** 600%	**9.** 25%	**10.** 68%
11. 800%	**12.** 150%	**13.** 500%	**14.** 75%	**15.** 160%
16. 175%	**17.** 900%	**18.** 12%	**19.** 50%	**20.** 1000%
21. 95%	**22.** 14%	**23.** 200%	**24.** 125%	**25.** 400%
26. 700%	**27.** 90%	**28.** 30%	**29.** 115%	**30.** 1200%

Write as a fraction or mixed number in simplest form.

Example.

Multiply by $\frac{10}{10}$ to get whole numbers.

$$37.5\% = \frac{37.5}{100} = \frac{375}{1000} = \frac{3}{8}$$

31. 7.5%	**32.** 12.5%	**33.** 0.1%	**34.** 0.5%
35. 6.25%	**36.** 93.75%	**37.** 2.5%	**38.** 18.75%
39. 87.5%	**40.** 8.25%	**41.** 262.5%	**42.** 137.5%

Write as a fraction or mixed number in simplest form.

Example.

$$16\tfrac{2}{3}\% = \frac{16\frac{2}{3}}{100} = \frac{\frac{50}{3}}{\frac{100}{1}} \leftarrow = \frac{50}{3} \times \frac{1}{100} = \frac{1}{6}$$

43. $14\frac{1}{7}\%$	**44.** $5\frac{1}{2}\%$	**45.** $2\frac{1}{2}\%$
46. $33\frac{1}{3}\%$	**47.** $41\frac{2}{3}\%$	**48.** $37\frac{1}{2}\%$
49. $62\frac{1}{2}\%$	**50.** $12\frac{1}{2}\%$	**51.** $66\frac{2}{3}\%$
52. $16\frac{2}{3}\%$	**53.** $81\frac{1}{4}\%$	**54.** $87\frac{1}{2}\%$
55. $6\frac{1}{4}\%$	**56.** $166\frac{2}{3}\%$	**57.** $137\frac{1}{2}\%$

Writing a fraction as a percent

Andrew got $\frac{3}{4}$ of the problems correct on a math test.

Here is how to write the fraction as a percent:

First write the equivalent fraction with a denominator of 100. Then write as a percent.

$$\frac{3}{4} = \frac{75}{100}$$
$$= 75\%$$

Andrew got 75% of the problems correct.

To write some fractions as percents, you may need to solve a proportion.

EXAMPLE 1.

$$\frac{7}{8} = \frac{?}{100} \qquad \frac{7}{8} = \frac{x}{100}$$
$$8x = 700$$
$$x = 87\tfrac{1}{2}$$

So $\frac{7}{8} = \frac{87\frac{1}{2}}{100}$

$= 87\frac{1}{2}\%$

EXAMPLE 2.

$$\frac{1}{240} = \frac{?}{100} \qquad \frac{1}{240} = \frac{x}{100}$$
$$240x = 100$$
$$x = \frac{100}{240}$$
$$x = \frac{5}{12}$$

So $\frac{1}{240} = \frac{\frac{5}{12}}{100}$

$= \frac{5}{12}\%$

EXERCISES

Write as a percent.

1. $\frac{1}{2}$	**2.** $\frac{9}{10}$	**3.** $\frac{3}{5}$	**4.** $\frac{5}{4}$	**5.** $\frac{9}{4}$	**6.** $\frac{2}{5}$
7. 1	**8.** $\frac{1}{5}$	**9.** $\frac{1}{10}$	**10.** $\frac{1}{4}$	**11.** $\frac{7}{10}$	**12.** $\frac{4}{5}$
13. 2	**14.** $\frac{3}{2}$	**15.** $\frac{6}{5}$	**16.** $\frac{7}{2}$	**17.** $\frac{3}{4}$	**18.** $\frac{7}{4}$
19. $\frac{9}{50}$	**20.** $\frac{13}{20}$	**21.** $\frac{3}{25}$	**22.** $\frac{9}{20}$	**23.** $\frac{7}{25}$	**24.** $\frac{13}{50}$

Write as a percent by solving a proportion.

25. $\frac{1}{3}$ 26. $\frac{1}{6}$ 27. $\frac{1}{8}$ 28. $\frac{9}{16}$ 29. $\frac{4}{3}$

30. $\frac{5}{9}$ 31. $\frac{5}{12}$ 32. $\frac{3}{8}$ 33. $\frac{2}{3}$ 34. $\frac{1}{9}$

35. $\frac{5}{3}$ 36. $\frac{5}{6}$ 37. $\frac{11}{8}$ 38. $\frac{17}{12}$ 39. $\frac{2}{7}$

Complete.

40. If $\frac{1}{8} = 12\frac{1}{2}\%$, then $\frac{2}{8} = \underline{?}\%$.
41. If $\frac{1}{9} = 11\frac{1}{9}\%$, then $\frac{5}{9} = \underline{?}\%$.
42. If $\frac{1}{11} = 9\frac{1}{11}\%$, then $\frac{3}{11} = \underline{?}\%$.
43. If $\frac{1}{6} = 16\frac{2}{3}\%$, then $\frac{2}{6} = \underline{?}\%$.
44. If $\frac{1}{100} = 1\%$, then $\frac{1}{200} = \underline{?}\%$.

Problem Solving

The class results of the mathematics test are shown on the bar graph.

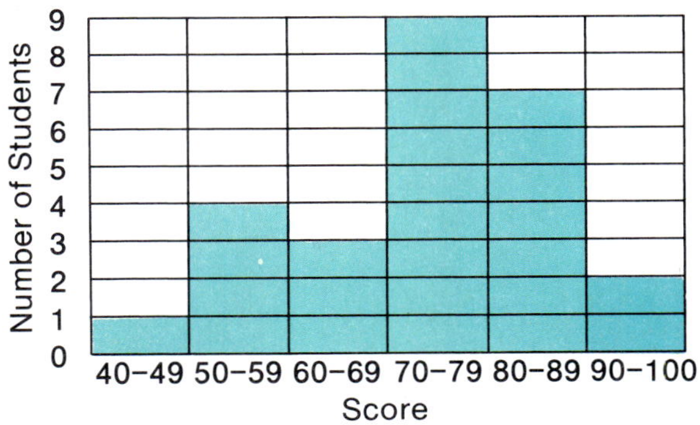

45. How many students took the test?
46. What percent of the students got a test score in the 80's?
47. What percent scored below 70?
48. What percent scored 80 or above?
49. What fraction of the students scored below 60?

KEEPING SKILLS SHARP

Give each sum in simplest form.

1. $\frac{1}{8} + \frac{3}{8}$
2. $\frac{1}{9} + \frac{1}{3}$
3. $\frac{3}{4} + \frac{1}{8}$
4. $\frac{3}{4} + \frac{3}{2}$
5. $\frac{5}{6} + \frac{2}{3}$
6. $\frac{7}{8} + \frac{5}{6}$
7. $\frac{3}{4} + \frac{4}{3}$
8. $1 + \frac{1}{3}$
9. $\frac{4}{5} + 2$
10. $\frac{5}{4} + \frac{5}{6}$
11. $\frac{2}{3} + 3$
12. $\frac{1}{2} + \frac{1}{3}$
13. $\frac{7}{8} + \frac{1}{4}$
14. $\frac{3}{5} + \frac{2}{3}$
15. $\frac{5}{6} + \frac{3}{4}$
16. $\frac{6}{5} + \frac{2}{3}$

Decimals and percents

These examples show how to write a decimal as a percent.

$0.24 = \frac{24}{100}$ $\qquad 0.6 = \frac{60}{100}$ $\qquad 0.66\frac{2}{3} = \frac{66\frac{2}{3}}{100}$

$= 24\%$ $\qquad = 60\%$ $\qquad = 66\frac{2}{3}\%$

SHORTCUT MENTAL MATH

To write a decimal as a percent, move the decimal point two places to the RIGHT and write a percent sign.

$0.525 = 52.5\%$

$0.8 = 80\%$

$0.37\frac{1}{2} = 37\frac{1}{2}\%$

These examples show how to write a percent as a decimal.

$42\% = \frac{42}{100}$ $\qquad 60\% = \frac{60}{100}$ $\qquad 87\frac{1}{2} = \frac{87\frac{1}{2}}{100}$

$= 0.42$ $\qquad = 0.6$ $\qquad = 0.87\frac{1}{2}$

SHORTCUT MENTAL MATH

To write a percent as a decimal, move the decimal point two places to the LEFT and omit the percent sign.

$42.4\% = 0.424$

$90\% = 0.90$

$33\frac{1}{3}\% = 0.33\frac{1}{3}$

EXERCISES

Write as a percent.

1. 0.18 **2.** 0.35 **3.** 0.92 **4.** 0.8 **5.** 0.3 **6.** 0.1

7. 1.5 **8.** 2.4 **9.** 3.5 **10.** 0.002 **11.** 0.023 **12.** 0.0475

13. $0.37\frac{1}{2}$ **14.** $0.62\frac{1}{2}$ **15.** $0.33\frac{1}{3}$ **16.** $0.66\frac{2}{3}$ **17.** $0.06\frac{1}{4}$ **18.** $0.08\frac{1}{3}$

Write as a decimal.

19. 15% **20.** 87% **21.** 4% **22.** 9% **23.** 175% **24.** 225%

25. 37.5% **26.** 54.2% **27.** 64.3% **28.** 4.2% **29.** 0.6% **30.** 0.5%

31. $16\frac{2}{3}\%$ **32.** $83\frac{1}{3}\%$ **33.** $37\frac{1}{2}\%$ **34.** $44\frac{1}{5}\%$ **35.** $133\frac{1}{3}\%$ **36.** $137\frac{1}{2}\%$

SKILL GAME

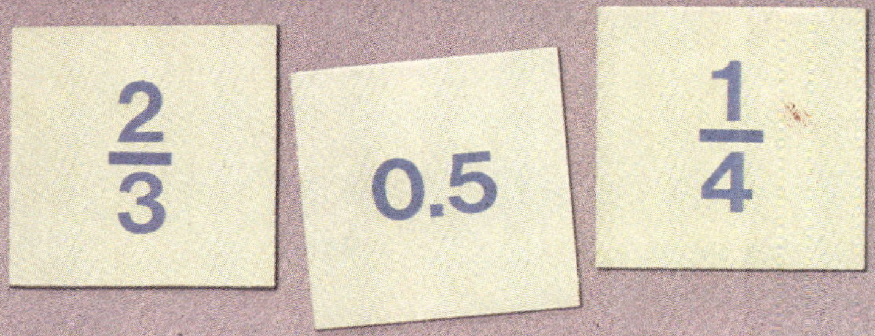

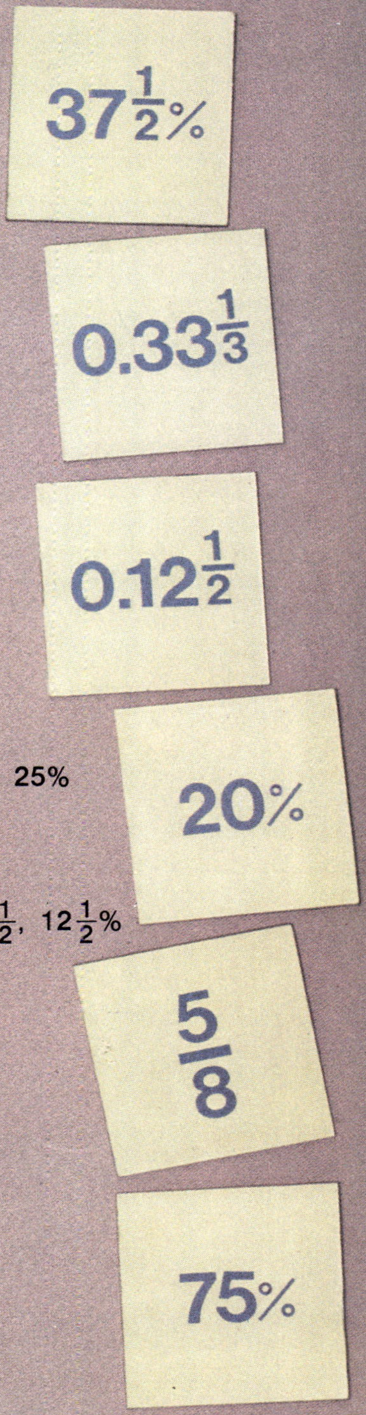

In this game you must be able to tell whether two numbers are equal. These exercises will give you some practice.

Equal (=) or not equal (≠)?

1. $\frac{1}{6}$ ● 0.2 **2.** $83\frac{1}{3}\%$ ● 0.8 **3.** $\frac{5}{6}$ ● $\frac{3}{4}$

4. 40% ● $\frac{2}{5}$ **5.** $\frac{1}{6}$ ● $0.12\frac{1}{2}$ **6.** 60% ● $\frac{5}{8}$

Here is how to play the game.

1. Make cards for these numbers.

   $\frac{1}{2}$, 0.5, 50%, $\frac{1}{3}$, $0.33\frac{1}{3}$, $33\frac{1}{3}\%$, $\frac{2}{3}$, $0.66\frac{2}{3}$, $66\frac{2}{3}\%$, $\frac{1}{4}$, 0.25, 25%

   $\frac{3}{4}$, 0.75, 75%, $\frac{1}{5}$, 0.2, 20%, $\frac{2}{5}$, 0.4, 40%, $\frac{3}{5}$, 0.6, 60%

   $\frac{4}{5}$, 0.8, 80%, $\frac{1}{6}$, $0.16\frac{2}{3}$, $16\frac{2}{3}\%$, $\frac{5}{6}$, $0.83\frac{1}{3}$, $83\frac{1}{3}\%$, $\frac{1}{8}$, $0.12\frac{1}{2}$, $12\frac{1}{2}\%$

   $\frac{3}{8}$, $0.37\frac{1}{2}$, $37\frac{1}{2}\%$, $\frac{5}{8}$, $0.62\frac{1}{2}$, $62\frac{1}{2}\%$, $\frac{7}{8}$, $0.87\frac{1}{2}$, $87\frac{1}{2}\%$

2. A game leader divides the class into two teams and picks a player from each team.
3. The game leader shuffles the deck and slowly places the cards on the chalktray, one at a time. A player who sees two numbers that are equal says "match" and identifies the cards. If the player is correct, that team earns a point. If incorrect, that team loses a point.
4. The matched cards are set aside, and play continues with two new players.
5. The team earning the greater number of points wins.

Finding a percent of a number

In a survey of a class of 36 students, it was found that 25% earned more than \$10 a week. To find how many students earned more than \$10, we can solve this equation:

$$25\% \text{ of } 36 = n$$

Here are 3 ways to find the solution:

Method 1: Write the percent as a fraction and multiply.

$$25\% \text{ of } 36 = n$$

$$\frac{1}{\cancel{4}_1} \times \overset{9}{\cancel{36}} = n$$

$$9 = n$$

Method 2: Write the percent as a decimal and multiply.

$$25\% \text{ of } 36 = n$$

$$0.25 \times 36 = n$$

$$9 = n$$

$$\begin{array}{r} 36 \\ \times 0.25 \\ \hline 180 \\ 72 \\ \hline 9.00 \end{array}$$

Method 3: Solve a proportion.

$$25\% \text{ of } 36 = n$$

$$\begin{array}{c} \textit{part} \rightarrow \\ \textit{whole} \rightarrow \end{array} \frac{25}{100} = \frac{n}{36} \begin{array}{c} \leftarrow \textit{part} \\ \leftarrow \textit{whole} \end{array}$$

$$100n = 900$$

$$n = 9$$

Look for the easiest method to solve percent problems. The method you pick will depend on the numbers in the problem.

EXERCISES

Solve by writing the percent as a fraction and multiplying.

1. 50% of $30 = n$
2. 25% of $24 = n$
3. 75% of $48 = n$
4. 40% of $45 = n$
5. 150% of $18 = n$
6. 100% of $32 = n$
7. 60% of $30 = n$
8. 200% of $14 = n$
9. 5% of $40 = n$
10. 10% of $80 = n$
11. $33\frac{1}{3}\%$ of $21 = n$
12. $66\frac{2}{3}\%$ of $27 = n$

Solve by writing the percent as a decimal and multiplying.

13. 14% of $32 = n$
14. 42% of $53 = n$
15. 23% of $75 = n$
16. 19% of $29 = n$
17. 18% of $64 = n$
18. 36% of $83 = n$
19. 5.2% of $95 = n$
20. 6.5% of $115 = n$
21. 7.3% of $142 = n$
22. 0.88% of $315 = n$
23. 0.92% of $250 = n$
24. 1.30% of $432 = n$

Solve by using a proportion. Round answers to the nearest hundredth.

25. $5\frac{1}{3}\%$ of $80 = n$
26. $3\frac{3}{4}\%$ of $64 = n$
27. $6\frac{1}{2}\%$ of $75 = n$
28. $7\frac{2}{3}\%$ of $152 = n$
29. $6\frac{3}{4}\%$ of $125 = n$
30. $5\frac{1}{2}\%$ of $248 = n$
31. $12\frac{1}{3}\%$ of $225 = n$
32. $10\frac{3}{8}\%$ of $358 = n$
33. $11\frac{2}{3}\%$ of $465 = n$

Problem Solving

Below are some more facts found in the survey of the 36 students. Write and solve an equation.

34. 50% had record collections. How many had record collections?
35. 75% had radios. How many had radios?
36. $33\frac{1}{3}\%$ had regular chores at home. How many had regular chores?
37. $16\frac{2}{3}\%$ had coin collections. How many had coin collections?
38. $11\frac{1}{9}\%$ had home computers. How many had home computers?
39. $77\frac{7}{9}\%$ received weekly allowances. How many received weekly allowances?

Finding the number when a percent is known

60% of the Sonoma Junior High School band are boys. There are 30 boys in the junior high band. How many are in the band altogether?

$$60\% \text{ of } n = 30$$

Here are three ways to solve the equation.

Method 1: Write the percent as a fraction and divide.

$$\frac{\frac{3}{5}n}{\frac{3}{5}} = \frac{30}{\frac{3}{5}}$$

$$n = 50$$

$$30 \div \frac{3}{5} = \frac{\overset{10}{\cancel{30}}}{1} \times \frac{5}{\underset{1}{\cancel{3}}} = 50$$

Method 2: Write the percent as a decimal and divide.

$$\frac{0.6n}{0.6} = \frac{30}{0.6}$$

$$n = 50$$

$$0.6\overline{)30.0} = 50.$$

Method 3: Solve a proportion.

$$\begin{matrix} \text{part} \rightarrow \\ \text{whole} \rightarrow \end{matrix} \frac{30}{n} = \frac{60}{100} \begin{matrix} \leftarrow \text{part} \\ \leftarrow \text{whole} \end{matrix}$$

$$60n = 3000$$

$$n = 50$$

There are 50 students in the school band.

EXERCISES

Solve.

1. 50% of $n = 12$
2. 25% of $n = 24$
3. 20% of $n = 18$
4. 60% of $n = 27$
5. 40% of $n = 46$
6. 75% of $n = 42$
7. 5% of $n = 80$
8. 10% of $n = 75$
9. 150% of $n = 86$
10. 125% of $n = 60$
11. 250% of $n = 120$
12. $33\frac{1}{3}\%$ of $n = 145$
13. $66\frac{2}{3}\%$ of $n = 248$
14. $16\frac{2}{3}\%$ of $n = 74$
15. $83\frac{1}{3}\%$ of $n = 205$

Solve.

16. 30% of $n = 15$
17. 45% of $n = 90$
18. 36% of $n = 9$
19. 72% of $n = 18$
20. 15% of $n = 45$
21. 48% of $n = 12$
22. 53% of $n = 53$
23. 56% of $n = 63$
24. 56% of $n = 56$
25. 78% of $n = 42.9$
26. 55% of $n = 40.7$
27. 79% of $n = 7.9$
28. 84% of $n = 96.6$
29. 90% of $n = 50.4$
30. 8.32% of $n = 6.24$
31. 8.5% of $n = 5.27$
32. 7.3% of $n = 4.672$
33. 6.25% of $n = 5.875$
34. $83\frac{1}{3}\%$ of $n = 60$
35. $62\frac{1}{2}\%$ of $n = 60$
36. $37\frac{1}{2}\%$ of $n = 60$

Problem Solving

37. A TV set sells for \$480. A down payment of $33\frac{1}{3}\%$ of the cost was made. How much was the down payment?

38. Dotty was at bat 256 times during the season. She got a hit 25% of the time. How many times did she get a hit?

39. Iona read 40% of a book having 350 pages. How many pages did she have left to read?

40. Bill bought a bicycle for \$126.90. This was $66\frac{2}{3}\%$ of the regular selling price. What was the regular selling price?

41. Henry spends 80% of his weekly allowance of \$6 for lunches. How much does he spend on lunch each week?

42. A bicyclist rode $37\frac{1}{2}\%$ of a 56-kilometer bicycle tour and then took a break. How many kilometers did the cyclist ride before taking a break?

43. Karen made a 25% down payment of \$130 on a stereo set. What was the cost of the stereo set?

44. In a school of 824 students, $62\frac{1}{2}\%$ ride a bus to school. How many students ride the bus to school?

45. 80% of the students in one class walk to school. 24 students in that class walk to school. How many students are in the class?

46. One type of home computer sold for \$345. This was 60% of the regular cost. What was the regular price?

47. 7.25% of $n = 35.4$
48. $9\frac{1}{2}\%$ of $n = \$57.60$
49. $12\frac{3}{4}\%$ of $n = \$160.50$

Mental computation with percents

SHORTCUT
MENTAL MATH

You can find 10% or 1% of a number by "moving" the decimal point.

1% of 168.3 = 1.683

10% of 168.3 = 16.83

SHORTCUT
MENTAL MATH

You can find some percents of numbers by simple division.

$33\frac{1}{3}\%$ of 90 $= \frac{90}{3}$
$= 30$

$\frac{1}{3}$

25% of 84 $= \frac{84}{4}$
$= 21$

$\frac{1}{4}$

Here are some other special cases:

$$\begin{aligned} 5\% \text{ of } 72 &= \tfrac{1}{2} \text{ of } 10\% \text{ of } 72 \\ &= \tfrac{1}{2} \text{ of } 7.2 \\ &= 3.6 \end{aligned}$$

$$\begin{aligned} 15\% \text{ of } 72 &= 10\% \text{ of } 72 + 5\% \text{ of } 72 \\ &= 7.2 + 3.6 \\ &= 10.8 \end{aligned}$$

EXERCISES

Complete. Do not use paper and pencil.

1. 10% of 58 = ?
2. 1% of 58 = ?
3. 10% of 6.23 = ?
4. 1% of 6.23 = ?
5. 20% of 85 = ?
6. 20% of \$9.20 = ?
7. 50% of 4.6 = ?
8. 50% of \$8.30 = ?
9. $33\frac{1}{3}\%$ of 123 = ?
10. $33\frac{1}{3}\%$ of 66 = ?
11. $33\frac{1}{3}\%$ of \$4.50 = ?
12. 25% of 0.44 = ?
13. 25% of \$10.80 = ?
14. 5% of 88 = ?
15. 5% of \$2.60 = ?
16. 15% of 60 = ?
17. 15% of \$8.20 = ?
18. $66\frac{2}{3}\%$ of \$12 = ?
19. 75% of 40 = ?
20. 1.5% of 80 = ?
21. 1.5% of 600 = ?

Problem Solving

22. When Carol is satisfied with the service in a restaurant, she tips the waiter 15% of the cost of her meal. What would be the tip for each of these amounts?

a. \$6.00 **b.** \$12.40 **c.** \$8.45

Practice Exercises

Write as a fraction in lowest terms.

1. 20%
2. 60%
3. 25%
4. 85%
5. 75%
6. 140%
7. $12\frac{1}{2}\%$
8. $33\frac{1}{3}\%$
9. $16\frac{2}{3}\%$
10. $37\frac{1}{2}\%$
11. $62\frac{1}{2}\%$
12. $83\frac{1}{3}\%$

Write as a percent.

13. $\frac{3}{4}$
14. $\frac{1}{2}$
15. $\frac{4}{5}$
16. $\frac{5}{4}$
17. $\frac{8}{5}$
18. $\frac{5}{2}$
19. $\frac{1}{8}$
20. $\frac{5}{8}$
21. $\frac{5}{6}$
22. $\frac{7}{8}$
23. $\frac{5}{12}$
24. $\frac{9}{16}$

Write as a decimal.

25. 30%
26. 90%
27. 65%
28. 5%
29. 0.8%
30. 16.5%
31. $8\frac{1}{3}\%$
32. $66\frac{2}{3}\%$
33. $91\frac{2}{3}\%$
34. $8\frac{2}{3}\%$
35. $7\frac{3}{4}\%$
36. $9\frac{2}{3}\%$

Solve.

37. 8% of $50 = m$
38. 25% of $64 = s$
39. $33\frac{1}{3}\%$ of $78 = p$
40. 20% of $25 = u$
41. 200% of $18 = w$
42. 40% of $56 = r$
43. $87\frac{1}{2}\%$ of $64 = x$
44. $112\frac{1}{2}\%$ of $100 = q$
45. 0.5% of $70 = v$
46. $66\frac{2}{3}\%$ of $96 = t$
47. $37\frac{1}{2}\%$ of $140 = y$
48. 150% of $66 = n$

Solve.

49. 50% of $x = 17$
50. 75% of $x = 18$
51. 20% of $x = 35$
52. 120% of $x = 42$
53. 45% of $x = 18$
54. $33\frac{1}{3}\%$ of $x = 18$
55. $37\frac{1}{2}\%$ of $x = 48$
56. 0.8% of $x = 20$
57. 0.1% of $x = 60$
58. $66\frac{2}{3}\%$ of $x = 44$
59. $44\frac{4}{9}\%$ of $x = 64$
60. $112\frac{1}{2}\%$ of $x = 54$

Finding the percent

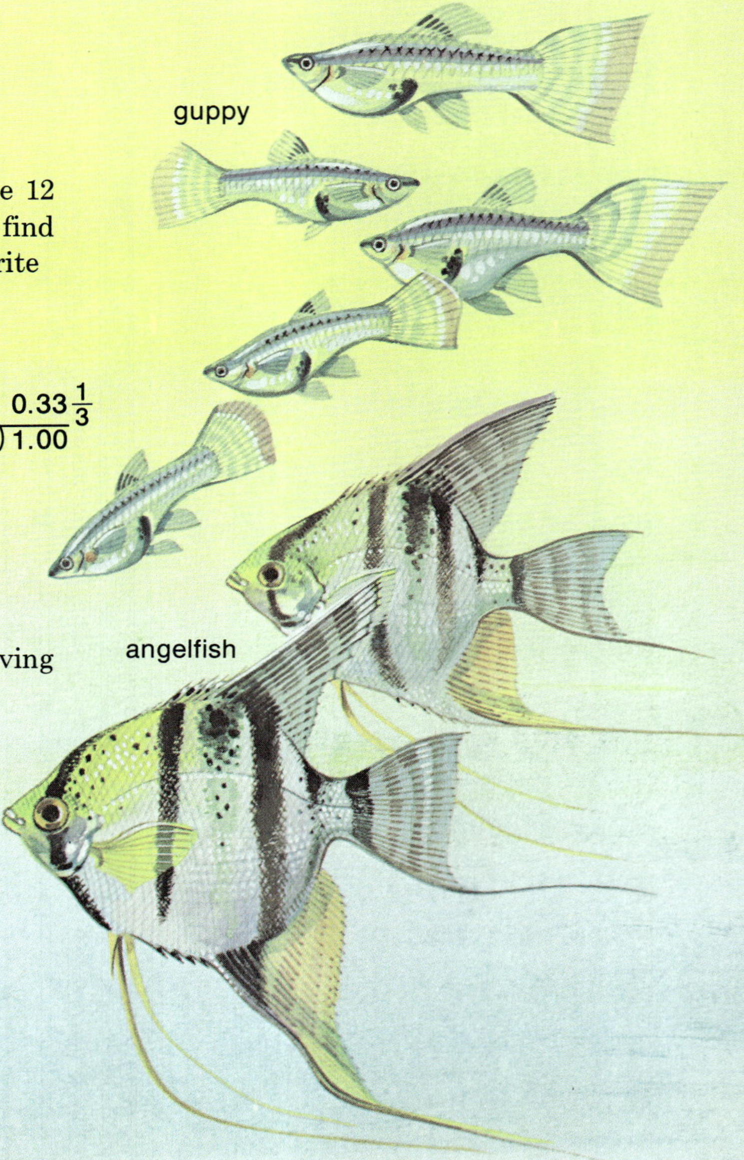

Mary set up this aquarium. 4 of the 12 fish she bought are swordtails. To find the percent of swordtails, we can write this fraction (ratio) as a percent:

swordtails ⟶ / *total fish bought* ⟶ $\frac{4}{12} = \frac{1}{3}$

$$= 0.33\frac{1}{3} \qquad 3\overline{)1.00} \text{ quotient } 0.33\frac{1}{3}$$

$$= 33\frac{1}{3}\%$$

So $33\frac{1}{3}\%$ of the fish are swordtails.

We could also find the percent by solving this proportion:

$$\frac{4}{12} = \frac{x}{100}$$

$$12x = 400$$

$$x = 33\frac{1}{3}$$

So $33\frac{1}{3}\%$ of the fish are swordtails.

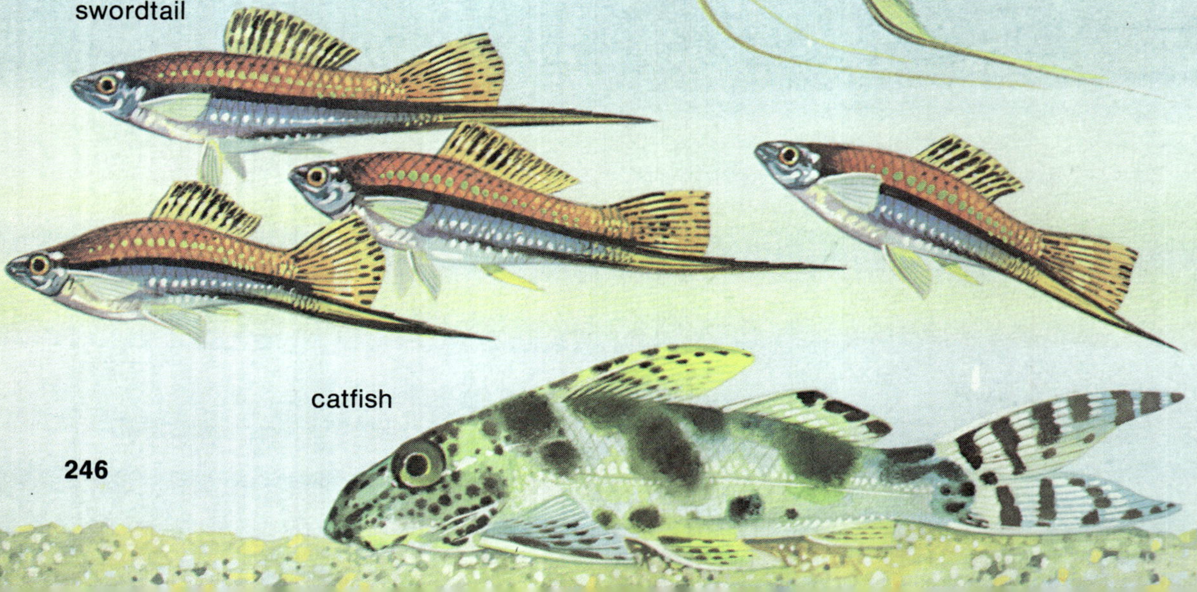

EXERCISES

Solve by writing a fraction as a percent.

whole *part*

1. $n\%$ of $18 = 12$

part → whole → $\frac{12}{18} = \frac{2}{3}$

$= 0.66\frac{2}{3}$

$= ?\%$

2. $n\%$ of $20 = 15$

$\frac{15}{20} = \frac{3}{4}$

$= 0.75$

$= ?\%$

3. $n\%$ of $16 = 10$
4. $n\%$ of $24 = 20$
5. $n\%$ of $42 = 21$
6. $n\%$ of $72 = 54$

Round answers to the nearest tenth of a percent.

7. $n\%$ of $65 = 18$
8. $n\%$ of $54 = 42$

Solve by using a proportion. Round answers to the nearest tenth of a percent.

9. $n\%$ of $45 = 23$
10. $n\%$ of $64 = 28$
11. $n\%$ of $38 = 42$
12. $n\%$ of $75 = 92$
13. $n\%$ of $67 = 50$
14. $n\%$ of $118 = 75$
15. $n\%$ of $93 = 40$
16. $n\%$ of $236 = 156$

Problem Solving

Give the percent of fish on page 246 that are

17. catfish.
18. guppies.
19. angelfish.
20. not swordtails.
21. Mary bought a filter and pump that regularly sold for \$18.75 for 20% off. What was the sale price?
22. Mary learned that for each 5 centimeters of fish (body length), 8 liters of water are needed. Her aquarium holds 38 liters of water. How many centimeters of fish are the limit?

KEEPING SKILLS SHARP

Give each difference in simplest form.

1. $\frac{7}{8} - \frac{3}{8}$
2. $\frac{5}{9} - \frac{1}{3}$
3. $\frac{5}{6} - \frac{2}{3}$
4. $\frac{7}{8} - \frac{3}{4}$
5. $\frac{7}{6} - \frac{5}{8}$
6. $\frac{6}{5} - \frac{5}{6}$
7. $1 - \frac{3}{8}$
8. $\frac{3}{2} - 1$
9. $\frac{5}{6} - \frac{3}{4}$
10. $\frac{7}{9} - \frac{1}{6}$
11. $\frac{3}{4} - \frac{2}{3}$
12. $\frac{3}{5} - \frac{1}{3}$
13. $\frac{3}{2} - \frac{5}{8}$
14. $\frac{7}{4} - 1$
15. $\frac{7}{6} - \frac{5}{9}$
16. $\frac{5}{3} - \frac{5}{4}$

Estimating with percents

Karen Daley won the election with about $\frac{2}{3}$ of the vote. The actual percentage of the vote was 65.2%.

$$65.2\% \approx \frac{2}{3}$$

We often use fraction approximations for percents.

EXERCISES

Choose the best approximation.

1. 34.8% of the voters voted no.
- **a.** About $\frac{1}{3}$ of the voters voted no.
- **b.** About $\frac{1}{4}$ of the voters voted no.
- **c.** About 1 out of 5 voters voted no.

2. Carol scored 47% of her team's points.
- **a.** Carol scored about $\frac{4}{7}$ of the points.
- **b.** Carol scored about $\frac{2}{5}$ of the points.
- **c.** Carol scored about 1 out of every 2 points.

3. 98% of all homes have a TV set.
- **a.** Almost every home has a TV set.
- **b.** About $\frac{4}{5}$ of the homes have a TV set.
- **c.** Some homes have 2 TV sets.

4. 12% of the students got an A.
- **a.** About $\frac{1}{6}$ of the students got an A.
- **b.** About 12 students got an A.
- **c.** About 1 out of every 10 students got an A.

5. Sale: 55% off
- **a.** Sale: prices reduced by almost half
- **b.** Sale: less than half price
- **c.** Sale: $55 off

6. 30% chance of rain
- **a.** There's 1 chance in 3 it will rain.
- **b.** Chances are 1 in 4 that it will rain.
- **c.** It is almost certain to rain.

Estimate.

Example.

19% of 80 = ?

19% is near 20%

20% of 80 = 16

So, 19% of 80 is a little less than 16.

7. 26% of 60 = ?
8. 48% of 168 = ?
9. 31% of $1.47 = ?
10. 12% of 286.52 = ?
11. 34% of $6.93 = ?
12. 53% of $268.40 = ?
13. 73% of 80 = ?
14. 97% of 53 = ?
15. 29% of $700 = ?
16. 65% of $900 = ?

Example.

?% of 26 is 12

12 is 50% of 24

So, 12 is a little less than 50% of 26.

17. ?% of 42 = 10
18. ?% of 59 = 6
19. ?% of 13 = 8
20. ?% of 31 = 3
21. ?% of 31 = 4
22. ?% of 49 = 11
23. ?% of $250 = $47
24. ?% of 61 = 19
25. ?% of $10.03 = $.99
26. ?% of 99 = 23

Estimate.

27. Altogether, 4572 voters voted in the special election and 65.2% voted yes. How many voted yes?
28. Only 28% of the eligible voters voted in the special election. How many eligible voters were there?
29. The mayor of a small city was elected with 57.2% of the votes. The total number of votes was 29,156. How many voted for the winner?
30. One of the candidates for office received only 9.6% of the 57,436 votes cast. How many votes did he receive?

Challenge!

Solve.

31. Leon sold 50% of his baseball cards. Then he gave away 60% of those that were left. He then had 8 baseball cards. How many baseball cards did Leon have to begin with?

Problem Solving

Percent of increase or decrease

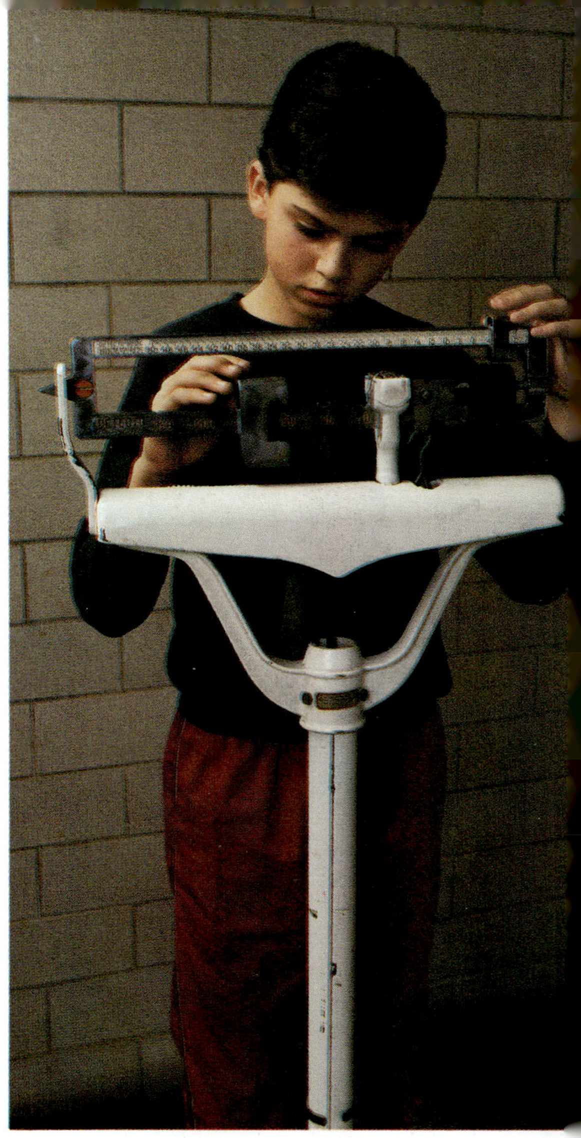

A year ago Mike weighed 102 pounds. His school sets 105 pounds as the minimum weight for joining the school wrestling team. Since then he has gained 8 pounds. To find the **percent of increase,** express the ratio given below as a percent:

$$\frac{\textit{amount of increase} \longrightarrow}{\textit{weight before increase} \longrightarrow} \quad \frac{8}{102} = \frac{4}{51} \qquad 51\overline{)4.0000} \ \ 0.0784$$

Rounded to the nearest tenth percent, his increase in weight was 7.8%.

Tom wants to maintain his weight at 127 pounds so that he can remain in his wrestling class. His doctor started him on a 2600-calorie diet, then reduced it to a 2250-calorie diet. Here is how to find the **percent of decrease:**

Subtract to find the amount of decrease.

$$\frac{\textit{amount of decrease} \longrightarrow}{\textit{calories before decrease} \longrightarrow} \quad \frac{350}{2600} = \frac{7}{52} \qquad 52\overline{)7.0000} \ \ 0.1346$$

Rounded to the nearest tenth percent, his decrease in calories was 13.5%.

Solve. Round each answer to the nearest percent.

1. Roberto's height a year ago was 58 inches. This year he grew 2 inches. What was the percent of increase?

2. One type of bicycle regularly sells for $156. Now it is on sale for $132. What is the percent of decrease?

3. Jeremiah earned $18 his first week on a job. On the second week he earned $21. What was the percent of increase in weekly earnings?

4. Geraldine studied 90 minutes one night and 20 minutes less the next night. By what percent was her study time decreased?

Solve.

City	Population in Census Year		
	1960	1970	1980
New York City	7,781,984	7,894,862	7,071,030
Chicago	3,550,404	3,369,359	3,005,072
Los Angeles	2,479,015	2,809,596	2,966,763
Philadelphia	2,002,512	1,950,098	1,688,210
Houston	938,219	1,232,802	1,594,086
Detroit	1,670,144	1,513,601	1,203,339
Dallas	679,684	844,401	904,078
Baltimore	939,024	905,759	786,775
Washington, D.C.	763,956	756,510	637,651

5. a. Which cities had a decrease in population from 1960 to 1970? From 1970 to 1980?

   b. Which cities had an increase in population from 1960 to 1970? From 1970 to 1980?

6. a. How much did the population of Detroit decrease from 1970 to 1980?

   b. What was the percent of decrease? Round answer to the nearest tenth percent.

7. a. How much did the population of Dallas increase from 1970 to 1980?

   b. What was the percent of increase? Round to the nearest tenth percent.

8. By what percent did the population of Houston increase from 1960 to 1970?

9. The price of a car was reduced from $7238 to $6743. What was the percent of reduction?

KEEPING SKILLS SHARP

Give each quotient in simplest form.

1. $8 \div \frac{1}{3}$
2. $\frac{2}{3} \div 2$
3. $\frac{5}{9} \div \frac{2}{3}$
4. $\frac{7}{4} \div \frac{7}{9}$
5. $\frac{0}{4} \div \frac{5}{4}$
6. $\frac{3}{2} \div \frac{2}{3}$
7. $\frac{9}{4} \div \frac{9}{4}$
8. $\frac{15}{16} \div \frac{5}{4}$
9. $\frac{7}{6} \div \frac{2}{3}$
10. $\frac{2}{3} \div \frac{7}{6}$
11. $5 \div \frac{3}{4}$
12. $\frac{7}{8} \div \frac{3}{4}$
13. $\frac{3}{2} \div \frac{9}{8}$
14. $\frac{15}{4} \div \frac{5}{6}$
15. $\frac{5}{8} \div \frac{10}{3}$
16. $\frac{3}{4} \div 4$

Practice exercises

Copy and complete.

	1.	2.	3.	4.	5.	6.	7.	8.
Fraction	$\frac{1}{4}$	?	$\frac{5}{4}$	?	$\frac{1}{3}$	?	$\frac{5}{6}$	?
Decimal	?	?	?	0.6	?	0.875	?	?
Percent	?	50%	?	?	?	?	?	62.5%

Solve.

9. 40% of 25 = n

10. 150% of 18 = n

11. 75% of 96 = n

12. 8.5% of 50 = n

13. 22.6% of 65 = n

14. 7.35% of 84 = n

15. $33\frac{1}{3}$% of 141 = n

16. $87\frac{1}{2}$% of 128 = n

17. $66\frac{2}{3}$% of 249 = n

18. 25% of n = 16

19. 60% of n = 48

20. 125% of n = 60

21. 6.2% of n = 12.4

22. 12.5% of n = 6.25

23. 58.4% of n = 93.44

24. $37\frac{1}{2}$% of n = 36.3

25. $8\frac{1}{3}$% of n = 304

26. 22.5% of n = 36.9

27. n% of 50 = 25

28. n% of 420 = 42

29. n% of 60 = 45

Challenge!

Find the missing input or output.

In order to find percents with a computer, you must input decimal numbers. To find 25% of 64, you would command the computer to PRINT .25 * 64.

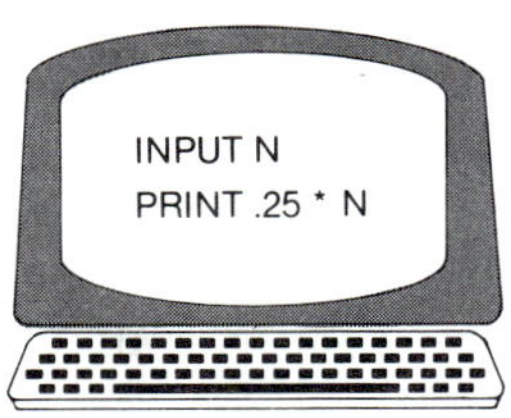

	Input	Output
	64	16
30.	360	?
31.	?	400
32.	?	382

	Input	Output
	24	18
33.	7.2	?
34.	?	36
35.	?	9.6

Problem Solving / Doing simpler problems

1. Study and understand.
2. Plan and do.
3. Answer and check.

A drive-in restaurant serves hamburgers plain or with any combination of toppings. How many kinds of toppings does the restaurant have in order to serve 256 varieties of hamburgers?

Let P represent the plain hamburger and A, B, C, etc., represent the different toppings.

Number of toppings	*Varieties of hamburgers*	*Number of varieties*	*Pattern*
0	P	1	2^0
1	P, A	2	2^1
2	P, A, B, AB	4	2^2
3	P, A, B, C, AB, AC, BC, ABC	8	2^3

Problem: What power of two is 256? $2^4 = 16$, $2^5 = 32$, $2^6 = 64$, $2^7 = 128$, $2^8 = 256$

Answer: The restaurant has 8 different hamburger toppings.

Solve by first solving some simpler but similar problems.

1. A pizza parlor serves pizza plain or with any combination of toppings. How many toppings are needed to offer 512 varieties of pizza?

2. A group of people met, and each person shook hands with everyone else. There were 21 handshakes in all. How many people were there?

Problem Solving Using percents

Work in groups. Get squared paper and a calculator.

1. Mr. Arg has planted pear trees on 30% of his land. Ms. Blick also has pear trees on 30% of her land.
   a. Make up sizes and shapes of land to show how it is possible that
      (1) the pear trees on Mr. Arg's land cover more area than the pear trees on Ms. Blick's land.
      (2) the pear trees on Ms. Blick's land cover more area than the pear trees on Mr. Arg's land.
      (3) pear trees cover the same area on the two pieces of land.
   b. For the condition in **a (3)** to be true, must the two pieces of land be the same shape? Show why or why not.

2. Mr. Cap and Ms. Dodd are two other pear growers. Mr. Cap's land is a rectangle 108 m by 100 m, and Ms. Dodd's land is a rectangle 180 m by 100 m. The pear orchard on each covers 30% of the land.
   a. Show how this could be possible.
   b. Mr. Cap and Ms. Dodd each allot an additional 540 m^2 of land to pear trees without increasing the size of their land.
      (1) By what percent does each person's orchard area increase?
      (2) By what percent does each person's non-orchard area decrease?

3. Discuss your answers. Are they reasonable? Explain why they are or are not what you expected.

4. A tree in Ms. Blick's orchard grew 21% in one year and 18% in the next year. What percent of its original height did it grow in these two years?

5. a. Suppose Mr. Cap sold 95% of his pears, but Ms. Dodd sold only 80% of her pears. Is it possible that
   (1) Ms. Dodd sold more pears than Mr. Cap?
   (2) they both sold the same number of pears?
   b. Defend your answers with examples.

6. Mr. Elk and Ms. Fawn are also pear growers.
   a. Suppose Mr. Elk made $2220 profit for his first year as a pear grower. If his profits increased by 10% each year, in how many years would his profits double?
   b. Suppose Ms. Fawn made $3450 profit for her first year and also wanted her profits to increase by 10% each year. Would it take more, less, or the same time to double her profits as it would take Mr. Elk to double his?

7. a. If Mr. Elk's profits decreased by 10% each year, in how many years would they be half as much?
   b. Did you get the answer you expected? Why or why not?

8. a. Suppose you bought some land, which increased the total amount of land you owned by 20% one year. The next year you sold some land, which decreased the amount of land you owned by 20%. What percent of the original amount of land you owned would you have left?
   b. Compare your answers with those of another group. Does the answer depend on the size of the original land? Why or why not?

9. Discuss the strategies used for solving the problems in this lesson. For which parts of the lesson was each of these methods useful: the calculator, estimation, mental math, paper-and-pencil computation? Why?

Chapter Checkup

Write as a fraction or mixed number in simplest form. [pages 234–235, 245, 252]

1. 50% **2.** 120% **3.** $33\frac{1}{3}\%$ **4.** $55\frac{5}{9}\%$ **5.** 1.2% **6.** 0.8%

Write as a percent. [pages 236–237, 245, 252]

7. $\frac{3}{10}$ **8.** $\frac{1}{5}$ **9.** $\frac{2}{3}$ **10.** $\frac{7}{4}$ **11.** $\frac{5}{8}$ **12.** $\frac{15}{16}$

Copy and complete. [pages 238, 252]

	13.	14.	15.	16.	17.	18.
Decimal	0.59	0.3	2.4	?	?	?
Percent	?	?	?	$37\frac{1}{2}\%$	183%	0.9%

Solve. [pages 240–241, 244–245, 248–249, 252]

19. 16% of 54 = n **20.** $66\frac{2}{3}\%$ of 171 = n **21.** 105% of 1240 = n

22. $6\frac{1}{4}\%$ of 256 = n **23.** 87.5% of 584 = n **24.** 0.6% of 1240 = n

Solve. [pages 242–243, 245, 248–249, 252]

25. 30% of n = 18 **26.** 75% of n = 54 **27.** 110% of n = 22

28. $33\frac{1}{3}\%$ of n = 17 **29.** $16\frac{2}{3}\%$ of n = 16 **30.** 9.4% of n = 6.11

Solve. [pages 246–249, 252]

31. n% of 60 = 6 **32.** n% of 120 = 30 **33.** n% of 150 = 22.5

34. n% of 57 = 19 **35.** n% of 40 = 6 **36.** n% of 48 = 54

Solve. [pages 250–251, 253–255]

37. The sales tax is 5.5%. How much is the sales tax on the reduced price?

38. The regular price is 120% of the price the dealer paid for the bike. What was the dealer's cost?

39. By what percent was the price decreased?

Chapter Project

CIRCLE GRAPH

Number of people in each vehicle	Tally	Percent	Number of people
1	𝍸 𝍸 ||	$\frac{12}{48} = \frac{1}{4} = 25\%$	12
2	𝍸 𝍸 𝍸 |||		
3	𝍸 |||		
4	𝍸 |		
5 or more	||||		24
Total number of vehicles:	48	Total number of people:	

Robert counted the number of people that rode in each vehicle that passed by his school. He used the results of his survey to construct a circle graph.

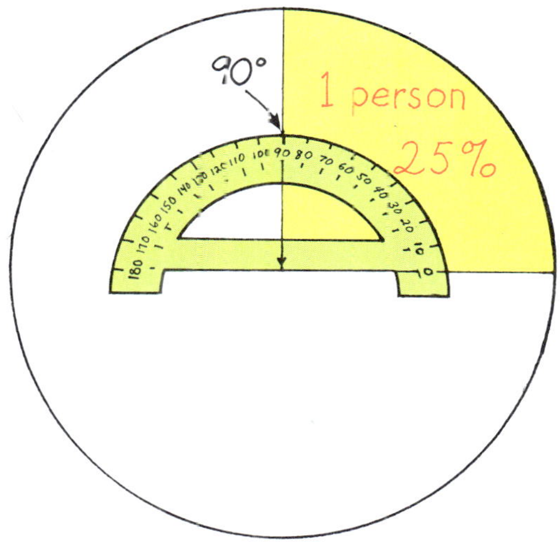

The survey shows that $\frac{1}{4}$, or 25%, of the vehicles had only 1 person. To find the angle for that part of the circle graph, you would need to find 25% of 360° (the total number of degrees in a circle).

$$
\begin{aligned}
25\% \times 360° &= 0.25 \times 360° \\
&= 90°
\end{aligned}
$$

1. Copy and complete the survey.
2. Use your percent column to complete the circle graph.
3. What is the total number of people in the survey?
4. If these people had ridden 3 to a vehicle, how many vehicles would have been needed?
5. By what percent would the traffic have been reduced?
6. Repeat exercises 4 and 5 assuming that 4 people rode in a vehicle.
7. Take your own traffic survey. Repeat exercises 1–6 using the results.

Chapter Review

$20\% = \frac{20}{100} = \frac{1}{5}$

$33\frac{1}{3}\% = \frac{33\frac{1}{3}}{100} = \frac{\frac{100}{3}}{\frac{100}{1}} = \frac{1}{3}$

Write as a fraction in lowest terms.

1. 15%
2. 85%
3. 75%
4. 0.5%
5. 17.5%
6. $62\frac{1}{2}\%$

$\frac{3}{5} = \frac{60}{100} = 60\%$

$\frac{5}{8} = \frac{n}{100}$

$8n = 500$

$n = 62\frac{1}{2}$

So, $\frac{5}{8} = 62\frac{1}{2}\%$

Write as a percent.

7. $\frac{1}{2}$
8. $\frac{3}{4}$
9. $\frac{8}{5}$
10. $\frac{1}{3}$
11. $\frac{2}{3}$
12. $\frac{5}{16}$

Decimal → Percent

$0.06\frac{1}{2} = 6\frac{1}{2}\%$

Percent → Decimal

$35.6\% = 0.356$

Copy and complete.

	13.	14.	15.	16.	17.
Decimal	0.23	?	0.055	?	$0.36\frac{1}{2}$
Percent	?	6%	?	$8\frac{1}{3}\%$	?

$40\% \text{ of } 32 = 0.40 \times 32 = 12.8$

$30\% \text{ of } n = 18$

$\frac{0.30n}{0.30} = \frac{18}{0.30}$

$0.3\overline{)18.0}$ = 60.

$n = 60$

Solve.

18. $25\% \text{ of } 48 = n$
19. $42\% \text{ of } 67 = n$
20. $9.4\% \text{ of } 35 = n$
21. $8.5\% \text{ of } 24 = n$
22. $33\frac{1}{3}\% \text{ of } 112 = n$
23. $8\frac{1}{2}\% \text{ of } 72 = n$
24. $50\% \text{ of } n = 18$
25. $36\% \text{ of } n = 27$
26. $9.6\% \text{ of } n = 9.6$
27. $6.25\% \text{ of } n = 25$
28. $37\frac{1}{2}\% \text{ of } n = 27$
29. $33\frac{1}{3}\% \text{ of } n = 18.2$

Chapter Challenge

NUMBER PUZZLE

Copy the grid. Then work each exercise and write the answer in the correct box in your grid.

1. $66\frac{2}{3}\%$ of what number is 24?
2. 23% of 100 is what number?
3. $\frac{2}{9} = \underline{?}\%$ (Round your answer to the nearest percent.)
4. What is 36 decreased by $8\frac{1}{3}\%$?
5. $62\frac{1}{2}\%$ of 40 $= \underline{?}$
6. What is 15% of 200?
7. What is 5 less that 9% of 400?
8. What is 4 times 14% of 50?
9. What is 30 decreased by $3\frac{1}{3}\%$?
10. 52% of 50 is what number?
11. 150% of 18 $= \underline{?}$
12. 2.08 is $6\frac{1}{2}\%$ of $\underline{?}$
13. What is 20 increased by 20%?
14. 8% of 325 is 9 less than what number?
15. $0.336 = \underline{?}\%$ (Round your answer to the nearest percent.)
16. 0.5% of 4200 is what number?

To check your work, add the 4 numbers in each row, column, and diagonal. If all the sums are the same, your answers are correct.

Cumulative Checkup

STANDARDIZED FORMAT

Choose the correct letter.

1. 58.0695 rounded to the nearest thousandth is

a. 58.071
b. 58.069
c. 58.070
d. none of these

2. Give the quotient to the nearest hundredth.

$5.7\overline{)11.59}$

a. 2.33
b. 2.03
c. 23.3
d. none of these

3. Solve.

$4y - 9 = 59$

a. $12\frac{1}{2}$
b. $14\frac{3}{4}$
c. 17
d. none of these

4. The prime factorization of 108 is

a. $2^3 \times 3^2$
b. $2^3 \times 3^3$
c. $2^2 \times 3^2$
d. none of these

5. Give the difference.

$15\frac{1}{2} - 3\frac{5}{6}$

a. $12\frac{1}{3}$
b. $12\frac{2}{3}$
c. $11\frac{2}{3}$
d. none of these

6. Give the quotient.

$8\frac{1}{2} \div 3\frac{1}{4}$

a. $27\frac{5}{8}$
b. $\frac{13}{34}$
c. $2\frac{8}{13}$
d. none of these

7. Find the surface area. Use 3.14 as an approximation for π.

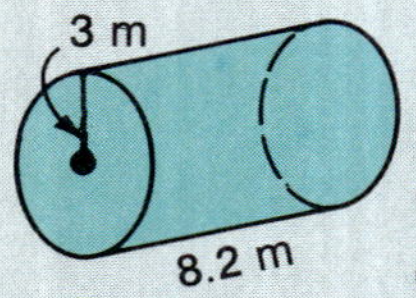

a. 231.73 m^2
b. 154.49 m^2
c. 77.25 m^2
d. none of these

8. Solve.

$\frac{6}{7} = \frac{8}{y}$

a. $9\frac{1}{3}$
b. $5\frac{1}{4}$
c. $6\frac{6}{7}$
d. none of these

9. Solve.

$16\frac{2}{3}\%$ of $x = 48$

a. 288
b. 80
c. 8
d. none of these

10. Solve.

$x\%$ of $57 = 38$

a. $33\frac{1}{3}\%$
b. 150%
c. $66\frac{2}{3}\%$
d. none of these

11. Bagels are 80¢ a dozen. What is the cost of 9 bagels?

a. 60¢ **b.** 72¢
c. \$1.20 **d.** none of these

12. Arthur was 60 in. tall. Now he is 2% taller. How tall is he now?

a. 62 in. **b.** 72 in.
c. 61.2 in. **d.** none of these

9

Consumer Mathematics

Corey has 75 customers. Each pays $3.85 per week for newspaper delivery. If Corey earns 15% of the money she collects, how much does she earn per week?

READY OR NOT?

Change to a percent.

1. $\frac{1}{4}$
2. $\frac{1}{3}$
3. $\frac{52.8}{66}$
4. $\frac{34.6}{51.9}$

Solve.

5. 20% of 76 = x
6. 6% of 38 = x
7. 9.5% of 50 = x
8. $33\frac{1}{3}$% of 57 = x

Solve.

9. 25% of x = 8
10. 8% of x = 36
11. 4.5% of x = 27
12. $37\frac{1}{2}$% of x = 51

Earning money

People are paid for their work in different ways.

A teacher is generally paid an annual (yearly) salary.

A hair stylist is often paid a fee for each haircut or style.

A real estate agent is usually paid a commission (a percent) of sales.

A factory worker is generally paid an hourly wage.

EXERCISES

1. A cookware company pays its salespeople a commission of 23% of all sales. How much would a person earn for selling \$2340 worth of cookware?

2. A welder is paid an hourly wage of \$12.35. How much does the welder earn for working five 8-hour days?

3. A clothing store pays its salespeople a combination salary and commission. It pays each salesperson \$105 a week plus 5% of all sales. How much would a person earn whose weekly sales were \$2375?

4. A jewelry store pays its salespeople a 10% commission on the first \$1500 of sales each week and a 12% commission on all sales over \$1500. How much would a salesperson make for selling \$2240 worth of jewelry?

5. A hairstyling shop pays its employees 72% of all the money they take in. One employee did 6 stylings and charged \$20.75 for each. How much did the employee earn that day?

6. A paint company pays its salespeople \$13,750 a year plus a bonus of $8\frac{1}{2}$% of all sales over \$290,000. How much would a person whose yearly sales were \$378,000 earn?

7. A microchip factory pays its employees an hourly wage for the first 40 hours. Each hour that an employee works more than the 40 hours is called "overtime." For overtime work an employee is paid "time and half" ($1\frac{1}{2}$ times the regular hourly wage). The factory also pays "double time" for work done on Sundays or holidays. Compute the weekly earnings for the hours shown on the time card.

8. List as many occupations as you can that are paid by fees, by an annual salary, by commission, or by an hourly rate.

Name: *Roberta Davis*

Employee Number: 348

Hourly Wage: \$5.60

	Hours Worked
Monday	8
Tuesday	9
Wednesday	8
Thursday	$9\frac{1}{2}$
Friday	11
Saturday	8
Sunday	0

Payroll and deductions

Toyland Industries pays Carol an hourly wage of $7.78 and time and a half for each hour over 40 hours per week. Each week she receives with her paycheck a *statement of withholdings* and *deductions*. This statement lists the amounts that her employer is required by law to deduct from her weekly earnings. This money is sent to the federal and state tax-collecting agencies. Toyland Industries also has optional deductions for employees who sign up for life insurance, savings bonds, and other group programs. One of Carol's weekly withholding statements is shown here:

Toyland Industries
Employee Record

Earnings

Employee Number	Hourly Wage	Total Hours Worked	Regular Earnings	Overtime Earnings	Total Earnings (Gross Pay)
3054	$7.78	48.5	$311.20	$99.20	$410.40

Withholdings and deductions

Social Security (F.I.C.A.)	Federal Income Tax	State Income Tax	Life Insurance	Net Pay
$27.57	$68.53	$18.36	$4.59	$291.35

EXERCISES

1. How much was Carol's net, or "take home," pay?
2. How much did she earn for her first 40 hours of work?
3. What was her overtime pay?
4. Social security withholding is for such things as retirement income and disability income if one should become disabled. How much was withheld for social security?
5. What percent of her gross pay was withheld for social security?
6. How much did the employer withhold for federal income tax? For state income tax?
7. The total income tax withheld amounted to what percent of the gross income?
8. What percent of her gross pay was her net pay?
9. A salesman for Toyland Industries is paid \$360 a month and 12% of his monthly sales over \$2300. What were his gross earnings for a month in which his sales totaled \$11,680?
10. For the salesman in exercise 9, Toyland withholds 6.95% for social security, 18% for federal income tax, 4% for state income tax, and \$6.35 for insurance. What is his net pay for the month?

KEEPING SKILLS SHARP

Solve.

1. 50% of $y = 17$
2. 30% of $153 = y$
3. 160% of $96 = y$
4. $33\frac{1}{3}\%$ of $y = 26$
5. $87\frac{1}{2}\%$ of $y = 182$
6. 39% of $68 = y$
7. 15% of $450 = y$
8. $66\frac{2}{3}\%$ of $y = 54$
9. $116\frac{2}{3}\%$ of $84 = y$
10. 0.8% of $y = 16$
11. 67% of $72 = y$
12. 9% of $y = 45$
13. 15% of $y = 30$
14. $37\frac{1}{2}\%$ of $y = 27$
15. 0.18% of $120 = y$
16. 5.2% of $48 = y$
17. $62\frac{1}{2}\%$ of $y = 52$
18. 110% of $150 = y$
19. $133\frac{1}{3}\%$ of $y = 76$
20. 175% of $84 = y$

Income tax

In the preceding lesson you learned that employers withhold federal income tax from an employee's pay. By April 15 of the following year, the employee has to file with the Internal Revenue Service an income tax form similar to the one shown below.

Use IRS label. Otherwise, please print or type.	Your first name and initial (if joint return, also give spouse's name and initial): John T. & Mary L.	Last name: Brown	Your social security number: 516 04 1492
	Present home address (Number and street, including apartment number, or rural route): 885 Scott Street		Spouse's social security no.: 575 10 1776
	City, town or post office, State and ZIP code: Hometown, Maryland 01234	Your occupation ▶ Clerk Spouse's occupation ▶ Bus Driver	

Presidential Election Campaign ▶ Do you want $1 to go to this fund? ✓ Yes ☐ No
If joint return, does your spouse want $1 to go to this fund? ✓ Yes ☐ No
Note: *Checking "Yes" will not increase your tax or reduce your refund.*

For Privacy Act and Paperwork Reduction Act Notice, see page 23 of Instructions

Filing Status — Check Only One Box.
1 ☐ Single
2 ✓ Married filing joint return (even if only one had income)
3 ☐ Married filing separate return. Enter spouse's social security no. above and full name here ▶
4 ☐ Head of household (with qualifying person). (See page 8 of Instructions.) If he or she is your unmarried child, enter child's name ▶

Exemptions — Always check the box labeled Yourself. Check other boxes if they apply.
5a ✓ Yourself ☐ 65 or over ☐ Blind
b ✓ Spouse ☐ 65 or over ☐ Blind — Enter number of boxes checked on 5a and b ▶ 2
c First names of your dependent children who lived with you ▶ James — Enter number of children listed on 5c ▶ 1

d Other dependents: (1) Name	(2) Relationship	(3) Number of months lived in your home.	(4) Did dependent have income of $1,000 or more?	(5) Did you provide more than one-half of dependent's support?

Enter number of other dependents ▶
6 Total number of exemptions claimed — Add numbers entered in boxes above ▶ 3

Please Attach Copy B of Forms W-2 Here

7	Wages, salaries, tips, etc. *(Attach Forms W-2. See page 10 of Instructions)*			7	19,558 00
8a	Interest income *(Complete page 2 if over $400 or you have any All-Savers interest)*	8a	345 00		
b	Dividends *(Complete page 2 if over $400)*	8b	135 00		
c	Total *(add lines 8a and 8b)*	8c	480 00		
d	Exclusion *(See page 11 of Instructions)*	8d	400 00		
e	**Subtract line 8d from line 8c** *(but not less than zero)*			8e	80 00
9a	Unemployment compensation (insurance). Total received from Form(s) 1099–UC				
b	Taxable amount, if any, from worksheet on page 12 of Instructions			9b	—
10	Adjusted gross income *(add lines 7, 8e, and 9b). If under $10,000, see page 13 of Instructions on "Earned Income Credit"*			10	19,638 00
11	Multiply $1,000 by the total number of exemptions claimed on line 6			11	3000 00
12	Taxable income *(subtract line 11 from line 10)*			12	16,638 00
13a	**Credit for contributions to candidates for public office.** *(See page 13 of Instructions)*	13a			
	IF YOU WANT IRS TO FIGURE YOUR TAX, PLEASE STOP HERE AND SIGN BELOW.				
b	**Total Federal income tax withheld** *(If line 7 is more than $29,700, see page 13 of Instructions)*	13b	2306 00		

EXERCISES

Refer to the completed form on page 266 to answer the following questions.

1. What is the filing status?
2. An exemption is someone who is supported by the taxpayer. How many exemptions are there? Who are they?
3. What does the entry on line 7 represent?
4. What was the Browns' taxable income from interest and dividends (line 8e)?
5. What was their adjusted gross income?
6. What was their taxable income?
7. How much income tax was withheld from their pay?

This table is like the one the Browns can use to find the amount of their income tax.

If line 12 (taxable income) is—		And you are—			
At least	But less than	Single	Married filing jointly	Married filing separately	Head of a household
		Your tax is—			
16,500	16,550	3,024	2,361	3,637	2,837
16,550	16,600	3,039	2,373	3,655	2,849
16,600	16,650	3,054	2,385	3,674	2,862
16,650	16,700	3,069	2,397	3,692	2,875
16,700	16,750	3,083	2,409	3,710	2,888
16,750	16,800	3,098	2,420	3,729	2,901
16,800	16,850	3,113	2,432	3,747	2,914
16,850	16,900	3,128	2,444	3,765	2,926
16,900	16,950	3,143	2,456	3,783	2,939
16,950	17,000	3,158	2,468	3,802	2,952
17,000	17,050	3,172	2,480	3,820	2,965
17,050	17,100	3,187	2,49[illegible]	3,838	2,978
17,100	17,150	3,202	2,503	3,856	2,991
17,150	17,200	3,217	2,515	3,875	3,003
17,200	17,250	3,232	2,527	3,893	3,016
17,250	17,300	3,246	2,539	3,911	3,029
17,300	17,350	3,261	2,55[illegible]	3,930	3,042
17,350	17,400	3,276	2,563	3,948	3,055
17,400	17,450	3,291	2,574	3,966	3,068
17,450	17,500	3,306	2,586	3,984	3,081

8. Use the table to find the Browns' income tax.
9. Was the amount of tax withheld from the Browns' pay greater or less than the income tax they should pay? How much?

Challenge!

10. Mr. and Mrs. Delano have 3 children and file a joint return. Their adjusted gross income (line 10) is $22,178. Find their taxable income (line 12) and then use the tax table to find the amount of their income tax.

Spending wisely

Mike is looking at some blue jeans. Before he decides to buy the jeans, he should ask himself such questions as

1. Is the price within my budget?
2. Do I really need or want another pair of jeans at this time?
3. Are they well made?
4. Can I buy the jeans at a better price?

Can you think of other questions that Mike should ask himself?

Mike decided to buy the jeans.
The regular price was \$22.60.
They were on sale for 25% off.
Here are two ways to find the sale price:

$$\begin{array}{rl} \$22.60 & \leftarrow \textit{regular price} \\ \times 0.25 & \\ \hline 11300 & \\ 4520 & \\ \hline \$5.6500 & \leftarrow \textit{discount} \end{array}$$

$$\begin{array}{rl} \$22.60 & \leftarrow \textit{regular price} \\ -5.65 & \leftarrow \textit{discount} \\ \hline \$16.95 & \leftarrow \textit{sale price} \end{array}$$

SHORTCUT

If the discount is 25%, then the sale price is 75% of the regular price.

$$\begin{array}{rl} \$22.60 & \leftarrow \textit{regular price} \\ \times 0.75 & \\ \hline 11300 & \\ 15820 & \\ \hline \$16.9500 & \leftarrow \textit{sale price} \end{array}$$

EXERCISES

Compute the sale price.

1. Blue jeans
   Regular price \$26.40
   25% off

2. Blue jeans
   Regular price \$32.80
   20% off

3. Casual cords
   Regular price \$22.40
   15% off

Round each sale price up to the next cent.

4. T-shirts
   \$7.98 each
   22% off

5. Flannel shirts
   \$12.79 each
   24% off

6. T-shirts
   \$11.89 each
   19% off

The state Mike lives in has a 5% sales tax. Here is how to find the total cost of the blue jeans he bought:

\$16.95	← *price*	**\$16.95**	← *price*
×0.05		**+.85**	← *sales tax*
\$.8475	← *sales tax*	**\$17.80**	← *total cost*

Compute the total cost, including a 5% sales tax. Round each answer to the nearest cent.

7. 1 pair of blue jeans for \$17.80

8. 2 pairs of blue jeans for \$21.80 each

9. 1 shirt for \$8.65 and 3 pairs of socks for \$1.60 a pair

10. 2 pairs of blue jeans for \$21.50 each and 3 shirts for \$9.28 each

11. David bought a \$24.30 pair of blue jeans that was on sale for 30% off.

12. Jan bought a \$5.60 belt that was on sale for 25% off and a \$24.30 pair of jeans that was on sale for 20% off.

Challenge!

This formula gives the total cost of buying a certain brand of blue jeans by mail.

Total cost		Number of pairs		Postage and handling
C	=	$\$18h$	+	\$2.50

Solve. Use the formula.

13. What is the total cost for 3 pairs of blue jeans?

14. Is \$40 enough money to buy 2 pairs of blue jeans?

Problem Solving / Using a pictograph

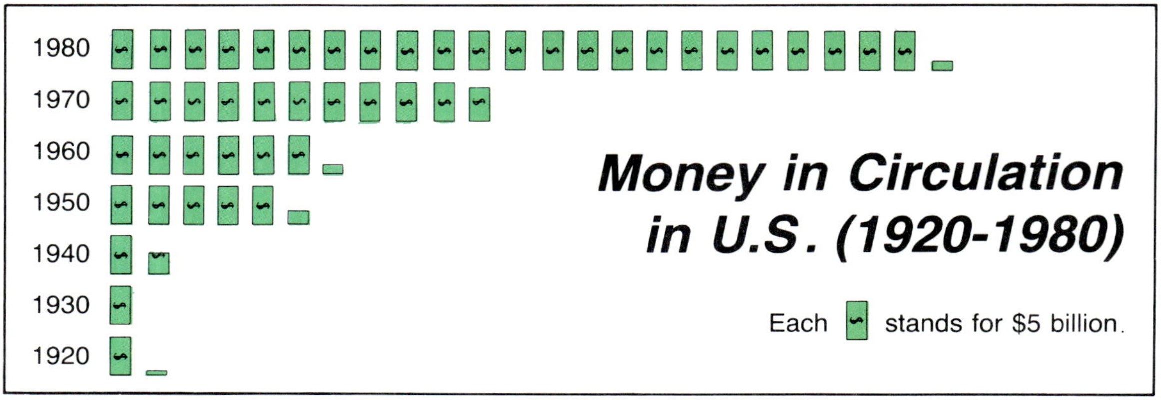

Solve.

1. About how much money was in circulation in 1940? In 1980?
2. In which year was there about $32 billion in circulation?
3. About how much more money was in circulation in 1950 than in 1920?
4. In what year was there a decrease in money circulation?
5. What was the approximate increase in money circulation from 1960 to 1970?
6. What was the approximate percent of increase from 1960 to 1980?
7. What was the approximate percent of increase from 1940 to 1950?
8. Was the approximate percent of increase from 1940 to 1950 less than or greater than the approximate percent of increase from 1970 to 1980?

Challenge!

9. The silver dollar was issued in 1795. By law, the ratio of the weight of pure gold in the gold dollar to the weight of pure silver in the silver dollar was 1 to 15. The gold dollar contained 24.75 grains of pure gold. How many grains of pure silver were in a silver dollar?

Problem Solving Choosing a strategy

1. Study and understand.
2. Plan and do.
3. Answer and check.

Doing systematic trials
Writing an equation
Making a list
Making a drawing
Making a table
Finding a pattern
Doing simpler problems

Solve.

1. A hot-air balloon was made of 24 panels (gores) that were 7 feet wide at the equator of the balloon. What was the diameter of the balloon at its equator?

2. The table shows the altitude of a balloon at various times during its ascent. If it continues to rise at the same rate, how high will it be after 10 minutes in the air?

Time in minutes	0	1	2	3	4
Altitude in feet	0	60	130	200	270

3. Two student tickets and 1 adult ticket to the takeoff of a balloon race cost \$17. One student ticket and 3 adult tickets cost \$26. What was the cost of each kind of ticket?

4. How much money did Paco take to the balloon race? He paid \$4.95 each for two T-shirts, and \$1.05 for a poster. He has \$17 now.

Borrowing money—simple interest

When you borrow money from a bank, credit union, or loan company, you pay for the use of it. The amount you pay is called **interest.** The interest paid depends upon the amount borrowed (**principal**), the length of time the money is kept (**time**), and the percent of interest charged (**rate**).

Here is how to compute the interest on an \$800 loan at 18% interest. It was repaid in 6 months.

Principal = \$800
Rate = 18% per year
Time = 6 months ($\frac{1}{2}$ *year*)

Simple Interest Formula: $I = prt$

$$= \$800 \times \frac{18}{100} \times \frac{1}{2}$$

$$= \$72$$

Important: When you use the simple interest formula, the time and the interest period of the rate must be expressed in the same unit. In the example above, the rate was a yearly rate, so the time had to be expressed in years.

EXERCISES

Compute the interest.

1. Principal = \$1000
   Rate = 13% per year
   Time = 1 year

2. Principal = \$700
   Rate = 19.5% per year
   Time = 3 years

3. Principal = \$950
   Rate = 16.5% per year
   Time = 6 months

4. Principal = \$500
   Rate = 12% per year
   Time = 3 months

5. Principal = \$700
   Rate = 12.5% per year
   Time = 15 months

6. Principal = \$450
   Rate = 21.5% per year
   Time = 1 year

Sometimes money is borrowed for a short period of time, for example, a few days.

Example. Principal = \$600

Rate = 15% per year

Time: from 1/1/85 to 1/25/85

Time = 24 days

$$I = 600 \times \frac{15}{100} \times \frac{24}{365}$$
$$= \$5.92$$

Compute the interest. Round to the nearest cent.

7. Principal: \$500
   Rate: 12% per year
   Time: from 1/1/85
   to 1/28/85

8. Principal: \$780
   Rate: 15% per year
   Time: from 3/1/85
   to 4/18/85

Problem Solving

9. Suppose that you borrowed \$500 at 12.5% per year. One month later you paid \$100 on your loan. How much of the \$100 was interest? How much principal would you still owe?

KEEPING SKILLS SHARP

Write as a percent.

1. 0.08
2. 0.065
3. 0.358
4. 2.4
5. 0.009
6. 0.0036
7. $\frac{1}{4}$
8. $\frac{3}{8}$
9. $\frac{2}{3}$
10. $\frac{7}{8}$
11. $\frac{3}{2}$
12. $\frac{9}{4}$
13. $1\frac{1}{4}$
14. $1\frac{3}{5}$
15. $2\frac{1}{2}$
16. $2\frac{3}{4}$
17. $3\frac{3}{8}$
18. $1\frac{5}{8}$
19. $4\frac{3}{8}$
20. $5\frac{7}{8}$

Problem Solving

Installment buying

Tomas has chosen a guitar that costs $160. He has decided to buy it on an **installment plan.** He will make a **down payment** and then pay the balance in regular **monthly payments.**

Since the merchant is really lending Tomas an amount of money, he charges Tomas interest on that amount.

CASH PLAN	INSTALLMENT PLAN
$160	$40 down payment and $11.40 per month for 1 year

$$
\begin{array}{rl}
\$11.40 & \leftarrow \textit{per month} \\
\times 12 & \leftarrow \textit{months} \\
\hline
2280 & \\
1140 & \\
\hline
\$136.80 & \leftarrow \textit{total monthly payments} \\
+40.00 & \leftarrow \textit{down payment} \\
\hline
\$176.80 & \leftarrow \textit{installment plan cost}
\end{array}
$$

The installment plan costs $16.80 more than paying cash for the guitar. The $16.80 is interest. Before buying something on an installment plan, check to see if you could pay less by borrowing the money from a bank or credit union.

EXERCISES

How much greater than the cash price is the installment price?

1. Drums: $175 cash or $50 down and $11.67 per month for 12 months.
2. Bass guitar: $280 cash or $70 down and $13.77 per month for 18 months.
3. Amplifier: $265.88 cash or $60 down and $37.40 per month for 6 months.
4. Guitar: $475 cash or $100 down and $34.37 per month for 1 year.

Solve.

5. Andrew bought a $627 stereo set on the installment plan. The down payment was $150.

   a. After paying the down payment, how much did he owe? This amount is the principal of the installment loan.

   b. The first monthly payment was $43.72. Of that payment, $3.98 was interest and the rest was principal. How much was paid on the principal?

   c. After the first payment was made, how much principal was still owed? This is called the **unpaid balance.**

   d. Since the principal during the second month was less, the interest was also less. Of the next $43.72 payment, $3.64 was interest. How much was principal?

   e. After the second payment was made, how much was the unpaid balance?

6. Sarah bought a microcomputer and software package for $3400. She paid $1000 down and 30 monthly payments of $120.

   a. What was the total amount of the monthly payments?

   b. What was the principal of the installment loan?

   c. What was the amount of interest?

   ★ d. What was the simple interest rate?

Checking accounts

The basic idea of a checking account is simple. You give the bank some money, called a **deposit.** When you need some of the money, you write a check, which tells the bank how much to take from your account and to whom it is to be paid.

Ⓑ Person or business to whom the money is *to be paid*

Ⓐ The *date* the check was written

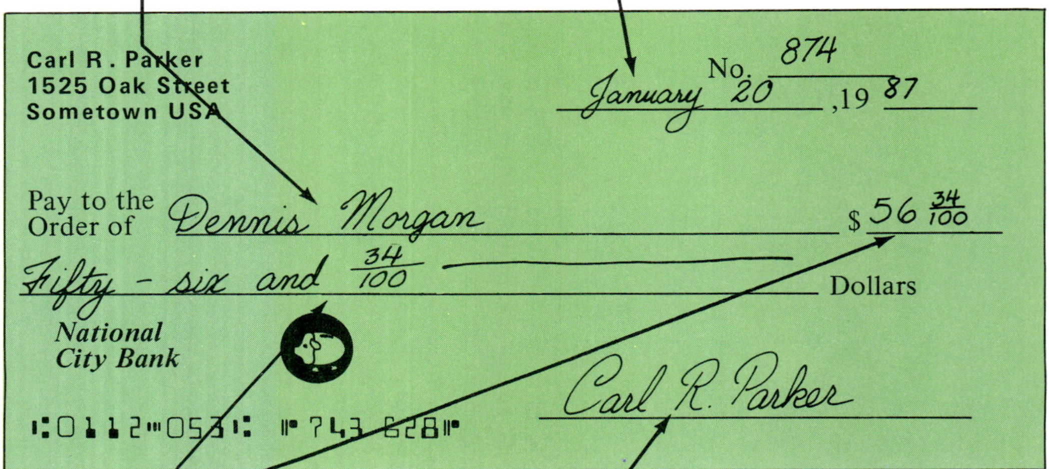

Ⓒ The *amount* of the check is written two ways to avoid error and possible alteration.

Ⓓ The *signature* assures the bank that the check is authentic. It can be checked against the signature on file.

The check also acts as a receipt, because the person who receives the money endorses the check (signs his name on the back of the check). This is proof that he received the money.

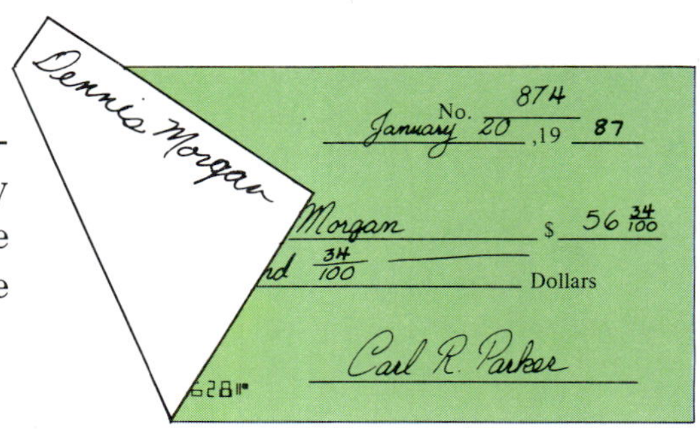

EXERCISES

Look at the check on page 276. Tell how you would fill in Ⓐ, Ⓑ, Ⓒ, and Ⓓ for each of the following transactions.

1. On January 24, 1987, Mr. Parker took his dog to the Central Animal Clinic for some shots. The shots cost $15.75.

2. On January 25, 1987, Mr. Parker bought a TV set for $695.20 from Nichols Furniture Company.

3. On January 26, 1987, Mr. Parker paid his electric bill by check. The amount was $50.25. The company was Northwest Power Company.

4. On January 26, 1987, Mr. Parker needed $40 in cash. He went to his bank and deposited $157.35. He wrote a check to "cash."

5. Mr. Parker's checkbook has a section in which he keeps a record of his account. Complete each balance.

Check No.	Date	Pay to	Amount	Deposit	Balance	
					$895	38
874	Jan. 20	Dennis Morgan	$56.34		839	04
	Jan. 20			200.00	1039	04
875	Jan. 24	Central Animal Clinic	15.75			
876	Jan. 25	Nichols Furniture Co.	695.20			
877	Jan. 26	Northwest Power	50.25			
878	Jan. 26	Cash	40.00	157.35		

6. The National City Bank charges Mr. Parker a service charge if his balance falls below $200. Did he have to pay a service charge during the time shown in the record above?

7. Why do you think that the bank does not charge for a checking account when the balance stays above $200?

Challenge!

8. Find out whether a local bank charges a service charge. If so, how do they determine how much is charged?

9. List some advantages to having a checking account. Can you think of any disadvantages?

Computers in careers

Often computer programmers write programs that involve branches and loops. Study this flow chart.

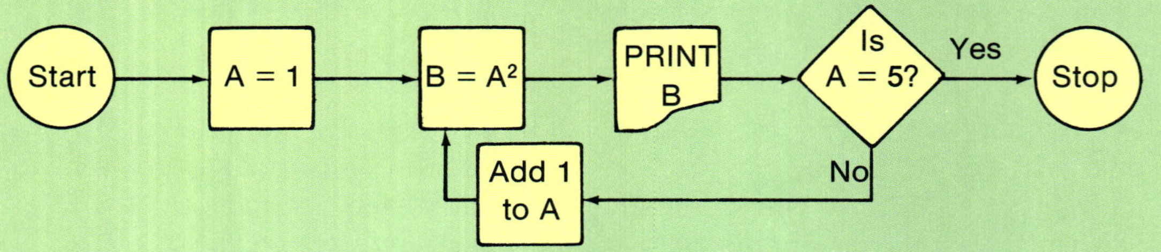

Now study this program to see how the program *branches* at a question, how *loops* are written, and how to tell the computer to skip steps.

```
1Ø  A = 1
2Ø  B = A ↑ 2
25  PRINT B
3Ø  IF A = 5 THEN 6Ø
4Ø  A = A + 1
5Ø  GO TO 2Ø
6Ø  END
```

At line 30 there is a **branch.** The statement tells the computer to skip to line 60 (END) if A = 5. If A is not 5, then the computer goes to the next line, line 40.

Line 40 is a little strange, since it is impossible for a number to *equal* a number 1 greater. This statement tells the computer to add 1 to the number stored in A and then to store that new number back in A. This means that the old value of A is lost, but that's OK since it has already served its purpose.

Line 50 tells the computer to go back to line 20 and do the computation with the new value of A. Lines 20, 25, 30, 40, and 50 form a **loop.** The computer repeats the steps over and over until A = 5. Then it skips to line 60 and stops.

Notice the new words that the computer recognizes:

```
IF . . . THEN
GO TO
```

EXERCISES

Be the computer. Follow the program.

1.
```
1Ø  A = 1
2Ø  C = 2*A
3Ø  PRINT C
4Ø  IF A = 1Ø THEN 7Ø
5Ø  A = A + 1
6Ø  GO TO 2Ø
7Ø  END
```

2.
```
1ØØ  G = 1
2ØØ  H = 1
3ØØ  I = G + H
4ØØ  PRINT I
5ØØ  IF H = 6 THEN 9ØØ
6ØØ  G = G + 1
7ØØ  H = H + 1
8ØØ  GO TO 3ØØ
9ØØ  END
```

3.
```
1ØØ  M = 1
11Ø  N = 2 ↑ M
12Ø  PRINT N
13Ø  IF N = 32 THEN 16Ø
14Ø  M = M + 2
15Ø  GO TO 11Ø
16Ø  END
```

4.
```
1ØØØ  A = 1
2ØØØ  B = A ↑ 2
3ØØØ  PRINT B
4ØØØ  IF A = 1Ø THEN 7ØØØ
5ØØØ  A = A + 1
6ØØØ  GO TO 2ØØØ
7ØØØ  END
```

5.
```
1Ø  A = 1
2Ø  C = 3*A
3Ø  PRINT C
4Ø  IF A = 2Ø THEN 7Ø
5Ø  A = A + 1
6Ø  GO TO 2Ø
7Ø  END
```

6.
```
1Ø  A = 1
2Ø  B = 3.14*(A ↑ 2)
3Ø  PRINT B
4Ø  IF A = 5 THEN 8Ø
6Ø  A = A + 1
7Ø  GO TO 2Ø
8Ø  END
```

7. Write flow charts for the programs in exercises 1–4.

Challenge!

8. Write a program for this flow chart. Give the output.

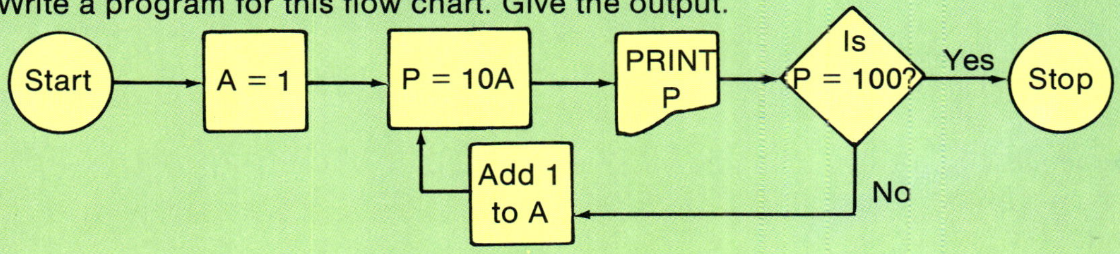

Savings accounts

No. 22357 VILLAGE SAVINGS BANK

Date	Withdrawal	Deposit	Interest	Balance
1/1/87		$500.00		$500.00
4/1/87			$7.50	$507.50
7/1/87			$7.61	$515.11
10/1/87			$7.73	$522.84
1/1/88			$7.84	$530.68

You may have a savings account at a bank (or a savings and loan company). If so, you receive interest on the money in your account. This interest is **compound interest** instead of simple interest. This means that the interest is added to the principal at regular intervals and that the principal gets larger and larger. In this example the interest is 6% per year compounded quarterly.

Jan. 1 $500 deposited

April 1 $7.50 interest paid $\left(6\% \times 500 \times \frac{1}{4}\right)$
$507.50 new principal

July 1 $7.61 interest paid $\left(6\% \times \$507.50 \times \frac{1}{4}\right)$
$515.11 new principal

Oct. 1 $7.73 interest paid $\left(6\% \times \$515.11 \times \frac{1}{4}\right)$
$522.84 new principal

Jan. 1 $7.84 interest paid $\left(6\% \times \$522.84 \times \frac{1}{4}\right)$
$530.68 new principal

How much interest was paid in 1 year?
How much interest would have been paid at 6% simple interest?

EXERCISES

1. Pretend that on January 1 you deposited $1000 in a savings account that pays 8% interest compounded quarterly.

   a. How much interest will your account earn the first quarter?
   b. What will the principal be for the second quarter?
   c. What will the interest be for the second quarter?
   d. What will the principal be for the third quarter?
   e. How much will the third-quarter interest be?
   f. What will be the fourth-quarter principal?
   g. How much will the fourth-quarter interest be?
   h. What will be the total interest earned during the year?
   i. What percent of the original $1000 deposit is the total interest earned?
   j. How much greater is the actual percent earned than the 8% rate used?

This table shows the total interest accumulated when $1000 is on deposit at 6% and at 7% interest, both compounded quarterly.

Total Amount of Interest When $1000 Is on Deposit		
Year	6% *compounded quarterly*	7% *compounded quarterly*
1	$ 61.36	$ 71.86
2	126.49	148.88
3	195.62	241.44
4	268.98	319.93
5	346.85	414.78
...	...	...
10	814.02	1001.60

2. How much interest is earned in the first year at 6%?

3. How much interest is earned in the first 2 years at 6%?

4. How much interest is earned at 6% in the 2nd year? 3rd year? 4th year? 5th year?

5. Why does the amount of interest earned increase from year to year?

6. How much more interest is earned in the first year at 7% than at 6%?

7. Suppose you invest $1000 at 7% *simple interest* for 10 years. How much interest would you earn? How much more would you earn at 7% compounded quarterly?

26637 26637

Investing money

There are many ways to invest money. You can earn interest by lending money to the federal government (buying savings bonds) or by lending money to a corporation (buying corporate bonds). You also can become a part-owner of almost any large company by buying shares of stock in the company. As an owner, you will share in the profits of the company.

EXERCISES

1. **a.** A \$100 United States Savings Bond cost \$50. At the end of 11 years the bond will be worth \$100. What is the amount of interest? What is the simple annual interest rate?
   b. If a savings bond is held for less than 5 years, a lower rate of interest is paid. At the end of 1 year a \$100 bond is worth \$53.04. What is the amount of interest? What is the simple annual interest rate?

2. Jean bought a \$1000 bond from a large company. The bond pays 9% interest each year.
   a. How much interest does she receive each year?
   b. The company pays the interest semiannually (every 6 months). How much is Jean paid each 6 months?
   c. The bond matures in 10 years. This means that the company will repay the loan 10 years after she bought the bond. How much interest will she earn in all?

1000 1000

3. Stock prices are given in dollars and eighths of dollars. (An eighth of a dollar is 12.5¢, or $.125.) Suppose one share of stock in General Electronics Company can be bought for $23\frac{3}{8}$ ($23.375). What will 100 shares cost?

4. In each exercise, you are given the price per share of a stock and a number of shares. Compute the total cost.

   a. 100 shares at $18\frac{1}{2}$ per share

   b. 1000 shares at $3\frac{3}{4}$ per share

   c. 50 shares at $47\frac{5}{8}$ per share

   d. 200 shares at $83\frac{7}{8}$ per share

5. Suppose that you own 20 shares of Cooper Oil Company stock. The company pays a dividend (part of the profits) of 22¢ per share each quarter. How much would you receive in a year?

6. Suppose that you bought 100 shares of General Electronics at $21\frac{3}{8}$ per share and sold them a year later at $29\frac{1}{8}$. How much did your stock increase in value? (Not all the gain is profit. You must pay a **stock broker** a commission both for buying and for selling stock.)

7. You bought 50 shares of International Motor Company stock at $93\frac{1}{4}$ per share. Two years later it was worth $74\frac{3}{8}$ per share. How much did the stock decrease in value?

Challenge!

8. Suppose that you bought 150 shares of International Industries for $17\frac{5}{8}$ per share plus a 1.5% commission to the stockbroker. One year later you sold the stock for $13\frac{3}{8}$ per share and paid a 1.5% commission on the sale. How much money did you lose on this investment?

KEEPING SKILLS SHARP

Write as a decimal or mixed decimal.

1. $\frac{7}{8}$
2. $\frac{3}{4}$
3. $\frac{3}{2}$
4. $\frac{1}{3}$
5. $\frac{5}{6}$
6. $\frac{5}{12}$
7. $1\frac{1}{4}$
8. $1\frac{3}{5}$
9. $2\frac{1}{2}$
10. $2\frac{3}{4}$
11. $3\frac{3}{8}$
12. $1\frac{2}{3}$
13. 18%
14. 9%
15. 0.5%
16. 0.06%
17. $66\frac{2}{3}\%$
18. $9\frac{1}{3}\%$
19. $16\frac{2}{3}\%$
20. $133\frac{1}{3}\%$

Chapter Checkup

Solve. [pages 262–263]

1. Robert earns $6.85 per hour for a 40-hour week, with time and a half for overtime. How much does he earn in a 44-hour week?

2. Sandy earns $9600 per year plus 7% of all sales above $150,000. What does she earn in a year if her total sales are $175,000?

Solve. [pages 268–269]

3. a. The radio regularly sells for $39.95. What is the sale price?
   b. Suppose that there is a 5% sales tax. Compute the total cost of the radio, including the sales tax.

Compute the simple interest for these loans. [pages 272–273]

4. $300 at 10% per year for 2 years

5. $500 at 8% per year for 3 months

Solve. [pages 270, 271, 274–277; 280–283]

6. Ms. Kendrick bought a $380 TV set for $100 down and $11.95 per week for 26 weeks. How much extra did it cost her to use the installment plan?

7. Jerry Black had a balance of $216.12 in his checking account. He then wrote checks for $85.15, $26.85, and $50.00 and deposited $89.90. What was his balance then?

8. Alice has $480 in a credit union that pays her $6\frac{1}{2}$% interest compounded semiannually. How much interest will she earn in 1 year?

9. Jon bought a $1000 corporate bond that paid $5\frac{3}{4}$% interest. How much interest was he paid every 6 months?

10. Mrs. Pitkin bought 20 shares of stock that paid a 40¢ quarterly dividend on each share. She paid $25\frac{1}{4}$ per share. How much was her total yearly dividend?

Chapter Project

INSTALLMENT BUYING

1. Look at the used-car ads in a newspaper and pick out a car you would like to buy.
2. What is the price?
3. Assume that you would have to pay 25% down. What would be your down payment?
4. How much would you still owe?
5. The table can be used to help you find how much the monthly payments would be if you paid 15% yearly interest.
   a. Pick the number of monthly payments you wish to make.
   b. Find the number in the table opposite the number of months you picked.
   c. Multiply that number by the balance you owe. This is the monthly payment.
6. How much would you actually pay for the car?
7. How much more than the cash price would you pay?

Monthly Payment to Repay $1 at an Annual Rate of 15%

Number of Monthly Payments	*Amount of Payment*
6	$0.17403
9	0.11816
12	0.09026
15	0.07352
18	0.06238
24	0.04848

Chapter Review

Find the pay for 40 hours and add the pay for the 6 hours overtime.

1. A company pays its employees time and a half for all hours over 40 that they work each week. How much should they pay a worker that earns $5.30 an hour and works 46 hours?

20% off $24.60 is the same as 80% of $24.60.

2. a. Compute the sale price of the tennis racket.

   b. Compute the total price if there is a 6% sales tax.

$I = prt$

Interest =
principal × rate × time

Find the simple interest for a loan of

3. $240 at 6% for one year.
4. $385 at 8.5% for 2 years.
5. $750 at 9% for 9 months.
   [*Hint:* 9 months is $\frac{3}{4}$ of a year.]

The total installment cost is the down payment plus the total of the monthly payments.

6. a. How much is the down payment?

   b. What is the total of the 12 monthly payments?

   c. Give the total installment cost.

Chapter Challenge

COMPOUND INTEREST

Consider an investment of $1000 at an interest rate of 6.5% compounded quarterly for a period of 2 years (8 quarters). Here is a BASIC program that will compute and print the principal at the end of each quarter. Notice that the value of A is used as a **counter.** Statements 10, 60, and 70 tell the computer to count the times that the interest is added to the principal and to stop when the number reaches 8.

```
1Ø  A = 1
2Ø  P = 1ØØØ
3Ø  I = P * Ø.Ø65/4
4Ø  P = P + I
5Ø  PRINT P
6Ø  IF A = 8 THEN 9Ø
7Ø  A = A + 1
8Ø  GO TO 3Ø
9Ø  END
```

1. Give the output for the program above.
2. Consider an investment of $1 at 7.75% compounded quarterly. Write a program that will compute and print the principal at the end of each quarter for a period of 10 years.
3. Write a program that will compute and print the number of quarters that it will take a principal of $1000 to double at 6.5% compounded quarterly.

Cumulative Checkup

STANDARDIZED FORMAT

Choose the correct letter.

1. Find the circumference. Use 3.14 as an approximation for π.

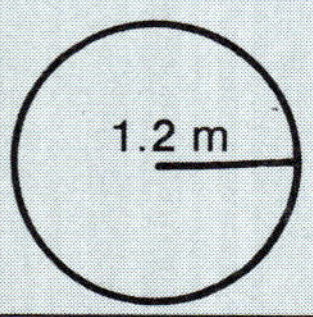

a. 3.768 m
b. 7.536 m
c. 4.5216 m
d. none of these

2. The least common multiple of 6 and 9 is

a. 36
b. 3
c. 18
d. none of these

3. Give the difference.

$3\frac{2}{3} - 2\frac{3}{4}$

a. $1\frac{11}{12}$
b. $1\frac{1}{12}$
c. $\frac{1}{12}$
d. none of these

4. Give the product.

$2\frac{1}{2} \times 1\frac{3}{5}$

a. 4
b. $\frac{1}{4}$
c. $1\frac{9}{16}$
d. none of these

5. Compute the area.

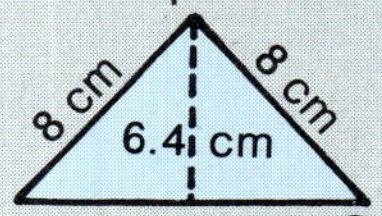

a. 61.44 cm²
b. 38.4 cm²
c. 30.72 cm²
d. none of these

6. Solve.

$\frac{5}{n} = \frac{3}{8}$

a. $4\frac{4}{5}$
b. $\frac{5}{24}$
c. $1\frac{7}{8}$
d. none of these

7. Solve.

6.5% of $n = 1.17$

a. 7.605
b. 18
c. 0.18
d. none of these

8. Solve.

$33\frac{1}{3}$% of $60 = n$

a. 40
b. 180
c. 20
d. none of these

9. The simple interest for a loan of $600 at 8% a year for 9 months is

a. $48
b. $24
c. $36
d. none of these

10. Give the total installment cost.
$38.95 down
$34.79 a month for 1 year

a. $456.43
b. $884.88
c. $502.19
d. none of these

11. Laura hiked 0.4 mile in 5 minutes. At that rate, how many miles could she hike in an hour?

a. 2 **b.** 2.4
c. 4.8 **d.** none of these

12. David earned $26.10 for working 6 hours. How much did he earn per hour?

a. $4.35 **b.** $4.40
c. $156.60 **d.** none of these

10

Rational Numbers and Algebra

Stock bought at $38\frac{1}{4}$ was sold at $38\frac{1}{2}$. The net change was $^{+}\frac{1}{4}$. These numbers represent dollars. What was the profit on 100 shares of the stock?

Rational numbers

We can use positive and negative numbers to show a stock's net change (the difference between yesterday's closing price and today's closing price).

If a stock had a day's net change of $^{+}\frac{5}{8}$, the price of each share of that stock went up $\frac{5}{8}$ of a dollar.

If a stock had a day's net change of $^{-}1\frac{1}{4}$, the price of each share of that stock went down $1\frac{1}{4}$ dollars.

DAILY STOCK QUOTATIONS

Stocks	Close	Net Change
DuPont	49	+5/8
Exxon	33 7/8	−1 1/4
GTE	43 1/8	+1/8
IBM	113	+7/8
Mobil	33 1/2	−3/8
Uniroyal	14 1/2	−1 1/2

A **rational number** is a number that can be written as a fraction in which

(a) the numerator is an integer and
(b) the denominator is an integer not equal to zero.

All the numbers shown on the number line below are rational numbers.

$^{-}3 \quad {}^{-}2\frac{5}{8} \quad {}^{-}2 \quad {}^{-}1.6 \quad {}^{-}1 \quad \frac{^{-}2}{5} \quad 0 \quad \frac{3}{4} \quad 1 \quad 1.41 \quad 2 \quad \frac{5}{2} \quad 3 \quad 3\frac{1}{5}$

On the number line, a number is less than every number to its right and greater than every number to its left.

$^{-}3 < {}^{-}2\frac{5}{8}$ $\qquad$ $^{-}1 > {}^{-}1.6$ $\qquad$ $0 < \frac{3}{4}$ $\qquad$ $2 > 1.41$

EXERCISES

Give the letter that names the point for each rational number.

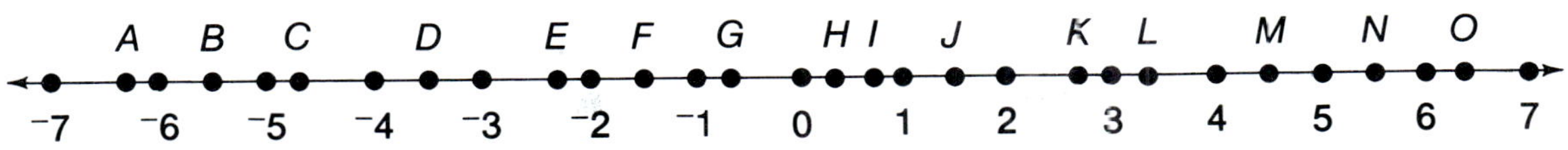

1. 1.5
2. $^-4\frac{3}{4}$
3. $^-1\frac{1}{2}$
4. $\frac{^-11}{2}$
5. $\frac{1}{4}$
6. 3.25
7. $^-3.5$
8. $5\frac{1}{2}$
9. $^-6.4$
10. $\frac{9}{2}$
11. $\frac{^-9}{4}$
12. $2\frac{7}{8}$
13. $\frac{^-3}{4}$
14. 0.8
15. $\frac{19}{3}$

<, =, or >?

16. $^-2$ ● $^-3$
17. 2 ● 3
18. $^-3$ ● 2
19. 3 ● $^-1$
20. 1.41 ● 0
21. $\frac{^-1}{4}$ ● $^-0.25$
22. $\frac{^-1}{2}$ ● $^-1$
23. 0 ● $\frac{^-2}{3}$
24. $^-0.8$ ● $^-0.80$
25. $\frac{1}{8}$ ● $\frac{3}{8}$
26. $\frac{^-1}{4}$ ● $\frac{^-3}{4}$
27. $\frac{^-7}{10}$ ● $\frac{^-9}{10}$
28. $^-1.6$ ● $^-1\frac{1}{2}$
29. $\frac{3}{5}$ ● $\frac{1}{5}$
30. $\frac{^-5}{3}$ ● $^-1\frac{2}{3}$
31. $^-3\frac{1}{2}$ ● $3\frac{1}{2}$
32. 0 ● $^-0.1$
33. $\frac{^-2}{5}$ ● $\frac{^-4}{5}$
34. $^-4\frac{1}{4}$ ● $^-4\frac{1}{2}$
35. $2\frac{1}{2}$ ● $\frac{5}{2}$
36. $1\frac{2}{3}$ ● $1\frac{1}{3}$
37. $\frac{3}{8}$ ● $\frac{^-1}{2}$
38. $^-6\frac{1}{2}$ ● $^-6.5$
39. $^-7$ ● $^-5\frac{1}{2}$
40. $\frac{^-7}{3}$ ● $^-2\frac{1}{3}$
41. $^-1.6$ ● $^-2.3$
42. 0 ● $\frac{1}{4}$
43. $^-0.5$ ● $\frac{^-3}{5}$
44. $\frac{1}{8}$ ● 0.125
45. $^-0.04$ ● $^-0.4$
46. $\frac{^-3}{5}$ ● $\frac{^-5}{3}$
47. $^-17$ ● 17

Problem Solving

Use the daily stock quotations on page 290.

48. Which stock had a net change of $^+87\frac{1}{2}$¢ per share?
49. Which stock had a net change of $^-$\$1.50 per share?
50. Which stocks went down more than 50¢ per share?
51. Which stocks went up more than 50¢ per share?

Adding and subtracting rational numbers

Integers	Rational numbers
$^{-}3 + 5 = 2$	$\frac{^{-}3}{8} + \frac{5}{8} = \frac{2}{8}$ $= \frac{1}{4}$
$^{-}1 + {}^{-}6 = {}^{-}7$	$^{-}0.1 + {}^{-}0.6 = {}^{-}0.7$

Remember: To subtract a number, add its opposite.

Integers	Rational numbers
$^{-}2 - 3 = {}^{-}2 + {}^{-}3$ $= {}^{-}5$	$\frac{^{-}2}{5} - \frac{3}{5} = \frac{^{-}2}{5} + \frac{^{-}3}{5}$ $= \frac{^{-}5}{5}$ $= {}^{-}1$
$^{-}4 - {}^{-}17 = {}^{-}4 + 17$ $= 13$	$^{-}0.4 - {}^{-}1.7 = {}^{-}0.4 + 1.7$ $= 1.3$

EXERCISES

Give each sum.

1. $^{-}6 + 7$ **2.** $^{-}6 + {}^{-}7$ **3.** $6 + 7$ **4.** $6 + {}^{-}7$

5. $^{-}23 + {}^{-}13$ **6.** $47 + {}^{-}17$ **7.** $^{-}55 + 55$ **8.** $^{-}103 + {}^{-}12$

9. $\frac{^{-}2}{5} + \frac{^{-}1}{5}$ **10.** $\frac{^{-}2}{5} + \frac{1}{5}$ **11.** $\frac{2}{5} + \frac{^{-}1}{5}$ **12.** $\frac{2}{5} + \frac{1}{5}$

13. $\frac{3}{4} + \frac{^{-}1}{2}$ **14.** $0 + \frac{^{-}7}{8}$ **15.** $\frac{^{-}1}{2} + \frac{^{-}1}{8}$ **16.** $\frac{7}{12} + \frac{^{-}7}{12}$

17. $^{-}3.8 + 6.9$ **18.** $3.4 + {}^{-}8.5$ **19.** $^{-}2.4 + 2.4$ **20.** $^{-}1.06 + {}^{-}2.7$

21. $4 + 8.6$ **22.** $^{-}1 + 0.4$ **23.** $^{-}3.6 + 0.9$ **24.** $^{-}5.08 + {}^{-}0.12$

Give each difference.

25. $7 - 10$
26. $6 - {}^{-}5$
27. ${}^{-}10 - {}^{-}4$
28. ${}^{-}6 - 3$
29. $3 - {}^{-}8$
30. $8 - 0$
31. ${}^{-}21 - {}^{-}9$
32. $18 - 21$
33. $\frac{3}{4} - \frac{1}{4}$
34. $\frac{{}^{-}3}{4} - \frac{1}{4}$
35. $\frac{{}^{-}3}{4} - \frac{{}^{-}1}{4}$
36. $\frac{3}{4} - \frac{{}^{-}1}{4}$
37. $\frac{3}{4} - \frac{{}^{-}1}{2}$
38. $\frac{{}^{-}5}{8} - \frac{3}{4}$
39. ${}^{-}2\frac{1}{2} - 3\frac{1}{2}$
40. $5\frac{3}{5} - {}^{-}2\frac{1}{5}$
41. $6.8 - 7.9$
42. ${}^{-}2.6 - 1.5$
43. ${}^{-}9.5 - 6$
44. ${}^{-}7.2 - 8.3$
45. ${}^{-}8.06 - {}^{-}7.2$
46. $5.88 - 3.06$
47. ${}^{-}12.25 - 1.5$
48. ${}^{-}10.76 - {}^{-}1.12$

<, =, or >?

49. ${}^{-}6 + {}^{-}3$ ● ${}^{-}10$
50. $9 - {}^{-}3$ ● 10
51. ${}^{-}8 - 3$ ● ${}^{-}11$
52. ${}^{-}5 + {}^{-}2$ ● ${}^{-}4$
53. ${}^{-}5 + 8$ ● ${}^{-}3$
54. ${}^{-}15 - {}^{-}17$ ● ${}^{-}2$
55. $\frac{{}^{-}3}{8} + \frac{{}^{-}1}{8}$ ● $\frac{{}^{-}1}{2}$
56. $\frac{3}{8} - \frac{1}{2}$ ● $\frac{1}{8}$
57. $\frac{{}^{-}7}{4} - \frac{3}{4}$ ● ${}^{-}2\frac{1}{2}$
58. $6.5 + {}^{-}3$ ● 3.5
59. $7.2 - 8$ ● ${}^{-}0.8$
60. ${}^{-}3.9 - 4.2$ ● ${}^{-}8.1$
61. $\frac{{}^{-}2}{3} + \frac{2}{3}$ ● $\frac{{}^{-}1}{8} - \frac{1}{8}$
62. $\frac{3}{10} - \frac{1}{5}$ ● $\frac{1}{5} + \frac{3}{10}$
63. $3\frac{1}{3} - 1\frac{1}{3}$ ● $1\frac{1}{2} + \frac{1}{2}$
64. $1\frac{3}{4} + \frac{{}^{-}1}{4}$ ● $2 - 2\frac{1}{2}$
65. ${}^{-}7.2 + 4.1$ ● ${}^{-}4.1 + 7.2$
66. $7.21 + {}^{-}7.21$ ● ${}^{-}8.16 + 8.17$
67. ${}^{-}8.9 + {}^{-}6.2$ ● ${}^{-}12 - 4$
68. ${}^{-}8.5 + 8.5$ ● $2.1 - {}^{-}2.1$

Challenge!

Substitute and simplify to find which expression has the greatest value.

m	n	r	s
${}^{-}1$	2	4	8

69. a. $\frac{m}{n} + \frac{r}{s}$ b. $\frac{m}{r} + \frac{n}{s}$ c. $\frac{n}{s} + \frac{r}{s}$ d. $\frac{m}{r} + \frac{s}{n}$

70. a. $\frac{m}{n} - \frac{r}{s}$ b. $\frac{n}{r} - \frac{m}{s}$ c. $\frac{n}{s} - \frac{r}{s}$ d. $\frac{m}{r} - \frac{s}{n}$

Multiplying and dividing rational numbers

You multiply and divide rational numbers in the same way that you multiply and divide integers.

> The product or quotient of two rational numbers with the same sign is positive.
>
> The product or quotient of two rational numbers with different signs is negative.
>
> The product of any rational number and 0 is 0.
>
> The quotient of 0 divided by any nonzero rational number is 0.

Study these examples.

Integers

$^-3 \times {}^-4 = 12$

$^-7 \times 3 = {}^-21$

$6 \div {}^-3 = {}^-2$

reciprocal of $^-4$

$$^-14 \div {}^-4 = {}^-14 \times \frac{^-1}{4} = \frac{14}{4} = 3\frac{1}{2}$$

Rational Numbers

$\frac{^-3}{5} \times \frac{^-4}{5} = \frac{12}{25}$

$^-0.7 \times 0.3 = {}^-0.21$

$$\frac{6}{5} \div \frac{^-3}{4} = \frac{6}{5} \times \frac{^-4}{3} = \frac{^-24}{15} = {}^-1\frac{3}{5}$$

Dividing by a number is the same as multiplying by its reciprocal.

$^-0.14 \div {}^-0.4 = 0.35$

$$\begin{array}{r} 0.35 \\ 0.4\overline{)\,0.140} \\ \underline{-12} \\ 20 \\ \underline{-20} \\ 0 \end{array}$$

EXERCISES

Give each product or quotient.

1. $^-6 \times {}^-7$	**2.** $^-6 \times 7$	**3.** $6 \times {}^-7$	**4.** 6×7
5. $^-0.4 \times 3$	**6.** $^-1.2 \times {}^-3$	**7.** $^-0.25 \times 0$	**8.** 8×4.3
9. $\frac{^-1}{4} \times \frac{1}{2}$	**10.** $\frac{^-1}{4} \times \frac{^-1}{2}$	**11.** $\frac{1}{4} \times \frac{^-1}{2}$	**12.** $\frac{1}{4} \times \frac{1}{2}$

13. $\frac{^-3}{5} \times \frac{^-3}{4}$

14. $\frac{2}{5} \times \frac{3}{4}$

15. $\frac{5}{9} \times \frac{^-3}{4}$

16. $^-16 \div {}^-4$

17. $^-24 \div 3$

18. $32 \div {}^-8$

19. $27 \div 3$

20. $^-36 \div {}^-9$

21. $72 \div {}^-8$

22. $^-5.6 \div 7$

23. $^-4.2 \div {}^-6$

24. $^-4.8 \div 6$

25. $0.63 \div {}^-9$

26. $^-0.25 \div {}^-5$

27. $4.9 \div 7$

28. $^-3.5 \div {}^-5$

29. $0.21 \div {}^-3$

30. $0.5 \div {}^-0.5$

31. $0 \div {}^-1.5$

32. $^-0.35 \div {}^-10$

33. $0.6 \div {}^-100$

Give each quotient. [*Hint:* Multiply by the reciprocal.]

34. $5 \div {}^-7$

35. $^-3 \div 4$

36. $^-9 \div 4$

37. $\frac{^-2}{3} \div \frac{^-1}{3}$

38. $\frac{2}{5} \div \frac{^-1}{4}$

39. $\frac{5}{9} \div \frac{^-2}{3}$

40. $\frac{3}{8} \div \frac{^-1}{2}$

41. $\frac{^-1}{2} \div \frac{3}{8}$

42. $\frac{^-5}{4} \div \frac{^-3}{2}$

43. $\frac{^-1}{2} \div 2$

44. $2 \div \frac{^-1}{2}$

45. $\frac{^-1}{3} \div \frac{1}{2}$

46. $\frac{3}{2} \div \frac{^-3}{2}$

47. $0 \div \frac{^-1}{4}$

48. $\frac{^-3}{8} \div \frac{^-1}{4}$

$<$, $=$, or $>$?

49. $^-6 \times 4$ ● $^-24$

50. $^-8 \div {}^-2$ ● $^-4$

51. $^-0.8 \times {}^-3$ ● $^-2.4$

52. $^-3.5 \times 7$ ● $^-0.4$

53. $\frac{^-1}{2} \times \frac{^-1}{2}$ ● $\frac{3}{8}$

54. $\frac{^-4}{3} \div \frac{1}{2}$ ● $^-2\frac{2}{3}$

55. $\frac{7}{3} \times \frac{^-1}{2}$ ● 1

56. $\frac{^-2}{3} \div \frac{2}{3}$ ● 1

57. $\frac{^-8}{5} \times \frac{^-5}{2}$ ● 4

58. $\frac{^-8}{5} \div \frac{^-2}{5}$ ● 4

KEEPING SKILLS SHARP

Choose the correct solution by estimating.

1. $34r = 918$
   a. 17 b. 27 c. 37 d. 47

2. $t + 23.5 = 117.06$
   a. 93.56 b. 103.56 c. 130.56 d. 140.56

3. $\frac{s}{5.7} = 9.2$
   a. 42.44 b. 52.44 c. 62.44 d. 72.44

4. $n - 6.53 = 49.5$
   a. 33.03 b. 43.03 c. 56.03 d. 66.03

5. $0.89f = 5.162$
   a. 5.8 b. 6.8 c. 7.8 d. 8.8

6. $\frac{m}{0.9} = 45$
   a. 40.5 b. 50.5 c. 60.5 d. 70.5

7. $3.2b = 22.72$
   a. 4.1 b. 5.1 c. 6.1 d. 7.1

8. $n - 1.75 = 4.2$
   a. 3.95 b. 4.95 c. 5.95 d. 6.95

Practice exercises

Compute.

1. $6 + {}^{-}3$
2. $7 - 9$
3. ${}^{-}3 \times {}^{-}7$
4. ${}^{-}24 \div 6$
5. ${}^{-}13 + {}^{-}11$
6. ${}^{-}8 - 6$
7. $12 \times {}^{-}3$
8. ${}^{-}88 \div {}^{-}8$
9. ${}^{-}65 + 83$
10. $25 - {}^{-}16$
11. $20 \times {}^{-}10$
12. $96 \div {}^{-}6$
13. $8.2 + {}^{-}5.8$
14. $7.4 - {}^{-}2.1$
15. $7.5 \times {}^{-}2$
16. $6.4 \div {}^{-}8$
17. ${}^{-}2.75 + 3.2$
18. ${}^{-}6.25 - 1.2$
19. ${}^{-}6.2 \times {}^{-}0.2$
20. ${}^{-}7.5 \div 5$
21. ${}^{-}3.14 + {}^{-}2.65$
22. $16.8 - {}^{-}4.2$
23. ${}^{-}39.1 \times 0.1$
24. ${}^{-}39.1 \div 0.1$
25. $\frac{1}{3} + \frac{{}^{-}1}{3}$
26. $\frac{1}{3} - \frac{{}^{-}1}{3}$
27. $\frac{1}{3} \times \frac{{}^{-}1}{3}$
28. $\frac{1}{3} \div \frac{{}^{-}1}{3}$
29. $\frac{{}^{-}3}{8} + \frac{{}^{-}1}{8}$
30. $\frac{{}^{-}7}{8} - \frac{{}^{-}1}{8}$
31. $\frac{{}^{-}3}{5} \times \frac{{}^{-}1}{3}$
32. $\frac{{}^{-}2}{3} \div \frac{1}{2}$
33. $\frac{{}^{-}7}{4} + \frac{1}{2}$
34. $\frac{{}^{-}6}{5} - \frac{1}{2}$
35. $\frac{{}^{-}8}{5} \times \frac{{}^{-}4}{5}$
36. $\frac{3}{4} \div \frac{{}^{-}1}{2}$

Simplify. Remember to work inside the grouping symbols first.

37. $({}^{-}8 \times {}^{-}3) \times 1$
38. $({}^{-}6 + 9) \div {}^{-}3$
39. $({}^{-}8 \div 2) + {}^{-}5$
40. $9 + ({}^{-}2 \times 0)$
41. ${}^{-}4 \times (5 - {}^{-}3)$
42. $(6 + {}^{-}6) \div 4$
43. $({}^{-}18 \div {}^{-}3) + (21 \div {}^{-}3)$
44. $({}^{-}24 \div {}^{-}4) \times (6 - {}^{-}2)$
45. $(3 \times {}^{-}4) - (8 \div {}^{-}4)$
46. $\left(\frac{{}^{-}1}{5} - \frac{{}^{-}3}{4}\right) \times \frac{1}{4}$
47. $\left(\frac{2}{3} + \frac{{}^{-}2}{3}\right) \times \frac{1}{2}$
48. $\frac{{}^{-}1}{2} \times \left(\frac{1}{5} - \frac{{}^{-}1}{5}\right)$
49. $\left(\frac{1}{3} + \frac{{}^{-}1}{3}\right) \times \frac{2}{3}$
50. $\left(\frac{{}^{-}3}{4} \times \frac{{}^{-}1}{4}\right) \div 2$
51. $\frac{3}{5} \div \left(\frac{1}{5} + \frac{2}{5}\right)$
52. $\left(\frac{1}{5} + \frac{1}{5}\right) \times \left(\frac{1}{2} - \frac{1}{2}\right)$
53. $\left(\frac{3}{4} \times \frac{{}^{-}5}{2}\right) + \left(\frac{5}{4} \times \frac{3}{2}\right)$
54. $\left(\frac{3}{4} - \frac{1}{4}\right) \div \left(\frac{{}^{-}1}{4} + \frac{{}^{-}1}{4}\right)$

Temperature formulas

Temperature can be measured in both degrees Celsius and degrees Fahrenheit.

On the Celsius scale, water freezes at 0° and boils at 100°.

On the Fahrenheit scale, water freezes at 32° and boils at 212°.

Sometimes it is necessary to change from one temperature unit to another. Study these formulas.

$$C = \frac{5}{9}(F - 32)$$

$$F = \frac{9}{5}C + 32$$

EXAMPLE 1.

Change 20°C to degrees Fahrenheit.

$$F = \frac{9}{5}C + 32$$

$$F = \frac{9}{\cancel{5}_{1}} \cdot \cancel{20}^{4} + 32$$

$$F = 36 + 32 = 68$$

So, 20°C = 68°F.

EXAMPLE 2.

Change 50°F to degrees Celsius.

$$C = \frac{5}{9}(F - 32)$$

$$C = \frac{5}{9}(50 - 32)$$

$$C = \frac{5}{9}(18) = 10$$

So, 50°F = 10°C.

EXERCISES

Change to degrees Fahrenheit.

1. 50°C
2. 18°C
3. 0°C
4. $^{-}12$°C
5. $^{-}40$°C

Change to degrees Celsius.

6. 86°F
7. 150°F
8. $^{-}13$°F
9. 0°F
10. $^{-}25$°F

11. The melting point of gold is 958°C. What is its melting point in degrees Fahrenheit?

12. The melting point of silver is 1761°F. What is its melting point in degrees Celsius?

Equivalent expressions

Look at these expressions:

$5n$ **$3n + 2n$** **$6n - 1$**

The table shows the values of these expressions for different values of n.

Substitute 1 for n.

n	$5n$	$3n + 2n$	$6n - 1$
1	$5(1) = 5$	$3(1) + 2(1) = 5$	$6(1) - 1 = 5$
$^{-}2$	$5(^{-}2) = {}^{-}10$	$3(^{-}2) + 2(^{-}2) = {}^{-}10$	$6(^{-}2) - 1 = {}^{-}13$
0	$5(0) = 0$	$3(0) + 2(0) = 0$	$6(0) - 1 = {}^{-}1$
0.5	$5(0.5) = 2.5$	$3(0.5) + 2(0.5) = 2.5$	$6(0.5) - 1 = 2$

Notice that $5n$ and $3n + 2n$ have the same value for each value given to n. In fact, the two expressions will have the same value for all values given to n. For this reason, $5n$ and $3n + 2n$ are called **equivalent expressions.**

$5n$ and $6n - 1$ do not have the same value for all the values given to n. They are *not* equivalent expressions.

$5n$ ← *equivalent expressions* → **$3n + 2n$**

$5n$ ← *not equivalent expressions* → **$6n - 1$**

EXAMPLE 1. Are $4a$ and $3a + 1$ equivalent expressions?

a	$4a$	$3a + 1$	
1	$4(1) = 4$	$3(1) + 1 = 4$	← same value
5	$4(5) = 20$	$3(5) + 1 = 16$	← different values

$4a$ and $3a + 1$ are not equivalent expressions.

EXAMPLE 2. Are $7t$ and $9t - 2t$ equivalent expressions?

t	$7t$	$9t - 2t$	
1	$7(1) = 7$	$9(1) - 2(1) = 7$	← same value
5	$7(5) = 35$	$9(5) - 2(5) = 35$	← same value
$^{-}3$	$7(^{-}3) = {}^{-}21$	$9(^{-}3) - 2(^{-}3) = {}^{-}21$	← same value

$7t$ and $9t - 2t$ are equivalent expressions.

EXERCISES

In each table, two of the expressions are equivalent. First copy and complete the table. Then tell which two of the expressions are equivalent.

1.

n	$7n$	$4n + 3n$	$5n + 2$
1	7	7	7
3	?	?	?
$^{-}2$	?	?	?
10	?	?	?

2.

r	$9r$	$10r - 1$	$12r - 3r$
1	?	?	?
3	?	?	?
$^{-}2$	?	?	?
10	?	?	?

3.

a	$6a$	$6 + a$	$7a - a$
1	?	?	?
4	?	?	?
$^{-}3$	?	?	?
10	?	?	?

4.

y	$5y$	$8y - 2 - y$	$7y - 2$
1	?	?	?
4	?	?	?
$^{-}3$	?	?	?
10	?	?	?

5.

b	$10b$	$8b + 2$	$(8+2)b$
1	?	?	?
6	?	?	?
$^{-}4$	?	?	?
10	?	?	?

6.

t	$10t$	$5t + 3 + 2t$	$7t + 3$
1	?	?	?
6	?	?	?
$^{-}4$	?	?	?
10	?	?	?

★ **7.**

d	$d - 4$	$d + 4$	$3d - (2d + 4)$
1	?	?	?
7	?	?	?
$^{-}6$	?	?	?
10	?	?	?

★ **8.**

p	$2p - 1$	$8p - 1$	$5p - (1 - 3p)$
1	?	?	?
7	?	?	?
$^{-}6$	?	?	?
10	?	?	?

Simplifying expressions

Study these examples that show how to simplify expressions.

EXAMPLE 1. Simplify $\mathbf{5}n + \mathbf{2}n$.

$5n + 2n = (5 + 2)n$ Use the distributive property.

$= 7n$

EXAMPLE 2. Simplify $\mathbf{8}t + t$.

$8t + t = 8t + 1t$ Use the multiplying by 1 property.

$= (8 + 1)t$ Use the distributive property.

$= 9t$

SHORTCUT

To simplify these expressions, combine the like terms by adding or subtracting.

EXAMPLE 3. Simplify $5r + 2 + 4r + 1$.

like terms

$5r + 2 + 4r + 1 = 9r + 3$

like terms

EXAMPLE 4. Simplify $6t + 7s - 2t + 3s$.

like terms

$6t + 7s - 2t + 3s = 4t + 10s$

like terms

EXERCISES

Copy and complete.

1. $4n + 5n = (4 + \underline{?})n$
   $= \underline{?}\,n$

2. $12s + {}^-7s = (12 + \underline{?})s$
   $= \underline{?}\,s$

3. $13t - 5t = (13 - \underline{?})t$
   $= \underline{?}\,t$

4. $7y - y = (7 - \underline{?})y$
   $= \underline{?}\,y$

Simplify. [*Hint:* Combine like terms.]

5. $5n + 3n$
6. $7t - 4t$
7. $^{-}7b + 2b$
8. $3a - 5a$
9. $4u + 2u + 5u$
10. $6r + 2r + r$
11. $11t + 9t + 2$
12. $8x + {}^{-}2x + 3x$
13. $4n + 3 + 5n - 1$
14. $12s + 9 - 4s + 5$
15. $r + r + 2r$
16. $15y - 6y - 2y$
17. $8t + 5t - 10t$
18. $a + 3 + 4a - 9$
19. $7x - 7x + 2$
20. $11d + 6 - 10d$
21. $9 + 2r + 4$
22. $^{-}2b - 10 + 6b$
23. $5m + 6t + 8t$
24. $9x + 2y - x + 3y$
25. $12f + 2e + 8f - 2e$
26. $4c + 8 - 3c - 8$
27. $3 - d - 7 + d$
28. $3x - 2x - x$
29. $9a + 2a + 1 - 6y$
30. $8r + 2t + 5t + 4r$
31. $^{-}7y + 9 - 3y + 6$
32. $4b + 9b - 6 + b$
33. $^{-}5 + n + 8 - d$
34. $a + a + a + a$
35. $c - c - c - c$
36. $8y - 2x - 5y - 4x$

Challenge!

First simplify the expression for each perimeter. Then find the perimeter P if $n = 2$ meters.

37. $P = n + 4n + 1 + 5n + 2n$

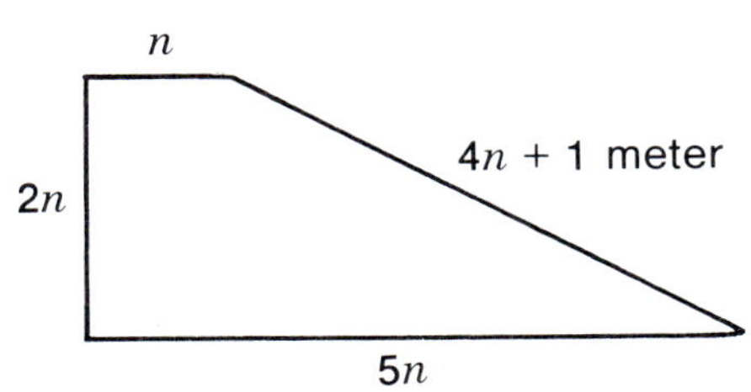

KEEPING SKILLS SHARP

Compute.

1. $^{-}6 + {}^{-}4$
2. $^{-}10 - 8$
3. $12 \times {}^{-}4$
4. $^{-}16 \div 2$
5. $^{-}1.5 + {}^{-}1.6$
6. $3.2 - {}^{-}1.5$
7. $^{-}0.3 \times {}^{-}0.8$
8. $7.2 \div {}^{-}9$
9. $\frac{1}{8} + \frac{^{-}5}{8}$
10. $\frac{1}{3} - \frac{2}{3}$
11. $\frac{4}{5} \times \frac{^{-}1}{2}$
12. $\frac{^{-}3}{10} \div \frac{^{-}1}{3}$
13. $({}^{-}6 + {}^{-}4) \div 5$
14. $(8 - 10) \times {}^{-}2$
15. $12 - ({}^{-}5 \times {}^{-}2)$
16. $({}^{-}6 + {}^{-}3) \div {}^{-}9$
17. $^{-}14 + ({}^{-}2 \times {}^{-}7)$
18. $(6 + {}^{-}3) \times (5 + {}^{-}5)$
19. $(12 - 10) \div (8 - 10)$
20. $(25 \div {}^{-}5) + ({}^{-}10 \div {}^{-}2)$

Solving equations

Karen spent \$2.65 for a photo album. She spent \$5 less for the album than she spent for film. How much did she spend for film?

Let f be the amount she spent for film. Then $f - 5$ is the amount she spent for the album.

Equation: $\overbrace{f - 5}^{\text{amount spent for the album}} = \overbrace{2.65}^{\text{amount spent for the album}}$

Add 5 to both sides. $f - 5 + 5 = 2.65 + 5$

$f = 7.65$

Check: $7.65 - 5 = 2.65$

So, she spent \$7.65 for film.

Andrew had 5 prints developed and his friend had 2 prints developed. Each print cost the same. The difference between the amounts they spent was \$1.08. How much did 1 print cost?

Let p be the cost of 1 print. Then $5p$ is the cost of 5 prints and $2p$ is the cost of 2 prints.

Equation: $\overbrace{5p - 2p}^{\text{difference between amounts spent}} = \overbrace{1.08}^{\text{difference between amounts spent}}$

Simplify by combining like terms. $3p = 1.08$

Divide both sides by 3. $\frac{3p}{3} = \frac{1.08}{3}$

$p = 0.36$

Check: $5(0.36) - 2(0.36) = 1.08$

So, each print cost \$.36.

EXERCISES

Copy and finish solving each equation. Check.

1. $y + 6 = 8.4$
$y + 6 - 6 = 8.4 - \underline{?}$
$y = \underline{?}$

2. $^{-}7 = \frac{n}{4}$
$^{-}7 \times \underline{?} = \frac{n}{4} \times 4$
$\underline{?} = n$

3. $7t + 5t = 6$
$\underline{?}\, t = 6$
$\frac{12t}{12} = \frac{6}{\underline{?}}$
$t = \underline{?}$

Solve and check.

4. $3a = ^-18$
5. $\frac{b}{^-7} = 2$
6. $c - 8 = ^-10$
7. $d + 6 = 1$
8. $^-40 = e + 5$
9. $6.8 = 2f$
10. $5n + 3n = 56$
11. $11p - 5p = ^-42$
12. $16 = 4r + 12r$
13. $8t - 7t = 6$
14. $150 = 3n + 12n$
15. $10p - 6p = ^-20$
16. $^-2n + 6n = 12$
17. $^-3y + 2y = 4$
18. $n - 8 = 0$
19. $8x - 3x = 30$
20. $8 = 5n - 7n$
21. $t + 2t = ^-3$
22. $35 = 11s - 4s$
23. $3r + r = 1$
24. $6 = 15 + t$
25. $9a - 3a = 5.4$
26. $8y - 6y = 3$
27. $0.24 = 3p + 5p$
28. $6n - 9n = 1$
29. $4x + 3x = ^-28$
30. $\frac{d}{2} = 1.38$
31. $10t - 4t = ^-28$
32. $r - 4r = 5.1$
33. $3 = a - 6a$

Problem Solving

Decide which equation would be used to solve each problem. Then solve the problem.

Equation A. $7n + 2n = 7.65$

Equation B. $7n - 2n = 7.65$

Equation C. $\frac{n}{5} = 7.65$

Equation D. $n + 5 = 7.65$

34. Steve bought 5 rolls of film. The average cost of a roll was $7.65. What was the total cost of the 5 rolls?

35. Rita bought 7 pages for her photo album and Cindy bought 2 pages. Each page cost the same. Together, they spent $7.65. How much did 1 page cost?

36. Greg bought a roll of film for $7.65. He also bought some flash cubes. He spent $5 more for the film than for the flash cubes. How much did he spend for the flash cubes?

37. Frank had 7 enlargements made and Holly had 2 made. Each enlargement cost the same. The difference between the amounts they spent was $7.65. How much did 1 enlargement cost?

Solving two-step equations

You can find the number by writing an equation and solving it.

$$5y - 23 = 14$$

To solve the equation, get an equation with y all by itself on one side of the equal sign.

First add 23 to both sides. $5y - 23 + 23 = 14 + 23$

$$5y = 37$$

Divide both sides by 5. $\frac{5y}{5} = \frac{37}{5}$

$$y = \frac{37}{5}$$

Check: $5\left(\frac{37}{5}\right) - 23 = 14$

The number is $\frac{37}{5}$ or $7\frac{2}{5}$.

EXAMPLE 1.

Equation: $8 = 5 + \frac{w}{^{-}4}$

Subtract 5. $8 - 5 = 5 - 5 + \frac{w}{^{-}4}$

$$3 = \frac{w}{^{-}4}$$

Multiply by $^{-}4$. $^{-}4 \times 3 = {^{-}4} \times \frac{w}{^{-}4}$

$$^{-}12 = w$$

Check: $8 = 5 + \frac{^{-}12}{^{-}4}$

EXAMPLE 2.

Equation: $5n + 3n - 2 = 3.6$

Simplify by combining like terms. $8n - 2 = 3.6$

Add 2. $8n - 2 + 2 = 3.6 + 2$

$$8n = 5.6$$

Divide by 8. $\frac{8n}{8} = \frac{5.6}{8}$

$$n = 0.7$$

Check: $5(0.7) + 3(0.7) - 2 = 3.6$

EXERCISES

Copy and finish solving each equation.
Check your solution.

1.

$${}^-6y - 7 = {}^-2$$
$${}^-6y - 7 + 7 = {}^-2 + \underline{?}$$
$${}^-6y = ?$$
$$\frac{{}^-6y}{{}^-6} = \frac{5}{?}$$
$$y = \underline{?}$$

2.

$$\frac{n}{6} + 3 = {}^-5$$
$$\frac{n}{6} + 3 - 3 = {}^-5 - \underline{?}$$
$$\frac{n}{6} = \underline{?}$$
$$6 \times \frac{n}{6} = \underline{?} \times {}^-8$$
$$n = \underline{?}$$

3.

$$3 = \frac{d}{{}^-4} - {}^-2$$
$$3 + \underline{?} = \frac{d}{{}^-4} - {}^-2 + {}^-2$$
$$\underline{?} = \frac{d}{{}^-4}$$
$$\underline{?} \times 1 = {}^-4 \times \frac{d}{{}^-4}$$
$$\underline{?} = d$$

4.

$$7t + 3t + 1 = 8$$
$$\underline{?}\,t + 1 = 8$$
$$10t + 1 - 1 = 8 - \underline{?}$$
$$10t = \underline{?}$$
$$\frac{10t}{10} = \frac{7}{?}$$
$$t = \underline{?}$$

5.

$$5r - 3r - 3 = {}^-1$$
$$\underline{?}\,r - 3 = {}^-1$$
$$2r - 3 + \underline{?} = {}^-1 + \underline{?}$$
$$2r = \underline{?}$$
$$\frac{2r}{2} = \frac{2}{?}$$
$$r = \underline{?}$$

6.

$$4w + 2w + 5 = 2.6$$
$$\underline{?}\,w + 5 = 2.6$$
$$6w + 5 - 5 = 2.6 - \underline{?}$$
$$6w = \underline{?}$$
$$\frac{6w}{6} = \frac{{}^-2.4}{?}$$
$$w = \underline{?}$$

Solve and check.

7. $9y + 10 = 1$

8. $\frac{x}{4} + 2 = 5$

9. $3y + 2y + 1 = 21$

10. $34 = {}^-8n + 2$

11. $\frac{m}{{}^-3} + 7 = 2$

12. $2t + 7t - 4 = 5$

13. $7y - 4 = {}^-5$

14. ${}^-1 = \frac{n}{3} - 4$

15. $20 = 9s - 5s + 6$

16. ${}^-8 = 12 + {}^-4n$

17. $\frac{t}{{}^-2} - 12 = 15$

18. $15y - 8y - 10 = 4$

19. $5x + 3 = {}^-1.5$

20. $\frac{w}{6} - 11 = {}^-10$

21. $2r + 8r - 10 = {}^-3$

22. $9y - 12.6 = 0$

23. $8 = 10 + \frac{n}{{}^-2}$

24. $23 = 15 + 7d - 6d$

25. $2 = 3x - 5$

26. $\frac{r}{{}^-8} + 6 = 0$

27. $4t + t - 3 = 0.5$

28. $123x - 1648 = 197$

29. $\frac{g}{{}^-26} + 57 = 110$

30. $1.31r - 0.57r + 3.8 = 0.1$

Exploring stairway patterns

Work in groups. Get squared paper.

PART A

Mr. Nail needs an outside stairway from the ground to a door on the third floor of his house. The third floor is 20 ft above the ground. The stairway will be along the wall. Each step will be 1 ft wide and $\frac{1}{2}$ ft high.

1. a. On squared paper, design the stairway. How far to the left of the door will the stairway begin?
   b. The first floor is 9 ft high. How far to the left of the door will the stairway cross the first-floor ceiling?
   c. Mr. Nail also wants to put a vertical drainpipe $7\frac{1}{2}$ ft to the left of the door. How far above the ground will the drainpipe cross the stairway?

2. In most stairways, each stair is the same size. Let's imagine a stairway for which this is not true. Suppose you need a stairway with a total height of 9 ft and a total width of 21 ft. Each step must be $\frac{1}{2}$ ft high, but the width of any step can be either 1 ft or $1\frac{1}{2}$ ft. Discuss different ways of designing this stairway. Compare your designs with those of another group.

PART B

We can use what we have learned about stairways to explore patterns on squared paper that resemble stairways.

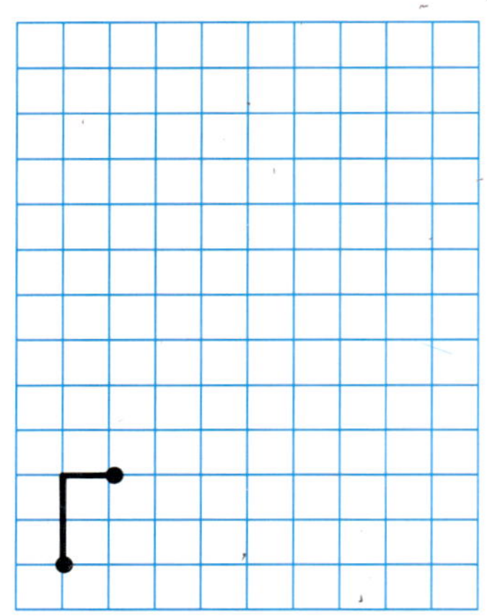

3. Near the lower left corner of squared paper, draw a step like this one. Draw a dot at each end of the step.

   Draw a stairway by drawing this step over and over until you reach the end of your paper. Draw a dot at the right end of each step.

Draw a straight line through the dots. Let's call this line the stair-line.

a. Discuss why your stair-line is straight, not curved.

b. Suppose you want another stair line that will meet the first one at a right angle. Find a way to draw such a stairway. Then draw the stair line. Check to see if the two stair lines form a right angle.

c. Compare the heights and widths of a step from each stairway. Look for a pattern. Use your pattern to make a rule about stair lines that meet at right angles.

d. Test your rule by trying it with a new stairway, made from a different-sized step.

4. Predict which of these steps will form a stairway with the steepest stair line. Also predict the least steep stair line.

a. 1 high and 3 wide

b. 0.5 high and 3 wide

c. 2 high and 7 wide

d. 2 high and 2.5 wide

Test your predictions.
How did you check?
Discuss ways to compare steepness.

5. a. Make a stairway with the steps shown in the chart. Complete the pattern. Is your stair line for this stairway straight or curved? Why?

	1st	2nd	3rd
Height	1	2	3
Width	1	1	1

b. Draw what you think the shape of the stair line will be for the pattern of steps in this chart. Test your prediction.

	1st	2nd	3rd
Height	1	1	1
Width	1	2	3

6. a. Make up a step pattern, but do not tell it to the group. Draw a stair line using your step pattern.

b. Show your stair line to the group. Ask them to try to find the mystery step pattern. If their guess of the step pattern is different, have them draw a stair line with the different pattern. Compare stair lines.

Graphing inequalities

These are examples of inequalities:

		is less than or equal to	*is greater than or equal to*	*is not equal to*
$x > 2$	$n < {}^-3$	$y \leq 1$	$t + 1 \geq 4$	$r \neq 5$

There are many numbers that are solutions of $x > 2$. In fact, every number greater than 2 is a solution. You can graph the solutions on a number line.

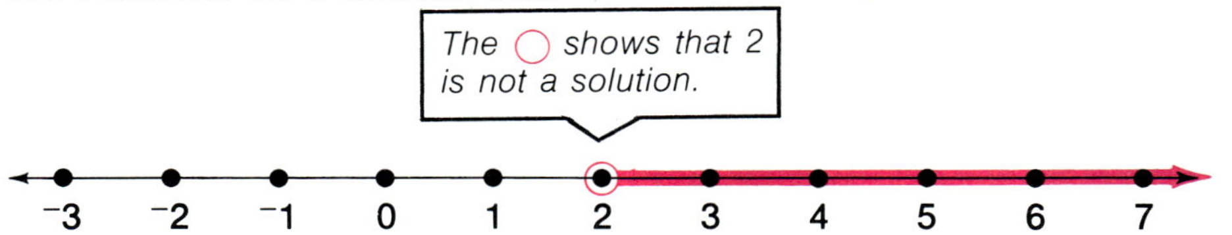

Here are other examples.

EXAMPLE 1.

Inequality: $n < {}^-3$ *Solution:* All numbers less than $^-3$

Graph:

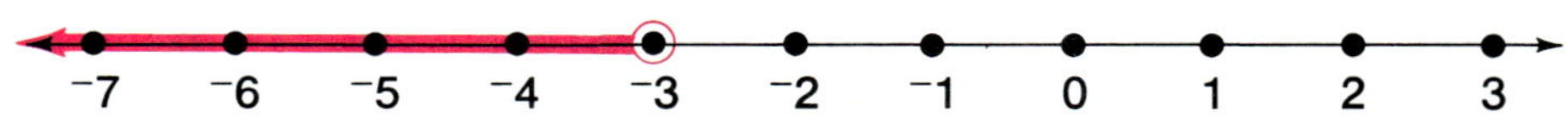

EXAMPLE 2.

Inequality: $y \leq 1$ *Solution:* 1 and all numbers less than 1

Graph:

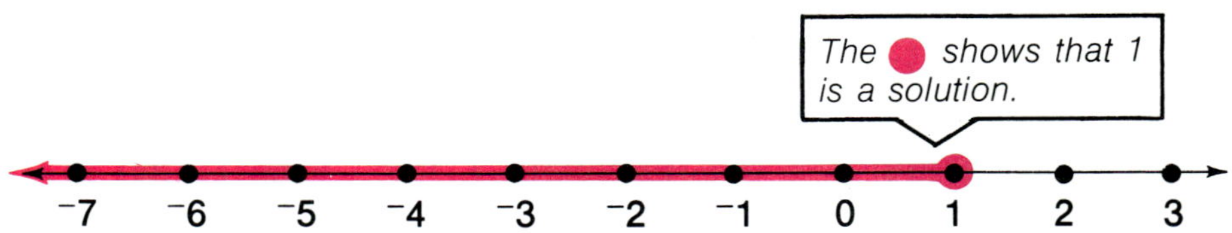

EXAMPLE 3.

Inequality: $t + 1 \geq 4$

Subtract 1 from both sides. $t + 1 - 1 \geq 4 - 1$

Simplify. $t \geq 3$ *Solution:* 3 and all numbers greater than 3

Graph:

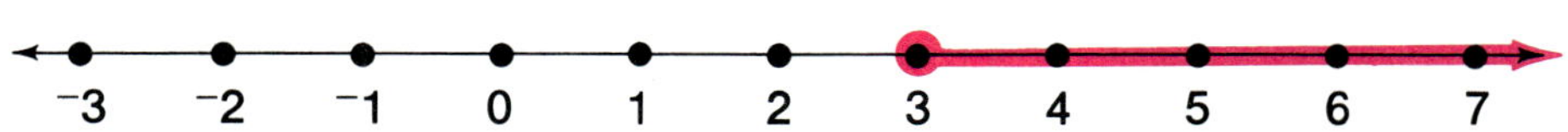

EXERCISES

Match each inequality with its graph.

1. $n < {}^-2$
2. $n > {}^-2$
3. $n \geq {}^-2$
4. $n \leq {}^-2$
5. $n \neq {}^-2$

a. 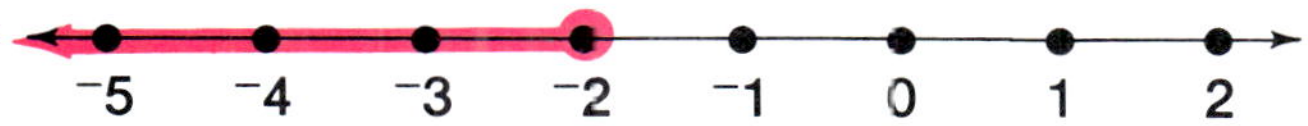

b. 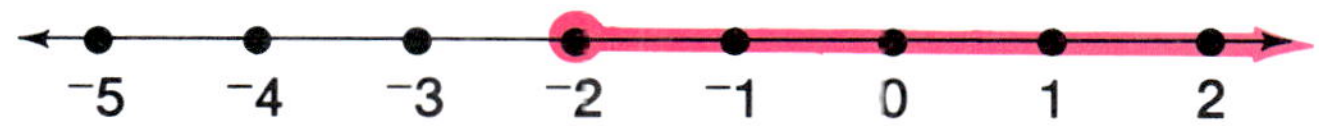

c. 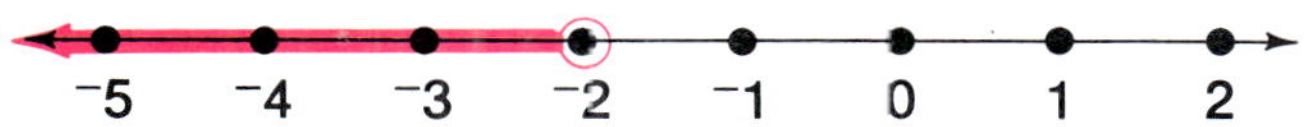

d. 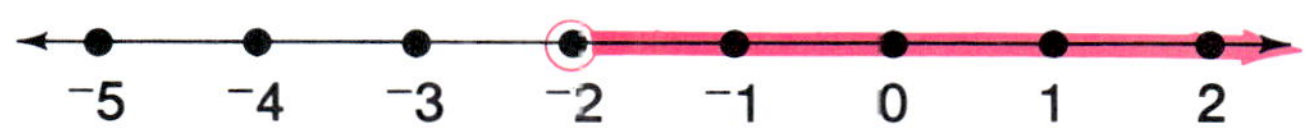

e. 

Match each inequality with its graph.

6. $t - 1 > 2$
7. $t - 1 < 2$
8. $t - 1 \geq 2$
9. $t - 1 \leq 2$
10. $t - 1 \neq 2$

a. 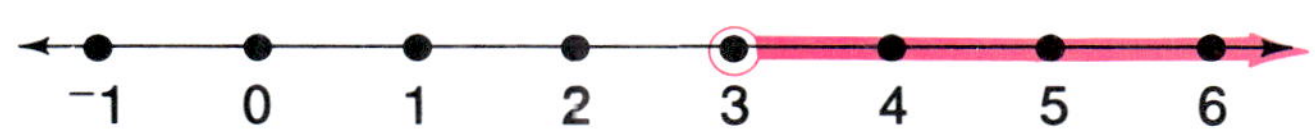

b. 

c. 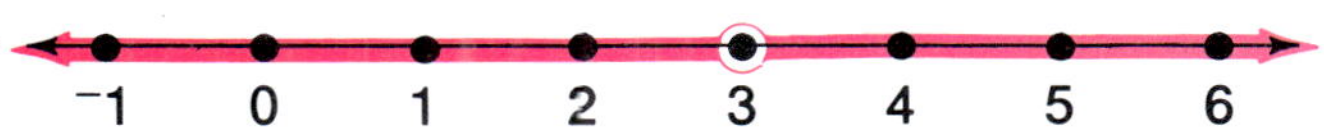

d. 

e. 

Graph the solutions of each inequality.

11. $n > {}^-1$
12. $r < 4$
13. $t \geq 0$
14. $x \leq {}^-1$
15. $s \neq 2$
16. $y \leq 3$
17. $n \geq {}^-3$
18. $y \neq {}^-3$
19. $t + 2 < 3$
20. $n + 1 \geq 2$
21. $r - 1 < 3$
22. $k - 3 \geq {}^-4$
23. $m + 3 \neq 5$
24. $n - 3 < 0$
25. $w + 4 > 0$
26. $t + 1 \geq 0$

The coordinate plane

Ordered pairs of rational numbers can be graphed on a coordinate plane. Notice that the first coordinate axis and the second coordinate axis each have negative numbers as well as positive numbers.

To graph the ordered pair ($^-4$, 3): Start at the origin and move along the first coordinate axis to the first coordinate of the ordered pair, $^-4$. Then move parallel to the second coordinate axis to the second coordinate, 3. We have labeled this point with the letter *P*. So, point *P* has coordinates ($^-4$, 3).

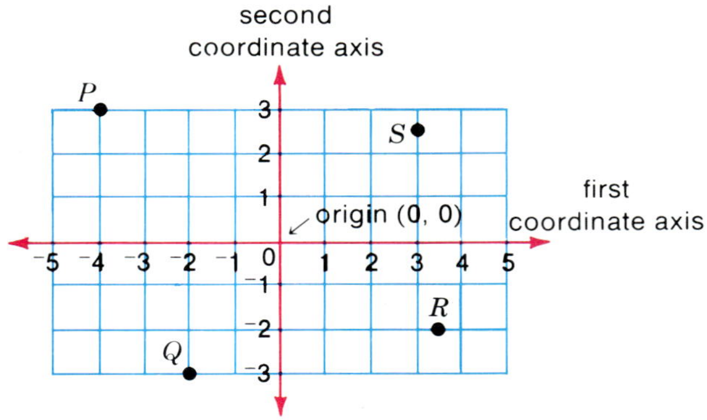

The coordinates of point *S* are $\left(3, 2\frac{1}{2}\right)$. What are the coordinates of point *Q*? of point *R*?

EXERCISES

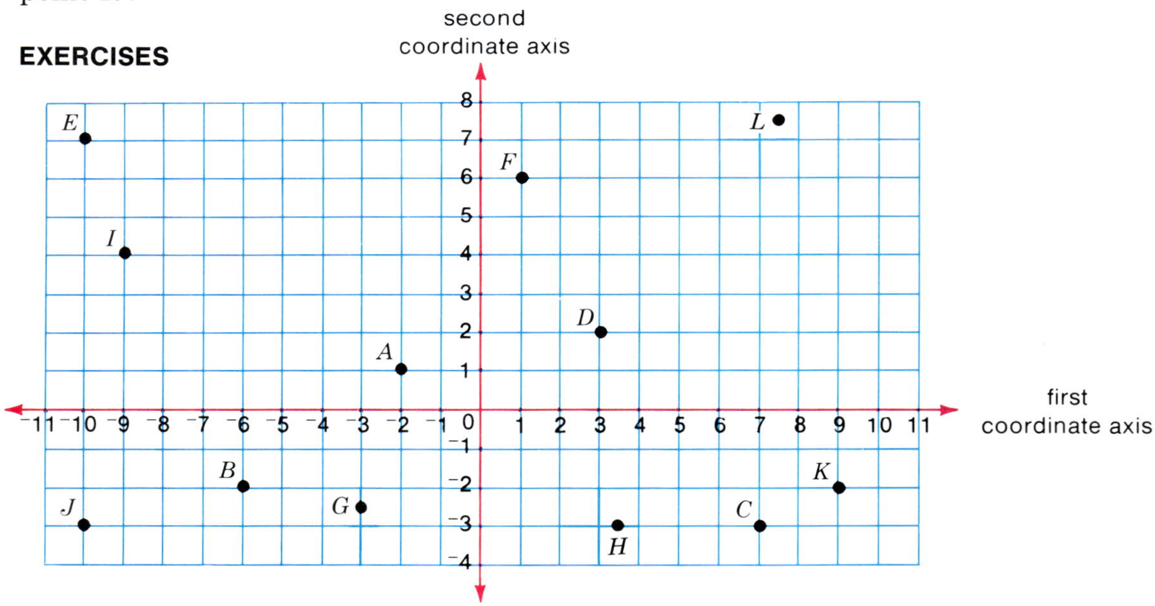

Give the coordinates of

1. *A* **2.** *B* **3.** *C* **4.** *D* **5.** *E* **6.** *F*

7. *G* **8.** *H* **9.** *I* **10.** *J* **11.** *K* **12.** *L*

Graph the following ordered pairs. Label each point with its coordinates.

13. $(3, {}^-6)$ **14.** $(4, {}^-2)$ **15.** $({}^-3, {}^-1)$ **16.** $(2, {}^-4)$

17. $({}^-3, 5)$ **18.** $({}^-2, {}^-2)$ **19.** $({}^-3, 0)$ **20.** $({}^-4, 4)$

21. $\left(4, \frac{1}{2}\right)$ **22.** $\left(\frac{1}{2}, {}^-3\right)$ **23.** $\left({}^-5, 2\frac{1}{2}\right)$ **24.** $\left(3\frac{1}{2}, {}^-1\right)$

25. Here are some ordered pairs in which the second coordinate is twice the first.
$({}^-2, {}^-4)$, $({}^-1, {}^-2)$, $\left(\frac{{}^-1}{2}, {}^-1\right)$
- **a.** List 7 more such ordered pairs.
- **b.** Graph all 10 ordered pairs.
- **c.** Do the points "line up"?

★ **26.** **a.** List 10 ordered pairs in which the second coordinate is the square of the first.
- **b.** Graph the ordered pairs.
- **c.** Do the points "line up"?

Challenge!

27. Copy the triangle on graph paper. Now draw the triangle that you get if you

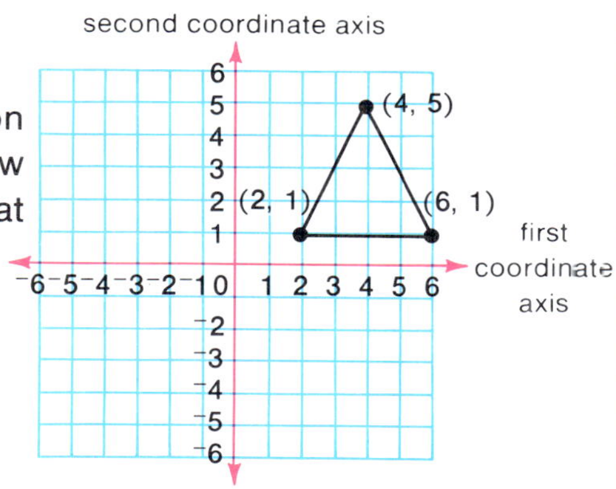

- **a.** multiply the first coordinate of each point by $^-1$ and keep the same second coordinate.
- **b.** multiply both coordinates of each point by $^-1$.
- **c.** multiply the second coordinate of each point by $^-1$ and keep the same first coordinate.
- **d.** add 6 to the first coordinate of each point and keep the same second coordinate.
- **e.** add $^-5$ to both coordinates of each point.
- **f.** add $^-4$ to the second coordinate of each point and keep the same first coordinate.

KEEPING SKILLS SHARP

Estimate to find the two incorrect answers.

1. a. $2.8 \times 7.5 = 21$

b. $5.6 \times 0.25 = 0.140$

c. $24.1 \div 3.6 = 66.9$

d. $8.67 \div 4.25 = 2.04$

Multiply or divide. Round each quotient to the nearest hundredth.

2. 7.6×0.43

3. 6.7×3.4

4. 0.82×0.4

5. 0.3×0.2

6. 1.23×4.06

7. $4.3\overline{)28.71}$

8. $0.6\overline{)5.33}$

9. $0.28\overline{)1.44}$

10. $6\overline{)38.1}$

11. $1.7\overline{)0.643}$

Equations with two variables

The difference between our ages is 20 years.

An equation with two variables shows how the two ages are related.

Let x be the mother's age and y be the son's age.

Equation: $x - y = 20$

The table of values shows some pairs of numbers that make the equation true. These pairs of numbers are solutions of the equation $x - y = 20$. There is an infinite number of solutions.

Notice that when the mother is 25 her son is 5. How old is the son when the mother is 45? 60?

$x - y = 20$

x	y
25	5
30	10
34	14
45	25
60	?

To find pairs of numbers that are solutions of the equation $2x + y = 10$, *choose* a value for x and solve the equation for y. Repeat for other chosen values for x.

If you choose 3 for x, then

$$2(3) + y = 10$$
$$6 + y = 10$$
$$6 - 6 + y = 10 - 6$$
$$y = 4$$

So, when $x = 3$, $y = 4$.

$2x + y = 10$

x	y
3	4
5	0
1	8
$^{-}2$	14
$^{-}5$	20
10	$^{-}10$

EXERCISES

Copy and complete the table of values for each equation.

1. $x + y = 0$

x	y
3	?
5	?
0	?
$^{-}2$	?
$^{-}5$	?

2. $3x + y = 12$

x	y
1	?
2	?
3	?
$^{-}1$	?
$^{-}2$	?

3. $2x + y = -12$

x	y
$^{-}2$	?
$^{-}1$	?
0	?
1	?
2	?

4. $\frac{x}{2} + y = 8$

x	y
$^{-}4$	?
$^{-}2$	?
0	?
2	?
4	?

5. $3x + 2y = 24$

x	y
0	?
1	?
$^{-}1$	?
2	?
$^{-}2$	?

6. $\frac{x}{3} + 3y = 6$

x	y
3	?
$^{-}3$	?
0	?
6	?
$^{-}6$	?

Make a table and find four pairs of numbers that are solutions of the equation.

7. $x + y = 16$
8. $2x + y = 12$
9. $3x + y = {}^{-}10$
10. $3x + y = 6$
11. $2x + 3y = {}^{-}12$
12. $3x + 4y = 24$

Challenge!

Write an equation with two variables for each table of values.

13.

x	y
0	8
1	6
2	4
$^{-}2$	12

14.

x	y
0	3
2	2
4	1
6	0

15.

x	y
0	6
1	$4\frac{1}{2}$
2	3
3	$1\frac{1}{2}$

Graphing equations with two variables

Here is how to graph an equation with two variables.

Step 1. Make a table of values.

$2x + y = 4$

x	y
$^-2$	8
0	4
2	0
4	$^-4$

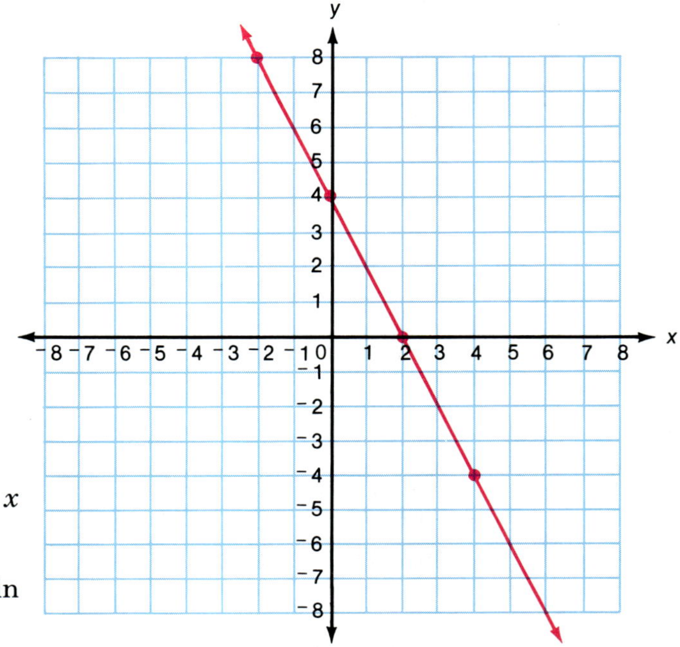

Step 2. Label the first coordinate axis x and the second coordinate axis y.

Step 3. Graph the ordered pairs listed in the table of values.

Step 4. Draw a line through the points.

The graphs of $x + y = 3$ and $2x + y = 5$ are shown on the same coordinate plane.

$x + y = 3$

x	y
0	3
$^-1$	4
1	2
$^-2$	5

$2x + y = 5$

x	y
0	5
1	3
3	$^-1$
4	$^-3$

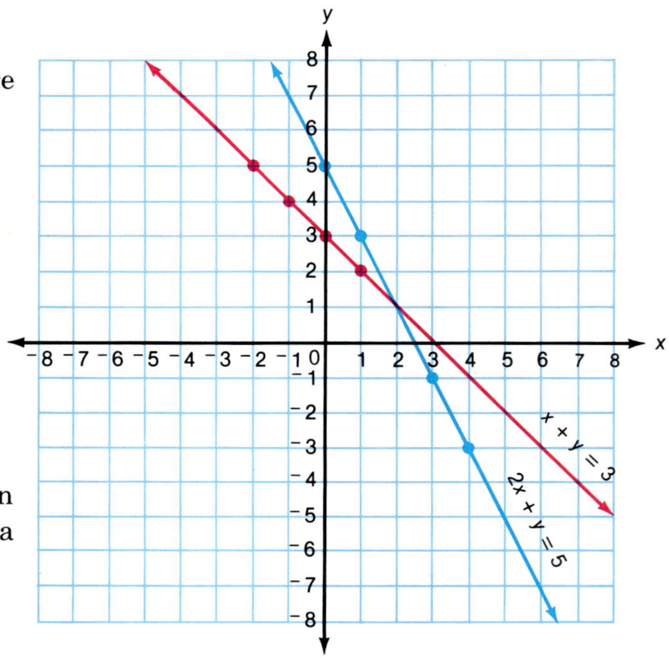

The point with coordinates (2, 1) is on both lines. So, the ordered pair (2, 1) is a solution of both equations.

EXERCISES

First complete the table of values. Then graph the equation.

1. $2x + y = 6$

x	y
0	?
1	?
$^-1$	?
2	?

2. $x + 2y = 6$

x	y
0	?
2	?
$^-2$	?
1	?

3. $2x + y = 8$

x	y
0	?
1	?
$^-1$	?
2	?

Graph the equation.

4. $x + y = 0$

5. $2x + y = {}^-4$

6. $3x + y = 6$

7. $4x + y = 12$

8. $x + 3y = {}^-6$

9. $x + 2y = 10$

Graph each pair of equations on the same coordinate plane. By looking at your graphs, give the ordered pair that is a solution of both equations.

10. $x + y = 0$
$2x + y = {}^-2$

11. $3x + y = 7$
$x + 2y = 9$

12. $2x + y = 9$
$x + 2y = 3$

Challenge!

Find the equation for each graph. [*Hint:* List some ordered pairs and look for a pattern.]

13.

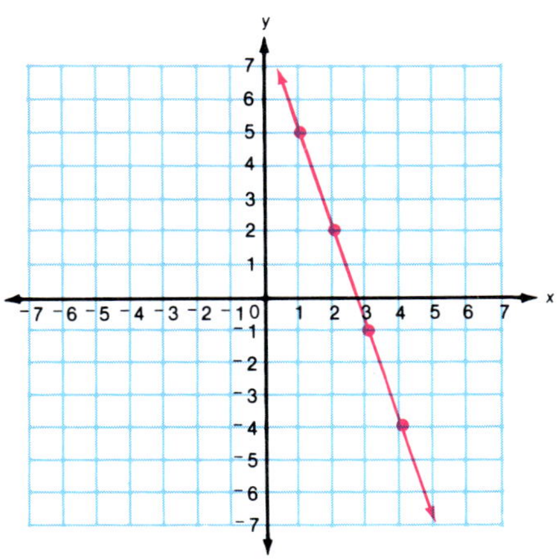

14.

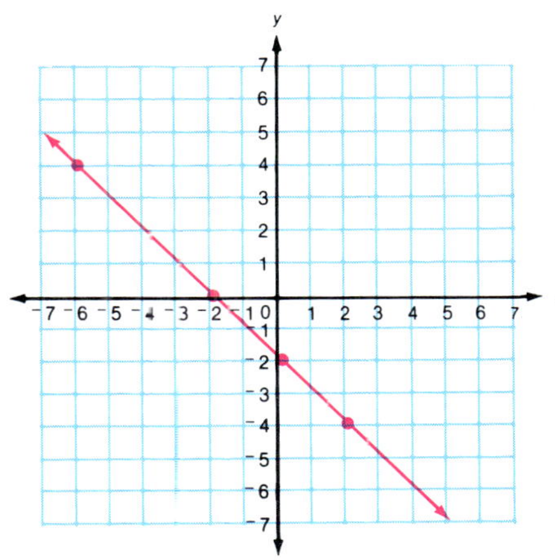

Problem Solving

Graphs

Imagine being on a ship that is sailing at a constant rate of 4 kilometers per hour. This graph shows the relationship between the sailing time in hours and the distance sailed in kilometers.

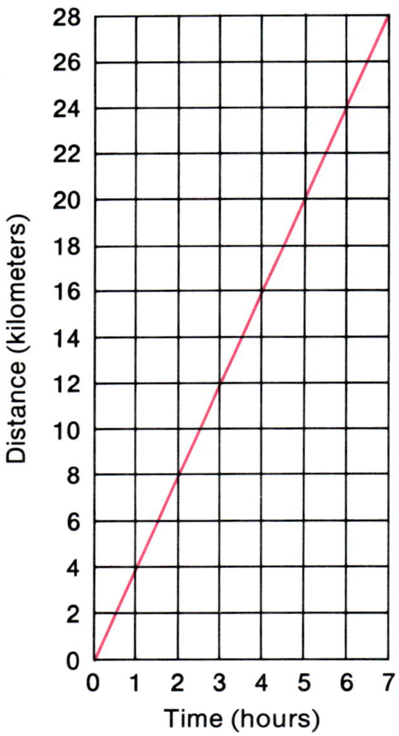

EXERCISES

Use the graph above to answer the questions.

1. How far would you sail in 3 hours?
2. How far would you sail in 6.5 hours?
3. How long would it take you to sail 10 km?
4. How long would it take you to sail 19 km?

The higher you are above the surface of the ocean, the farther away the horizon is. This graph shows how the two distances are related.

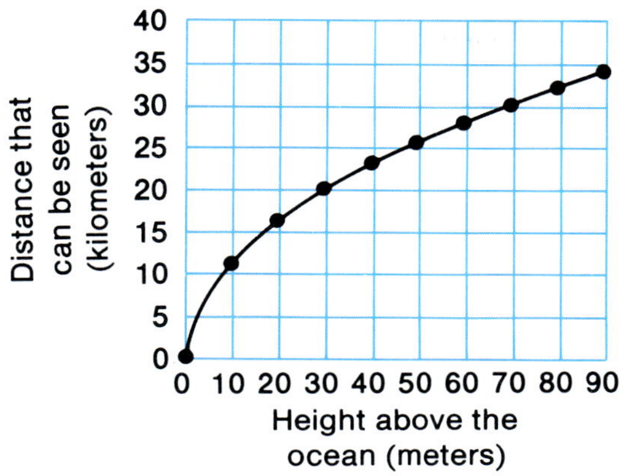

5. How far could you see if you were 20 meters above the ocean?

6. If the top of the mast of the Santa Maria was 30 meters above the ocean, how far could a sailor at the top of the mast see?

7. How many meters above the ocean would you have to be to be able to see 30 kilometers?

If an object is dropped from the top of the mast of the ship, it falls faster and faster until it strikes the water. This graph shows how far an object will fall in a given number of seconds.

8. How far will an object fall in 1 second? In 2 seconds? How much farther does it fall in the second second than in the first second?

9. How far will an object fall in 6 seconds? In 7 seconds? How far will it fall during the 7th second? How much farther does it fall during the 7th second than it does during the 1st second?

10. About how long does it take an object to fall 80 meters? 160 meters?

11. A sailor is in the rigging of a ship when his hat drops off and falls 30 meters to the deck. How long does the hat take to reach the deck?

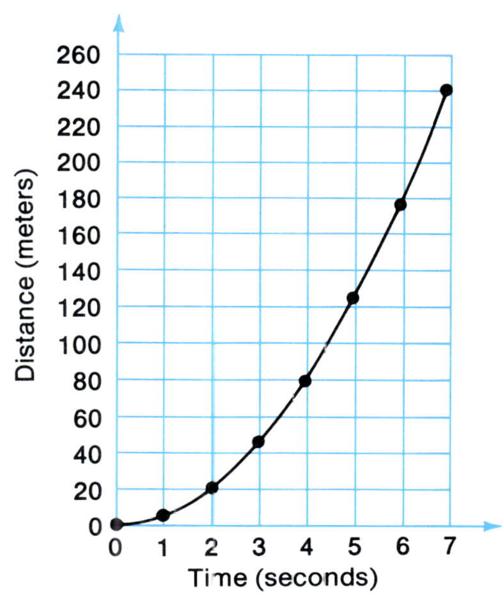

Chapter Checkup

Give each sum. [pages 292–293, 296]

1. $\frac{^-3}{8} + \frac{1}{4}$
2. $\frac{5}{6} + \frac{^-2}{3}$
3. $2\frac{1}{2} + 1\frac{2}{3}$
4. $^-3.2 + {^-4.6}$

Give each difference. [pages 292–293, 296]

5. $\frac{3}{5} - \frac{1}{10}$
6. $\frac{3}{4} - \frac{^-3}{8}$
7. $^-2\frac{1}{2} - {^-1}\frac{1}{4}$
8. $^-5.6 - 0.8$

Give each product. [pages 294–296]

9. $\frac{^-5}{8} \times \frac{3}{4}$
10. $\frac{^-3}{5} \times \frac{^-4}{5}$
11. $1\frac{1}{5} \times {^-1}\frac{1}{5}$
12. 5.3×4.8

Give each quotient. [pages 294–296]

13. $\frac{3}{4} \div \frac{^-1}{4}$
14. $\frac{^-5}{8} \div \frac{2}{3}$
15. $1\frac{1}{4} \div 2\frac{1}{2}$
16. $^-8.4 \div {^-0.03}$

Solve. [pages 302–306]

17. $4x + 2x = 30$
18. $5 = y - 7y$
19. $8 = {^-4}a + 20$
20. $\frac{n}{6} - 4 = {^-3}$
21. $6 = 5t + 3t - 1.2$
22. $6b - b + 8 = {^-7}$

Write an equation and solve. [page 307]

23. John has 76 stamps in his collection. He has 4 times as many stamps as Marcia. How many stamps does Marcia have in her collection?

24. Bill bought 101 stamps. He put the same number of stamps on each of 8 pages of his album and had 5 left over. How many stamps did he put on each page?

Give the coordinates of [pages 310–311]

25. point A.
26. point B.
27. point C.
28. point D.

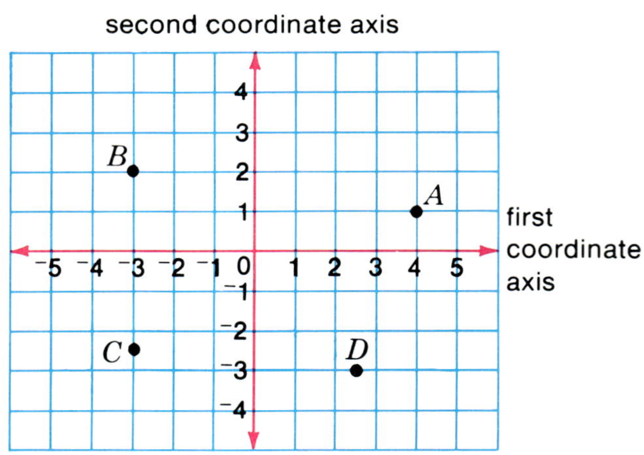

Chapter Project

STOCK QUOTATIONS

DAILY STOCK QUOTATIONS

Year High	Year Low	Stocks	Div.	Day High	Day Low	Close	Net Chg.
36½	22	Bk of NY	2.20	24½	24	24½	+¼
27¼	8¾	Bk of Va	.88	9⅛	9	9⅛	+⅜
57¼	29½	BankTr	3	31	30⅝	30⅞	+⅛
34	17¾	BarbOil	.80e	19	19	19	− ¼
24¼	9⅝	BardCR	.20	10	9⅝	9⅝	− ¼
22¾	2¼	BarnM	2.79e	4½	4⅛	4½	+⅜
10¼	6⅜	BasicInc	.40	6½	6¼	6½	− ⅛
30	24¾	Basic pf	2.50	24¾	24¾	24¾	. . .
15¼	10¾	BatesMf	.20	12¼	12⅛	12⅛	− ¼
19½	14¾	BatesMf pf	1	15⅞	15¾	15¾	. . .
22¼	4⅝	BathInd	.40	6⅜	6	6⅛	− ⅜
45	17¾	BauschL	.60	26¼	25¼	25⅜	− ⅝
48⅝	24⅛	BaxtLab	.17	26	24¾	25	− ¾
6	4½	BaykCig	.32	4⅝	4½	4½	. . .
28⅞	10	Bearing	.32	11⅛	11	11⅛	+⅛
23⅜	12¾	BeatFds	.72	12⅞	12⅛	12⅝	− ¼
40	16	Beckmn	.50	19⅛	18	18½	− ⅝
21¾	16¼	Beker	.28	17⅞	17	17¾	+¼
18½	9½	BelcoP	.77t	11	10¾	11	+⅛

The year's high for a share of Bank of Virginia stock was $\$27\frac{1}{4}$, or \$27.25.

The year's low for a share of this stock was \$8.75.

The day's net change (the difference between yesterday's closing price and today's closing price) was $+\frac{3}{8}$. That means that the price of each share of stock gained, or went up, $\frac{3}{8}$ of a dollar, or $37\frac{1}{2}$¢. Notice that some stocks had a negative net change. That means that the price of the stock lost, or went down, that day.

1. What was the day's high for a share of Bank of Virginia? The day's low? At what price was the stock selling when the market closed?
2. Select eight stocks from the list above. Pretend that you own 100 shares of each of the eight stocks. Compute your loss or gain from the quotations given above.
3. Select 10 stocks that are listed on the New York Stock Exchange. Pretend that you own 100 shares of each stock. For one week, compute the total dollar net change of your 1000 shares of stock.

Chapter Review

$^{-}7 + 2 = {}^{-}5$

To subtract a number, add its opposite.

$^{-}8 - 3 = {}^{-}8 + {}^{-}3$

$= {}^{-}11$

If the signs are the same, the product or quotient is positive.

$6 \times 3 = 18$

$^{-}16 \div {}^{-}2 = 8$

If the signs are different, the product or quotient is negative.

$^{-}9 \times 4 = {}^{-}36$

$21 \div {}^{-}7 = {}^{-}3$

$4y + 2y - 3 = {}^{-}27$

$6y - 3 = {}^{-}27$

$6y = {}^{-}24$

$y = {}^{-}4$

To graph an ordered pair:

Start at (0, 0). Move right or left for the first number. Move up or down for the second number.

Give each sum.

1. $6 + {}^{-}8$
2. $^{-}5 + {}^{-}8$
3. $\frac{^{-}2}{3} + \frac{1}{6}$
4. $\frac{3}{4} + \frac{^{-}1}{5}$
5. $^{-}2.8 + 1.6$
6. $7.4 + 3.9$

Give each difference.

7. $9 - 12$
8. $^{-}6 - {}^{-}7$
9. $\frac{3}{4} - \frac{^{-}1}{5}$
10. $\frac{^{-}5}{9} - \frac{2}{3}$
11. $5.3 - {}^{-}0.7$
12. $^{-}2.4 - {}^{-}6.9$

Give each product.

13. $^{-}6 \times {}^{-}4$
14. $9 \times {}^{-}7$
15. $\frac{^{-}2}{3} \times \frac{4}{5}$
16. $\frac{5}{8} \times \frac{3}{4}$
17. $^{-}6.5 \times {}^{-}2.8$
18. $0.72 \times {}^{-}0.6$

Give each quotient.

19. $18 \div {}^{-}3$
20. $^{-}30 \div {}^{-}6$
21. $\frac{3}{5} \div \frac{2}{3}$
22. $\frac{5}{4} \div \frac{^{-}2}{3}$
23. $^{-}7.2 \div 0.9$
24. $^{-}6.3 \div {}^{-}0.03$

Solve.

25. $w - 2 = {}^{-}6$
26. $3 = y + 8$
27. $^{-}2k = 15$
28. $8 = \frac{n}{2}$
29. $22 = 4a - 18$
30. $\frac{b}{^{-}5} + 2 = 3.4$
31. $12m - 3m - 7 = 11$
32. $2t + t + 6 = {}^{-}1$

Give the coordinates of

33. point A.
34. point B.
35. point C.
36. point D.

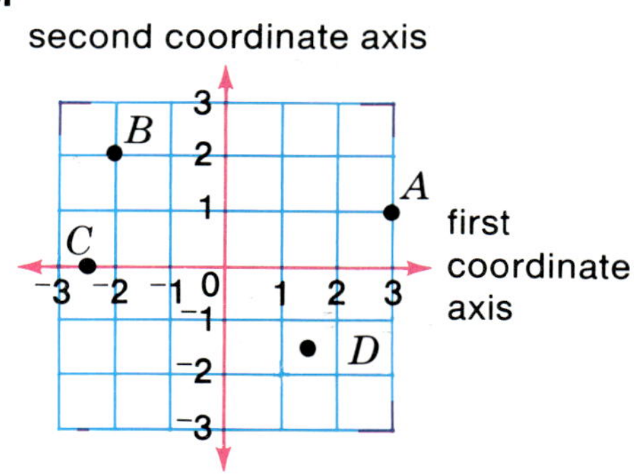

Chapter Challenge

GRAPHING EQUATIONS

In this chapter you graphed solutions to equations with two variables. In each case the graph was a straight line. Now you will work with graphs that are not straight lines.

EXAMPLE. $y = 2x^2$

x	y
$^-1\frac{1}{2}$	$4\frac{1}{2}$
$1\frac{1}{2}$	$4\frac{1}{2}$
$^-1$	2
1	2
$^-\frac{1}{2}$	$\frac{1}{2}$
$\frac{1}{2}$	$\frac{1}{2}$
0	0

$2\left(\frac{^-3}{2}\right)^2 = 2\left(\frac{9}{4}\right) = 4\frac{1}{2}$

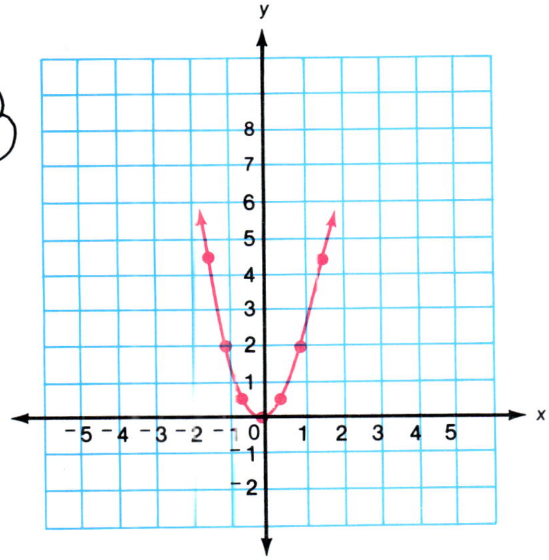

EXERCISES

Graph the solutions of these equations.

Use a different pair of axes for each exercise.

1. $y = x^2$ **2.** $y = \frac{1}{2}x^2$ **3.** $y = x^2 - 3$ **4.** $x = y^2$

Complete the table for the equation $y = x^3$.

5.

x	0	$\frac{1}{2}$	$^-\frac{1}{2}$	1	$^-1$	2	$^-2$
y	0	$\frac{1}{8}$	$^-\frac{1}{8}$	?	?	?	?

6. Graph the solutions of the equation, $y = x^3$

Cumulative Checkup

STANDARDIZED FORMAT

Choose the correct letter.

1. Divide. $0.36\overline{)8.294}$
Round the quotient to the nearest hundredth.

a. 23.03
b. 23.04
c. 230.49
d. none of these

2. Give the prime factorization for 72.

a. 9×8
b. $2^2 \times 3^3$
c. $2^4 \times 3$
d. none of these

3. Give the sum.

$2\frac{3}{4} + 1\frac{1}{6}$

a. $3\frac{11}{12}$
b. $4\frac{11}{12}$
c. $3\frac{2}{5}$
d. none of these

4. Give the quotient.

$\frac{3}{5} \div \frac{5}{3}$

a. 1
b. $2\frac{7}{9}$
c. $\frac{9}{25}$
d. none of these

5. Find the surface area.

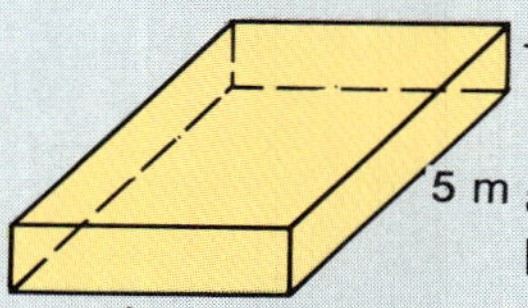

a. 20 m^2
b. 29 m^2
c. 58 m^2
d. none of these

6. Solve.

$\frac{7}{x} = \frac{10}{16}$

a. 4.375
b. 22.85
c. 11.2
d. none of these

7. Solve.

$16\frac{2}{3}\%$ of $n = 36$

a. 6
b. 96
c. 216
d. none of these

8. A parka that regularly sells for \$65.79 is on sale for 25% off. What is the sale price?

a. \$16.45
b. \$43.86
c. \$49.34
d. none of these

9. Give the difference.

$\frac{^-3}{4} - \frac{1}{2}$

a. $\frac{^-1}{4}$
b. $\frac{1}{4}$
c. $\frac{^-5}{4}$
d. $\frac{5}{4}$

10. Solve.

$^-2n + 4 = 10$

a. $^-3$
b. $^-7$
c. $^-12$
d. none of these

11. How many minutes long is 1 week?

a. 2520 **b.** 5040
c. 7200 **d.** none of these

12. Dody is 52 km from Elba. Dody is 3.7 km closer to Elba than Faro is. How far is Faro from Elba?

a. 48.3 km **b.** 49.7 km
c. 55.7 km **d.** none of these

11

Geometry

How many of these geometric figures can you find in the photograph: rectangle, triangle, trapezoid, square, parallelogram? Can you find any others?

READY OR NOT?

Match.

1. point *A* (*A*)
2. angle *ABC* ($\angle ABC$)
3. segment *AB* ($\overline{AB}$)
4. ray *AB* ($\overrightarrow{AB}$)
5. line *AB* ($\overleftrightarrow{AB}$)
6. angle *C* ($\angle C$)
7. ray *BA* ($\overrightarrow{BA}$)

a. 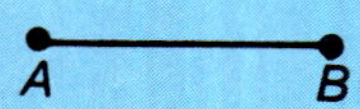

b. 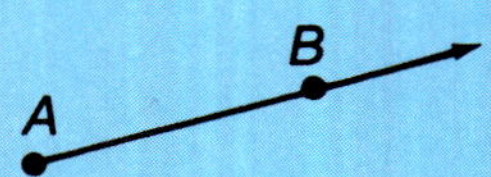

c. 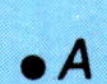

d. 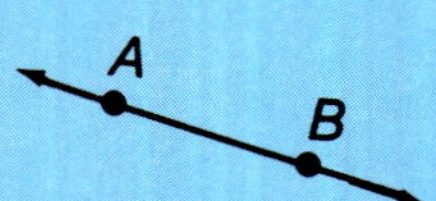

e. 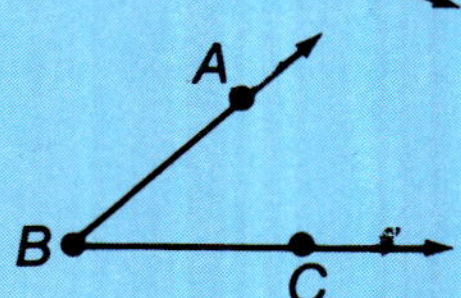

f. 

g. 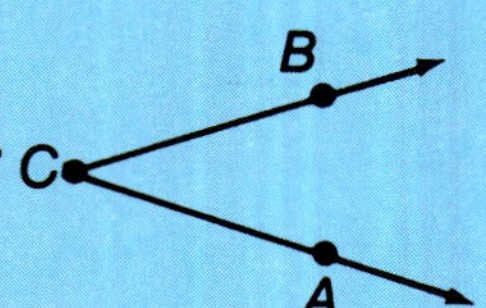

Measuring angles

Look at $\angle ABC$. $\overrightarrow{BA}$ and $\overrightarrow{BC}$ are the **sides** of the angle, and point *B* is the **vertex.**

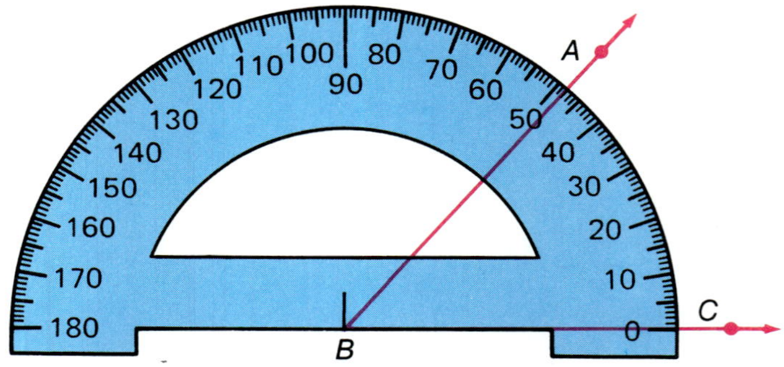

$m\angle ABC = 48°$

Read as "The measure of angle *ABC* is 48 degrees."

An **acute angle** is an angle that measures between 0° and 90°.

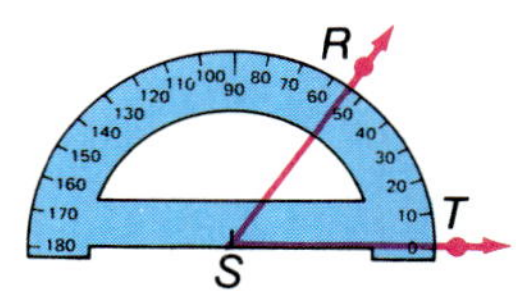

A **right angle** is an angle that measures 90°.

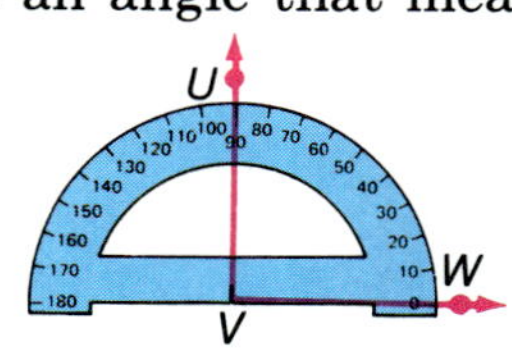

An **obtuse angle** is an angle that measures between 90° and 180°.

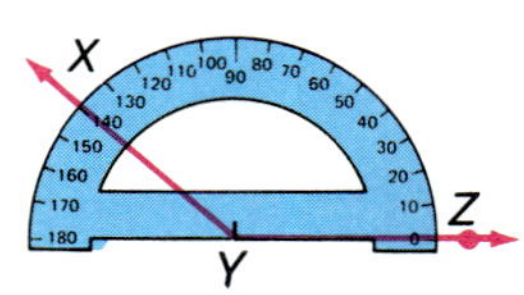

What kind of angle is $\angle ABC$?

EXERCISES

First measure the angle. Then tell whether it is acute, right, or obtuse.

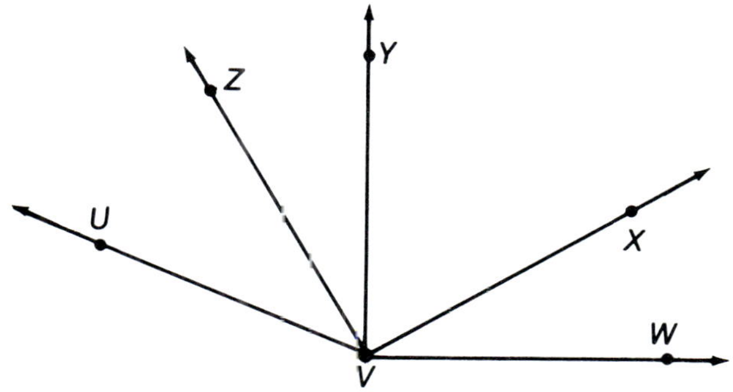

1. $\angle WVX$
2. $\angle UVZ$
3. $\angle YVZ$
4. $\angle XVZ$
5. $\angle WVY$
6. $\angle UVY$
7. $\angle WVZ$
8. $\angle UVX$
9. $\angle XVY$

Draw angles having these measures.

10. 25°
11. 72°
12. 132°
13. 90°
14. 168°

15. Follow these steps to **construct** a right angle.
   a. Draw $\overleftrightarrow{AB}$. Using A and B as centers and using the same compass setting, draw crossing arcs above and below $\overleftrightarrow{AB}$.
   b. Using a straightedge, draw a line through the points where the arcs cross.

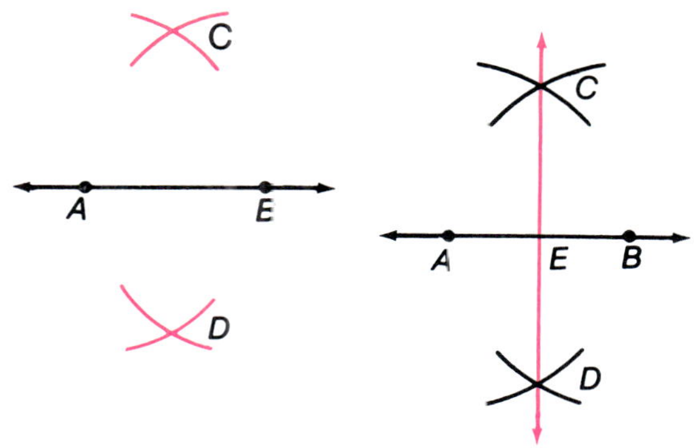

16. The construction in exercise 15 gives you four right angles. One angle is $\angle CEB$. Name the other right angles.

17. To **bisect** means to "cut into halves." Follow these steps to bisect an angle.
   a. Draw $\angle V$. Using V as the center, draw an arc that crosses the sides.
   b. Using U and W as centers and using the same compass setting, draw crossing arcs.
   c. Draw the ray from V through the point where the arcs cross.

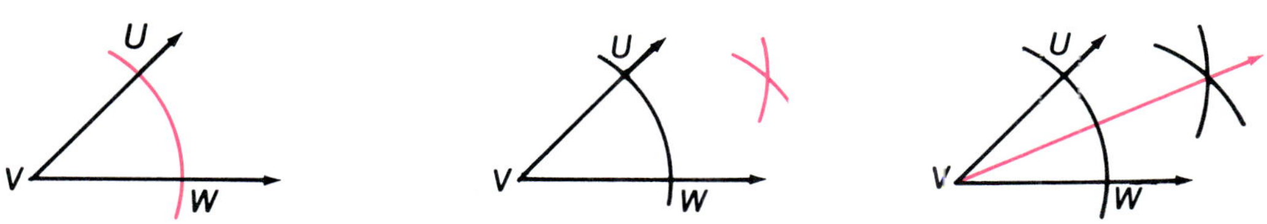

★ 18. Using only a compass and straightedge, construct a 45° angle.

★ 19. Construct a 135° angle.

Angles of polygons

The angles of the triangle were colored, torn off, and placed under the protractor.

Look at the pictures. What is the sum of the measures of the angles? The sum would have been the same regardless of the triangle we used.

For any triangle, the sum of the measures of the angles is 180°.

EXERCISES

1. Draw a triangle. Use a protractor to measure each angle.
2. What is the sum of the measures of the angles?

The measures of two angles are given. Find the measure of the remaining angle.

3.

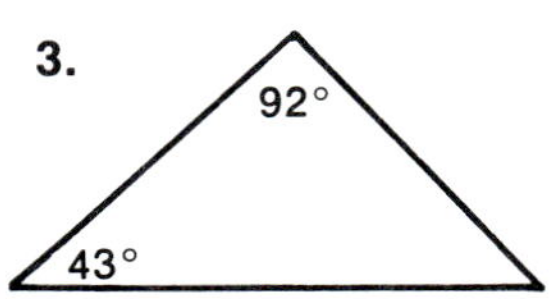

4.

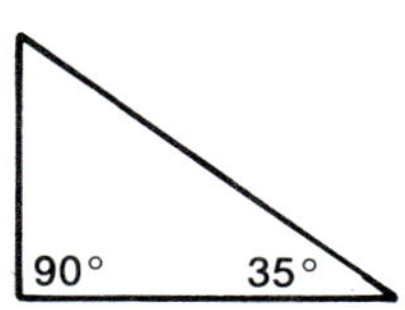

5.

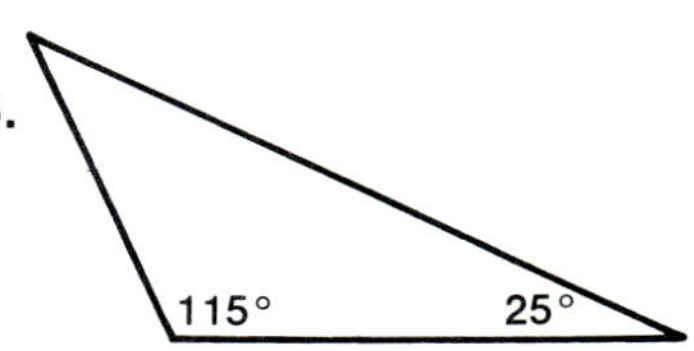

6.

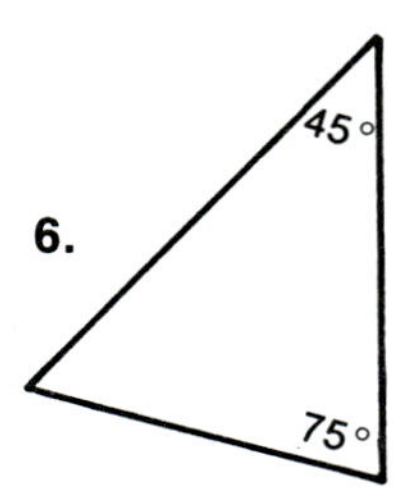

7. The red figure is called a **quadrilateral.**

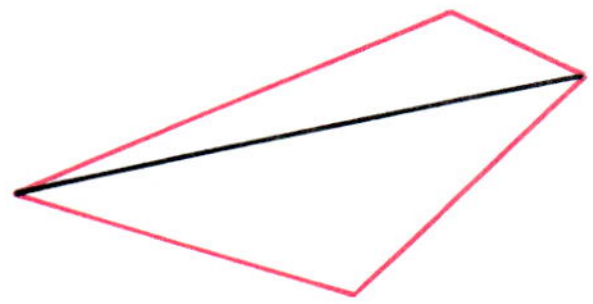

a. How many sides does a quadrilateral have?

b. How many triangles make up the quadrilateral?

c. What is the sum of the measures of the angles? [*Hint:* What is the sum of the measures of the angles of the two triangles?]

8. Copy and complete this table.

Polygon		*Number of sides*	*Number of triangles*	*Sum of the measures of the angles*
triangle	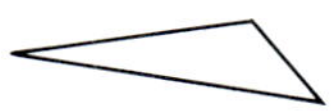	3	1	180°
quadrilateral	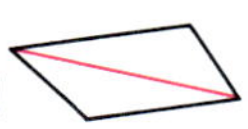	4	2	360°
pentagon	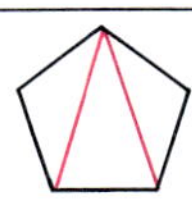	5	?	?
hexagon	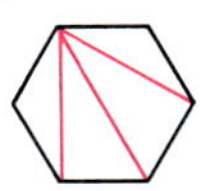	6	?	?
heptagon	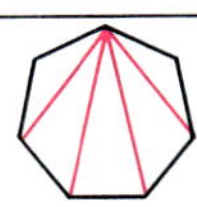	7	?	?
octagon	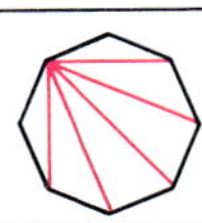	8	?	?

★ 9. Look for a pattern in your completed table. What is the relationship between the number of sides and the number of triangles?

★ 10. Give the sum of the measures of the angles for a polygon having 30 sides.

KEEPING SKILLS SHARP

Give the sum in simplest form.

1. $8\frac{2}{5} + 3\frac{3}{4}$
2. $9\frac{1}{6} + 8\frac{1}{3}$
3. $14\frac{5}{9} + 6\frac{2}{3}$
4. $6\frac{7}{8} + 9\frac{3}{8}$
5. $3\frac{2}{5} + 6\frac{1}{2}$
6. $31\frac{1}{3} + 24\frac{2}{9}$
7. $22\frac{5}{6} + 18\frac{3}{4}$
8. $15\frac{2}{3} + 17\frac{4}{5}$

Give the difference in simplest form.

9. $9\frac{1}{2} - 4\frac{3}{4}$
10. $12 - 9\frac{3}{5}$
11. $8\frac{1}{4} - 6\frac{3}{5}$
12. $4\frac{5}{9} - 2\frac{2}{3}$
13. $15 - 12\frac{5}{8}$
14. $15\frac{1}{6} - 8\frac{1}{8}$
15. $25\frac{3}{4} - 6\frac{9}{10}$
16. $17 - 16\frac{3}{8}$

Pairs of angles

Certain pairs of angles have special names.

Adjacent angles have a common side between them and a common vertex.

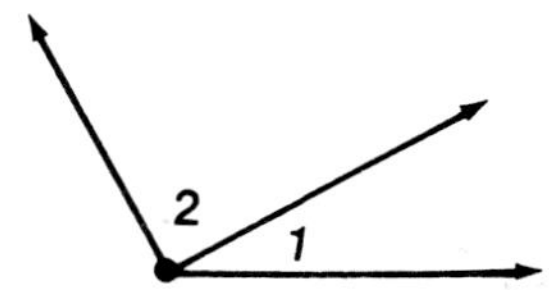

∠1 and ∠2 are adjacent angles.

Vertical angles are formed by two intersecting lines. They have a common vertex but no common side.

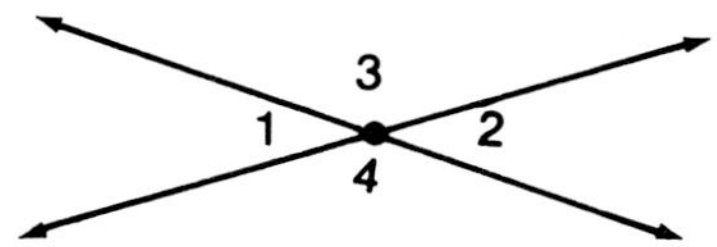

∠1 and ∠2 are vertical angles.
∠3 and ∠4 are vertical angles.

Two angles are **complementary** if the sum of their measures is 90°.

Note: Two angles need not be adjacent to be complementary.

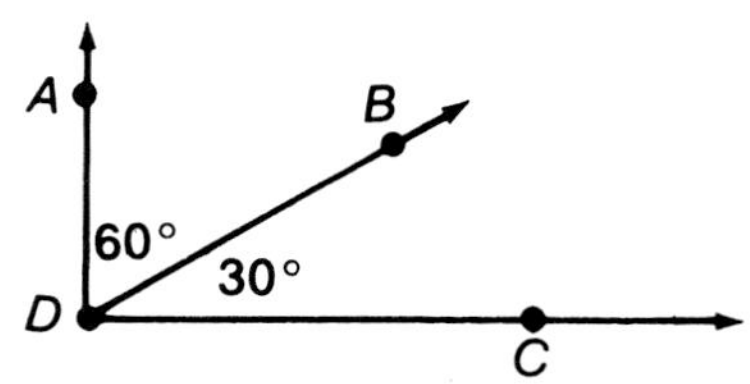

∠*ADB* and ∠*BDC* are complementary.

Two angles are **supplementary** if the sum of their measures is 180°.

Note: Two angles need not be adjacent to be supplementary.

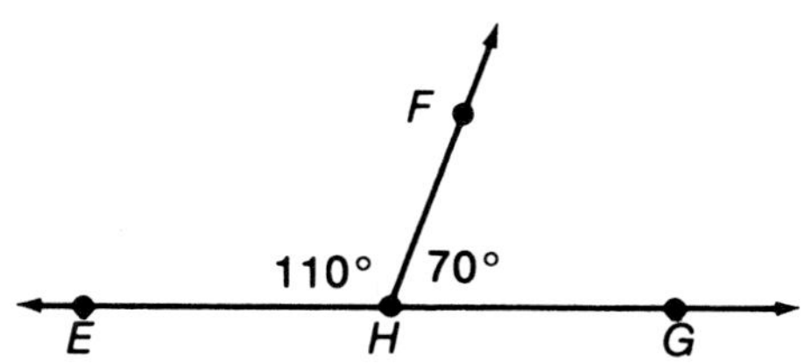

∠*EHF* and ∠*FHG* are supplementary.

EXERCISES

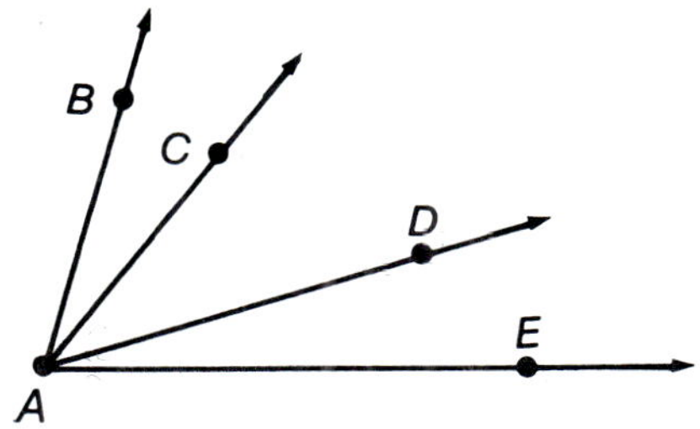

Give an angle that is adjacent to

1. ∠*BAC*
2. ∠*DAE*
3. ∠*BAD*
4. ∠*EAC*

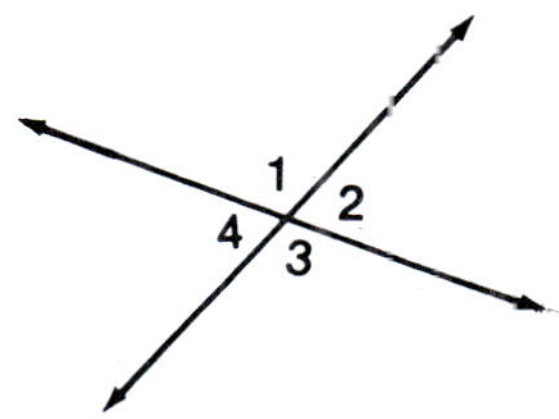

5. Find the measures of ∠1 and ∠3.
6. Find the measures of ∠2 and ∠4.
7. Do vertical angles have the same measure?

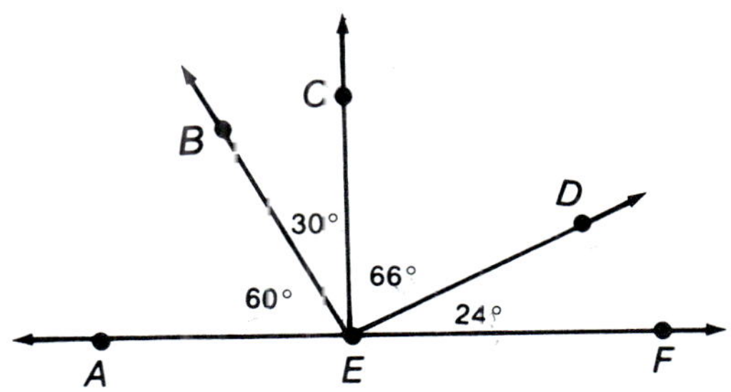

8. What angle is complementary to ∠*AEB*?
9. What angle is supplementary to ∠*AEB*?
10. What angle is complementary to ∠*CED*?
11. What angle is supplementary to ∠*FED*?

Draw a figure that satisfies the given conditions.

12. ∠*ABC* and ∠*DBC* are adjacent.
13. ∠*RST* and ∠*VST* are adjacent and complementary.
14. ∠*XYZ* and ∠*VYZ* are adjacent and supplementary.
15. ★ ∠*RST* and ∠*OSU* are vertical angles, and ∠*OSR* and ∠*RST* are supplementary.

Perpendicular lines and parallel lines

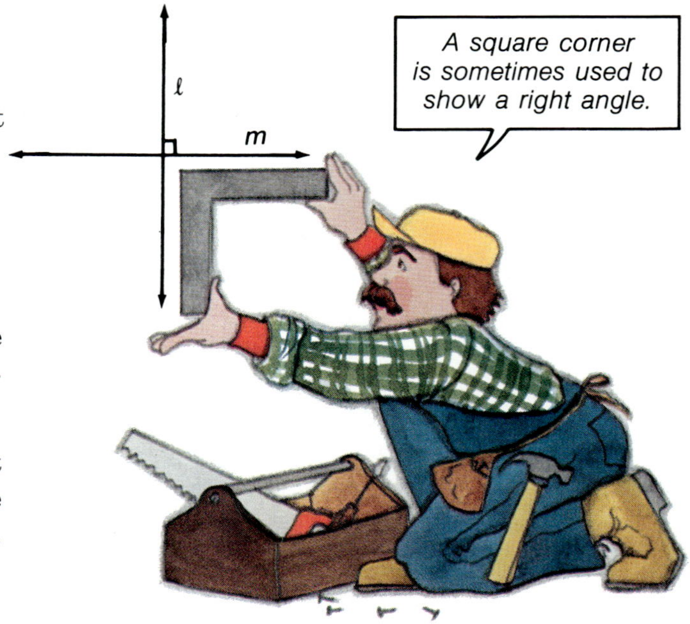

Two lines that intersect to form right angles are called **perpendicular lines.**

We say: "Line ℓ is perpendicular to line m."

We write: $\ell \perp m$.

Two segments are perpendicular if the lines that contain them are perpendicular.

The perpendicular bisector of a segment is the line that is perpendicular to the segment at the midpoint of the segment.

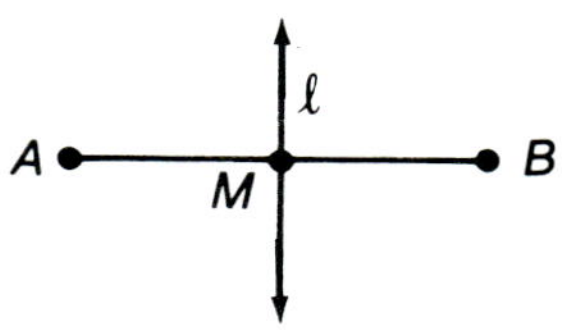

ℓ is the perpendicular bisector of $\overline{AB}$.

Two lines in a plane (a flat surface) that do not intersect are called **parallel lines.**

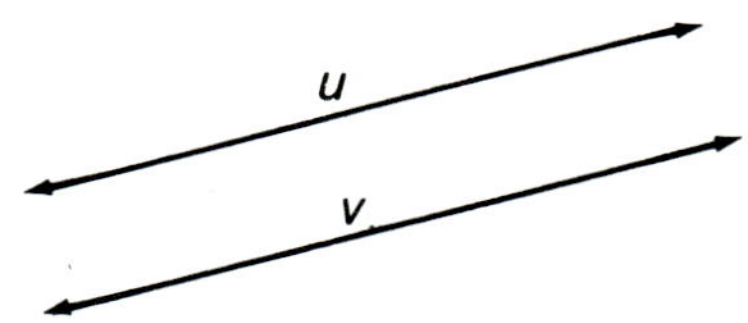

We say: "Line u is parallel to line v."
We write: Line $u \parallel$ line v.

Two segments are parallel if the lines that contain them are parallel.

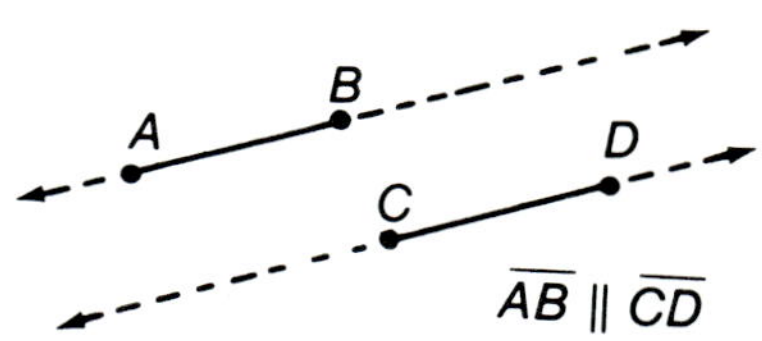

A line that intersects two lines is called a **transversal.**

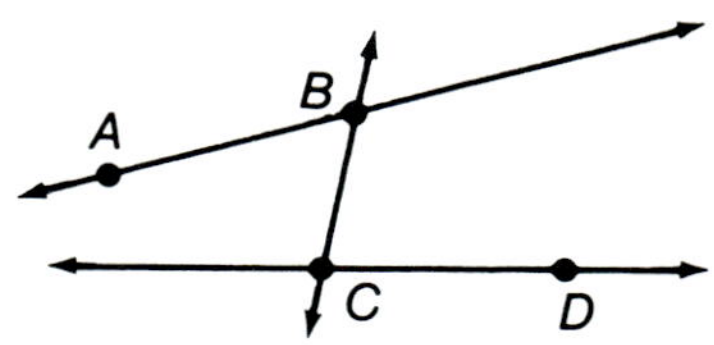

$\overleftrightarrow{CB}$ is a transversal of $\overleftrightarrow{AB}$ and $\overleftrightarrow{CD}$.

EXERCISES

True or false?

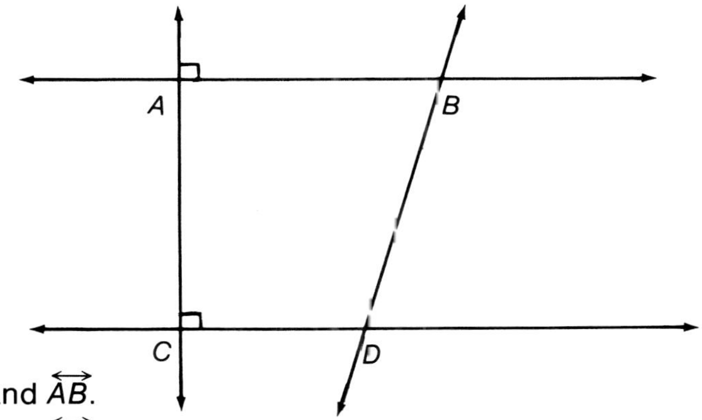

1. $\overleftrightarrow{AB} \perp \overleftrightarrow{BD}$
2. $\overleftrightarrow{CD} \parallel \overleftrightarrow{AB}$
3. $\overline{AC} \perp \overline{CD}$
4. $\overline{BD} \parallel \overline{AC}$
5. $\overleftrightarrow{CD} \perp \overleftrightarrow{AC}$
6. $\overleftrightarrow{CD} \perp \overleftrightarrow{BD}$
7. $\overleftrightarrow{AC}$ is a transversal of $\overleftrightarrow{CD}$ and $\overleftrightarrow{AB}$.
8. $\overleftrightarrow{CD}$ is a transversal of $\overleftrightarrow{AC}$ and $\overleftrightarrow{BD}$.

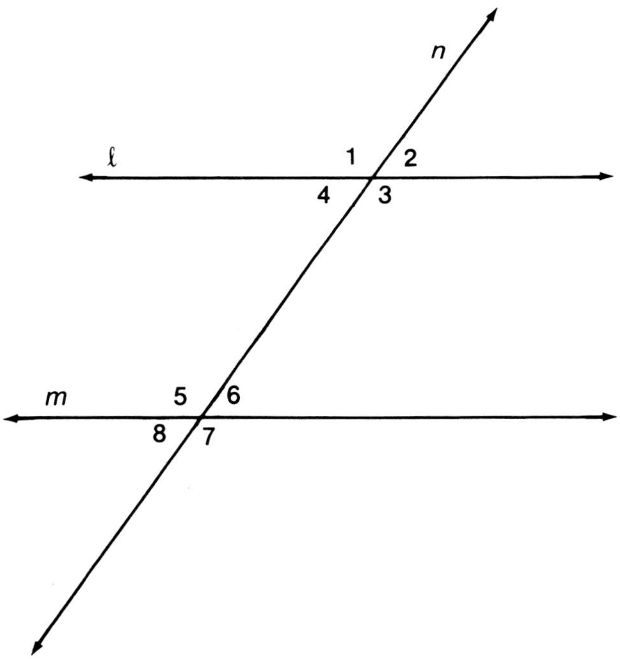

Line ℓ and line *m* are parallel. Line *n* is a transversal. Find the measure of each angle.

9. $\angle 1$ 10. $\angle 2$ 11. $\angle 3$ 12. $\angle 4$

13. $\angle 5$ 14. $\angle 6$ 15. $\angle 7$ 16. $\angle 8$

17. $\angle 4$ and $\angle 6$ are called **alternate interior angles.** $\angle 3$ and $\angle 5$ are also alternate interior angles. What can you say about the measures of these alternate interior angles?

Use a compass and straightedge to construct

18. a pair of perpendicular lines. [*Hint:* See exercise 15 on page 325.]

19. a pair of parallel lines.

Congruent figures

Figures that have the same size and shape are called **congruent figures.** Since a tracing of triangle *ABC* fits on triangle *STR*, the two triangles are congruent.

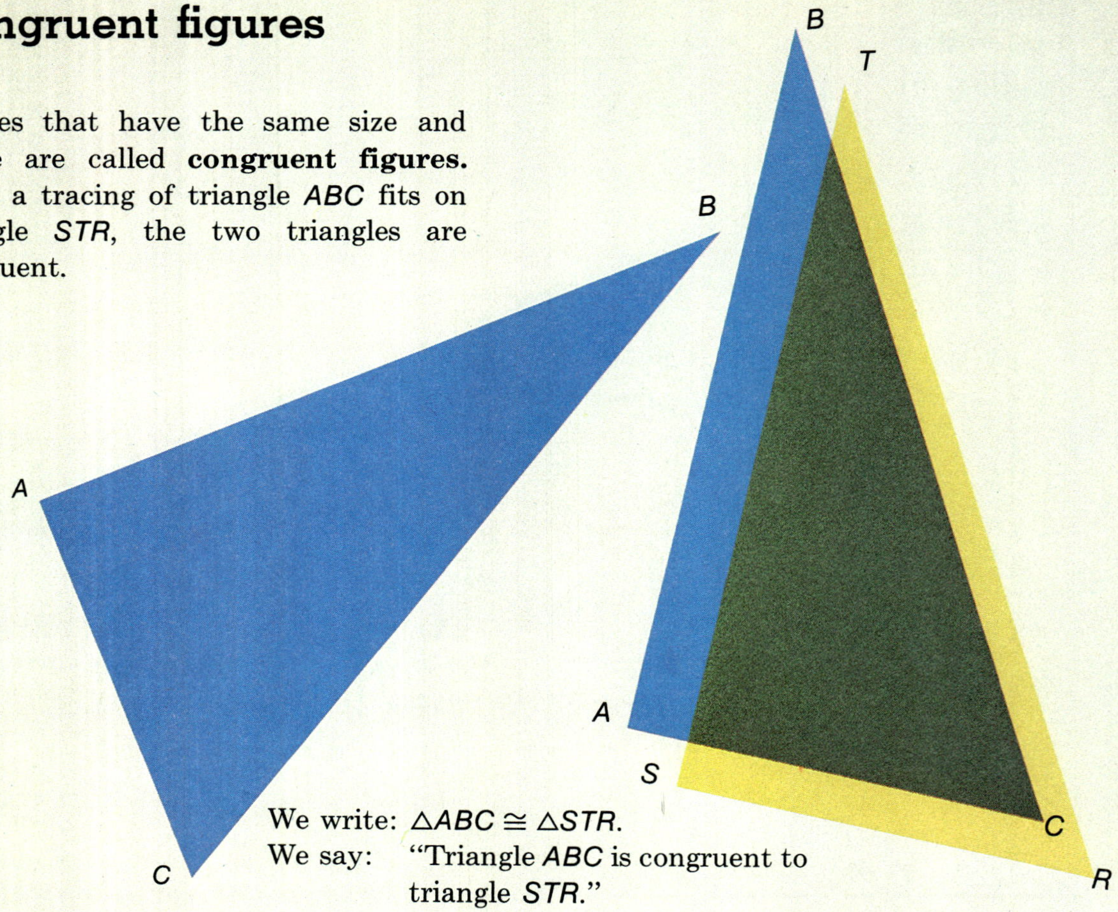

We write: $\triangle ABC \cong \triangle STR$.
We say: "Triangle *ABC* is congruent to triangle *STR*."

By looking at the picture above, we see that for this fitting the vertices are matched in this way:

$$A \longleftrightarrow S$$
$$B \longleftrightarrow T$$
$$C \longleftrightarrow R$$

The sides and angles that match are called **corresponding parts.** For a congruent fitting, the corresponding parts are congruent.

Congruent Angles	Congruent Sides
$\angle A \cong \angle S$	$\overline{AB} \cong \overline{ST}$
$\angle B \cong \angle T$	$\overline{BC} \cong \overline{TR}$
$\angle C \cong \angle R$	$\overline{CA} \cong \overline{RS}$

EXERCISES

Use a tracing to tell whether or not the two figures are congruent.

1.

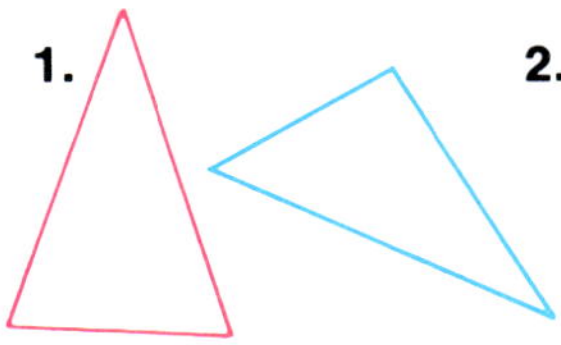

2.

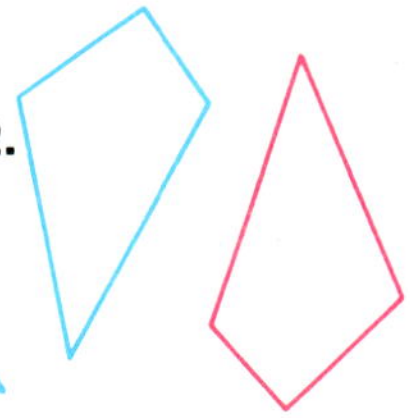

3.

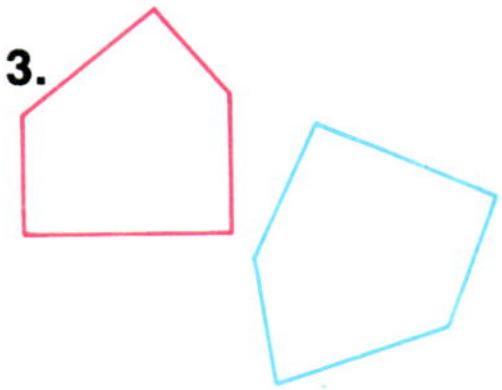

The two figures are congruent. Complete.

4. $\angle A \cong$?
5. $\angle C \cong$?
6. $\angle E \cong$?
7. $\overline{BC} \cong$?
8. $\overline{EH} \cong$?
9. $\overline{FE} \cong$?

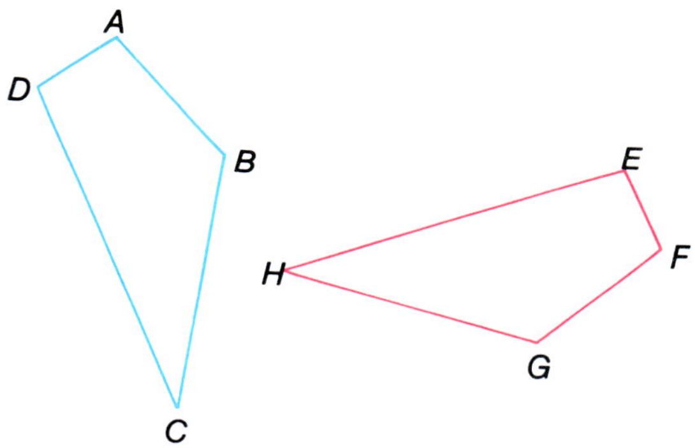

Complete.

10. $\angle R \cong$?
11. $\angle RSU \cong$?
12. $\overline{RS} \cong$?
13. $\angle RUS \cong$?

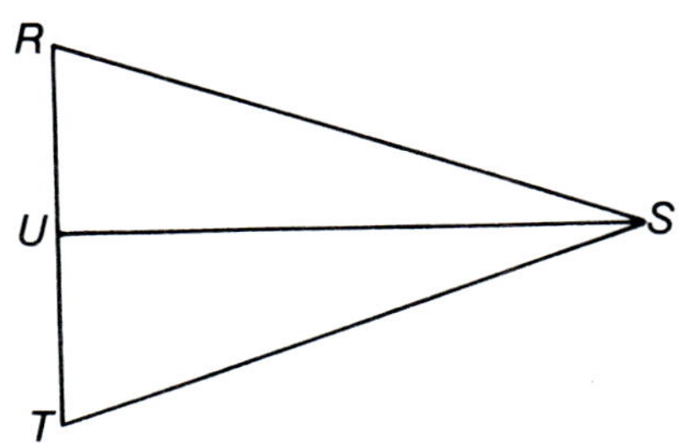

14. $\angle MNO \cong$?
15. $\overline{QN} \cong$?
16. $\angle P \cong$?
17. $\angle Q \cong$?
18. $\overline{MO} \cong$?

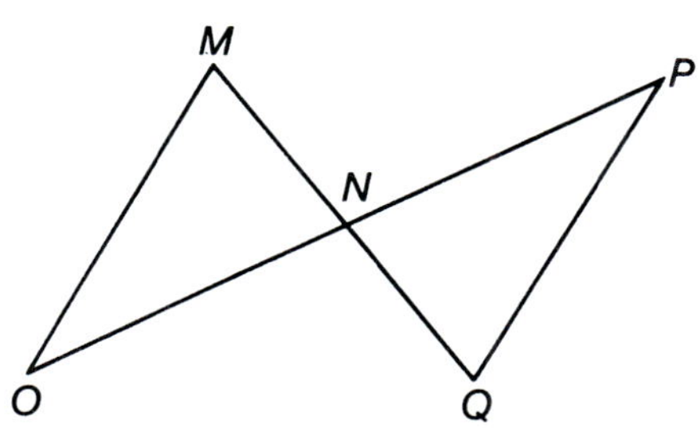

KEEPING SKILLS SHARP

Give each product in simplest form.

1. $2\frac{1}{4} \times 1\frac{1}{2}$
2. $7\frac{2}{3} \times 6\frac{1}{3}$
3. $4\frac{3}{4} \times 5$
4. $3\frac{3}{4} \times 5\frac{1}{8}$
5. $4\frac{2}{5} \times 5\frac{5}{8}$
6. $7 \times 9\frac{1}{2}$
7. $2\frac{1}{2} \times 3\frac{2}{3}$
8. $4\frac{1}{3} \times 3\frac{1}{4}$

Give each quotient in simplest form.

9. $8\frac{1}{2} \div 4\frac{1}{8}$
10. $7 \div 5\frac{2}{3}$
11. $10\frac{3}{5} \div 1\frac{4}{5}$
12. $12\frac{3}{4} \div 2\frac{1}{3}$
13. $9\frac{7}{8} \div 3\frac{1}{4}$
14. $14\frac{5}{9} \div 3\frac{2}{3}$
15. $8\frac{3}{4} \div 2\frac{1}{2}$
16. $12\frac{2}{3} \div 3\frac{3}{8}$

Constructions and triangles

Here is how to use just a compass and a straightedge to construct a segment congruent to $\overline{AB}$.

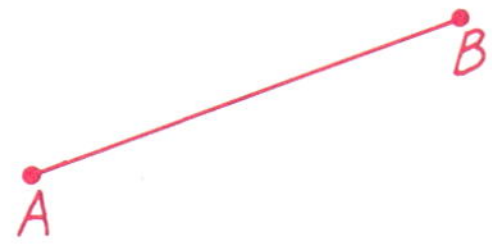

Step 1. Use a straightedge to draw $\overleftrightarrow{CX}$.

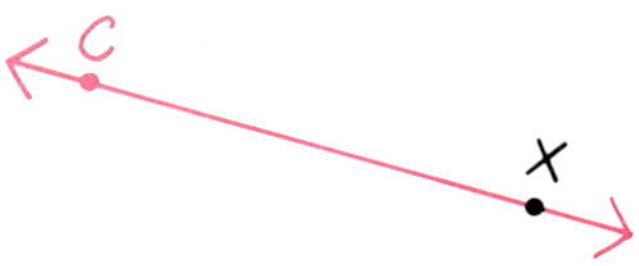

Step 2. Set the compass legs on points A and B.

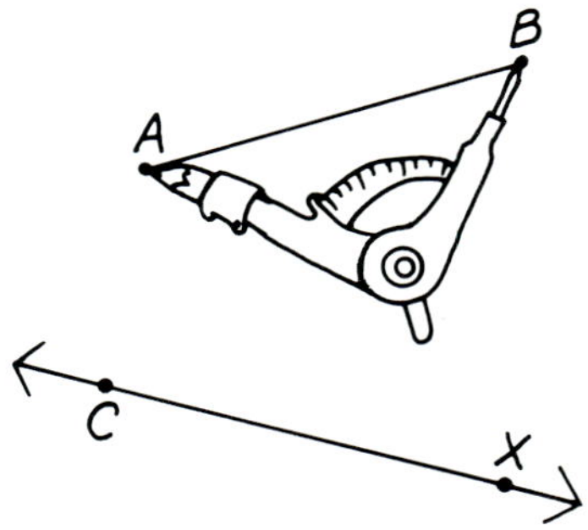

Step 3. Using point C as the center and the same compass setting, draw an arc as shown. $\overline{CD} \cong \overline{AB}$.

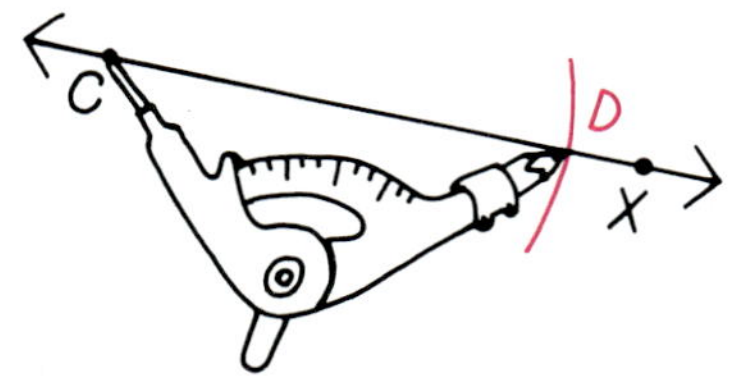

Here is how to construct an angle congruent to $\angle M$.

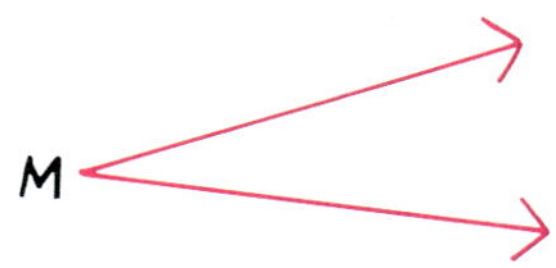

Step 1. Use a straightedge to draw $\overrightarrow{NP}$.

Step 2. Using point M as the center, draw an arc that intersects the sides of $\angle M$.

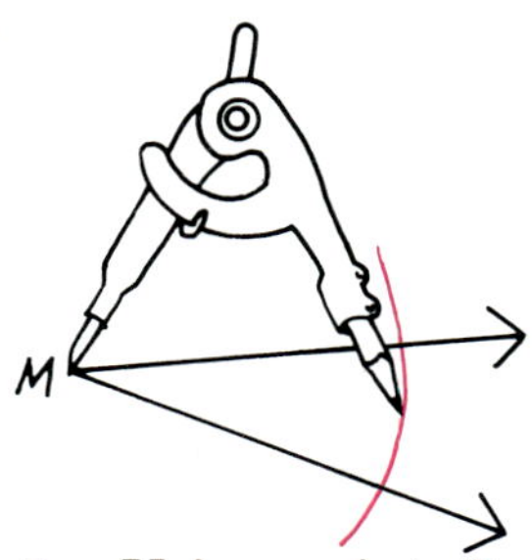

Step 3. Using point N as the center and the same compass setting, draw an arc. Label the intersection point O.

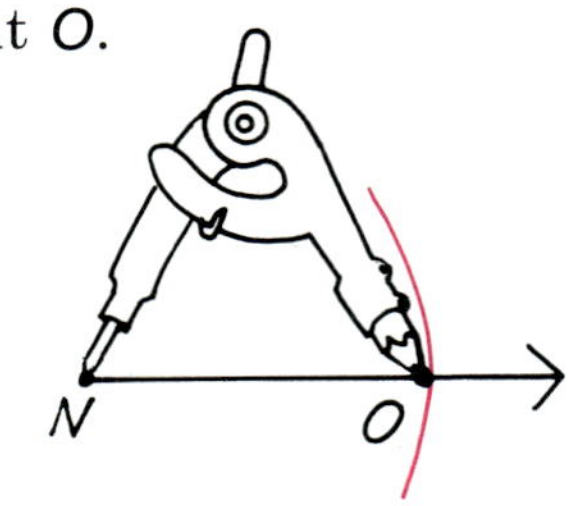

Step 4. Set the compass points on the two intersections.

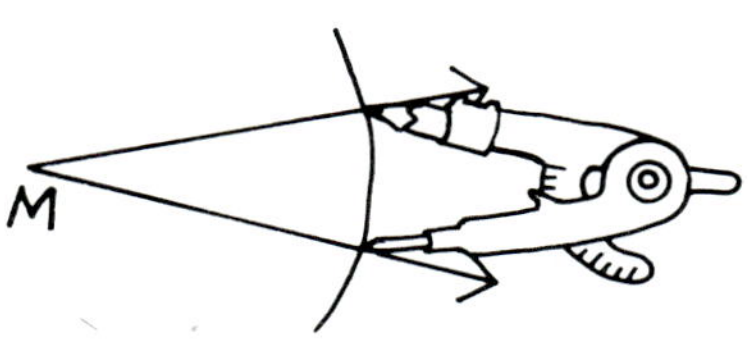

Step 5. Using point O as the center and the same compass setting, draw an arc as shown.

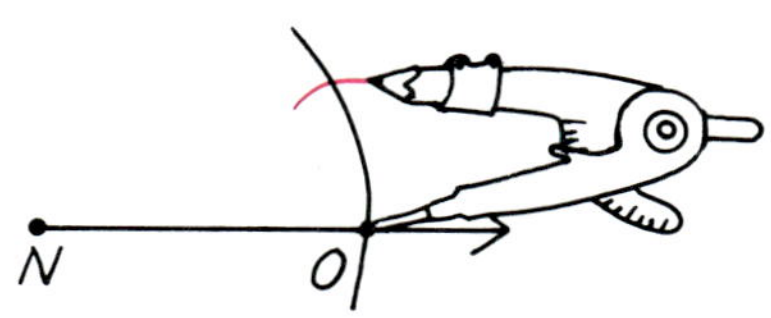

Step 6. Use a straightedge to draw the other ray. $\angle N \cong \angle M$.

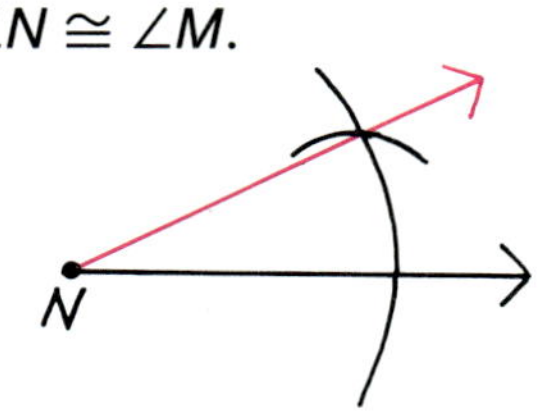

EXERCISES

1. Draw a segment and then use a straightedge and compass to construct a congruent segment.

2. Draw an acute angle and then construct a congruent angle.

3. Draw an obtuse angle and construct a congruent angle.

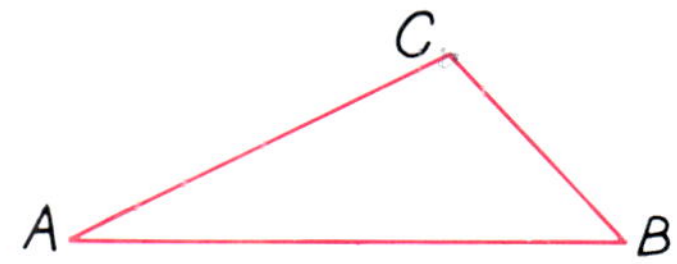

4. Draw a triangle ABC and then follow these steps to construct a triangle congruent to it.

   a. Construct a segment congruent to $\overline{AB}$.

   b. Measure $\overline{AC}$ with your compass. Put the compass point at R and draw an arc as shown.

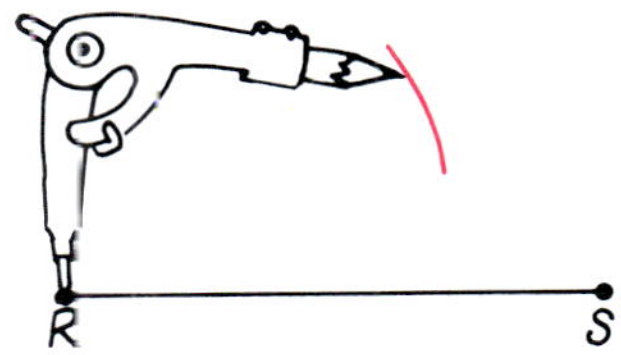

   c. Measure $\overline{BC}$ with your compass. Put the compass at point S and draw an arc as shown. Label the point of intersection T.

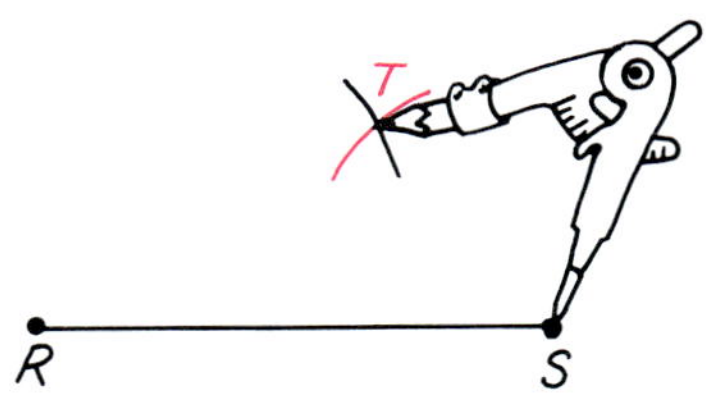

   d. Draw $\overline{RT}$ and $\overline{ST}$.

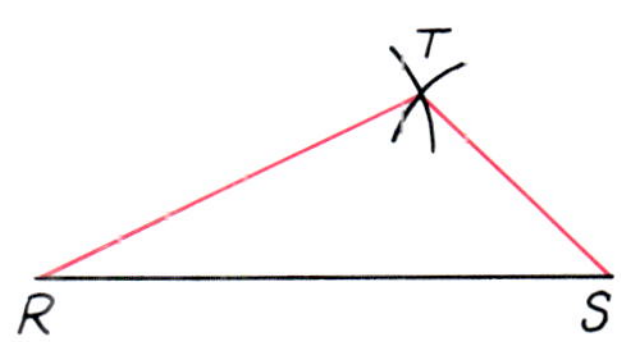

5. You did the construction in exercise 4 by ignoring the angles and just copying the three sides. Complete this statement:

   If the 3 sides of one triangle are congruent to the 3 sides of another triangle, the triangles are __?__.

6. Now construct a triangle congruent to $\triangle ABC$ by doing this:

   a. Use compass and straightedge and copy $\overline{AB}$.

   b. Copy $\angle A$.

   c. Copy $\overline{AC}$.

   d. Draw in the third side.

7. Construct a triangle congruent to $\triangle ABC$ by copying $\angle A$, $\overline{AB}$, and $\angle B$.

Line of symmetry

The dotted line divides the triangle "in half." If you fold the triangle along the dotted line, then the two halves match.

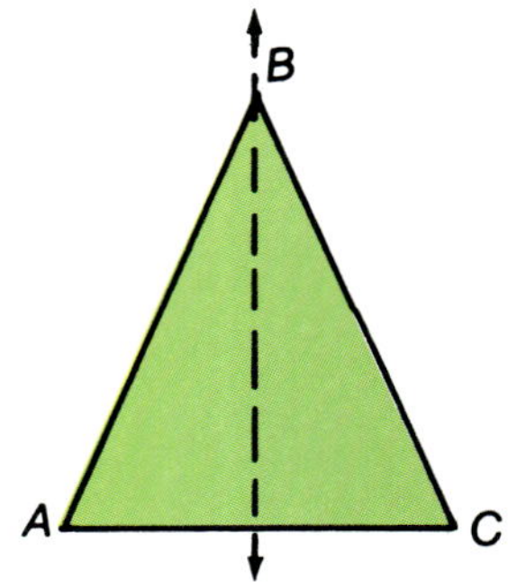

The dotted line is called a **line of symmetry.**

First we traced $\triangle ABC$ and the line of symmetry. Then we flipped the tracing about the line of symmetry. Notice that the flipped tracing fits exactly on $\triangle ABC$.

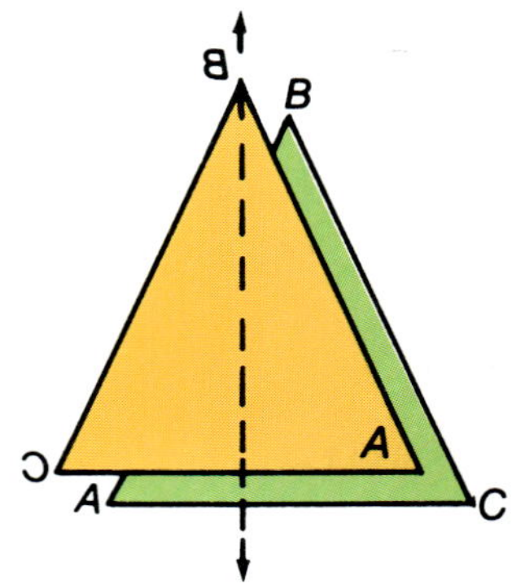

Matching of vertices

$B \leftrightarrow B$

$A \leftrightarrow C$

$C \leftrightarrow A$

Which angle is congruent to $\angle A$?
Which side is congruent to $\overline{AB}$?

EXERCISES

The dotted line is a line of symmetry. First make a tracing and flip it about the line of symmetry. Then answer each question.

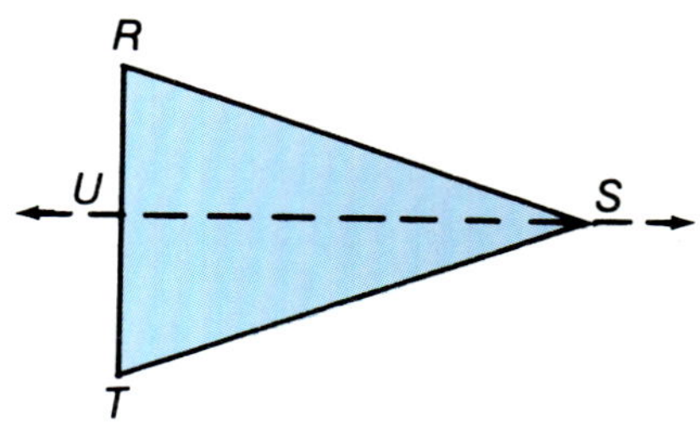

1. Which side is congruent to $\overline{RS}$?
2. Which angle is congruent to $\angle R$?
3. Which angle is congruent to $\angle RUS$?
4. Are $\angle RUS$ and $\angle TUS$ right angles?
5. Is $\overline{SU}$ perpendicular to $\overline{RT}$?
6. Which segment is congruent to $\overline{RU}$?
7. Is U the midpoint of $\overline{RT}$?
8. Is $\overleftrightarrow{SU}$ the perpendicular bisector of $\overline{RT}$?

True or false?

9. If a line of symmetry of a polygon goes through a side of the polygon, it is a perpendicular bisector of that side.

10. If a line of symmetry of a polygon goes through the vertex of an angle of the polygon, it bisects the angle.

This figure has two lines of symmetry. Think about flipping a tracing about $\overleftrightarrow{EG}$.

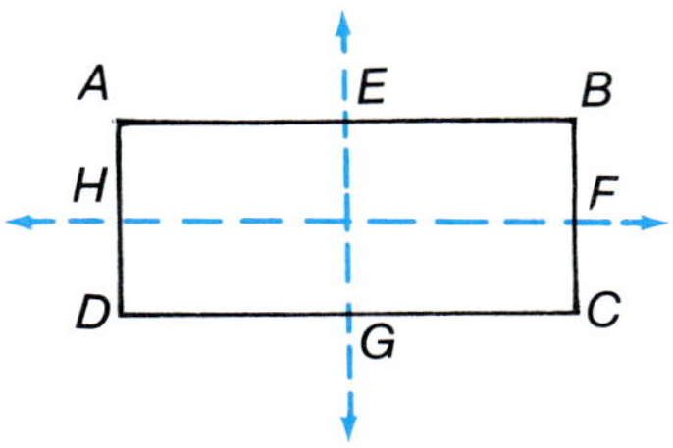

11. Is $\angle A \cong \angle B$?

12. Is $\angle D \cong \angle C$?

13. Is $\overline{AD} \cong \overline{BC}$?

Think about flipping a tracing about $\overleftrightarrow{HF}$.

14. Is $\angle A \cong \angle D$?

15. Is $\angle B \cong \angle C$?

16. Is $\overline{AB} \cong \overline{DC}$?

17. What can you say about the four angles of the figure?

18. What can you say about the opposite sides $\overline{AB}$ and $\overline{DC}$ of the figure?

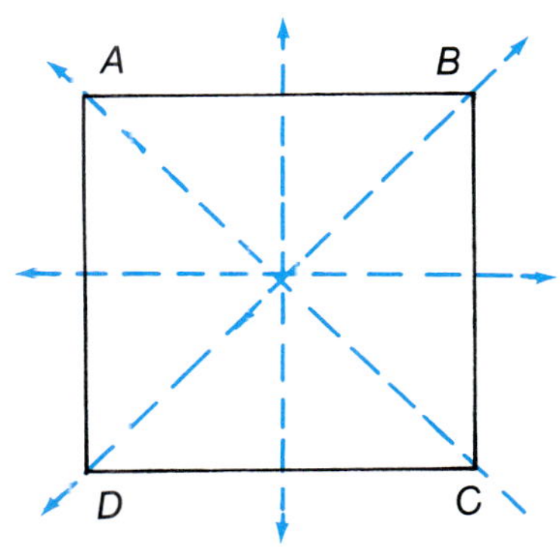

This figure has four lines of symmetry.

19. What can you say about the four angles?

20. What can you say about the four sides?

Challenge!

Solve.

21. Draw a 4-sided polygon with

   a. only 1 line of symmetry, going through midpoints of sides.

   b. only 2 lines of symmetry, going through vertices.

   c. only 2 lines of symmetry, going through midpoints of sides.

Triangles

There are three basic types of triangles.

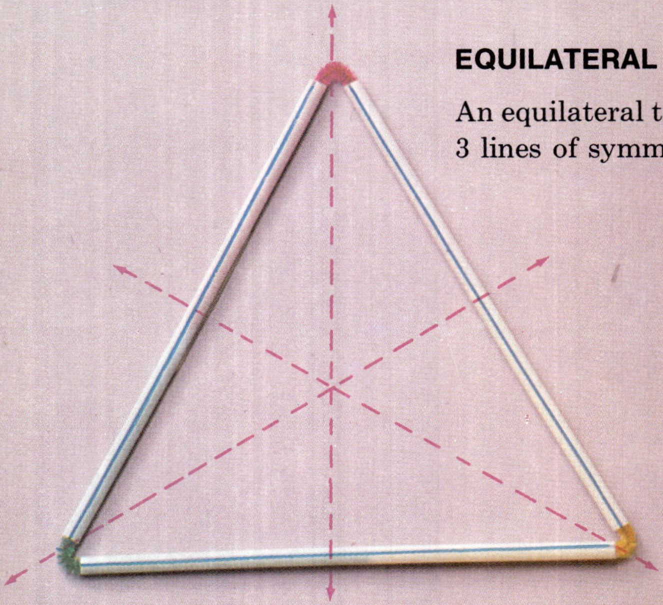

EQUILATERAL

An equilateral triangle has 3 lines of symmetry.

ISOSCELES

An isosceles triangle has 1 line of symmetry.

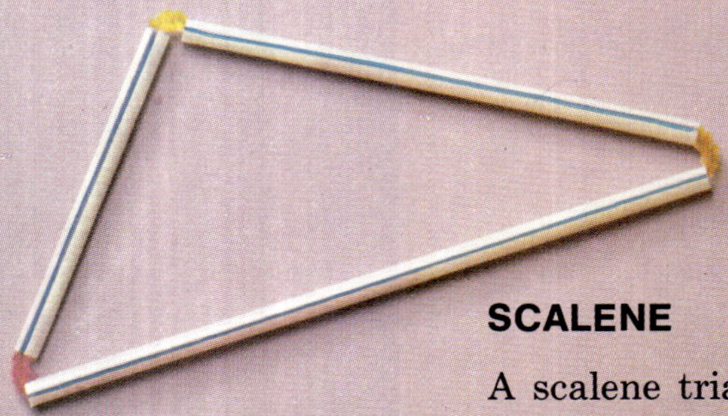

SCALENE

A scalene triangle has no lines of symmetry.

EXERCISES
True or false?

An equilateral triangle has

1. 3 congruent sides.

2. 3 congruent angles.

An isosceles triangle has

3. 2 congruent sides.

4. 2 congruent angles.

A scalene triangle has

5. no congruent sides.

6. no congruent angles.

Complete.

7. An equilateral triangle has ? congruent sides.
8. An isosceles triangle has ? congruent sides.
9. A scalene triangle has ? congruent sides.

Remember that the sum of the measures of the angles of a triangle is 180°. Give the length and angle measure of each marked side and angle.

10. Equilateral

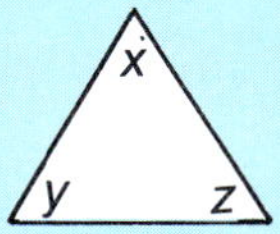

11. Equilateral

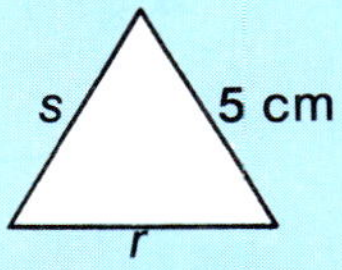

12. Isosceles

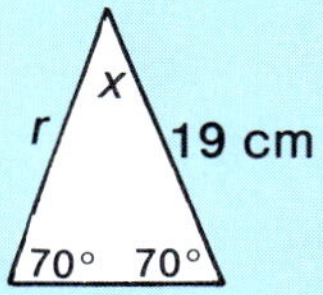

13. Isosceles

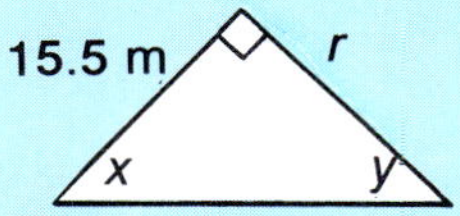

14. Can a triangle have 3 100° angles?
15. Can a triangle have 2 50° angles and 1 80° angle?
16. Can a triangle have 2 89° angles and 1 2° angle?
17. Can a triangle have 2 right angles?
18. Can a triangle have 1 right angle?

Use a compass and straightedge to construct

19. an equilateral triangle.
20. an isosceles triangle.

A **right triangle** is a triangle that has a 90° angle. Construct a right triangle that is

★ 21. isosceles.
★ 22. scalene.
★ 23. Is it possible for a right triangle to be an equilateral triangle?

KEEPING SKILLS SHARP

Solve. Round answers to the nearest hundredth.

1. 7% of 29 = x
2. 23% of 56 = x
3. 15% of 92 = x
4. 18% of 74 = x
5. 31% of 19 = x
6. 43% of 78 = x
7. 7.2% of 26 = x
8. 15.4% of 53 = x
9. 3.8% of 74 = x
10. 21.7% of 19 = x
11. 34.5% of 62 = x
12. 18.7% of 49 = x
13. 22.5% of 74 = x
14. 30.4% of 53 = x
15. 8.6% of 68 = x
16. 12.6% of 92 = x
17. 6.2% of 58 = x
18. 123% of 94 = x
19. 86.5% of 73 = x
20. 106.8% of 125 = x

Point of symmetry

If a figure fits itself after a half-turn about a point, the point is a **point of symmetry.**

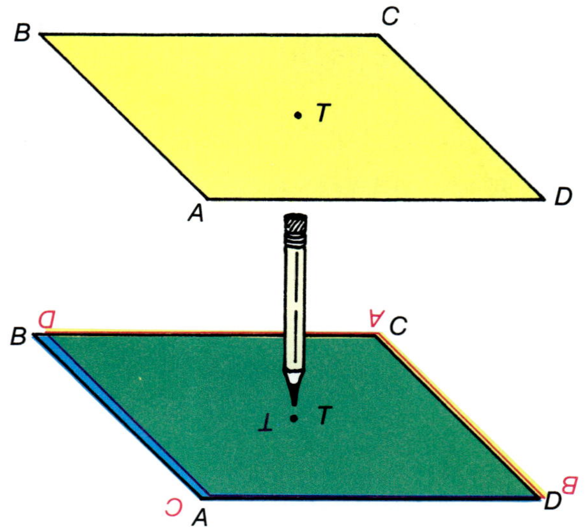

Here a tracing was given a half-turn about point T.
The tracing fits! So point T is a point of symmetry.

Matching of vertices

$A \longleftrightarrow C$ $\quad B \longleftrightarrow D$

$C \longleftrightarrow A$ $\quad D \longleftrightarrow B$

Which angle is congruent to $\angle A$? To $\angle B$?

Which side is congruent to $\overline{AB}$? To $\overline{AD}$?

$\overline{AD}$ and $\overline{CB}$ are called **opposite sides.** Which side is opposite $\overline{AB}$? When the figure has a point of symmetry, the opposite sides are parallel. $\angle A$ and $\angle C$ are called **opposite angles.** Which angle is opposite $\angle B$?

EXERCISES

Is point T a point of symmetry of the figure?

1.

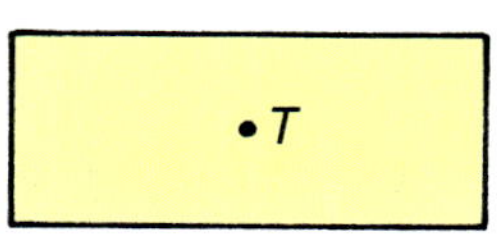

2.

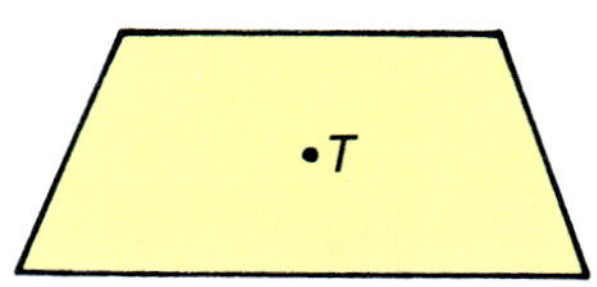

3.

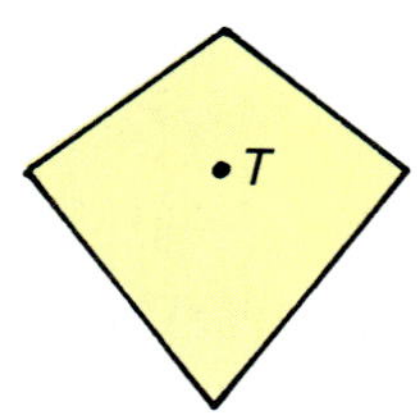

4.

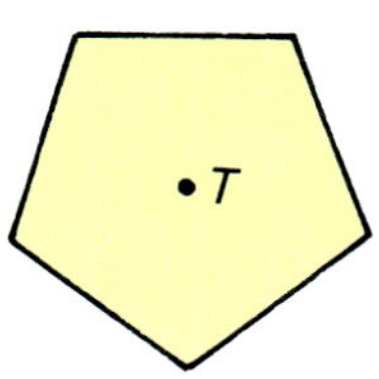

5.

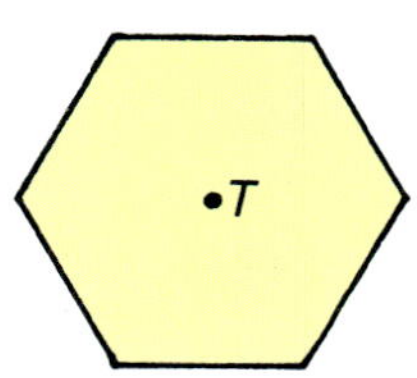

6.

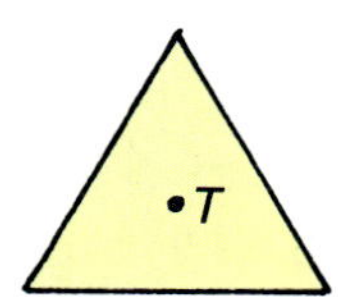

In each exercise half the figure is covered. In some of the figures there is a dashed line, which is a line of symmetry of the figure. In others there is a point *T*, which is a point of symmetry. Trace and complete each figure.

7.

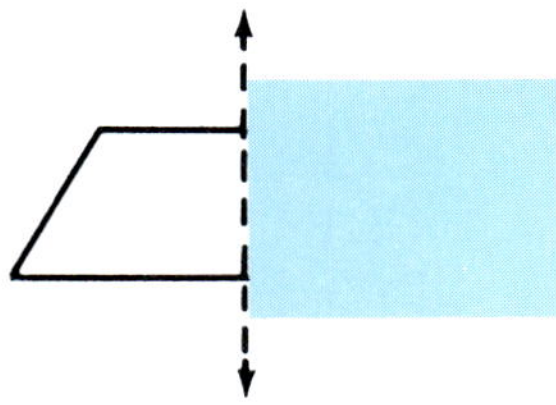

8.

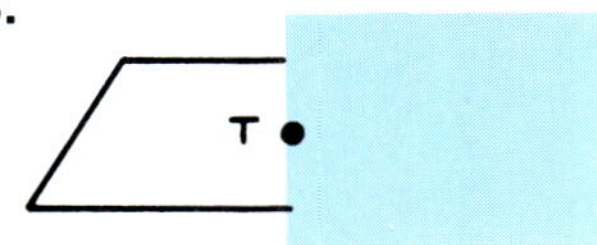

9.

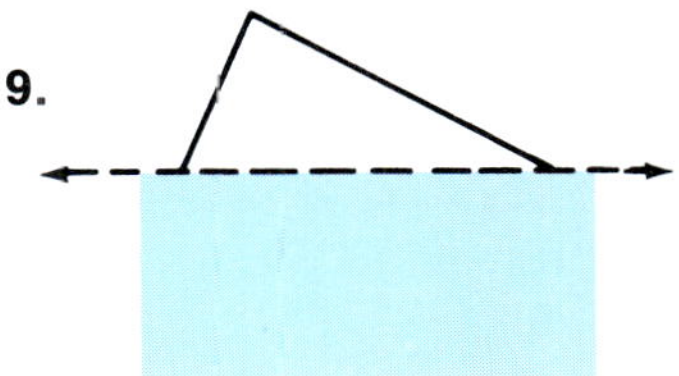

10.

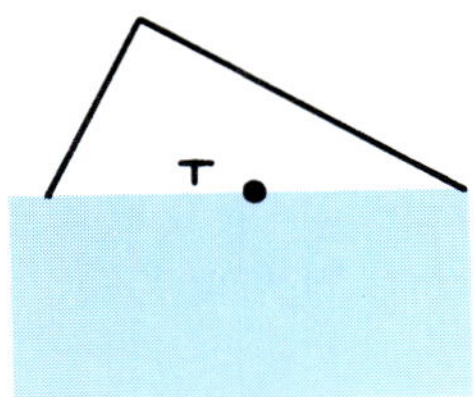

11.

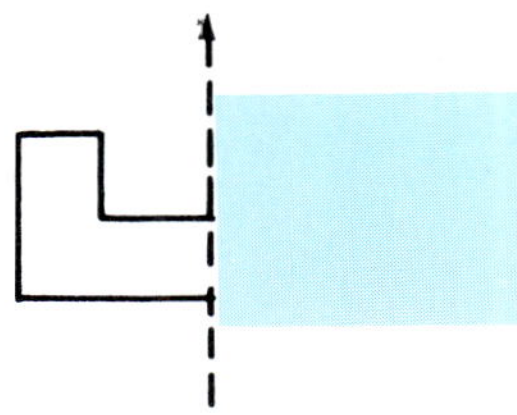

12.

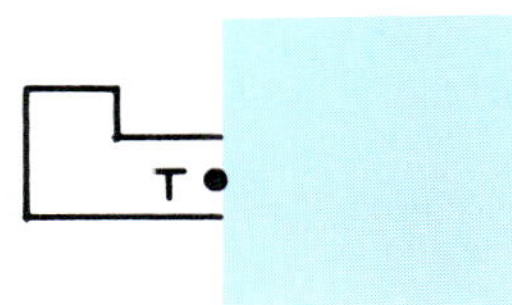

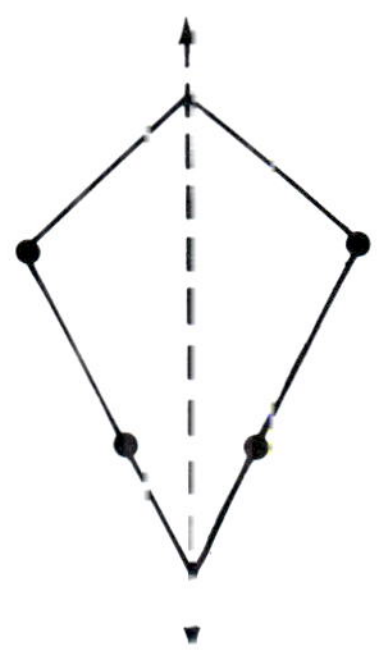

True or false?

13. Points of a figure directly opposite each other with respect to a line of symmetry are the same distance from the line of symmetry.

14. Points of a figure directly opposite each other with respect to the point of symmetry are the same distance from the point of symmetry.

Draw these polygons.

15. A quadrilateral (a 4-sided figure) that has a point of symmetry but no line of symmetry

16. A quadrilateral that has a line of symmetry but no point of symmetry

★ **17.** A hexagon (a 6-sided figure) that has a line of symmetry but no point of symmetry

★ **18.** A hexagon that has no line of symmetry but does have a point of symmetry.

Special quadrilaterals

Parallelogram

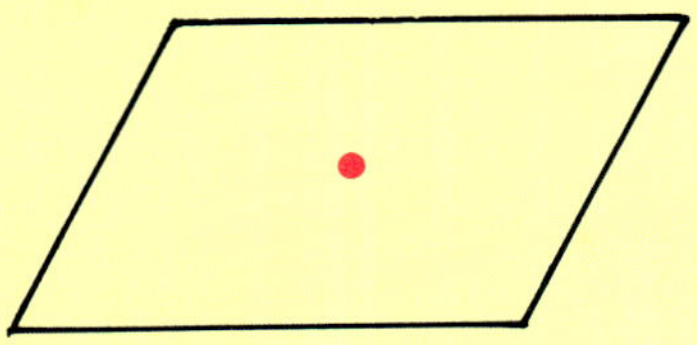

A parallelogram has a point of symmetry.

Rhombus

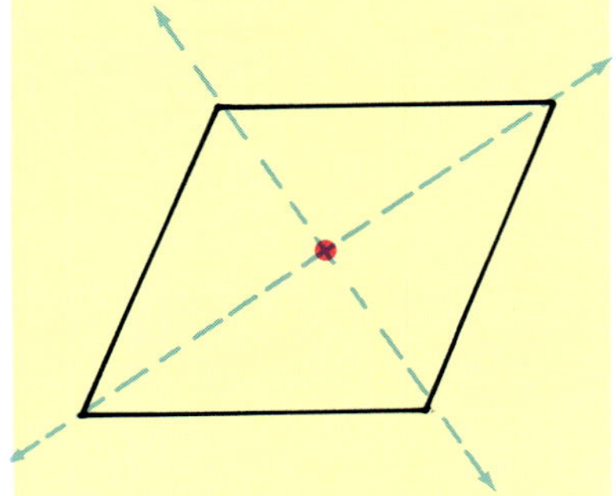

A rhombus has a point of symmetry and two lines of symmetry through opposite vertices.

Rectangle

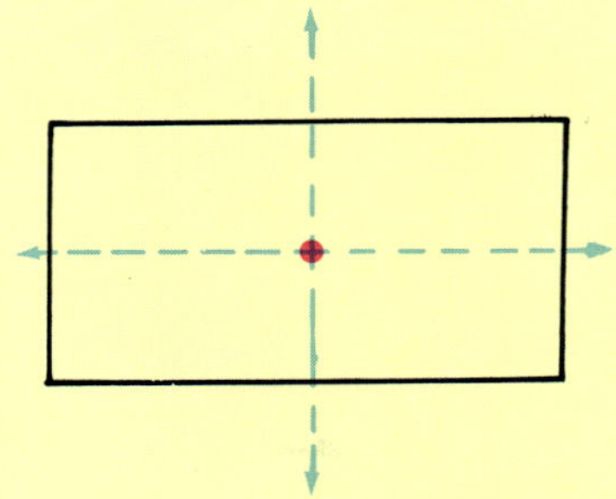

A rectangle has a point of symmetry and two lines of symmetry through opposite sides.

Square

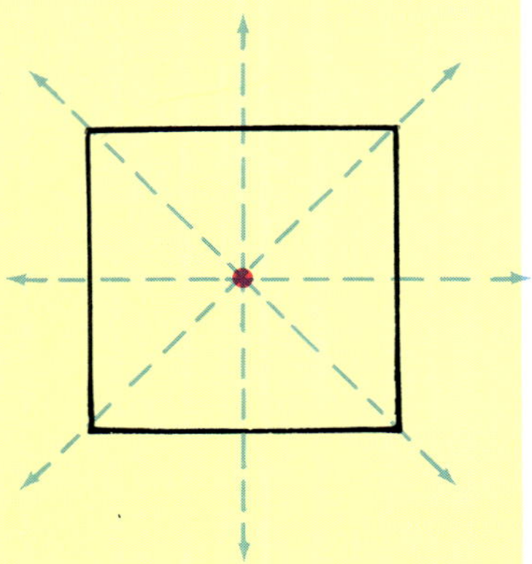

A square has a point of symmetry and four lines of symmetry.

Isosceles Trapezoid

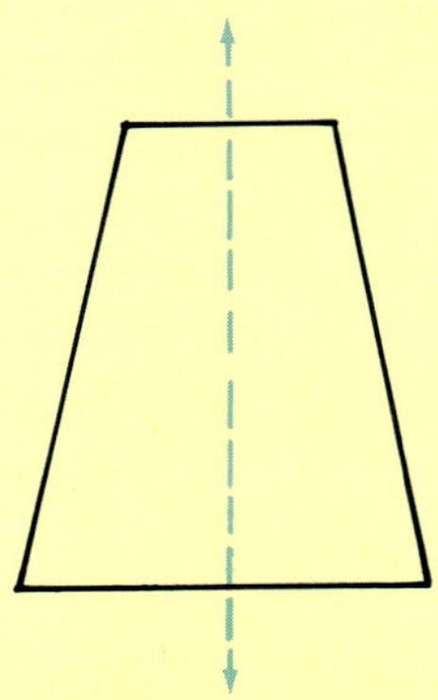

An isosceles trapezoid has exactly one line of symmetry through opposite sides.

Kite

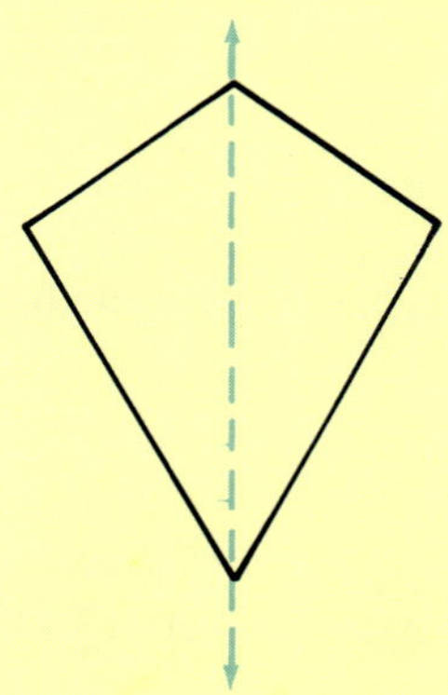

A kite has one line of symmetry through opposite vertices.

Notice that a rhombus, a rectangle, and a square are also parallelograms. Notice that a square is also a rectangle.

EXERCISES

1. What quadrilaterals have two pairs of opposite sides that are parallel?
2. What quadrilateral has exactly one pair of opposite sides that are parallel?

Two sides that have a common endpoint are called **consecutive sides.**

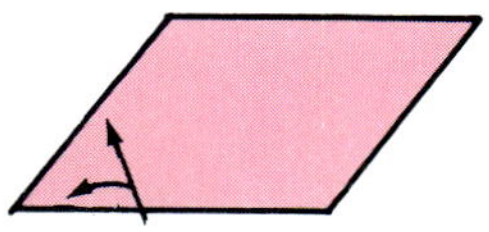

consecutive sides

Two angles that contain the same side of a polygon are called **consecutive angles.**

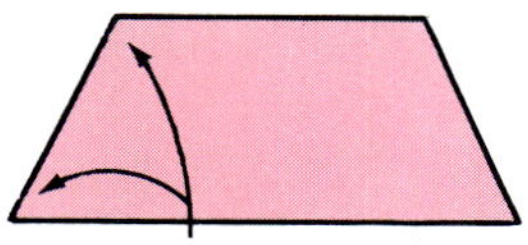

consecutive angles

3. What quadrilaterals have all pairs of consecutive sides congruent?
4. Which have all pairs of opposite sides congruent?
5. What quadrilaterals have all pairs of consecutive angles congruent?
6. Which have all pairs of opposite angles congruent?
7. What quadrilateral has exactly two pairs of consecutive angles that are congruent?

Give the measure of each side and each angle.

8.
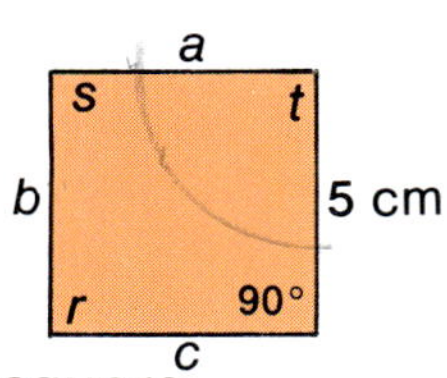

square

9.
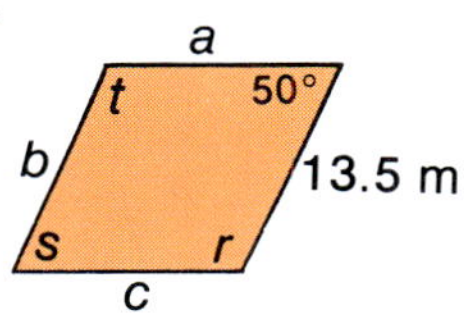

rhombus

10.
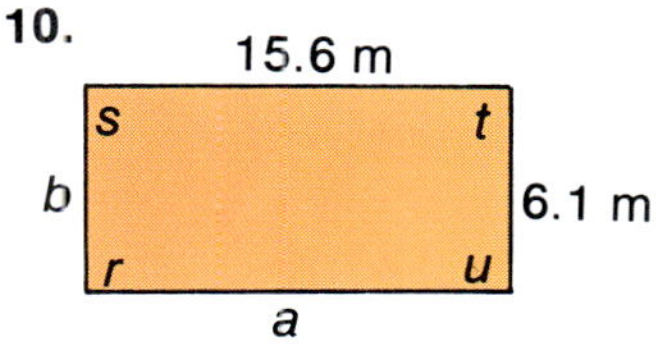

rectangle

11.
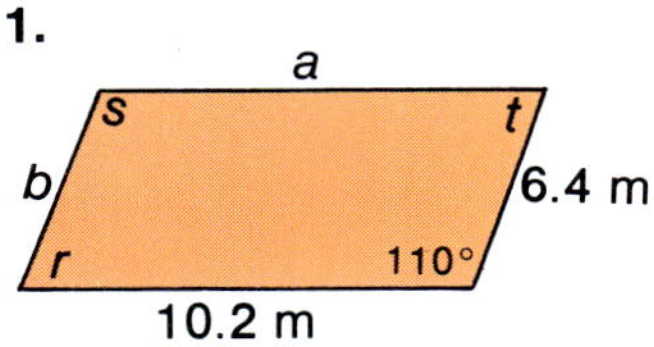

parallelogram

12.
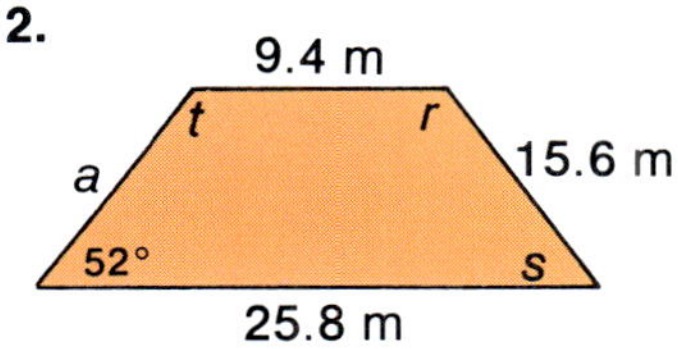

isosceles trapezoid

13.
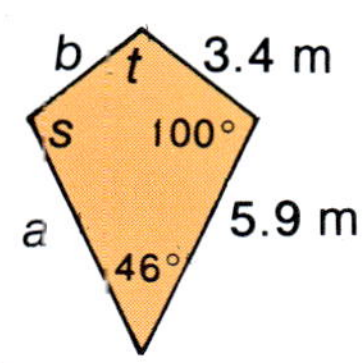

kite

Challenge!

Solve.

14. Construct each of the quadrilaterals that you studied in this lesson.

Regular polygons

A polygon having as many lines of symmetry as sides is called a **regular polygon.**

Regular Polygons

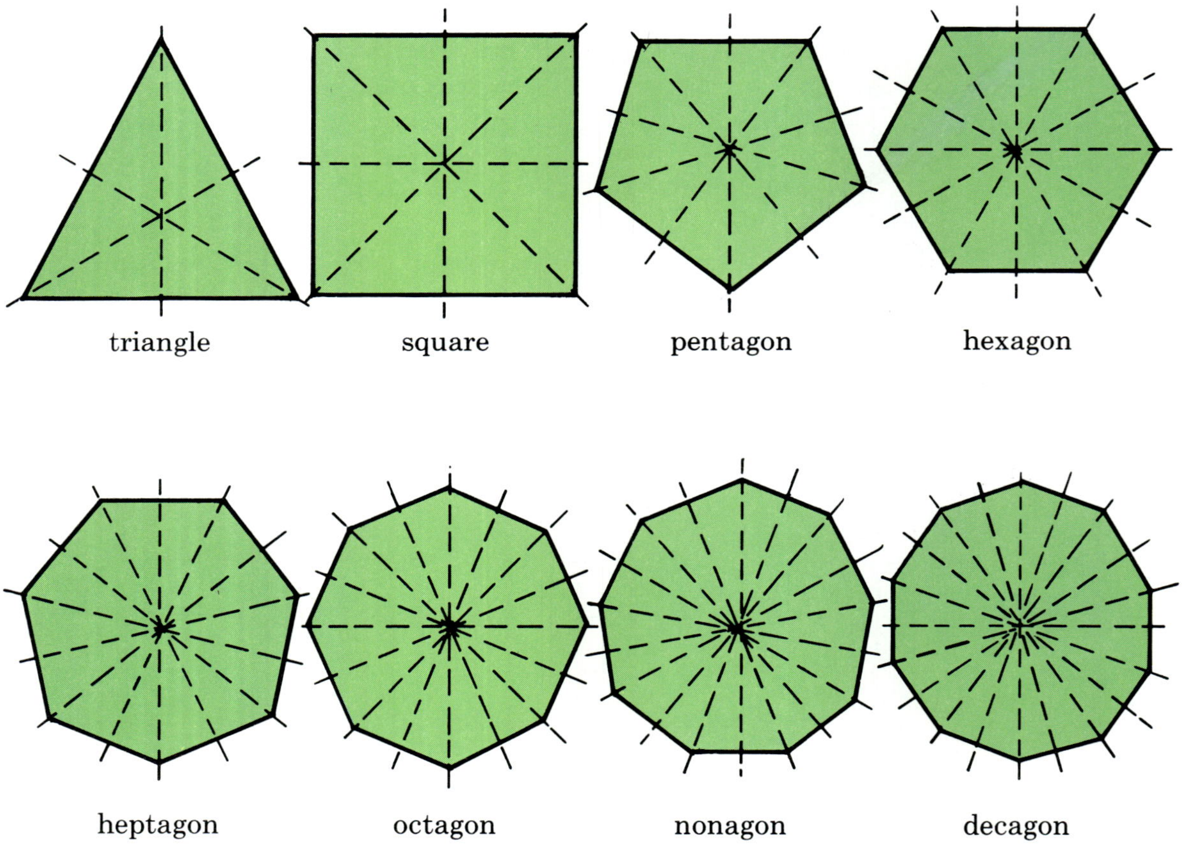

EXERCISES

True or false?

1. A polygon has the same number of sides as vertices.
2. A regular polygon with an odd number of sides has a point of symmetry.
3. A regular polygon with an even number of sides has a point of symmetry.
4. All sides of a regular polygon are congruent.
5. All angles of a regular polygon are congruent.

Complete this table for regular polygons. You may first need to review the table on page 327.

	Name	*Sum of the measures of the angles*	*Number of angles*	*Measure of each angle*
6.	Equilateral triangle	?	?	?
7.	Square	?	?	?
8.	Regular pentagon	?	?	?
9.	Regular hexagon	?	?	?
10.	Regular heptagon	?	?	?
11.	Regular octagon	?	?	?
12.	Regular nonagon	?	?	?
13.	Regular decagon	?	?	?

Challenge!

Solve.

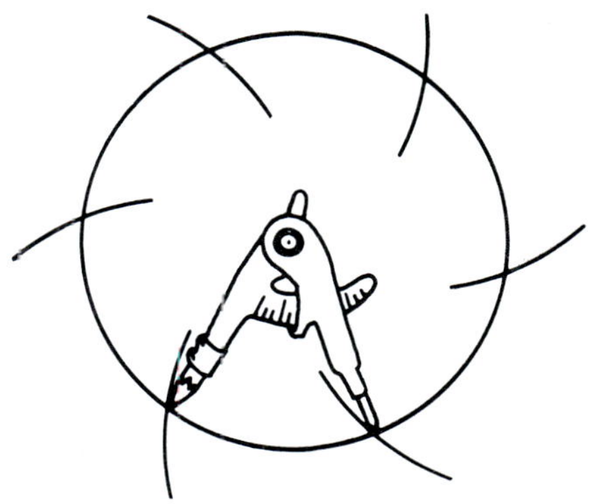

14. Draw a circle. With the same compass setting, draw six arcs as shown. Connect the crossing points to get a regular hexagon.
15. Use a similar method to construct an equilateral triangle.

Chapter Checkup

Acute, right, or obtuse? Complete. [pages 324–325]

1. $\angle RXS$ is ?.
2. $\angle RXU$ is ?.
3. $\angle TXR$ is ?.

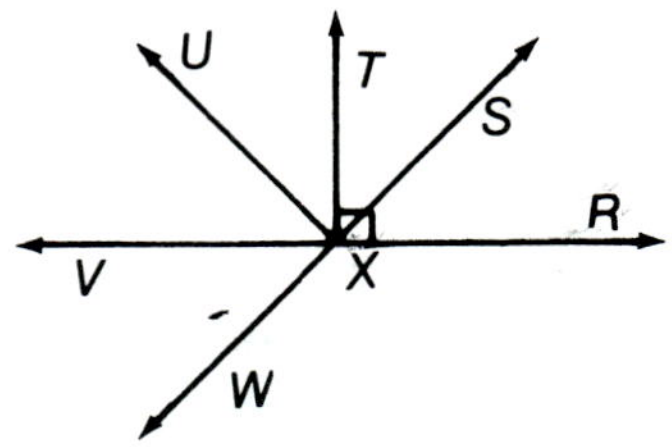

Adjacent, vertical, complementary, or supplementary? Complete. [pages 328–329]

4. $\angle UXV$ and $\angle VXW$ are ?.
5. $\angle SXR$ and $\angle VXW$ are ?.
6. $\angle VXU$ and $\angle UXR$ are adjacent and ?.

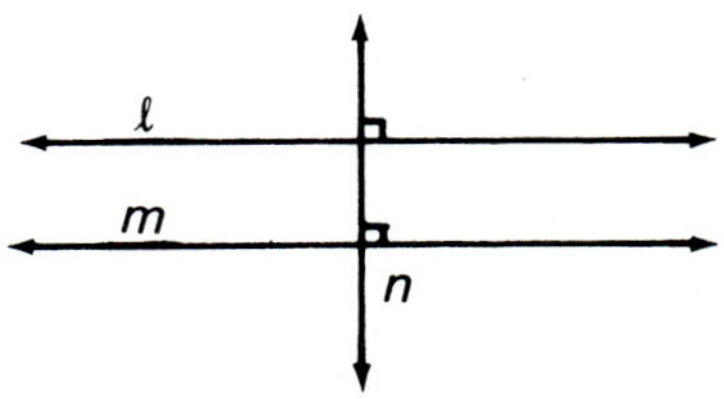

Answer. [pages 330–331]

7. Which lines are parallel?
8. Which line is a transversal?

The two quadrilaterals are congruent. Complete. [pages 332–333]

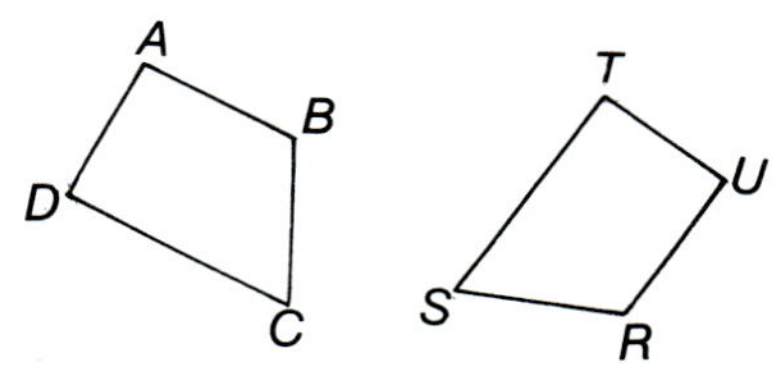

9. $\overline{AD} \cong$? 10. $\overline{BC} \cong$? 11. $\overline{DC} \cong$?
12. $\angle A \cong$? 13. $\angle C \cong$? 14. $\angle D \cong$?

Give the missing angle measures. [pages 338–339]

15.

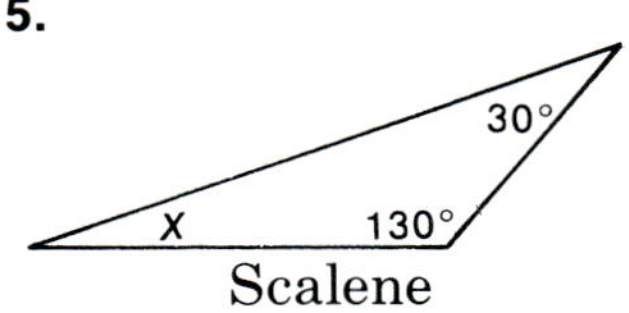

16.

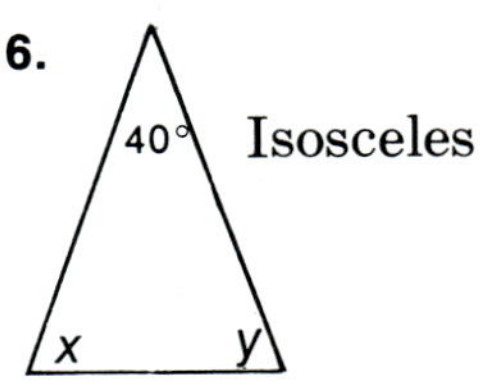

17.

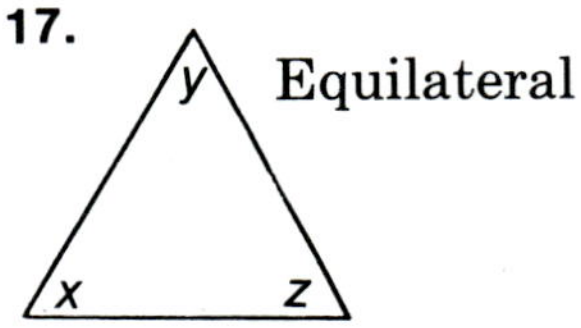

True or false? [pages 342–343]

18. The opposite sides of a parallelogram are parallel.
19. The four sides of a rhombus are congruent.
20. The four angles of a square are acute angles.
21. A rectangle is also a square.

Give the number of sides of a [pages 344–345]

22. quadrilateral. 23. hexagon. 24. octagon. 25. pentagon.

Chapter Project

GEOMETRIC DESIGNS

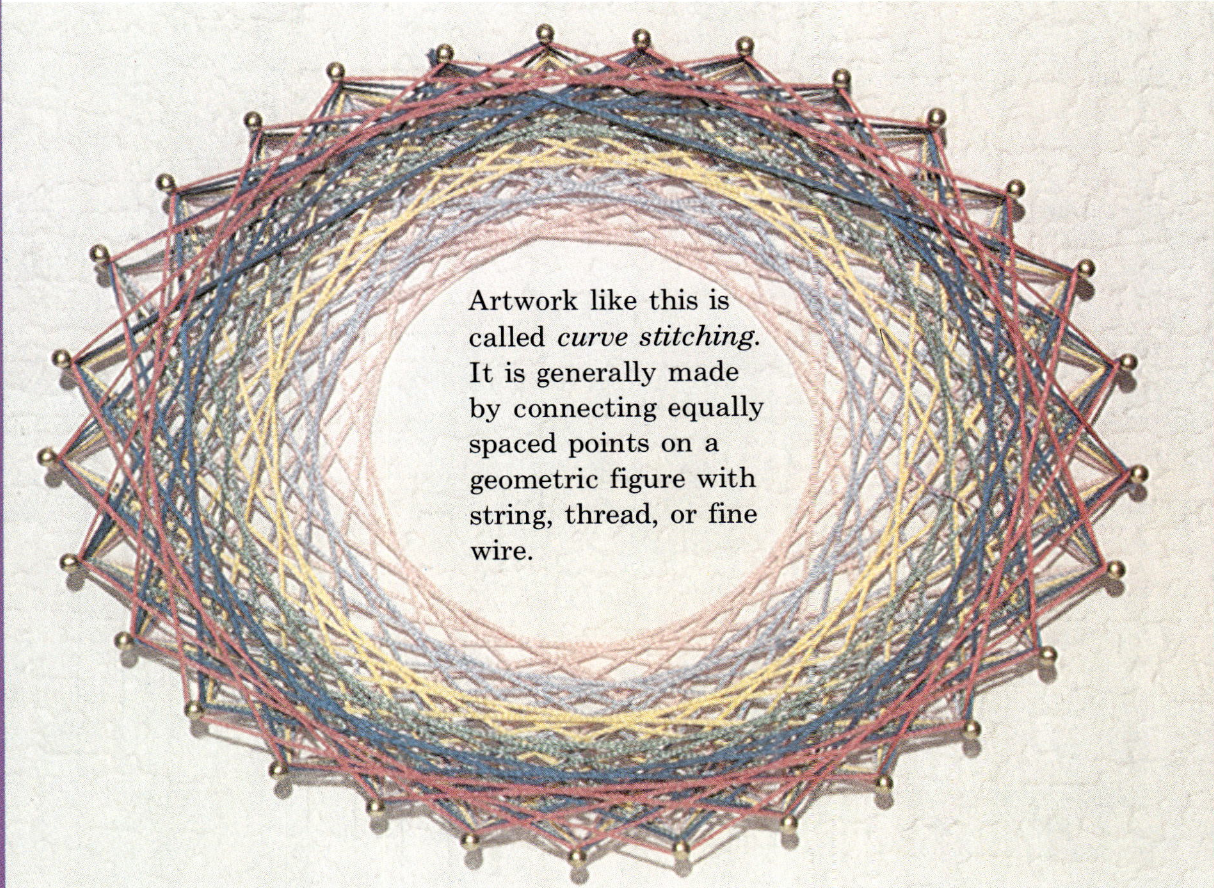

Artwork like this is called *curve stitching.* It is generally made by connecting equally spaced points on a geometric figure with string, thread, or fine wire.

1. Draw two intersecting lines and mark off equally spaced points as shown. Use segments to connect the points on the sides of the four angles. Connect them in the order shown by the red segments.

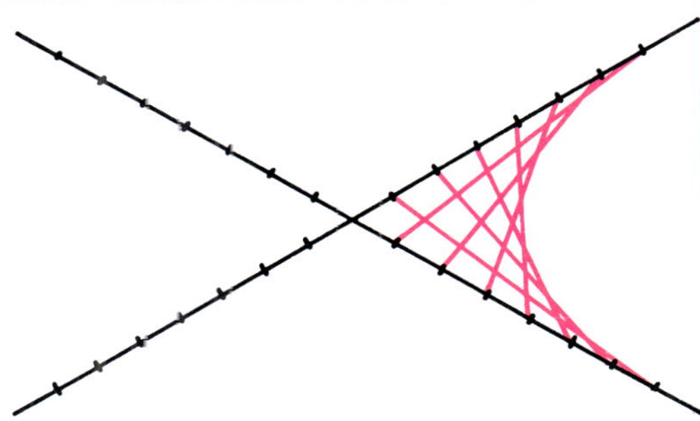

2. Make a design of your own. You may wish to work with geometric figures other than angles. Your design could consist of several geometric figures.

Chapter Review

acute angle 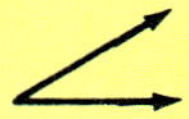

right angle 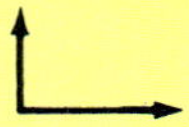

obtuse angle 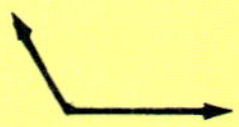

Complete.

1. An angle whose measure is between 0° and 90° is called an ? angle.
2. An angle whose measure is 90° is called a ? angle.
3. An angle whose measure is between 90° and 180° is called an ? angle.

Vertical angles are formed by 2 intersecting lines.

Complementary angles
The sum is 90°.

Supplementary angles
The sum is 180°.

4. $\angle AEF$ and ? are vertical angles.
5. $\angle BEC$ and ? are complementary angles.
6. $\angle AEC$ and ? are supplementary angles.

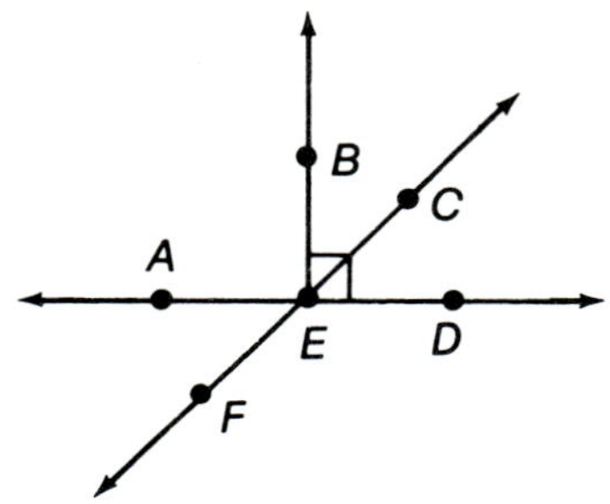

The corresponding parts of congruent figures are congruent.

The triangles are congruent.

7. $\angle A \cong$?
8. $\angle S \cong$?
9. $\overline{AC} \cong$?
10. $\overline{SR} \cong$?

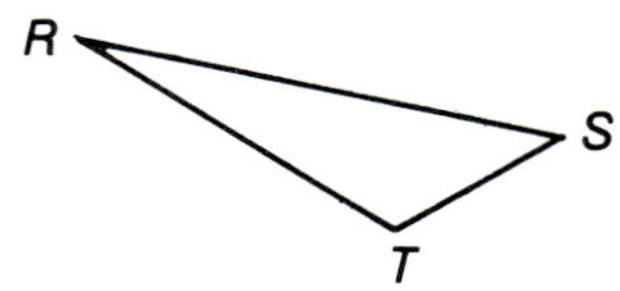

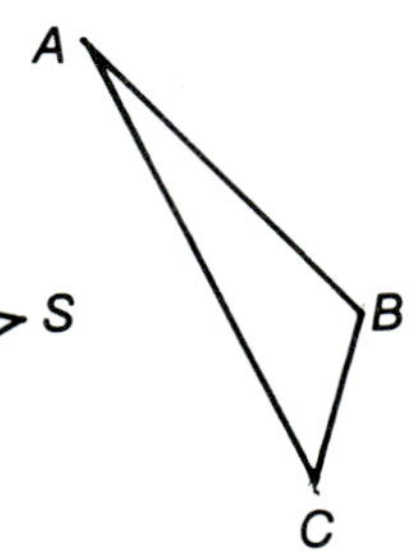

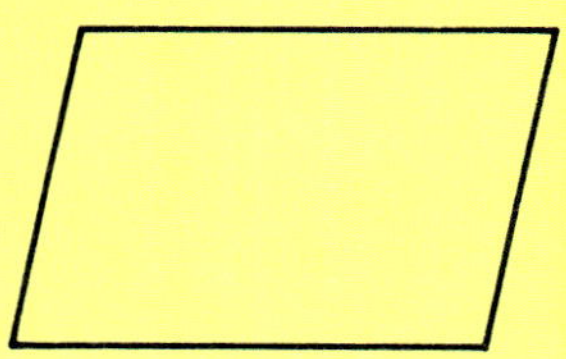

parallelogram

True or false?

11. A parallelogram has a point of symmetry.
12. A parallelogram has 2 lines of symmetry.
13. A rectangle is a parallelogram.

Chapter Challenge

TRANSLATIONS AND REFLECTIONS

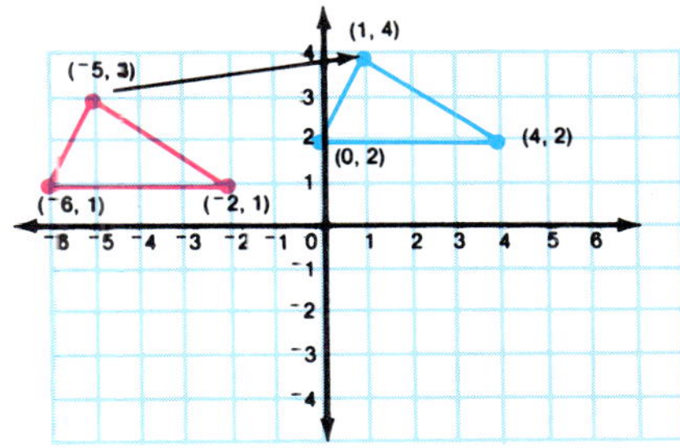

If you slide the red triangle in the direction and distance shown by the arrow, you get the blue triangle. The blue triangle is called the slide image or translation image of the red triangle.

Note: If you start with the red triangle and add 6 to the first coordinate of each point and add 1 to the second coordinate of each point, you will get the blue triangle.

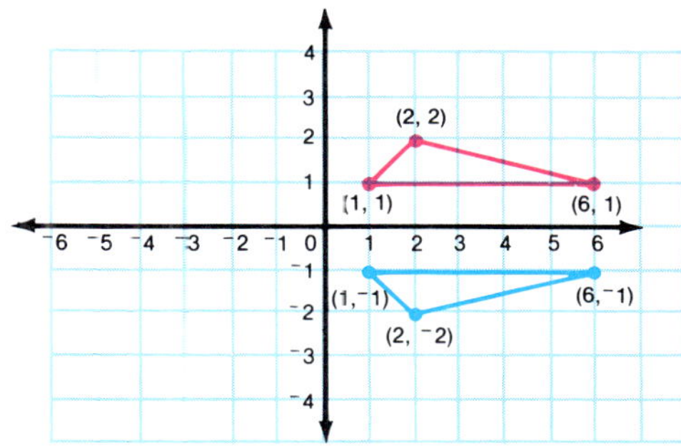

The first coordinate axis is a line of symmetry for these two triangles. If you flip the red triangle about the first coordinate axis, you get the blue triangle. The blue triangle is called the flip image or reflection image of the red triangle.

Note: If you start with the red triangle and keep the first coordinate of each point the same and multiply the second coordinate of each point by $^{-}1$, you will get the blue triangle.

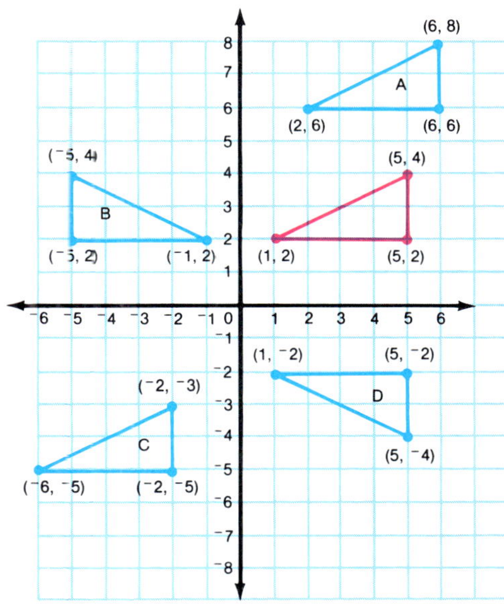

EXERCISES

Complete.

1. Triangle A is a ____ image of the red triangle.
2. Triangle B is a ____ image of the red triangle.
3. Triangle C is a ____ image of the red triangle.
4. Triangle D is a ____ image of the red triangle.

What would you have to do to the coordinates of the points of the red triangle to get

5. triangle A?
6. triangle B?
7. triangle C?
8. triangle D?

Cumulative Checkup

STANDARDIZED FORMAT

Choose the correct letter.

1. Give the difference.

4.03 − 2.974

a. 1.056
b. 2.056
c. 1.156
d. none of these

2. Give the product.

3.77 × 20.5

a. 9.425
b. 77.285
c. 7.7285
d. none of these

3. Which unit would most likely be used to measure the weight of a grain of sand?

a. milligram
b. kilogram
c. liter
d. centimeter

4. Give the standard numeral for $3^3 \times 5^2$.

a. 90
b. 270
c. 225
d. none of these

5. Give the difference.

$2\frac{1}{4} - 1\frac{5}{6}$

a. $\frac{5}{12}$
b. $\frac{7}{12}$
c. $1\frac{7}{12}$
d. none of these

6. Find the volume.

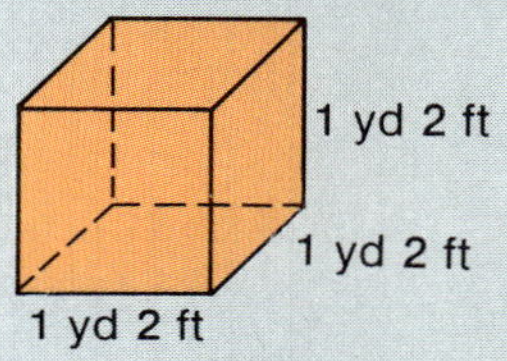

a. 5 yd^3
b. 125 ft^3
c. 1 yd^3 8 ft^3
d. none of these

7. Solve.

$\frac{4}{x} = \frac{15}{7}$

a. $8\frac{4}{7}$
b. $26\frac{1}{4}$
c. $1\frac{13}{15}$
d. none of these

8. Change $\frac{1}{6}$ to a percent.

a. 6%
b. 17%
c. $16\frac{2}{3}$%
d. none of these

9.

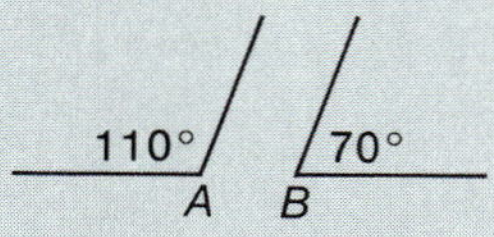

$\angle A$ and $\angle B$ are

a. complementary angles
b. supplementary angles
c. vertical angles
d. none of these

10. A quadrilateral that has 4 lines of symmetry is

a. rectangle
b. parallelogram
c. rhombus
d. none of these

11. Slacks were on sale for 30% off the regular price of $24. What was the sale price?

a. $16.80
b. $7.20
c. $23.28
d. none of these

12. Mio bought $1\frac{1}{2}$ dozen eggs for $1.26. How much did 1 dozen of these eggs cost?

a. $1.89
b. $1.05
c. $.84
d. none of these

12

Probability and Statistics

Weather forecasters use a wide variety of data to report and predict the weather. What numerical data can you find in the photograph?

READY OR NOT?

Give the ratio.

1. red marbles to marbles
2. blue marbles to marbles
3. yellow marbles to marbles
4. red marbles to not-red marbles
5. blue marbles to not-blue marbles
6. not-yellow marbles to yellow marbles

Equally likely outcomes

Event
Spinning this spinner

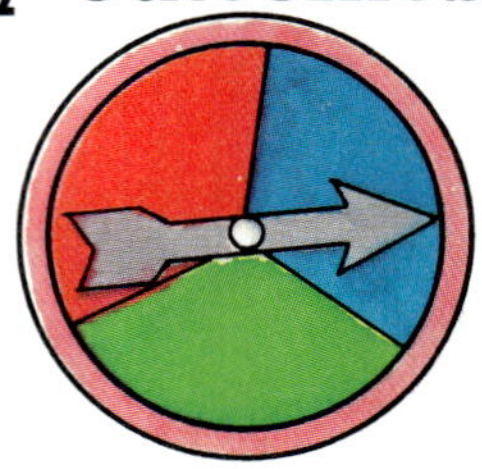

If you spin this spinner, there are 3 possible outcomes. Since the "chance" of getting each outcome is the same, we say that the outcomes are **equally likely.**

Event
Tossing this coin

There are two equally likely outcomes, "heads" and "tails."

Now think about combining the events, that is, first spinning the spinner and then flipping the coin. We can use a **tree diagram** to show the outcomes.

Event
Spinning the spinner
and then
flipping the coin

Tree Diagram of Outcomes

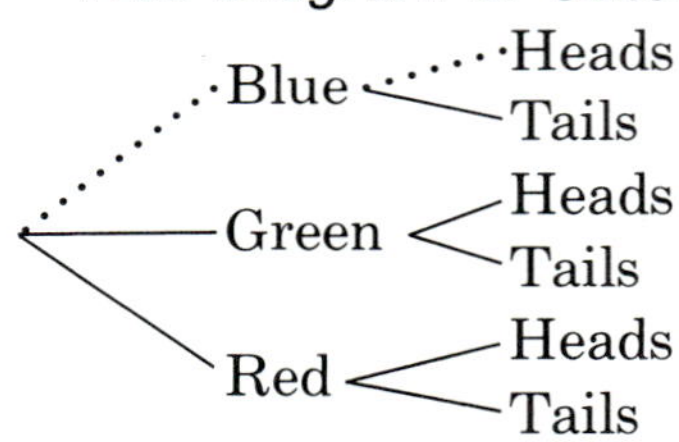

Notice that each "branch" of the tree diagram shows a possible outcome. For example, the dotted branch shows the outcome "spinning blue" **and then** "flipping heads."

A Basic Counting Principle

If a first event has m outcomes and if a second event has n outcomes, then the first event followed by the second event has $m \times n$ outcomes.

From the example we have

Outcomes for spinning spinner		Outcomes for flipping coin		Total number of outcomes
3	×	**2**	=	**6**

EXERCISES

First tell how many outcomes there are for an event. Then tell whether the outcomes are equally likely.

1. Tossing the number cube (The outcome is the number of dots on the top face.)

2. Spinning the spinner

3. Without looking, picking a marble from the bag

4. Tossing the paper cup

Draw a tree diagram that shows all possible outcomes. Then tell whether the outcomes are equally likely.

5. *Event*
   Picking a card and then tossing the number cube

6. *Event*
   Picking a card and then flipping the coin

7. How many outcomes are there for the event flipping a coin and then tossing a number cube? Are the outcomes equally likely?

8. Sarah has 3 sweaters and 4 pairs of slacks. How many different sweater-slacks outfits does she have?

★ 9. How many 5-digit license plates can be made? Assume that 0 can be the first digit.

Probability

When you spin the spinner, there are 4 equally likely outcomes. The **probability** of each outcome is $\frac{1}{4}$.

$P(\text{red}) = \frac{1}{4}$ ← *Number of red outcomes* ← *Total number of outcomes*

Read as "probability of red equals $\frac{1}{4}$."

Here are some other probabilities:

$P(\text{blue}) = \frac{1}{4}$ $\quad$ $P(\text{not blue}) = \frac{3}{4}$

$P(\text{red } \textit{or} \text{ green}) = \frac{2}{4} = \frac{1}{2}$

The probability of an impossible event is 0.

$P(\text{brown}) = 0$

> If an event has m equally likely outcomes, then the probability of one of the outcomes is $\frac{1}{m}$.

Notice the words *equally likely* in the sentence above. Let's consider an event where the outcomes are not equally likely.

Event

Picking a marble from the bag

$P(\text{blue}) = \frac{1}{8}$ $\quad$ $P(\text{not blue}) = \frac{7}{8}$

$P(\text{green}) = \frac{3}{8}$ $\quad$ $P(\text{not green}) = \frac{5}{8}$

$P(\text{red}) = \frac{2}{8} = \frac{1}{4}$ $\quad$ $P(\text{not red}) = \frac{6}{8} = \frac{3}{4}$

$P(\text{black}) = \frac{0}{8} = 0$ $\quad$ $P(\text{not black}) = \frac{8}{8} = 1$

Notice that the probability of the last outcome is 1. If an outcome is *certain* to occur, then the probability of that outcome is 1.

EXERCISES

Give each probability.

Event
Picking a marble from the bag

1. P(blue) **2.** P(not blue) **3.** P(not green)
4. P(red) **5.** P(yellow) **6.** P(not yellow)

Event
Tossing a number cube

7. P(1) **8.** P(6) **9.** P(odd number)
10. P(4 *or* 6) **11.** P(number greater than 6)

Event
Picking a card from a deck of cards of the first 50 counting numbers

12. P(even number)
13. P(number greater than 40)
14. P(multiple of 5)
15. P(number less than 36)
16. P(1-digit number)
17. P(square number)

Complete.

18. If the probability of an outcome is $\frac{5}{8}$, then the outcome is ___more/less___ likely to happen than not to happen.

19. If the probability of an outcome is $\frac{1}{4}$, then the outcome is ___more/less___ likely to happen than not to happen.

Problem Solving

20. Suppose that 12 red marbles and 18 white marbles are placed in a bag and thoroughly mixed. If one marble is picked, what is the probability that it is red? That it is white?

KEEPING SKILLS SHARP

Give each sum or difference in simplest form.

1. $2 + 4\frac{1}{3}$

2. $3\frac{1}{3} - 2\frac{1}{4}$

3. $4\frac{1}{2} + 5\frac{1}{3}$

4. $4\frac{7}{8} - 2\frac{3}{4}$

5. $5\frac{3}{5} + 2\frac{4}{5}$

6. $9\frac{1}{2} - 7\frac{2}{3}$

7. $8\frac{3}{8} + 3\frac{3}{4}$

8. $9\frac{3}{4} - 4\frac{2}{3}$

9. $12\frac{5}{8} + 10\frac{5}{6}$

10. $14\frac{3}{4} - 11\frac{2}{5}$

11. $15\frac{1}{8} + 11\frac{3}{4}$

12. $20\frac{2}{3} - 13\frac{5}{6}$

13. $24\frac{3}{4} + 19\frac{3}{4}$

14. $22\frac{4}{5} - 15\frac{1}{2}$

15. $17\frac{3}{4} + 2\frac{5}{6}$

16. $19\frac{3}{8} - 14\frac{5}{8}$

Probability—independent and dependent events

Now let's consider a first event followed by a second event.

First Event
Picking a marble

Second Event
Flipping a coin

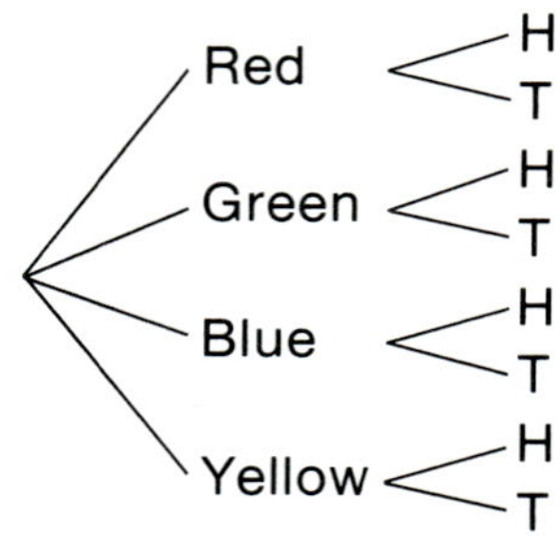

Notice that $P(\text{red}) = \frac{1}{4}$, $P(\text{H}) = \frac{1}{2}$,
and $P(\text{red and then H}) = \frac{1}{8}$

> If the probability of an outcome of a first event is $\frac{1}{m}$ and the probability of an outcome of a second event is $\frac{1}{n}$, then the probability of the first outcome followed by the second outcome is $\frac{1}{m} \times \frac{1}{n}$, or $\frac{1}{mn}$.

From the example above, we have

$$P(\text{red and then H}) = \frac{1}{4} \times \frac{1}{2} = \frac{1}{8}$$

The two events described above are **independent** events, since the outcome of the first event does not affect the outcome of the second event.

Two events such that the outcome of the first affects the outcome of the second are called **dependent** events.

First Event
Picking a marble

Second Event
Without replacing the first marble, picking a second marble

Since the first marble was not replaced, there are only three marbles left in the bag.

$$P(\text{red and then blue}) = \frac{1}{4} \times \frac{1}{3} = \frac{1}{12}$$

EXERCISES

Give each probability.

First Event
Picking a marble

Second Event
Tossing a number cube

1. P(red and then 5)
2. P(not red and then 4)
3. P(blue and then not 2)
4. P(not green and then not odd)

First Event
Spinning this spinner

Second Event
Spinning this spinner

5. P(green and then green)
6. P(blue and then not blue)
7. P(not yellow and then red)
8. P(yellow and then not red)

First Event: Picking a card

Second Event: Without replacing the first card, picking a second card

9. P( and then )
10. P(a four-sided figure and then )
11. P(a red card and then a black card)
12. P(a non-black card and then )

Challenge!

Give each probability.

First Event
Picking a card from a full deck of 52 playing cards

Second Event
Without replacing the first card, picking a second card

13. P(heart and then heart)
14. P(4 and then 6)
15. P(king and then king)
16. P(heart and then club)

Sample space

Every customer who makes a purchase is given a Sneaker Match card. When the card is rubbed, a picture of a green or red sneaker appears on each of the three boxes. The pictures are equally likely to appear.

What is the probability of getting a winning card? To find the probability of getting a winning card, it is helpful to first list the **sample space** (the set of all possible outcomes of an event is called the sample space of the event). The sample space is shown in this tree diagram.

Look at the sample space. Two of the 8 outcomes have colors that are all the same.

$P(\text{all colors the same}) = \frac{2}{8} = \frac{1}{4}$

So the probability of getting a winning card is $\frac{1}{4}$.

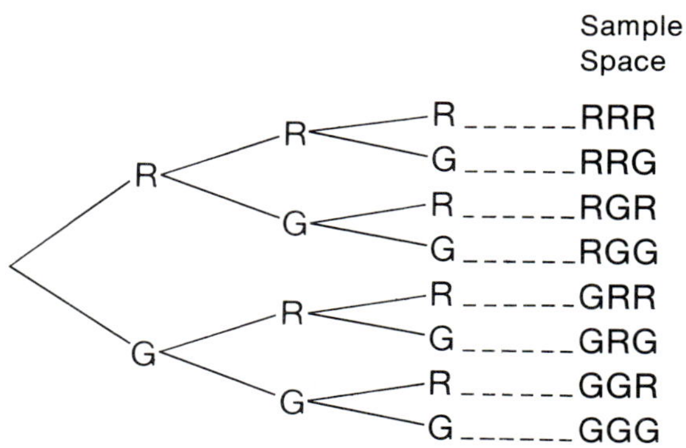

EXERCISES

Refer to the sample space on the preceding page. What is the probability of getting

1. red sneakers exactly once?

2. green sneakers exactly twice?

3. red sneakers at least once?

4. green sneakers in boxes *1* and *2*?

5. Think about tossing these number cubes.

The outcome shown can be represented by the number pair (1, 4).

a. Copy and complete this table.

Number on blue cube \ *Number on red cube*	1	2	3	4	5	6
1				(1, 4)		
2						
3						
4						
5						
6						

b. How many outcomes are in the sample space?

6. Refer to your sample space in exercise 5 to find the following probabilities.

a. *P*(doubles) **b.** *P*(not doubles) **c.** *P*(sum of 2)

d. *P*(sum of 12) **e.** *P*(sum of 7) **f.** *P*(not sum of 7)

7. Think about a family that has 3 children.

a. List the sample space. Use *BGG* to represent a boy as the oldest child, a girl as the middle child, and a girl as the youngest child.

b. Assume that the probability of the birth of a boy is the same as the probability of the birth of a girl. What is the probability that there are 2 boys and 1 girl in the family?

Challenge!

8. These five marbles are placed in a bag.

Then two marbles are picked, one after the other.

a. Make a tree diagram of the sample space

b. What is the probability that the red marble will be picked?

c. What is the probability that the green marble will not be picked?

Odds

The ratio of the number of ways the outcome *can* occur to the number of ways that the outcome *cannot* occur is called the **odds** in favor of the outcome.

Event

Spinning this spinner

The odds in favor of spinning a 2 are

$\frac{1}{7}$ ← 1: *number of ways that outcome can occur*; 7: *number of ways that outcome cannot occur*

Read as "one to seven."

The odds against spinning a 2 are

$\frac{7}{1}$ ← 7: *number of ways that outcome cannot occur*; 1: *number of ways that outcome can occur*

If the odds in favor of an outcome are $\frac{a}{b}$, then the odds against the outcome are $\frac{b}{a}$.

Other examples:

The odds in favor of spinning green are $\frac{3}{5}$.

The odds in favor of spinning red are $\frac{4}{4} = \frac{1}{1}$.

The odds against spinning a multiple of 3 are $\frac{6}{2} = \frac{3}{1}$.

The odds against spinning a number less than 4 are $\frac{5}{3}$.

EXERCISES

Compute the probability or the odds.

Event

Picking one marble

1. P(red)
2. P(not red)
3. Odds in favor of picking red
4. P(blue)
5. P(not blue)
6. Odds in favor of picking blue
7. Odds against picking blue

Event

Tossing the number cube

8. P(5)
9. P(not 5)
10. P(2 or 3)
11. Odds in favor of 5
12. Odds against 5
13. Odds in favor of an even number
14. Odds against an even number

Event

Picking one card

15. P(green)
16. P(triangle)
17. Odds in favor of a square
18. Odds in favor of a circle
19. Odds against red

Event

Picking one card

(whole numbers 1 through 24)

★ 20. Odds in favor of an even number

★ 21. Odds against a prime number

★ 22. Odds in favor of a multiple of 3 or a multiple of 4

★ 23. Odds in favor of a number less than 7

★ 24. Odds against a number greater than 16

KEEPING SKILLS SHARP

Give each product or quotient in simplest form.

1. $2 \times 4\frac{1}{2}$
2. $2\frac{1}{3} \div 2$
3. $1\frac{1}{3} \times 1\frac{1}{2}$
4. $3\frac{1}{3} \div 3$
5. $2\frac{2}{3} \times 3\frac{1}{2}$
6. $6 \div 3\frac{1}{3}$
7. $3\frac{3}{4} \times 4\frac{1}{4}$
8. $2\frac{1}{2} \div 2\frac{1}{2}$
9. $1\frac{2}{3} \times 4\frac{3}{5}$
10. $1\frac{4}{5} \div 2\frac{1}{4}$
11. $2\frac{1}{2} \times 3$
12. $3\frac{2}{3} \div 4\frac{1}{2}$
13. $4\frac{2}{3} \times 3\frac{1}{3}$
14. $5\frac{1}{4} \div 3\frac{1}{2}$
15. $4\frac{3}{4} \times 2\frac{2}{3}$
16. $2\frac{2}{3} \div \frac{3}{8}$

Sampling to estimate probabilities

Some students were asked to find what percent of the entire student body play a musical instrument. Of course, they could have checked with each student. But that would have taken too long. Instead, they decided to take a **sample** of the student population and then estimate the percent from their sample.

Their sample had to be representative of the population (the entire student body). It had to be large enough and had to be **unbiased.** For their sample to be unbiased, each member of the student body had to have an equally likely chance of being selected. For their sample they chose the first 80 students to enter the school.

From their sample they predicted the percent of the student body that play a musical instrument.

number who play an instrument → 49

number in sample → 80

$$\frac{49}{80} = \frac{x}{100}$$

$$80x = 4900$$

$$x = 61.25$$

They estimated that about 61% of the student body play a musical instrument. Of a student population of 480, they predicted that 61% of 480, or about 293, play a musical instrument.

EXERCISES

Use this sample to solve. Remember to round your answer to the nearest whole number.

1. How many students in the sample like WMS best?
2. Which station is the most popular?
3. How many students are in the sample?
4. What percent like WGB best? Round your answer to the nearest percent.
5. On the basis of the sample, how many students of a student population of 480 like WGB best?
6. On the basis of the sample, predict how many of the 480 students like WMS best.
7. Predict how many students of the student body do not like WBW best.
8. Predict how many students in the school like either WGB or WBW best.
9. If they had asked one more student for his favorite radio station, what would you guess the answer would have been? Can you be sure?

★ 10. Take a sample of your student body to find out what percent prefer your favorite radio station. Use your sample to estimate how many students in your school agree with your choice.

Presenting data—graphs

Statistics is a branch of mathematics that involves collecting data, analyzing the data, and then making inferences (drawing probable conclusions) from the data.

Graphs provide a visual organization of data. The data listed in the table below are represented on two different kinds of graphs: a bar graph and a broken-line graph.

Population Density of the United States	
Year	*Population per square mile*
1900	25.6
1910	31.0
1920	35.6
1930	41.2
1940	44.2
1950	42.6
1960	50.6
1970	57.4
1980	64.0

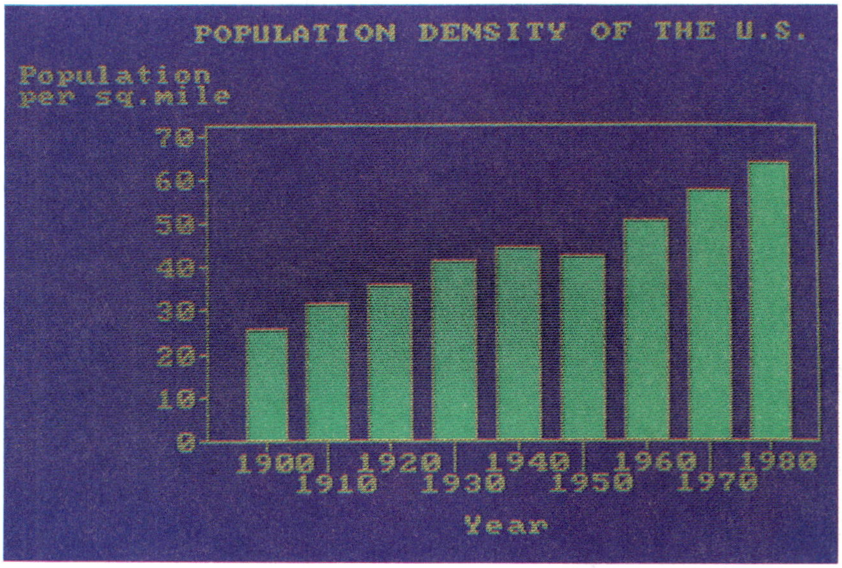

EXERCISES

Study the three ways in which the population density was presented.

1. Which is the most accurate?
2. Which is probably the least accurate?
3. Which do you think is least attractive?
4. Which do you think is the most attractive?
5. Could the bar graph have been made so that the population density could be read to the nearest tenth of a person? How?

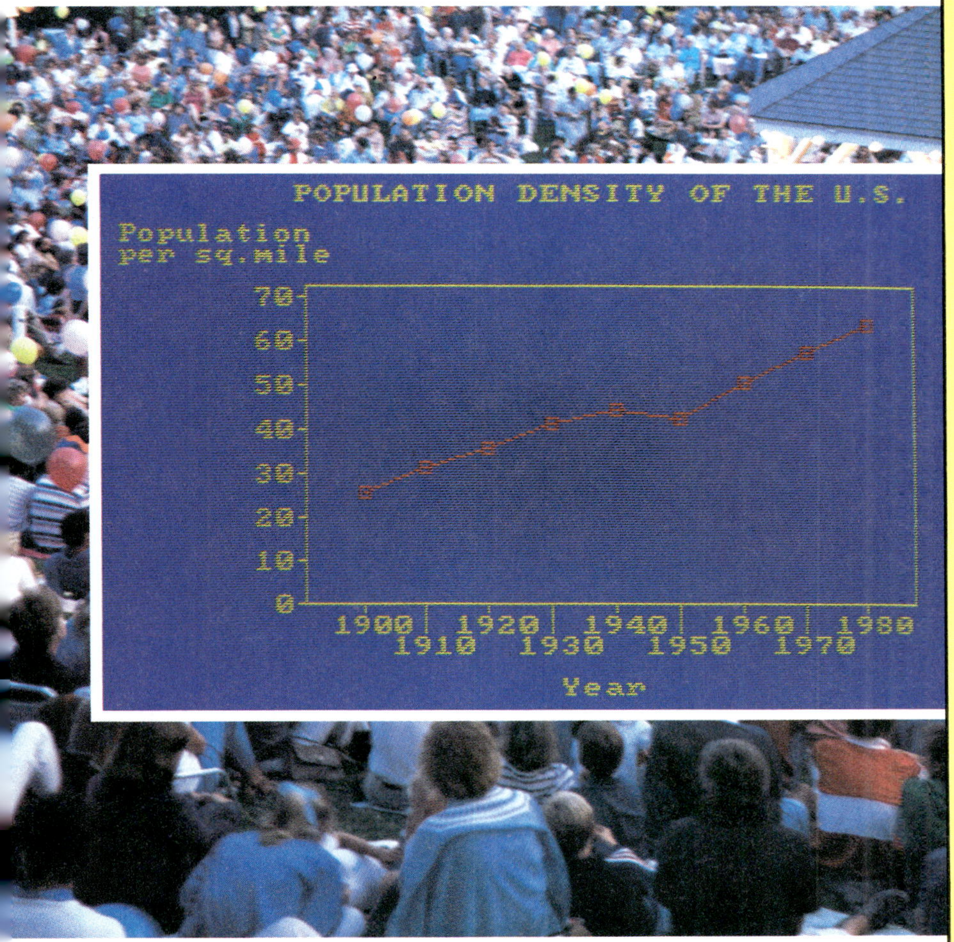

6. Look at the broken-line graph. Is it possible that the population density in 1965 was 60 persons per square mile? Why or why not? Does the broken-line graph really tell you anything about the population density of any years except those that are listed?

7. From a magazine or newspaper, cut out a graph. Construct a different kind of graph that shows the same information.

★ 8. Find out your school's daily attendance for each day of the past two full weeks. Show the daily attendance on a bar graph and on a broken-line graph.

KEEPING SKILLS SHARP

Solve and check.

1. $2x + 3 = 7$
2. $^{-}3a - 4 = 9$
3. $7 + 9y = 18$
4. $^{-}6c - 6 = {}^{-}2$
5. $3n + 9 = {}^{-}8$
6. $4 + 12b = 7$
7. $^{-}8w - 6 = {}^{-}5$
8. $^{-}4r - 3 = 8$
9. $^{-}1 = 14d + 5$
10. $9z + 6 = 9$
11. $\frac{t}{3} + 3 = 0$
12. $\frac{r}{5} - 5 = {}^{-}2$
13. $7 = \frac{g}{^{-}2} - 9$
14. $\frac{s}{7} + {}^{-}2 = {}^{-}6$
15. $\frac{m}{^{-}8} + {}^{-}8 = 9$
16. $\frac{k}{^{-}3} - 7 = 12$
17. $\frac{k}{6} - {}^{-}1 = 2$
18. $16 + \frac{s}{10} = {}^{-}1$
19. $\frac{k}{5} - 5 = 7$
20. $^{-}10 + \frac{v}{^{-}12} = {}^{-}4$

Presenting data—graphs

The **tally** below shows the results of the election for the student council.

Election Tally	
Andrews	𝍸 𝍸 𝍸 𝍸 |
Carter	𝍸 𝍸 𝍸 𝍸 𝍸 |
Howey	𝍸 𝍸 𝍸 𝍸 𝍸 𝍸 𝍸 |
Manning	𝍸 𝍸 𝍸
Sanchez	𝍸 𝍸 𝍸 𝍸 ||

The school election committee made this **pictograph** of the election results and placed it on a school bulletin board.

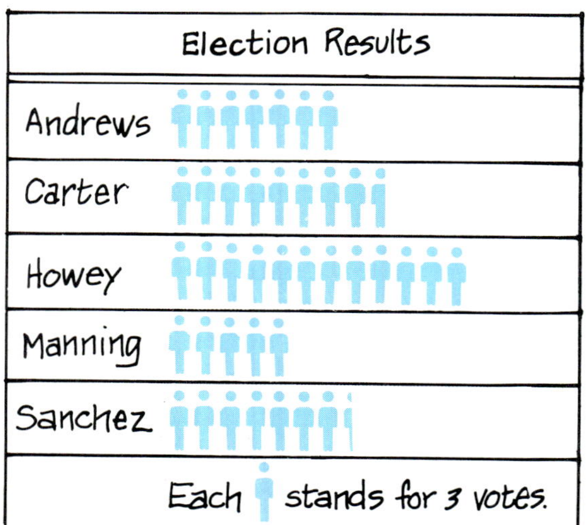

On another bulletin board they posted this **circle graph** of the results. The example below shows how they constructed the circle graph.

number of votes for Andrews → *total number of votes* →

$$\frac{21}{120} = \frac{x}{100}$$

$$120x = 2100$$
$$x = 17.5$$

Andrews got 17.5% of the votes. 17.5% of 360° is 63°. They marked off an angle of 63° in the circle.

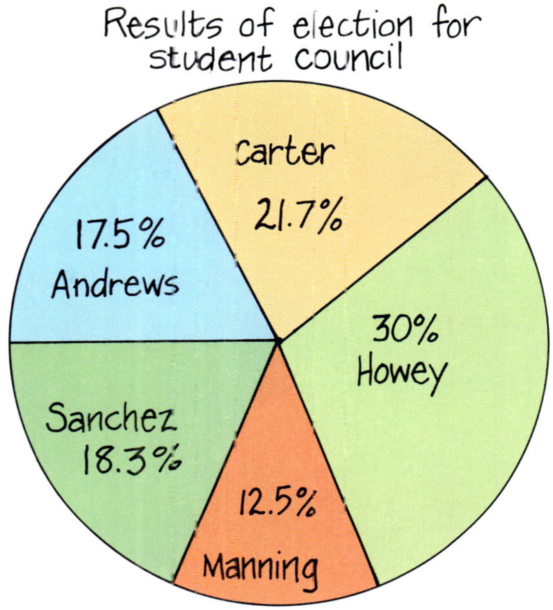

EXERCISES

Refer to the pictograph to answer the following questions.

1. How many votes does each  stand for?
2. Who got the most votes?
3. Who got the fewest votes?
4. How many votes did Andrews get?
5. Estimate the number of votes that Carter got.
6. How many students voted in the election?

Refer to the circle graph to answer the following questions.

7. How many candidates were there?
8. Who got the most votes?
9. Who got the fewest votes?
10. What percent of the votes did Sanchez get?
11. What percent of the votes did the top two "vote getters" get?
12. Can you tell just by looking at the circle graph how many students voted for Manning?

★ 13. **a.** List your six favorite songs.
b. Take a survey. Have each of your friends pick the song from your list that he or she likes best.
c. Show the results on a circle graph.

Mean, median, and mode

The table shows the pulse rate (heartbeats per minute) of some students. To help analyze the data, the pulse rates have been listed from least to greatest.

NAME	PULSE RATE
ADAMS, A.	81
ALLEN, J.	71
BORWICK, A.	76
BURNER, G.	80
CALLON, A.	82
COOKE, C.	86
DAVIS, F.	83
DRIVERS, R.	78
FALLON, C.	62
HOWARDS, F.	78
JACKSON, H.	70
JOHNSON, N.	72
LAVENTI, B.	81
LOGAN, D.	68
MARSHALL, G.	77
NEEDHAM, R.	78
OSLEY, J.	75
PARKER, N.	61
ROGERS, D.	69
SWEETMAN, W.	84
TROTTER, L.	78
TULMAN, M.	74
WALLACE, K.	65

ORDERED RATES
61
62
65
68
69
70
71
72
74
75
76
77
78
78
78
78
80
81
81
82
83
84
86

The **range** is the difference between the largest and smallest numbers.

$86 - 61 = 25$

The range is 25.

The **mean,** or average, is the sum of all the numbers divided by the number of numbers.

$$23\overline{)1729.00} = 75.17$$

Rounded to the nearest tenth, the mean is 75.2.

The **median** is the middle number. To find the median of an even number of numbers, average the two middle numbers.

The median is 77.

The **mode** is the number (or numbers) that occur most often.

The mode is 78.

EXERCISES

Find the range of each set of numbers.

1. 16, 23, 18, 19, 17, 21

2. 83.4, 86.9, 81.5, 90.3, 87.4

3. Compute the mean to the nearest tenth.

   59, 62, 71, 53, 74, 60, 57, 68, 75

4. Compute the mean to the nearest cent.

   \$42.15, \$38.75, \$46.29, \$50.18, \$51.65, \$43.83

Find the median. Remember that you average the two middle numbers if you have an even number of numbers.

5. 18, 14, 23, 19, 20, 17, 16, 24, 20, 21, 20

6. 152, 157, 150, 162, 159, 163, 158, 160

7. What is the mode of the data in exercise 5?

Mean, median, or mode?

8. Half the numbers in a set of data are less than the ? and half are greater.

9. The ? is always a number in the set.

10. The median and ? may not be numbers in the set.

11. Some sets of numbers do not have a ?.

12. a. You and your classmates take your pulse rates. Record your name along with your pulse rate on the chalkboard.
    b. Order the data.
    c. Determine the range, the mean, the median, and the mode.

★ 13. Find seven whole numbers that range from 17 to 25 and have a median of 22, a mode of 23, and a mean of 21.

Challenge!

Solve.

14. Ms. Brown drove 100 kilometers from A to B at 60 kilometers per hour. Then she drove the 100 kilometers from B to A at 40 kilometers per hour. What was her average speed for the round trip? (The answer is not 50 kilometers per hour.)

Frequency distribution

The scores of a mathematics test can be recorded in a table. Such a table is called a **frequency distribution.** It tells how many students received each of the listed scores.

From the table we can compute the mean. First multiply each test score by its frequency. Adding all these products gives the total number of points. To find the number of students, add the frequencies. Then divide the total number of points by the number of students.

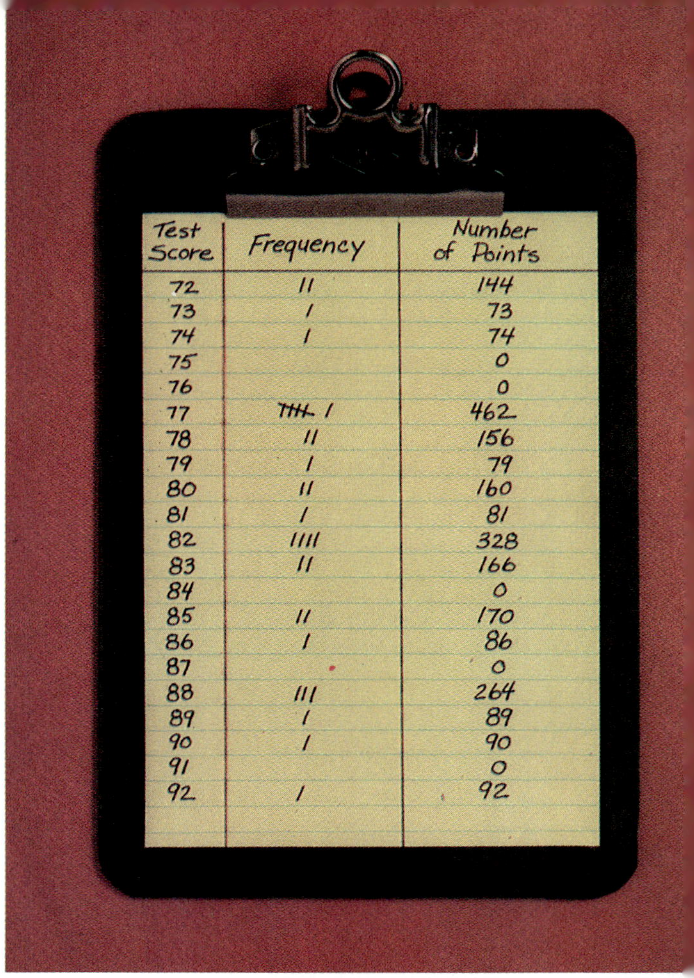

Test Score	Frequency	Number of Points
72	//	144
73	/	73
74	/	74
75		0
76		0
77	𝍸 /	462
78	//	156
79	/	79
80	//	160
81	/	81
82	////	328
83	//	166
84		0
85	//	170
86	/	86
87		0
88	///	264
89	/	89
90	/	90
91		0
92	/	92

A **histogram** is a special kind of bar graph that is used to show frequency. The data given above are shown in this histogram. Notice that all the intervals (72–74, 75–77, 78–80, etc.) are equal.

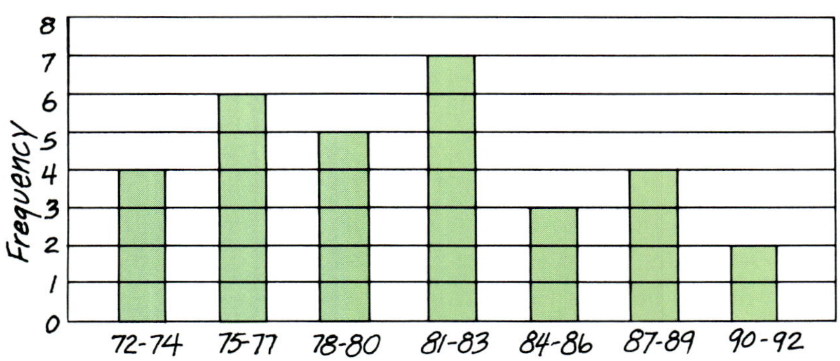

EXERCISES

For exercises 1–8, use the data on page 370.

1. How many students got a score of 82?
2. How many scored 80 or above?
3. Compute the mean score to the nearest tenth. [*Hint:* Divide the total number of points by the number of students.]
4. What is the median?
5. Find the mode.
6. Give the range.
7. Can you tell from the histogram alone whether any student got a score of 75? Why or why not?

Use the data below for exercises 8–13.

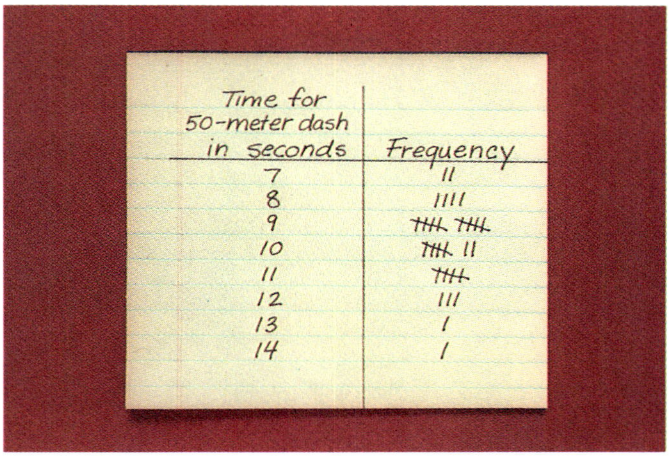

Time for 50-meter dash in seconds	Frequency
7	||
8	||||
9	𝍸 𝍸
10	𝍸 ||
11	𝍸
12	|||
13	|
14	|

8. How many ran the 50-meter dash?
9. Which time is the mode?
10. What is the range of times?
11. What is the mean time?
12. What is the median time?
13. Make a histogram of the frequency distribution.

KEEPING SKILLS SHARP

Solve and check.

1. $2x + 3x = 10$
2. $4x - x = 24$
3. $7x - 3x = 32$
4. $^-48 = 2x - 8x$
5. $3x + x = 0$
6. $8x - 2x = {}^-6$
7. $7 = 4x + 6x$
8. $9x - 3x = 11$
9. $6x - x = 8$
10. $^-3x + x = 16$
11. $5x + x - 3 = 4$
12. $2x + 3x + 8 = 7$
13. $5x - x - 3 = 28$
14. $4x + 3x - 6 = {}^-9$
15. $6x + x - 11 = 15$
16. $7x + x + 7 = {}^-8$
17. $10x - 2x - 3 = 5$
18. $9x + 3x - 2 = {}^-12$
19. $6x - x + 5 = {}^-14$
20. $5x + 2x + 6 = 20$

Misleading statistics

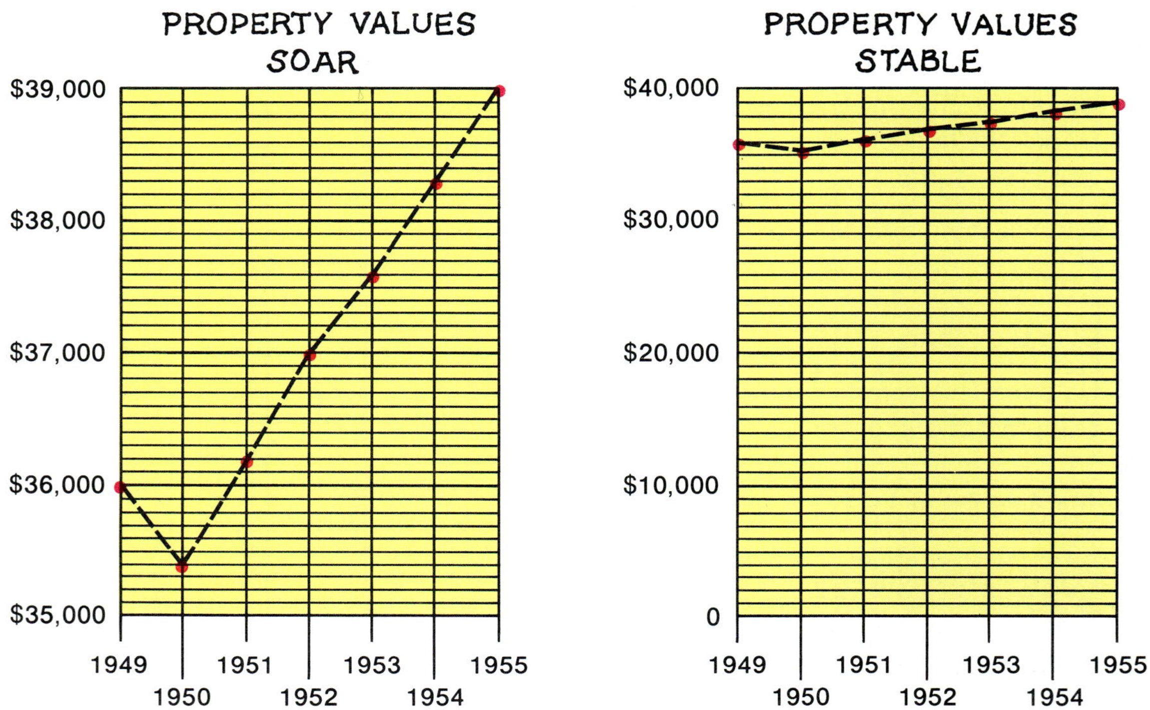

Average Price of a Home

EXERCISES

1. Answer these questions using the first graph.
   a. What was the average price of a home in 1949?
   b. What was the average price of a home in 1955?
   c. What was the increase in price over that period?
   d. What was the percent of increase?
2. Answer the 4 questions in exercise 1 using the second graph.
3. Do the graphs show the same information?
4. Do the pictures suggest the same idea?
5. Can the first graph be misleading?

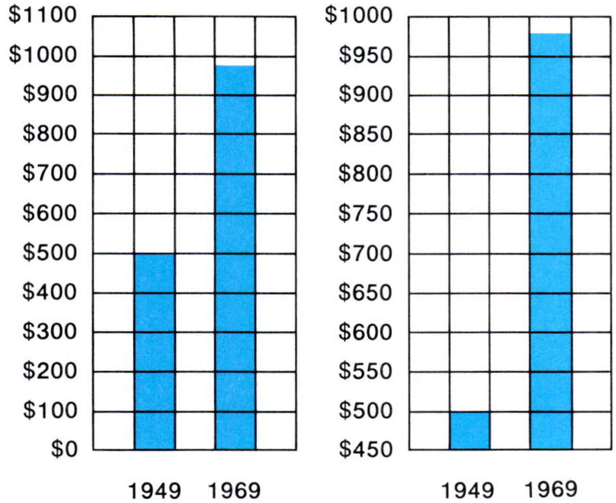

Average Tax on a Home

6. Do both bar graphs to the left show the same information?

7. Which graph do you think shows better the relationship between average tax on a home in 1949 and average tax on a home in 1969? Why?

Oak-Circle Residents' Yearly Incomes	
Smith	$200,000
Jones	$20,000
White	$15,000
Perez	$15,000
Chin	$15,000
Cooper	$15,000
Shultz	$12,000

8. The mean, median, and mode are all used as averages. Which one is the salesman using?

9. Do you believe that the salesman's claim describes the neighbors' incomes accurately?

10. What would be a better average to use? Why?

11. If you had to pick an average salary for this group, would you pick the mean, median, or mode?

Yearly Salary	
Sanchez (owner)	$175,000
Porter (foreman)	$24,000
Pride (laborer)	$13,000
Goza (laborer)	$12,000
Lee (laborer)	$11,500
Starz (laborer)	$10,000
Hyde (laborer)	$10,000

Chapter Checkup

Complete. [pages 352–355]

Event
Picking a marble from the bag

1. P(green) = ?
2. P(blue) = ?
3. P(red) = ?
4. P(not green) = ?
5. P(red or blue) = ?
6. P(yellow) = ?

Give the probability. [pages 356–357]

First Event
Picking a card

Second Event
Flipping the coin

7. P(square and then heads)
8. P(circle and then tails)
9. P(not a triangle and then heads)

Answer these questions about the circle graph. [pages 364–367]

10. What is the greatest percent spent for?
11. What percent is spent on taxes?
12. The family income is $24,000 per year. How much is spent per year on clothing?

Family Budget

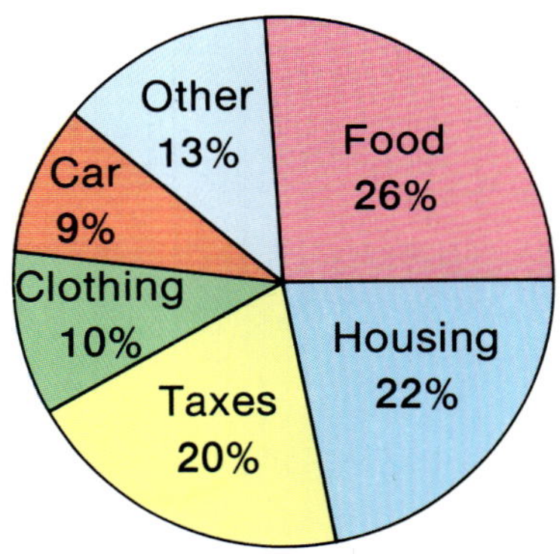

Give the range, the mean, the median, and the mode. [pages 368–369]

13. 5, 6, 7, 8, 9
14. 13, 8, 11, 13, 17, 19
15. 10.3, 8.4, 9.6, 8.4, 7.5, 11.4, 8.2

Chapter Project

DOING A SIMPLER PROBLEM

About how many students would it take to reach from Los Angeles to New York City?

1. Select 20 students at random.
2. Measure their arm spans to the nearest centimeter.
3. Compute the average arm span.
4. Use your average to answer the question above. [*Hint:* The distance from Los Angeles to New York City is 4683.8 kilometers.]

Chapter Review

The probability of spinning green is $\frac{1}{2}$.

$P(\text{green}) = \frac{1}{2}$

$P(\text{green and then red})$

$= \frac{1}{2} \times \frac{1}{6}$

$= \frac{1}{12}$

Give each probability.

1. P(green)
2. P(red)
3. P(not green)
4. P(not red)
5. P(green or blue)
6. P(green or red)

Remember that the probability of a sure outcome is 1. The probability of an impossible outcome is 0.

7. P(yellow)
8. P(not yellow)
9. P(red and then green)
10. P(green and then blue)

The **range** is the difference between the greatest and the smallest numbers.

The **mean** is the average.

The **median** is the middle number.

The **mode** is the number that occurs most often.

The table shows the weekly allowance (to the nearest dollar) of some students.

WEEKLY ALLOWANCE (rounded to the nearest dollar)			
5	4	5	3
3	6	2	4
5	3	1	2
2	4	0	5
1	2	8	1
3	3	3	0
3	5	7	2
2	7	2	6
2	4	4	3
0	0	2	6

11. Order the data from smallest to largest.
12. What is the range?
13. What is the mean allowance?
14. What is the median allowance?
15. What is the mode?

Chapter Challenge

RANDOM NUMBERS

Suppose that this spinner is a fair spinner—that is, that the probability of its stopping on any one of the digits is $\frac{1}{10}$.

We could generate random digits by spinning the spinner and recording (in order) the digits on which the spinner stops.

Since a fair spinner is difficult to construct, we will use the following random digits, which have been generated by a computer:

4	9	4	8	7	5	2	8	Ø	2	2
8	6	6	7	6	2	Ø	5	8	8	7
8	3	5	8	8	2	4	3	1	3	9
4	2	2	1	2	4	Ø	9	8	2	6
7	6	1	Ø	9	1	8	8	6	6	2
5	1	6	7	4	6	2	2	1	9	9

This list of random digits can be used to select unbiased samples. For example, suppose that 83 students want to go to the state basketball final game, but their school is allowed only 15 tickets. The school could randomly select 15 of the 83 students by assigning a 2-digit number to each student (10, 11, . . . , 90, 91, 92) and then, starting in the first row of the table, selecting 2-digit numbers. Students with the first 15 numbers may buy tickets.

49 48 75 28 Ø2 28 66 76 2Ø 58 87 . . .

↑ *duplicate* (under the second 28)

How could a table of random digits be used to make these selections?

1. Randomly divide a group of 80 students into 2 equally large groups.
2. Randomly assign 54 students to 6 groups of 9 students.
3. Randomly pick 60 students from a group of 720 students.
4. Randomly select 100 homes from a city of 12,250 homes for a TV survey.
5. The selection in exercise 1 also could have been made by tossing a coin. What is another way of making the selection in exercise 2?

Cumulative Checkup

STANDARDIZED FORMAT

Choose the correct letter.

1. Divide. Round the quotient to the nearest hundredth.

$0.19\overline{)3.74}$

a. 19.684
b. 19.68
c. 196.8
d. none of these

2. 2.84 kg = ? g

a. 284
b. 28.4
c. 2840
d. none of these

3. Multiply.

$\frac{4}{9} \times \frac{3}{8}$

a. $1\frac{5}{27}$
b. $\frac{27}{32}$
c. 6
d. none of these

4. The area of this trapezoid is

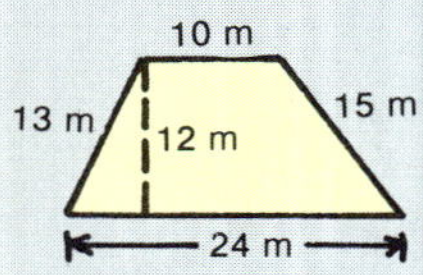

a. 204 m^2
b. 288 m^2
c. 62 m
d. none of these

5. The ratio of prime-number cards to composite-number cards is

a. 1 : 4
b. 2 : 3
c. 2 : 5
d. none of these

6. What is the percent of discount?

a. $37\frac{1}{2}\%$
b. 60%
c. $62\frac{1}{2}\%$
d. none of these

7. The simple interest for a loan of \$800 at 12% a year for 3 years is

a. \$288
b. \$28.80
c. \$24
d. none of these

8. Solve.

$^-3n + 6 = {}^-9$

a. 1
b. $^-1$
c. 5
d. none of these

9. The probability of spinning a number less than 6 and then flipping heads is

a. $\frac{3}{8}$
b. $\frac{9}{8}$
c. $\frac{5}{16}$
d. none of these

10. The median of 4, 9, 5, 8, and 9 is

a. 7
b. 8
c. 9
d. none of these

11. What fraction of a yard is 8 inches?

a. $\frac{2}{9}$ **b.** $\frac{2}{3}$
c. $\frac{1}{4}$ **d.** none of these

12. Leon bought 3 pens for 19¢ each and 3 pads for \$1.05 each. How much more did he spend for pads than pens?

a. \$3.72 **b.** \$2.58
c. 86¢ **d.** none of these

13

Indirect Measurement—Similar and Right Triangles

There are many congruent shapes in this photograph. Name some objects whose shapes are congruent.

READY OR NOT?

These two triangles are congruent.

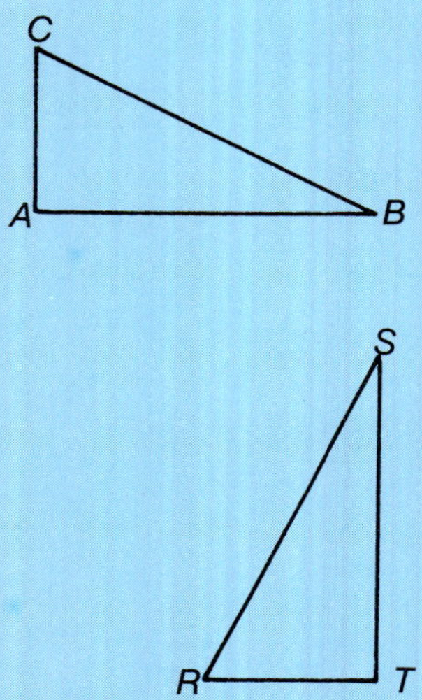

Complete each pair of corresponding parts for the congruent fitting.

1. $\overline{AB} \leftrightarrow$?
2. $\overline{BC} \leftrightarrow$?
3. $\overline{CA} \leftrightarrow$?
4. $\angle A \leftrightarrow$?
5. $\angle B \leftrightarrow$?
6. $\angle C \leftrightarrow$?

Solve.

7. $\frac{3}{n} = \frac{12}{20}$
8. $\frac{n}{6} = \frac{8}{10}$
9. $\frac{7}{3} = \frac{n}{5}$
10. $\frac{4}{7} = \frac{9}{n}$

Similar figures

These two photographs have the same shape. They are **similar.**

Two figures that have the same shape are called **similar figures.**

These two triangles are similar.

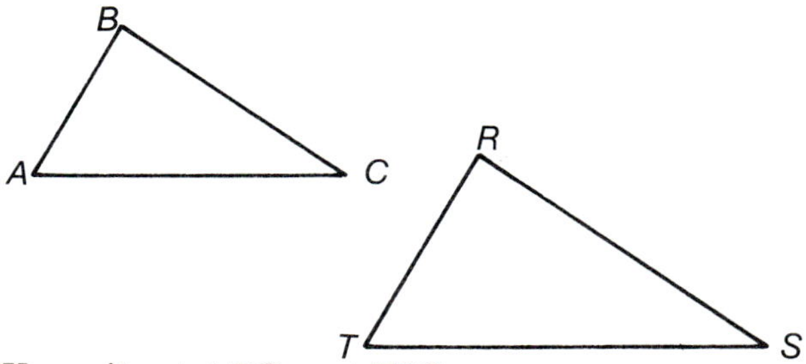

We write $\triangle ABC \sim \triangle TRS$
and say "Triangle ABC is similar to triangle TRS."

EXERCISES

Do the two figures look similar?

1. 

2. 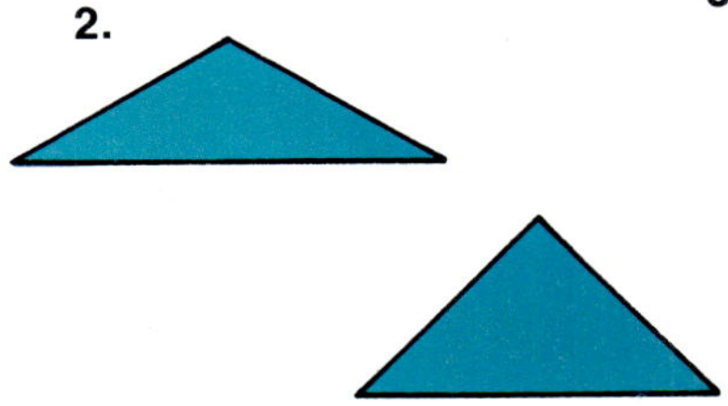

3. 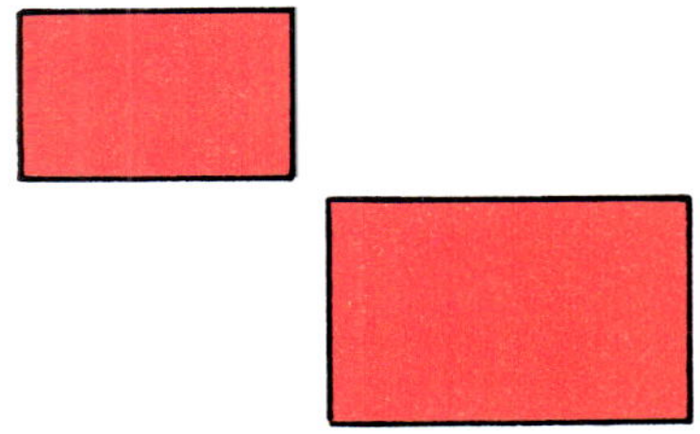

4. 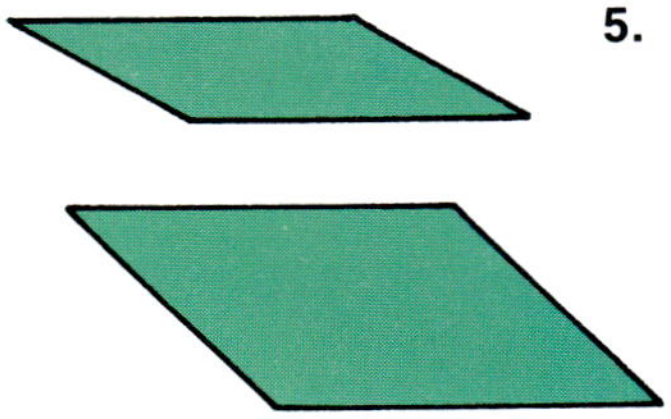

5. 

6. 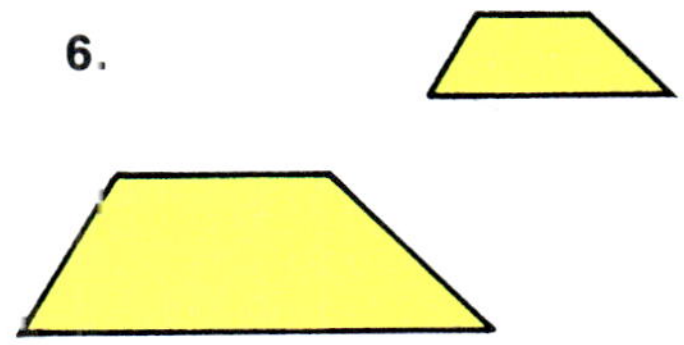

These two triangles are similar.

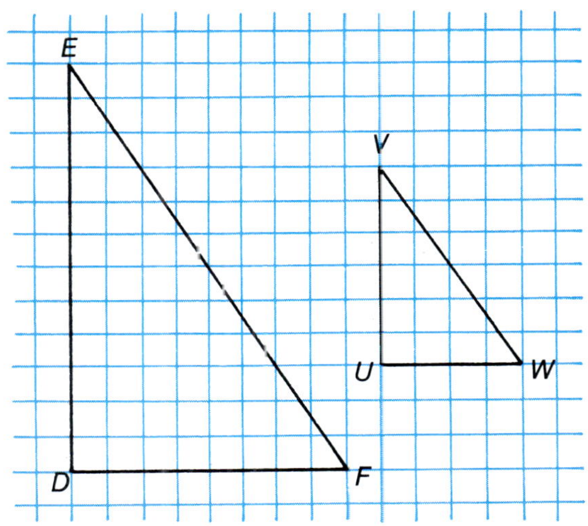

7. Is $\angle D \cong \angle U$?
8. Is $\angle E \cong \angle V$?
9. Is $\angle F \cong \angle W$?

What is the ratio of

10. the length of $\overline{DE}$ to the length of $\overline{UV}$?
11. the length of $\overline{DF}$ to the length of $\overline{UW}$?
12. **Guess** the ratio of the length of $\overline{EF}$ to the length of $\overline{VW}$.

13. **a.** On graph paper, draw a pair of similar triangles.
    b. List all pairs of congruent angles.
    c. Compare the lengths of the sides that are opposite congruent angles. Are the ratios equal?

Similar polygons

Here are two similar triangles.

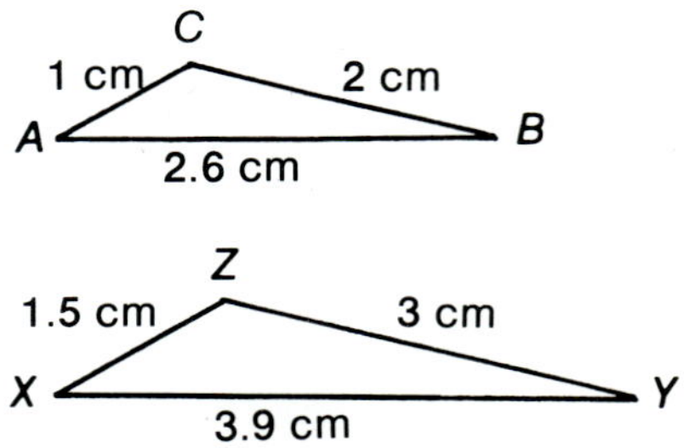

corresponding angles	*corresponding sides*
$\angle A \leftrightarrow \angle X$	$\overline{AB} \leftrightarrow \overline{XY}$
$\angle B \leftrightarrow \angle Y$	$\overline{BC} \leftrightarrow \overline{YZ}$
$\angle C \leftrightarrow \angle Z$	$\overline{CA} \leftrightarrow \overline{ZX}$

The corresponding angles are congruent and the corresponding sides are in the same ratio.

This symbol stands for the length of $\overline{AB}$.

$$\frac{AB}{XY} = \frac{2.6}{3.9} = \frac{2}{3} \qquad \frac{BC}{YZ} = \frac{2}{3} \qquad \frac{CA}{ZX} = \frac{1}{1.5} = \frac{2}{3}$$

Since all three ratios are equal, the corresponding sides are said to be **proportional.**

In similar polygons, the corresponding angles are congruent and the corresponding sides are proportional.

These two triangles are similar. Since similar triangles have proportional sides, we can solve a proportion to find the length x.

$$\frac{x}{20} = \frac{9}{15}$$

$$15x = 180$$

$$x = 12$$

The length x is 12 m.

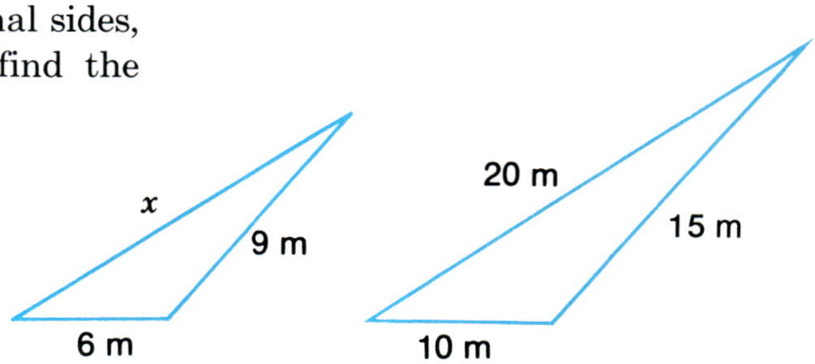

What are some other proportions that you could solve to find the length x?

EXERCISES

The two polygons in each exercise are similar. Solve a proportion to find each missing length. Round each answer to the nearest tenth unit.

1.

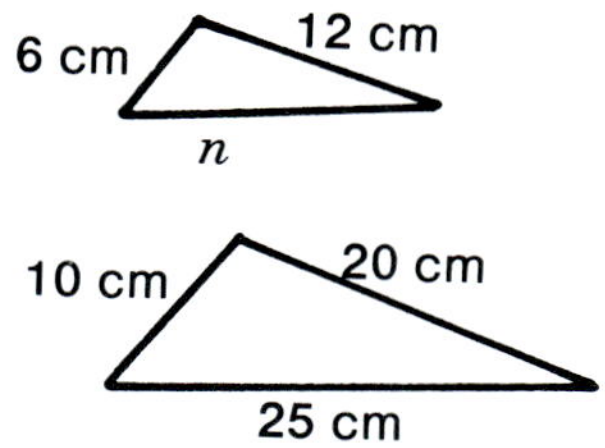

2.

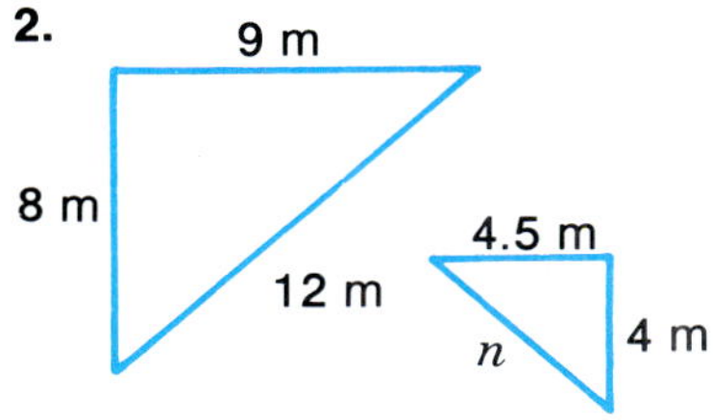

3.

4.

★ **5.**

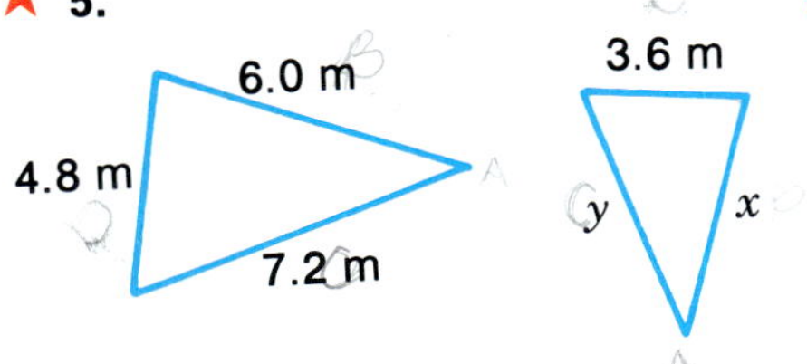

★ **6.**

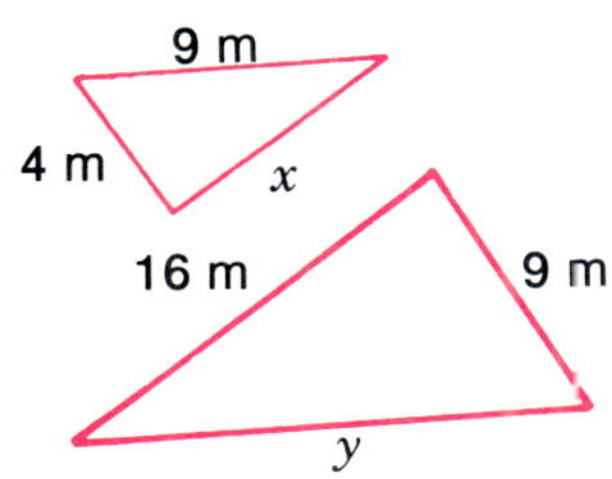

Challenge!

Solve.

7. a. Get a protractor and draw two triangles that have congruent corresponding angles.

b. Measure the sides. Are they proportional?

KEEPING SKILLS SHARP

Solve.

1. $2x + 6 = {}^{-}15$
2. $9 = \frac{x}{3} - 4$
3. $3x = {}^{-}12$
4. $\frac{x}{^{-}6} + 7 = {}^{-}2$
5. $3 + 2x = 13$
6. $\frac{x}{5} + 4 = 11$
7. ${}^{-}4 = 5x - 19$
8. $\frac{x}{^{-}3} - {}^{-}2 = {}^{-}26$
9. $5x + 5x = 0$
10. $3x - 2x = {}^{-}10$
11. $x + 9x = {}^{-}23$
12. $4x - x = {}^{-}20$
13. $3x + 2x - 3 = 8$
14. $-2x + 4x + 7 = {}^{-}10$
15. $4x + 6x - 5 = 12$
16. $8x - 5x + 9 = 22$

More about similar polygons

Here is a quick way to check whether two triangles are similar:

If the corresponding angles of two triangles are congruent, then the triangles are similar.

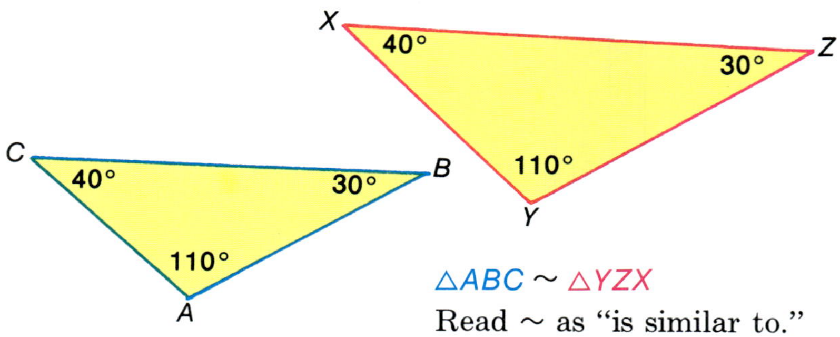

△*ABC* ~ △*YZX*

Read ~ as "is similar to."

EXERCISES

Decide whether the two triangles are similar.

1. 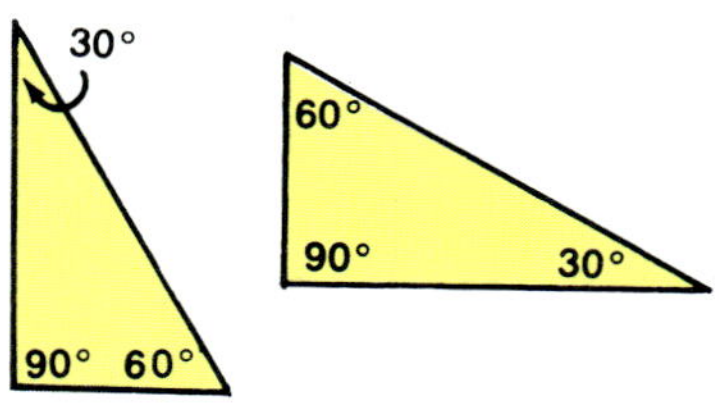

2. 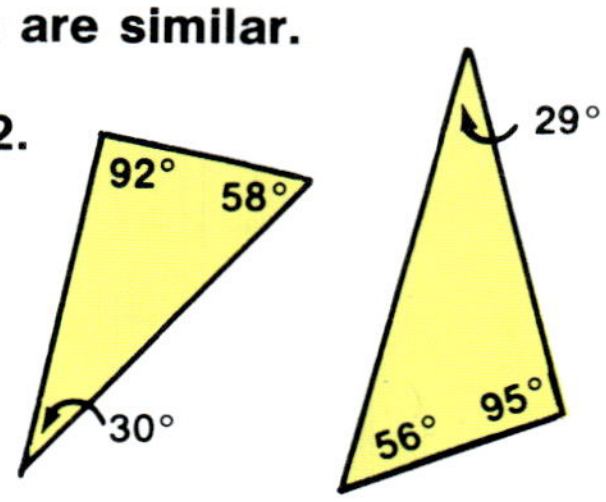

3. 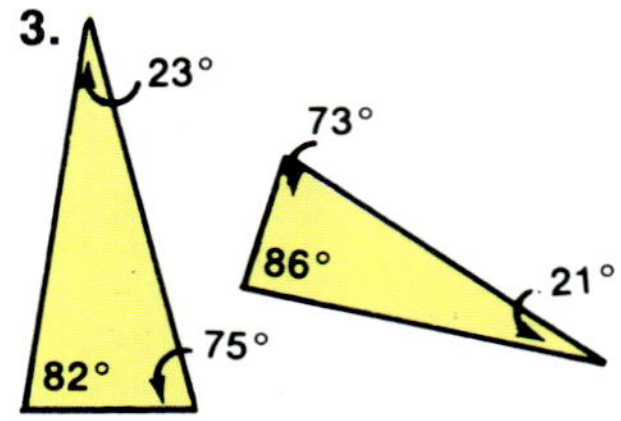

Remember that the sum of the angle measures of a triangle is 180°. In these exercises you are given two of the angles. Find the third. (There are 60 minutes in 1 degree; 60′ = 1°.)

4. 90°, 30°
5. 90°, 45°
6. 60°, 60°
7. 50°, 38°
8. 110°, 35°
9. 145°, 27°
★ 10. 27°18′, 59°17′
★ 11. 93°16′, 46°19′

12. △*ABC* has a 50° angle and a 70° angle. △*XYZ* also has a 50° angle and a 70° angle. Are the triangles similar?

13. a. Suppose that two right triangles also have 45° angles. Are they similar?

    b. Suppose that two right triangles also have 30° angles. Are they similar?

Find the missing angle measures.

14.

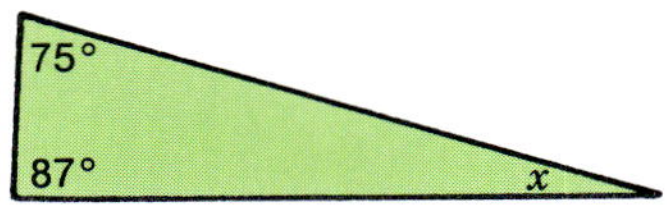

15.

16.

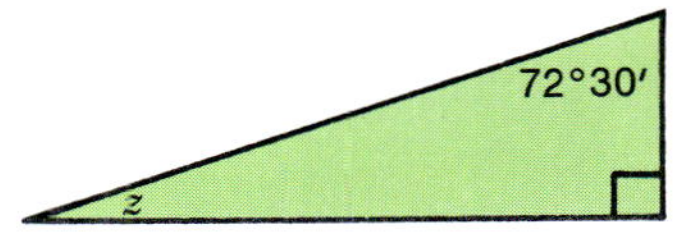

The two right triangles are similar.
Find the missing angle measures.

17.

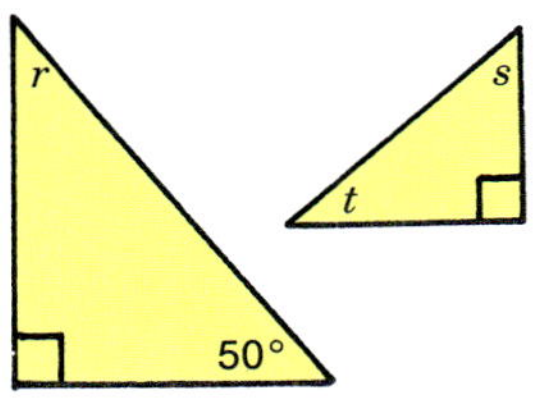

18.

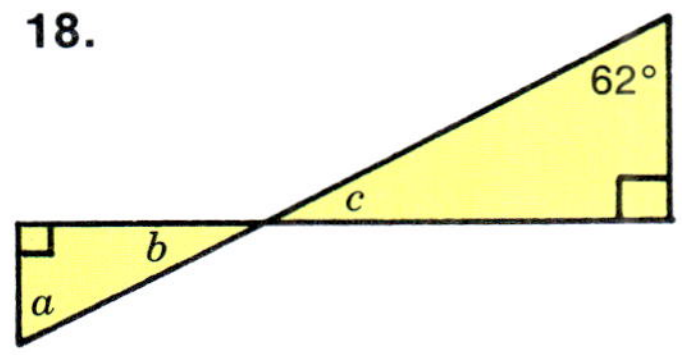

19.

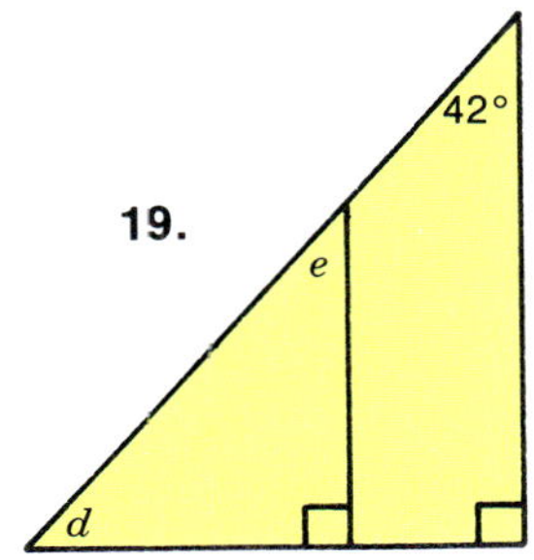

Challenge!

Solve.

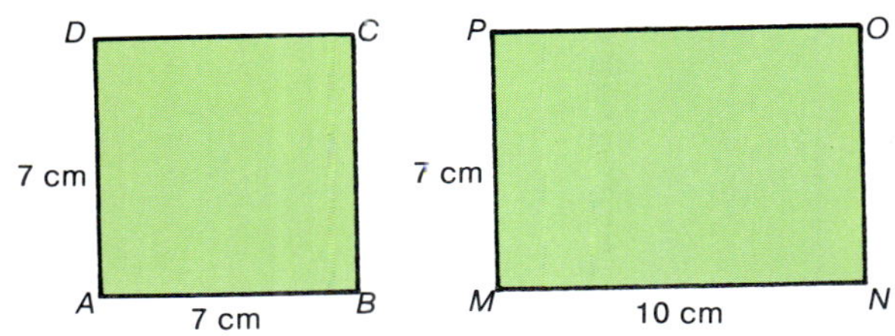

20. a. Are corresponding angles of these two rectangles congruent?

b. Are the corresponding sides proportional?
(Are these ratios equal: $\frac{AB}{MN}$, $\frac{BC}{NO}$, $\frac{CD}{OP}$, $\frac{AD}{MP}$?)

c. Are the rectangles similar?

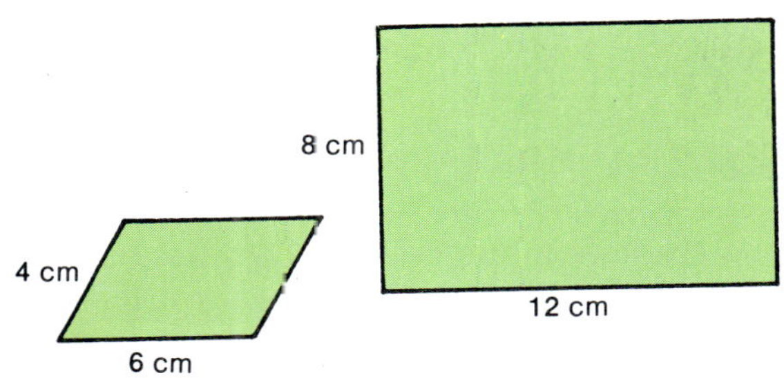

21. a. Are corresponding sides of the two parallelograms proportional?

b. Are the corresponding angles congruent?

c. Are the parallelograms similar?

22. Are all equilateral triangles similar?

23. Are all right triangles similar?

24. Are all rectangles similar?

25. Are all parallelograms similar?

26. Are all squares similar?

27. Are all isosceles triangles similar?

Problem Solving

Indirect measurement

Similar triangles can be used to find lengths that would be difficult to measure directly. Below, you are shown how indirect measurement can be used to find the height of the flagpole.

The triangle made by the meterstick and its shadow is similar to the triangle made by the flagpole and its shadow. Jane found the height of the flagpole by solving this proportion.

$$\frac{h}{100} = \frac{306}{68}$$

$$68 \times h = 100 \times 306$$

$$h = 450$$

So the height of the flagpole is 450 cm, or 4.50 m.

EXERCISES

First read the problem carefully and make a sketch. Then solve the problem.

1. A building casts a 150-meter shadow. At the same time, a man 2 meters tall casts a 3-meter shadow. How tall is the building?

2. A tree casts a 30.5-meter shadow. At the same time a meterstick casts a 1.6-meter shadow. How tall is the tree?

3. A meterstick casts a shadow that is 0.35 meter long. At the same time, a water tank casts a shadow that is 24.80 meters long. What is the height of the water tank?

4. A utility pole casts a shadow that is 24.50 meters long. The shadow of a meterstick that is stood upright 22.50 meters from the pole ends at the same point. Find the height of the utility pole.

★ 5. A backpacking club wants to measure the distance across a lake. The club made the measurements recorded in the sketch and used similar triangles to find the distance across the lake. How do you know the triangles are similar? How far is it across the lake?

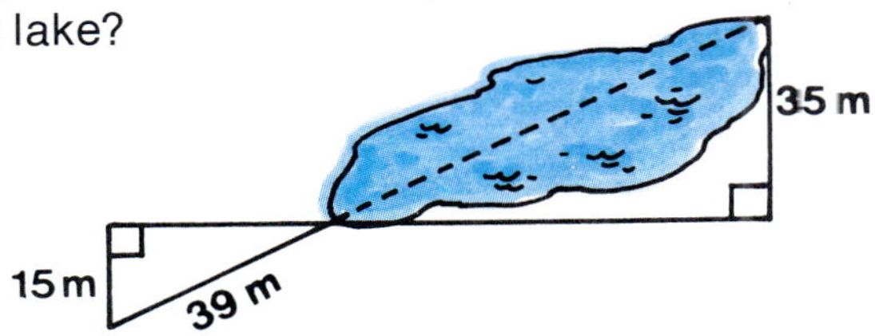

Challenge!

Solve.

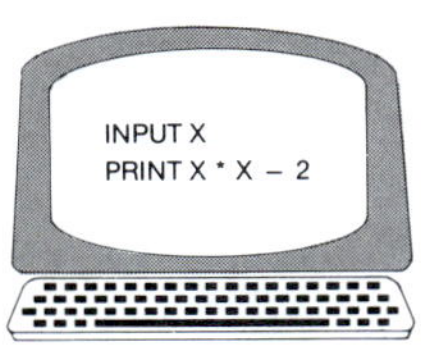

	Input	Output
	5	23
6.	7	?
7.	?	14
8.	9	?

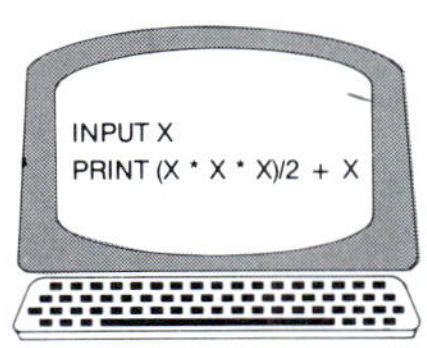

	Input	Output
	4	36
9.	2	?
10.	6	?
11.	?	510

KEEPING SKILLS SHARP

Solve.

1. 25% of $x = 11$
2. 20% of 130 $= x$
3. 150% of 84 $= x$
4. $33\frac{1}{3}$% of $x = 17$
5. $62\frac{1}{2}$% of $x = 65$
6. 43% of 74 $= x$
7. 12% of 350 $= x$
8. $66\frac{2}{3}$% of $x = 46$
9. $116\frac{2}{3}$% of 91 $= x$
10. 0.4% of $x = 16$
11. 79% of 80 $= x$
12. 7% of $x = 35$
13. 13% of $x = 39$
14. $37\frac{1}{2}$% of $x = 24$
15. 0.16% of 150 $= x$
16. 9.1% of 36 $= x$
17. $87\frac{1}{2}$% of $x = 56$
18. 120% of 240 $= x$
19. $133\frac{1}{3}$% of $x = 56$
20. 225% of 96 $= x$

Tangent ratios

All three of these right triangles are similar. Why?

The red side is called the **side opposite** the 30° angle. The blue side is called the **side adjacent** to the 30° angle. Since the triangles are similar, their ratios of

$$\frac{\text{side opposite (30° angle)}}{\text{side adjacent (30° angle)}}$$

are equal. The ratio of the side opposite an acute angle to the side adjacent to the acute angle in a *right* triangle is called the **tangent** of the angle. (The longest side of the triangle, shown here in black, is not used in the tangent ratio.) We could measure the sides and divide to find the approximate ratio. However, to save time and to be more accurate, we can use a table of tangent ratios.

$$\tan 30° \approx 0.58$$

We can use the tangent ratios to help us compute the lengths of sides of right triangles.

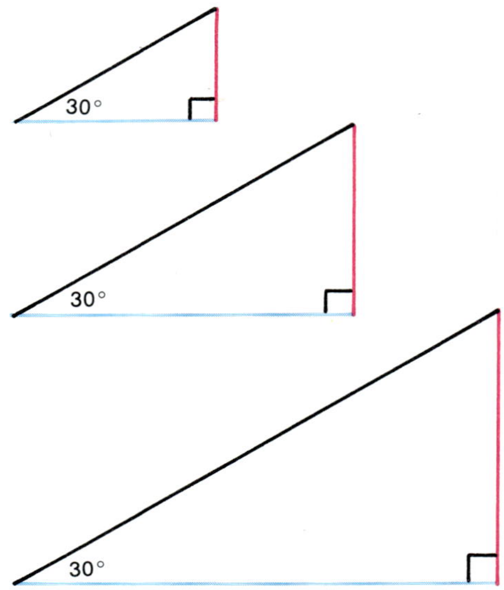

Some tangent ratios
tan 5° ≈ 0.09
tan 10° ≈ 0.18
tan 15° ≈ 0.27
tan 20° ≈ 0.36
tan 25° ≈ 0.47
tan 30° ≈ 0.58
tan 35° ≈ 0.70
tan 40° ≈ 0.84
tan 45° = 1.00
tan 50° ≈ 1.19
tan 55° ≈ 1.43
tan 60° ≈ 1.73
tan 65° ≈ 2.14
tan 70° ≈ 2.75
tan 75° ≈ 3.73
tan 80° ≈ 5.67
tan 85° ≈ 11.43

EXAMPLES.

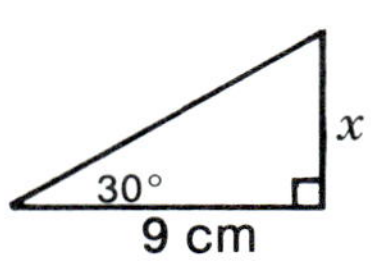

$$\tan 30° = \frac{x}{9} \approx 0.58$$

$$x \approx 5.22 \text{ cm}$$

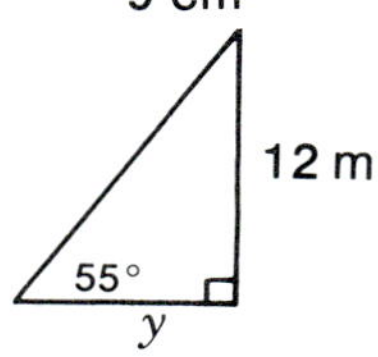

$$\tan 55° = \frac{12}{y} \approx 1.43$$

$$y \approx 8.39 \text{ m}$$

EXERCISES

Name the side opposite the given angle and the side adjacent to the given angle.

1. 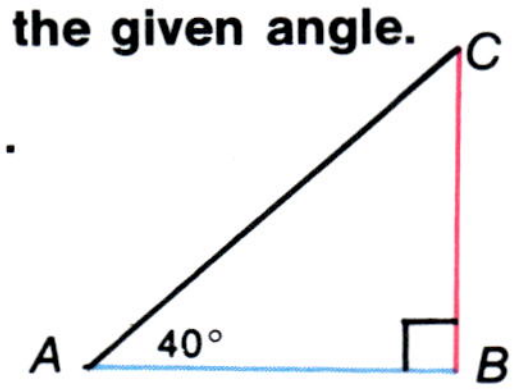

2. 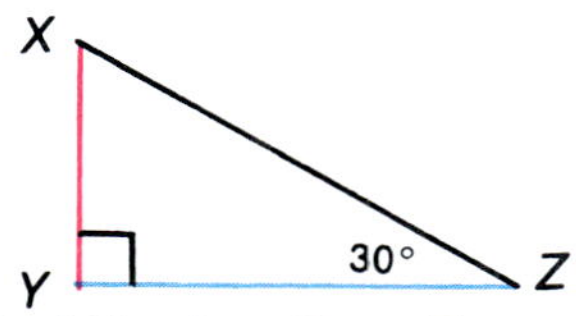

3. 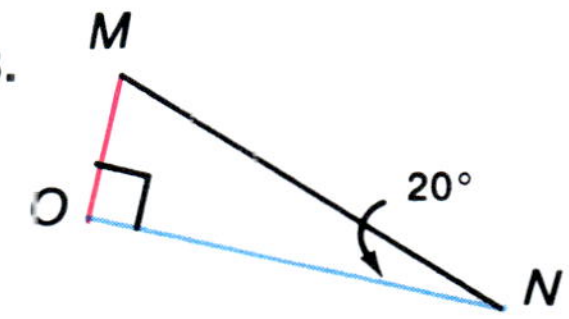

Use the table of tangent ratios to find the length x. Round each answer to the nearest hundredth.

4. 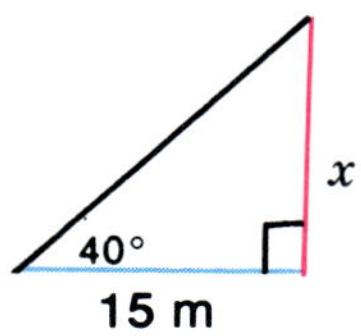

Hint: $\tan 40° = \frac{x}{15} \approx 0.84$

$x \approx 15 \times 0.84$

$x \approx ?$

5. 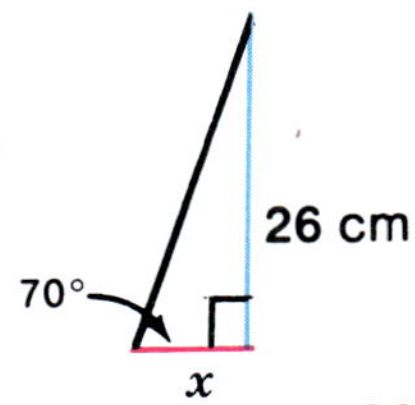

Hint: $\tan 70° = \frac{26}{x} \approx 2.75$

$2.75x \approx 26$

$x \approx ?$

6. 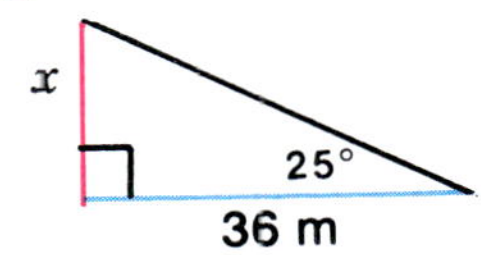

Hint: $\tan 25° = \frac{x}{36} \approx ?$

$x \approx ?$

7. 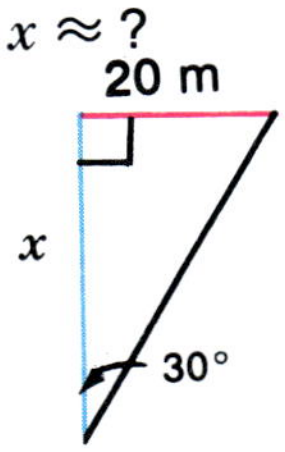

8. 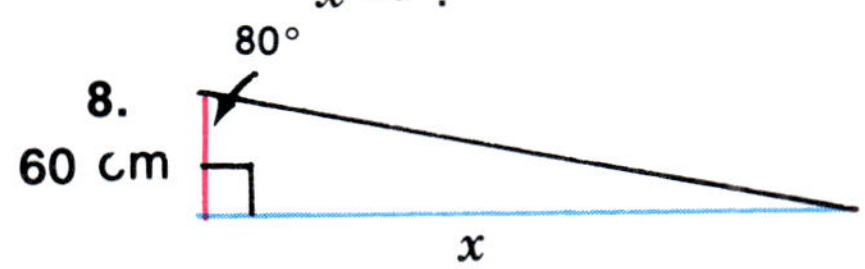

9. 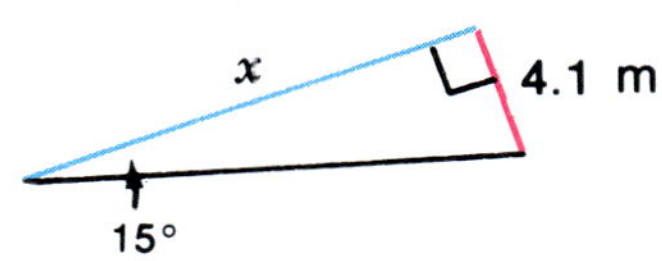

10. 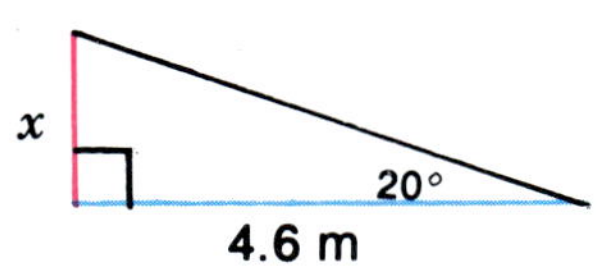

11. 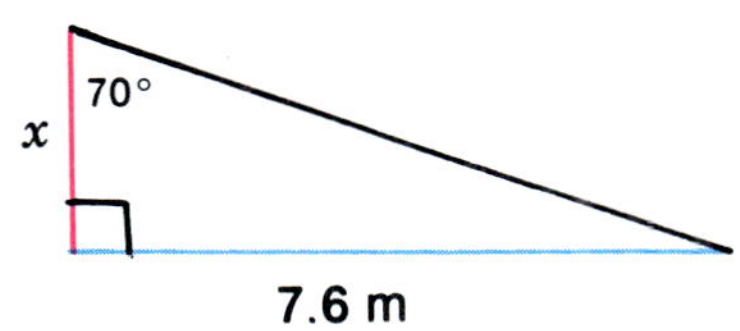

12. 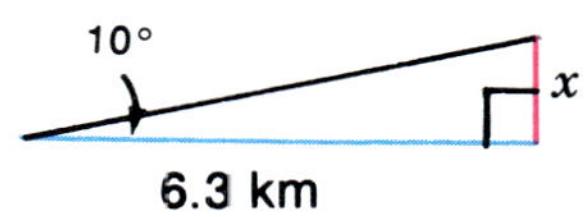

Challenge!

Solve. [*Hint:* First draw a picture. Round each answer to the nearest hundredth.]

13. A hot-air balloon is directly over a ballpark that is 10 kilometers from where the balloon was launched. A sighting from the point where it was launched shows the balloon to be 30° above horizontal. How high is the balloon over the ballpark?

14. How far from a 300-meter cliff would you have to stand in order to look directly at the top by sighting 70° above horizontal?

Practice exercises

Use similar triangles to find the length x. Round to the nearest tenth.

1.

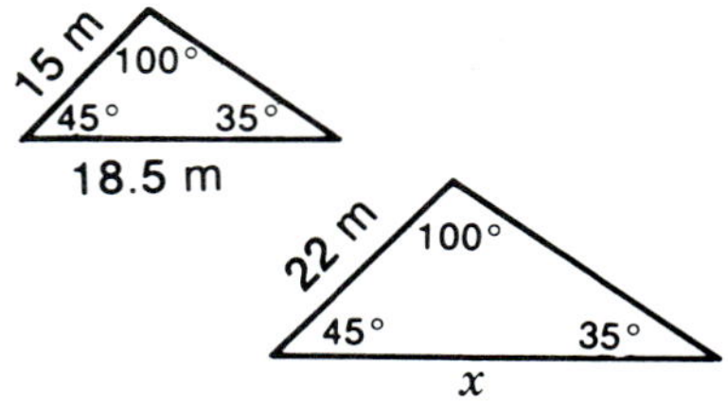

2.

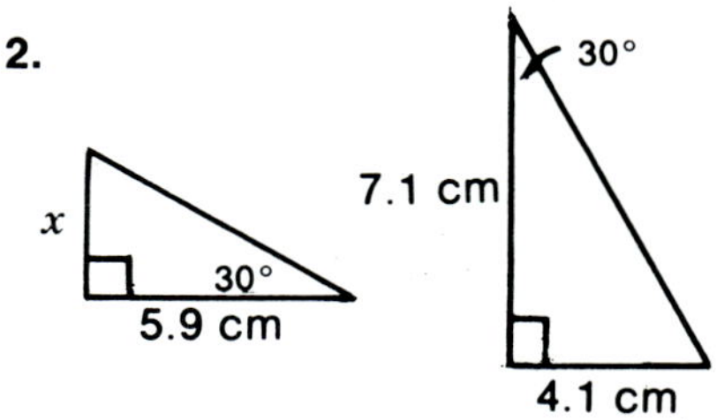

Use the table of tangent ratios to find the length x. Round your answer to the nearest hundredth.

3.

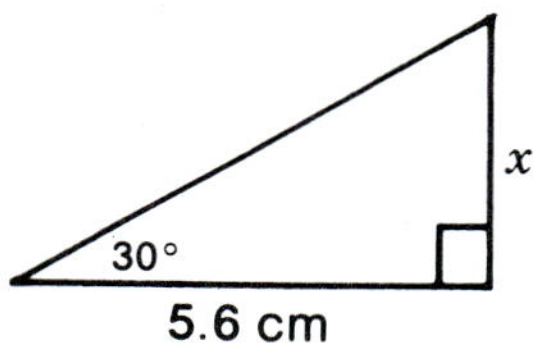

4.

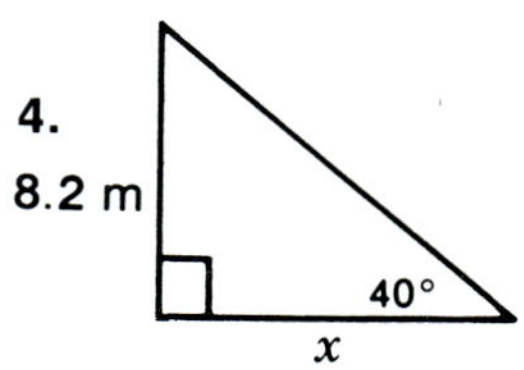

5.

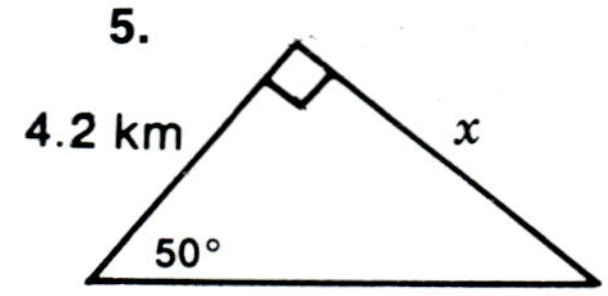

6.

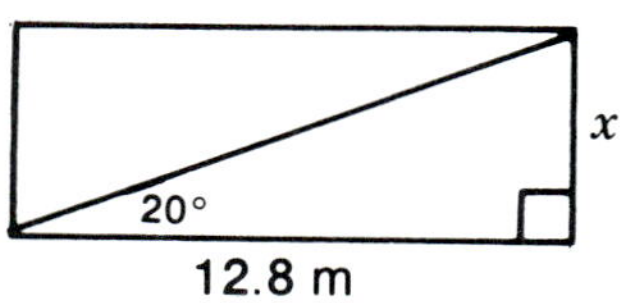

7.

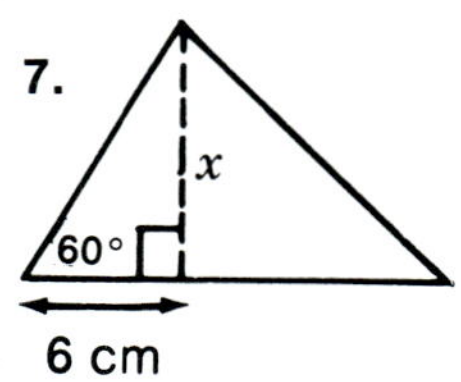

8.

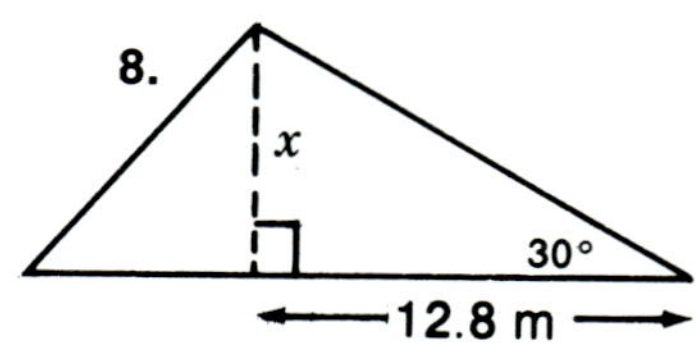

Problem Solving

9. An airplane is sighted from a control tower. The airplane is 18.2 km from a point directly over the tower. The tower is 80 m high. The angle of elevation is 25°. What is the altitude of the airplane?

Problem Solving Application

Grace is 16 years old and has her driver's license. Now she is thinking about buying a car. She listed the following information to help her decide whether she could afford to buy this car.

1. Study and understand.
2. Plan and do.
3. Answer and check.

- The car she wants will cost $85 per month for 3 years.
- Grace works 20 hours a week and earns $3.75 per hour.
- About $15 is deducted from her weekly pay for Social Security and withholding tax.
- Insurance will cost $960 per year.
- Gasoline costs $1.219 per gallon.
- Grace thinks that she will drive about 100 miles per week.
- She was told that the car averaged about 25 miles per gallon of gasoline.

Problem: Can she afford to buy the car?

The answers to these questions will help you decide.

1. About how many miles per month will she drive?
2. How many gallons of gasoline will she need in a week?
3. What will gasoline cost per month?
4. Including car payments and insurance, how much will she spend on her car each month?
5. What is her monthly take-home pay?
6. Are there other expenses that are not listed above?

Square roots

Ancient mathematicians thought about numbers in geometric terms. So when they multiplied a number by itself, they thought about finding the area of a square. That is the reason we still talk about **squaring** numbers.

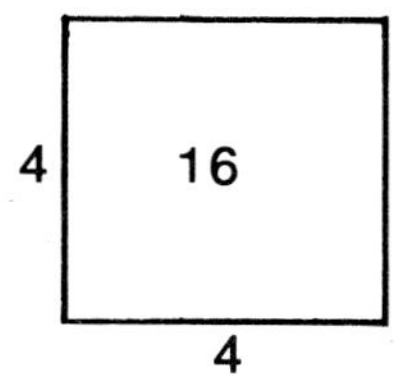

$$4^2 = 16$$

We say, "4 squared is 16."

We also say, "The square root of 16 is 4."

A short way of writing this is $\sqrt{16} = 4$.

Here are some more examples:

$7^2 = 49$	$10^2 = 100$	$1.2^2 = 1.44$
$\sqrt{49} = 7$	$\sqrt{100} = 10$	$\sqrt{1.44} = 1.2$

EXERCISES

Simplify.

1. $\sqrt{36}$	**2.** $\sqrt{9}$	**3.** $\sqrt{25}$	**4.** $\sqrt{64}$	**5.** $\sqrt{1}$
6. $\sqrt{81}$	**7.** $\sqrt{196}$	**8.** $\sqrt{256}$	**9.** $\sqrt{4}$	**10.** $\sqrt{361}$
11. $\sqrt{144}$	**12.** $\sqrt{400}$	**13.** $\sqrt{169}$	**14.** $\sqrt{225}$	**15.** $\sqrt{121}$
16. $\sqrt{2.25}$	**17.** $\sqrt{2.89}$	**18.** $\sqrt{1.69}$	**19.** $\sqrt{1.96}$	**20.** $\sqrt{3.24}$

Some square roots have only decimal approximations. The divide-and-average method is one way to find a decimal approximation of a square root. Study this example.

EXAMPLE. Find $\sqrt{32}$ to the nearest tenth.

Step 1. Estimate $\sqrt{32}$.

$\sqrt{25} = 5$

$\sqrt{32} = ?$

$\sqrt{36} = 6$

$\sqrt{32}$ is between 5 and 6. We will try 5.6.

Step 2. Divide 32 by 5.6.

$$\begin{array}{r} 5.71 \\ 5.6\overline{)32.000} \\ -28\,0 \\ \hline 4\,00 \\ -3\,92 \\ \hline 80 \end{array}$$

$\sqrt{32}$ is between 5.6 and 5.7.

Step 3. Average 5.6 and 5.7.

$$\frac{5.6 + 5.7}{2} = 5.65$$

Step 4. Divide 32 by 5.65.

$$\begin{array}{r} 5.66 \\ 5.65\overline{)32.0000} \\ -28\,25 \\ \hline 3\,750 \\ -3\,390 \\ \hline 3600 \end{array}$$

$\sqrt{32}$ is between 5.65 and 5.66.

$\sqrt{32} = 5.7$ to the nearest tenth.

True or false?

21. $\sqrt{2}$ is between 1 and 2.

22. $\sqrt{3}$ is between 1 and 2.

23. $\sqrt{5}$ is between 1 and 2.

24. $\sqrt{5}$ is between 2 and 3.

25. $\sqrt{5}$ is between 2.2 and 2.3.

26. $\sqrt{5}$ is between 2.3 and 2.4.

27. $\sqrt{5}$ is between 2.22 and 2.23.

28. $\sqrt{5}$ is between 2.23 and 2.24.

Use the divide-and-average method.
Give an approximation to the nearest tenth.

29. $\sqrt{7}$ **30.** $\sqrt{12}$ **31.** $\sqrt{50}$ **32.** $\sqrt{68}$ **33.** $\sqrt{75}$

34. $\sqrt{90}$ **35.** $\sqrt{110}$ **36.** $\sqrt{160}$ **37.** $\sqrt{218}$ **38.** $\sqrt{236}$

Right triangles

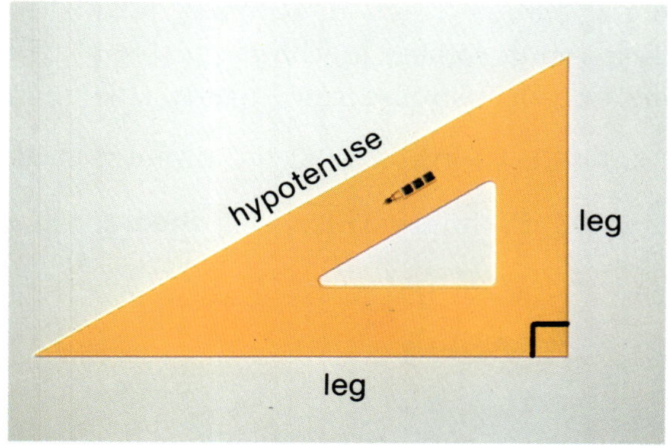

The sides of right triangles have special names. The side opposite the right angle, the longest side, is called the **hypotenuse.** The other two sides are called the **legs.**

In this lesson you will explore an interesting property of right triangles.

EXERCISES

Which side is the hypotenuse?

1.

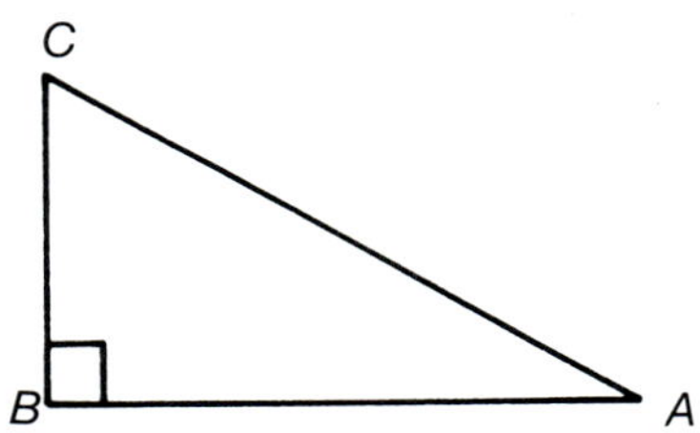

2.

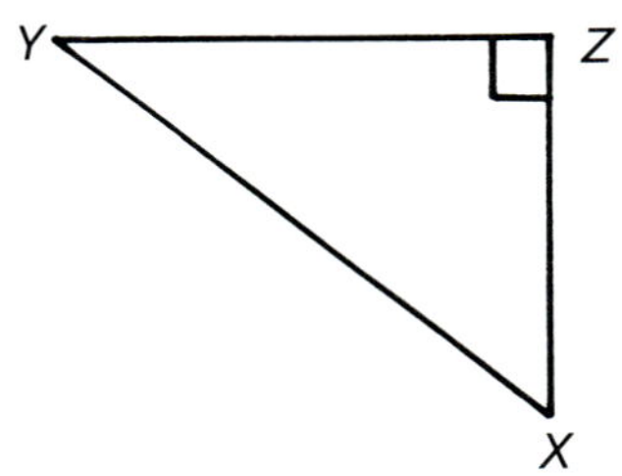

3.

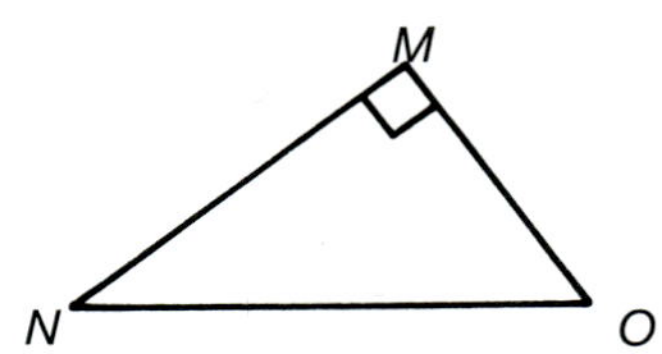

Which sides are the legs?

4.

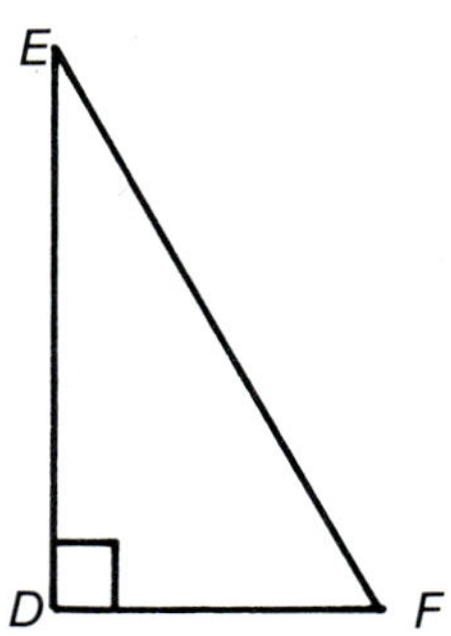

5.

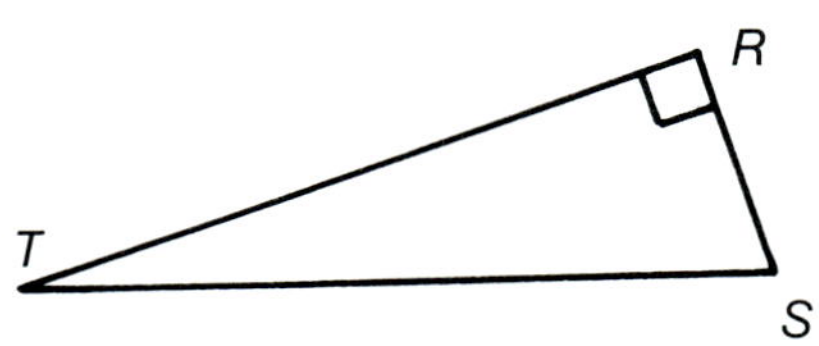

6.

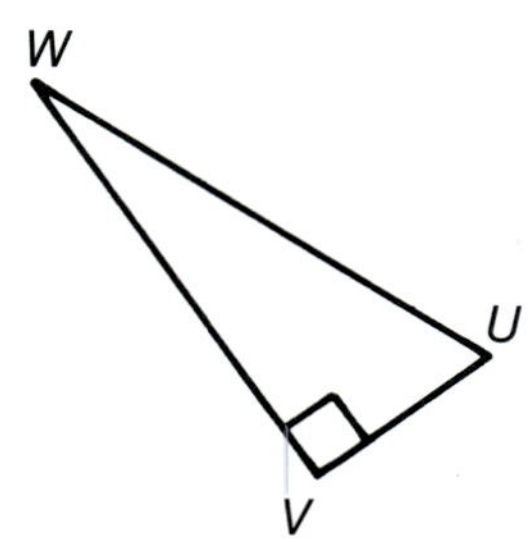

Complete.
Look for a pattern.

7\.

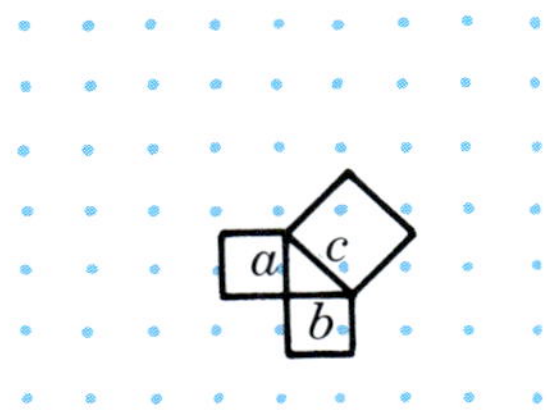

The area of the square on

a. leg a is __?__.
b. leg b is __?__.
c. the hypotenuse is __?__.

8\.

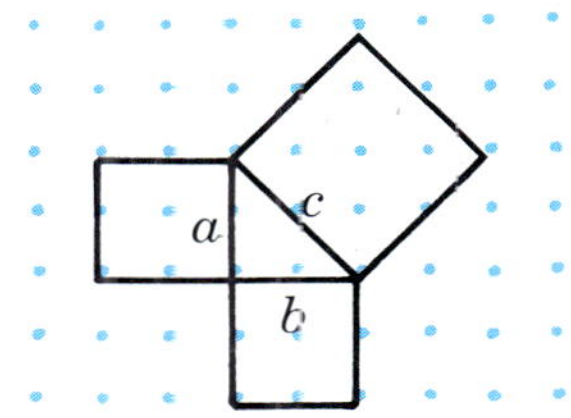

The area of the square on

a. leg a is __?__.
b. leg b is __?__.
c. the hypotenuse is __?__.

9\.

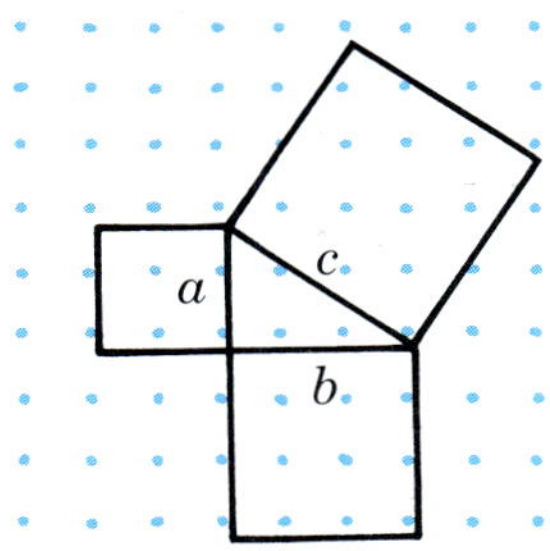

The area of the square on

a. leg a is __?__.
b. leg b is __?__.
c. the hypotenuse is __?__.

10\.

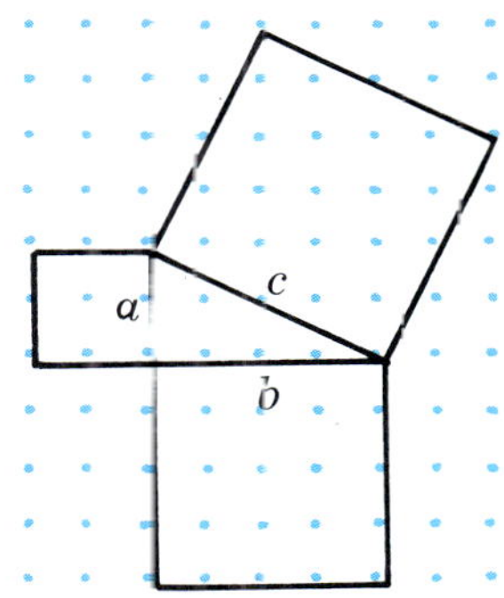

The area of the square on

a. leg a is __?__.
b. leg b is __?__.
c. the hypotenuse is __?__.

Challenge!

11. **a.** Use a protractor and carefully draw a right triangle.
b. Measure the sides to the nearest millimeter.
c. Square the length of the hypotenuse.
d. Square the lengths of the legs.
e. Add the squares of the legs.
f. How does the sum compare to the square of the hypotenuse?
g. Draw another right triangle and repeat steps a–f.

The Rule of Pythagoras

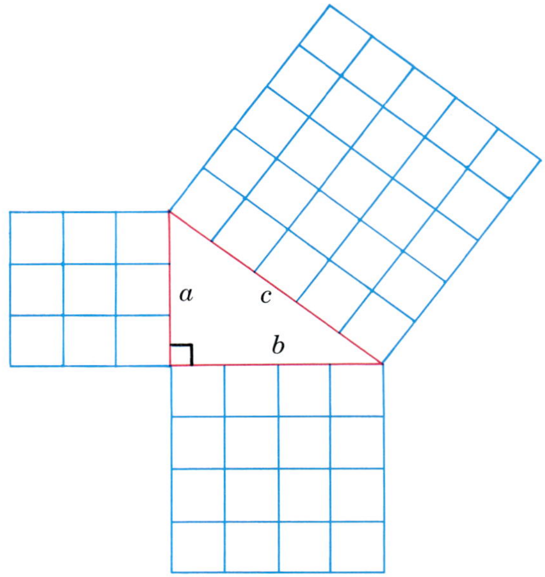

Ancient mathematicians discovered this remarkable fact about right triangles:

The area of the square drawn on the hypotenuse is equal to the sum of the areas of the squares drawn on the legs.

This fact is called the **Rule of Pythagoras.** It is often stated like this:

> In a right triangle the square of the hypotenuse is equal to the sum of the squares of the legs:
>
> $$a^2 + b^2 = c^2$$

We can use the Rule of Pythagoras to compute the length of a side of a right triangle when we know the lengths of the other two sides.

EXAMPLE 1.

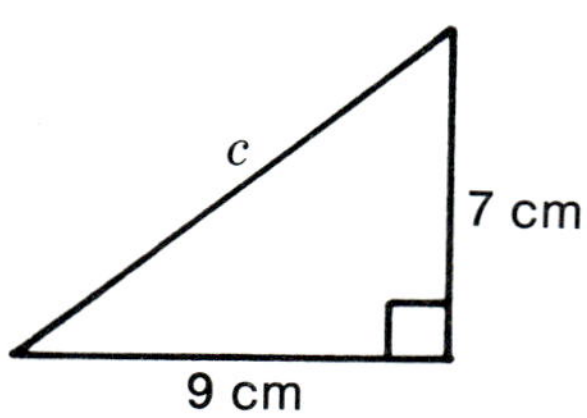

$$a^2 + b^2 = c^2$$
$$7^2 + 9^2 = c^2$$
$$49 + 81 = c^2$$
$$130 = c^2$$
$$\sqrt{130} = c$$

EXAMPLE 2.

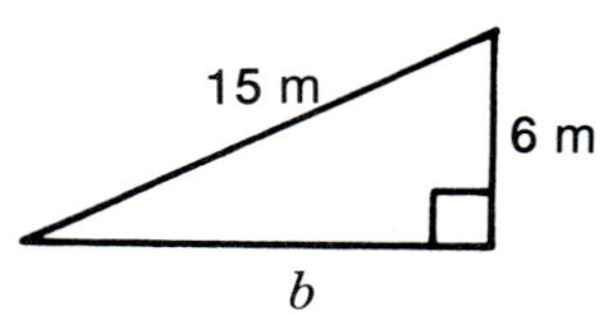

$$a^2 + b^2 = c^2$$
$$6^2 + b^2 = 15^2$$
$$36 + b^2 = 225$$
$$b^2 = 189$$
$$b = \sqrt{189}$$

In Example 1 we have found that the length of the hypotenuse is exactly $\sqrt{130}$ cm. When we need a decimal approximation, we can use a calculator or the divide-and-average method.

Guess and divide.	Average divisor and quotient.	Divide.	
$11\overline{)130}$ = 11.8	$\frac{11 + 11.8}{2} = 11.4$	$11.4\overline{)130.00}$ = 11.40	The answer to the nearest tenth is 11.4.

EXERCISES

Find the missing length.

1.

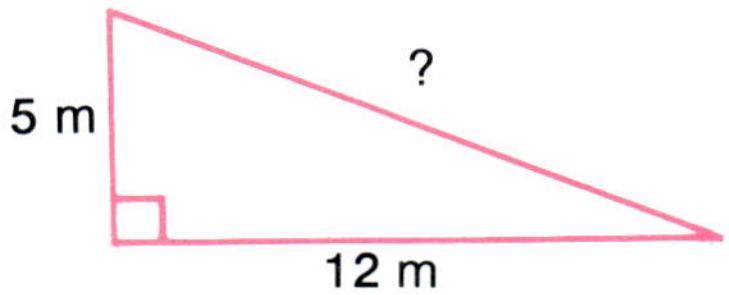

2.

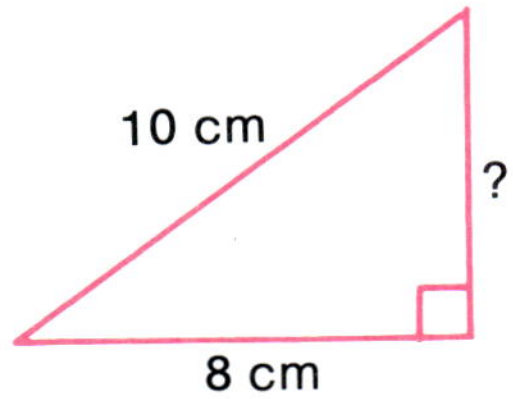

3.

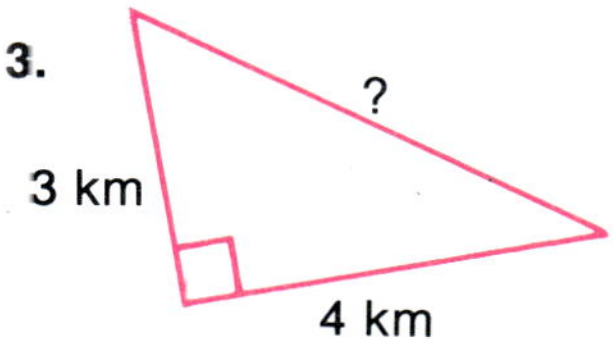

Find the missing length. Leave square roots in exact form.

4.

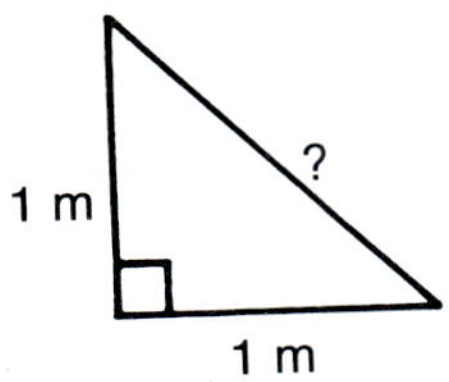

5.

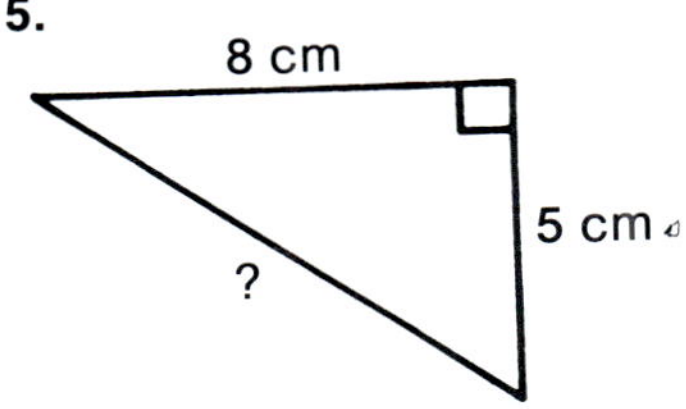

6.

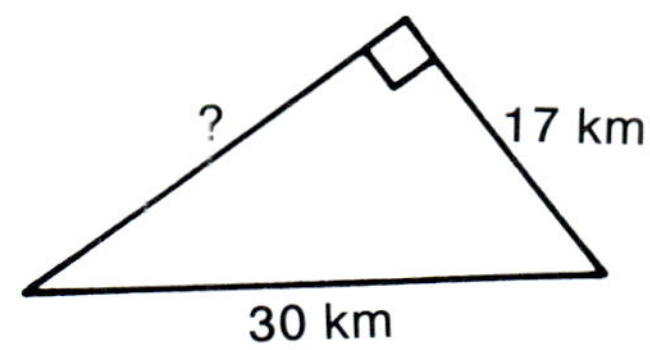

Find the missing length. Give the decimal value correct to the nearest tenth unit.

7.

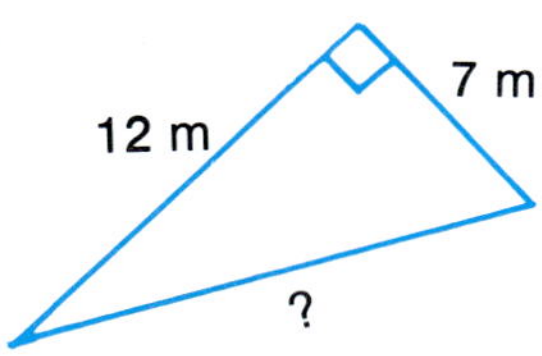

8.

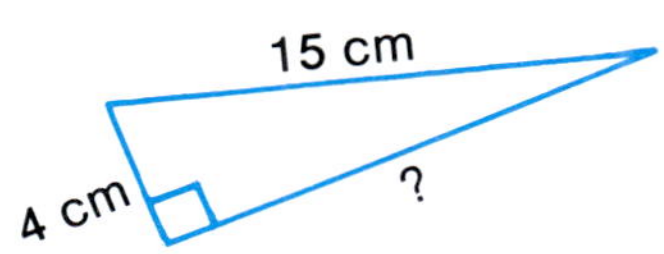

9.

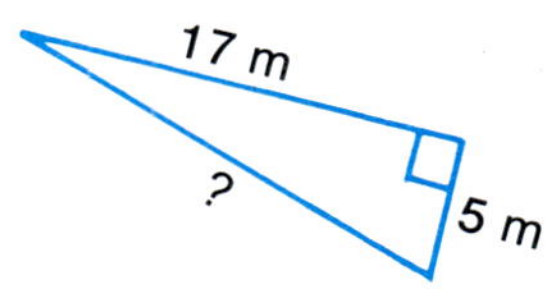

Problem Solving

[*Hint:* First draw a picture.]

10. A ladder that is 13 meters long is placed against a wall. The base of the ladder is 5 meters from the wall. How high on the wall does the ladder reach?

11. A plane flies over a control tower at a height of 8 km. An observer on the ground is 15 km from the control tower. How far is the plane from the observer?

12. An empty lot that Al uses as a short-cut is a 30-meter square. How much shorter is his walk diagonally across the lot compared to what his walk along two of its sides would be?

★ **13.** A flower garden is shaped like an equilateral triangle. It measures 12 meters on a side. What is its area? [*Hint:* First find its height to the nearest tenth meter.]

Exploring patterns in right triangles

Work in groups.
Get a calculator and a ruler.

When you used the Rule of Pythagoras, you may have noticed that in some right triangles all three sides have lengths that are whole numbers. In this lesson, you will explore these sets of whole numbers, which we will call Pythagorean triples.

PART A

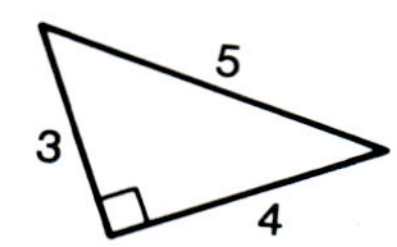

1. The chart shows the 1st Pythagorean triple (3-4-5). It also shows the leg lengths for the 2nd and 3rd triples. Find the missing hypotenuse lengths by calculating. Then find them by measuring.

	Short Leg a	Long Leg b	Hypotenuse c
1st	3	4	5
2nd	5	12	
3rd	7	24	
4th	9	40	

2. Did you get the same results from calculating as you did from measuring? Which method did you prefer? Why?

3. What patterns do you see in the chart? Find the next few Pythagorean triples. Compare your results with the results of other groups.

4. Why do you think the combinations 6-8-10 and 9-12-15 are not included in the chart?

5. Find the sum of b and c for the first few triples. What pattern do you see?

6. Use your pattern from problem 5 to make up a formula or rule for finding the rest of a Pythagorean triple when you know a.

7. Test your formula or rule. Use it to generate the 12th Pythagorean triple. Check it with the Rule of Pythagoras. Did it work?

8. How many Pythagorean triples are there? How did you get your answer?

9. Discuss your formula or rule and the ways you tested it with other groups.

PART B

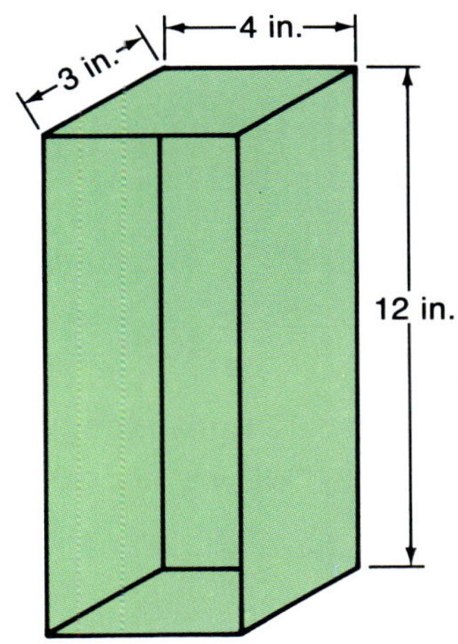

10. What is the length of the longest rigid pole that can fit in the box shown at the right?

11. Let's call the box shown a whole-number box because its dimensions are whole numbers and so is its longest diagonal. Find some other whole-number boxes that are not similar to the one shown.

12. How many non-similar whole-number boxes are there? How did you get your answer?

13. Discuss your results and the strategies you used to solve problems 10–12 with other groups.

PART C

14. The perimeter of a 3-4-5 right triangle is 12. What is the perimeter of each of the triangles formed by the next three Pythagorean triples?

Pythagorean triple	Perimeter
3-4-5	12

15. What pattern do you see in your list of perimeters? Use that pattern to predict the perimeter of a right triangle. Check your prediction.

16. Discuss your results and your methods for making and checking predictions with other groups.

17. For which parts of this lesson was a calculator useful? Why?

Chapter Checkup

Do the two figures look similar? [pages 380–381]

1.

2.

3.

These two figures are similar.
Complete each sentence. [pages 384–385]

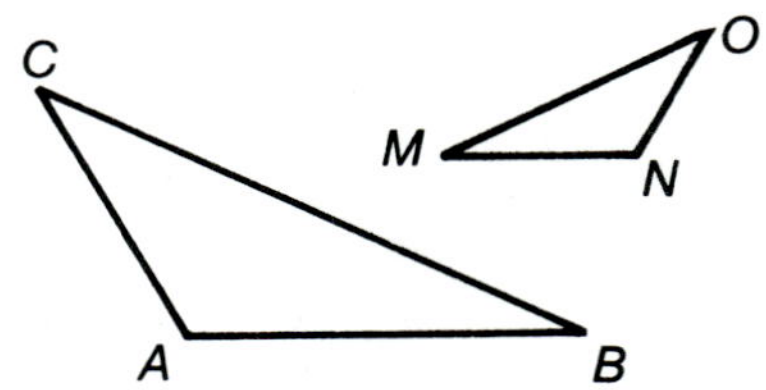

4. $\angle A \cong \underline{?}$ **5.** $\angle B \cong \underline{?}$ **6.** $\angle C \cong \underline{?}$

7. $\frac{AB}{MN} \cong \frac{?}{NO}$ **8.** $\frac{AB}{MN} \cong \frac{?}{OM}$ **9.** $\frac{NO}{AC} \cong \frac{MN}{?}$

The two right triangles are similar.
Find the length x. [pages 386–387, 390]

10.

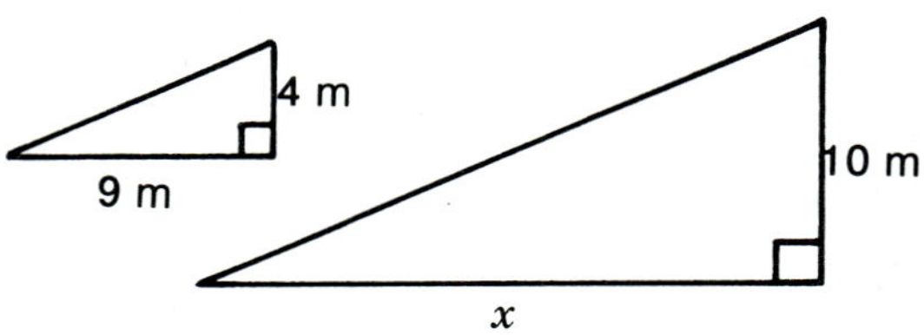

Use the table of tangents to find the length x. [pages 388–389]

11.

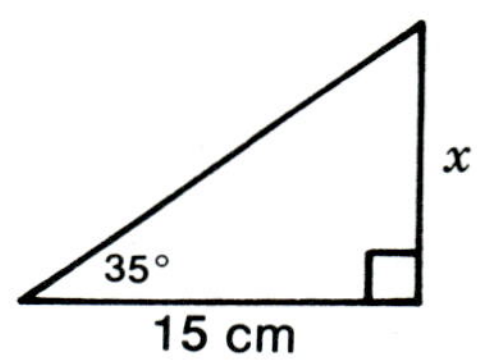

tan 30° ≈ 0.58
tan 35° ≈ 0.70
tan 40° ≈ 0.84

Simplify. [pages 392–393]

12. $\sqrt{16}$ **13.** $\sqrt{81}$ **14.** $\sqrt{0.64}$

Give an approximation to the nearest tenth. [pages 392–393]

15. $\sqrt{17}$ **16.** $\sqrt{54}$

Use the Rule of Pythagoras to find the missing length.
[pages 396–397]

17.

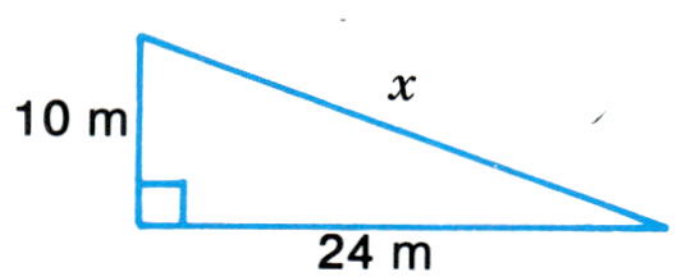

18.

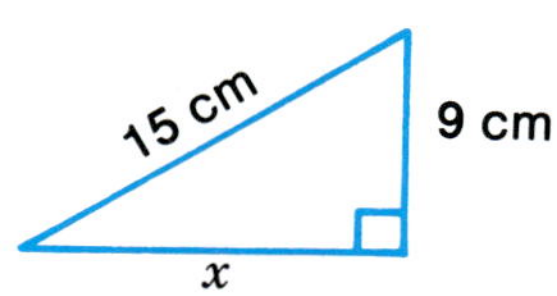

19.

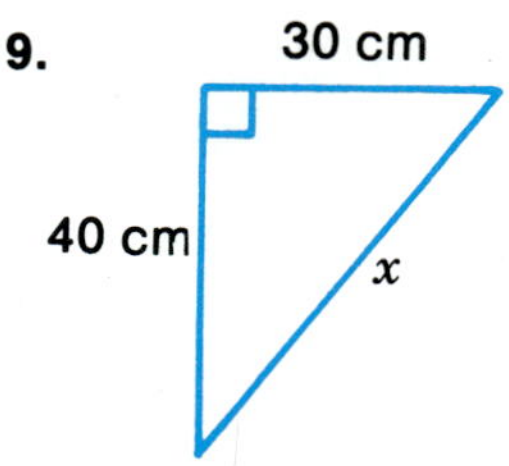

Chapter Project

INDIRECT MEASUREMENT

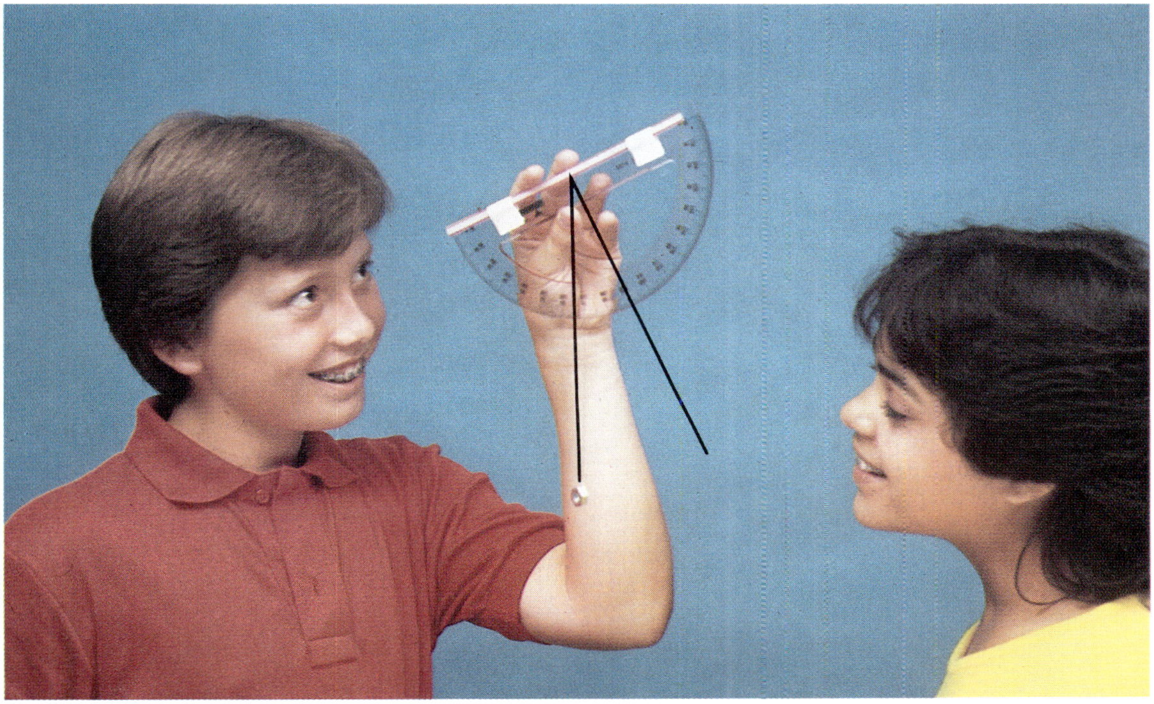

1. Choose a partner. Get a large protractor, a drinking straw, some tape, a string, and a weight. Make the instrument pictured above.

2. The following steps tell you how to use your instrument and a table of tangent ratios to measure the heights of trees, flagpoles, buildings, etc. (You can find the table of tangents in many high school mathematics textbooks.)

   a. Sight through the straw and have your partner read the angle of elevation (shown above in red).

   b. Measure the distance from the base of the object to the point where you sighted.

   c. First use a table of tangents to find the distance x shown in the sketch. Then add the distance d from the ground that you took the sighting.

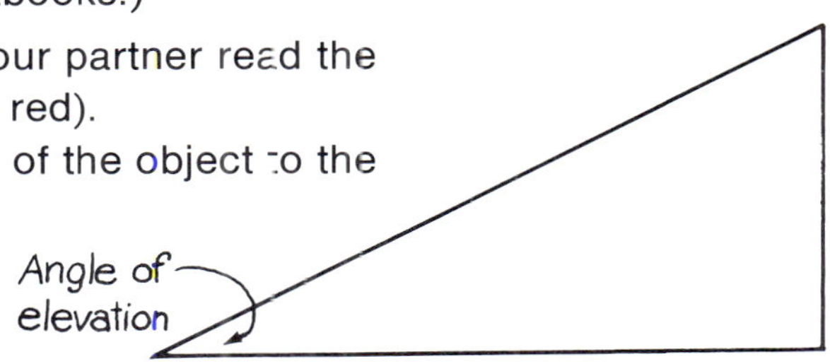

Chapter Review

In similar polygons the corresponding angles are congruent and the corresponding sides are proportional.

If the corresponding angles of two triangles are congruent, then the triangles are similar.

The tangent of an acute angle in a right triangle is the ratio of the side opposite the angle to the side adjacent to the angle.

Rule of Pythagoras

In a right triangle the square of the hypotenuse is equal to the sum of the squares of the legs.

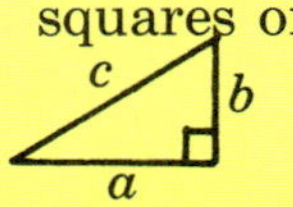

$a^2 + b^2 = c^2$

Here are two similar polygons.

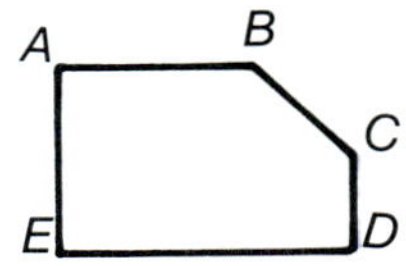

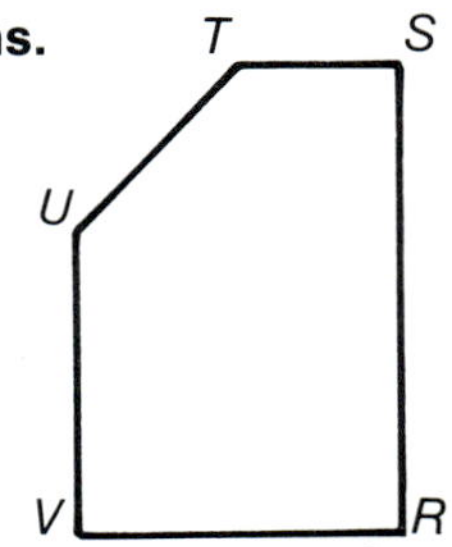

Complete.

1. $\angle A \cong$?

2. $\angle C \cong$?

3. $\frac{AB}{VU} \cong \frac{BC}{?}$

4. $\frac{AE}{VR} \cong \frac{DE}{?}$

$\triangle ABC \sim \triangle RST$

5. $x =$?

6. $y =$?

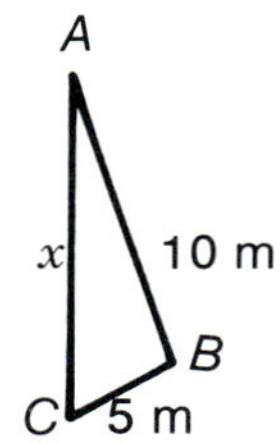

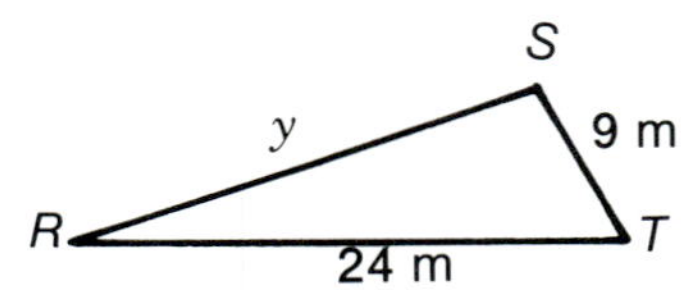

Use the table of tangent ratios to find the length x. Round answers to the nearest hundredth.

7.

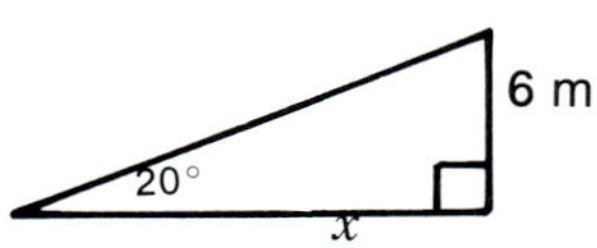

8.

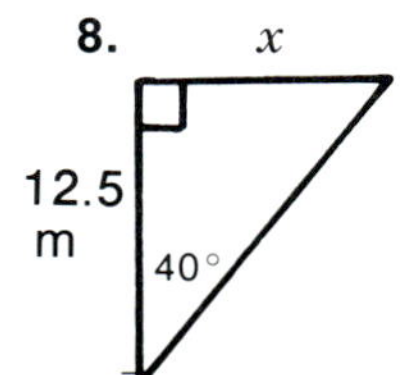

tan 20° ≈ 0.36
tan 30° ≈ 0.58
tan 40° ≈ 0.84

[*Hint:* tan 20° $= \frac{6}{x} \approx 0.36$]

Find the missing length.

9.

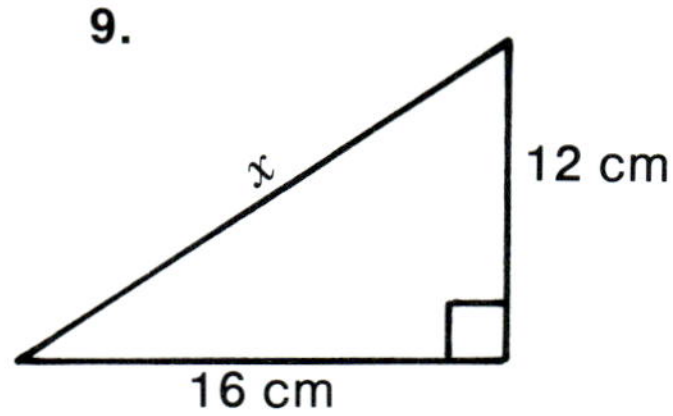

10.

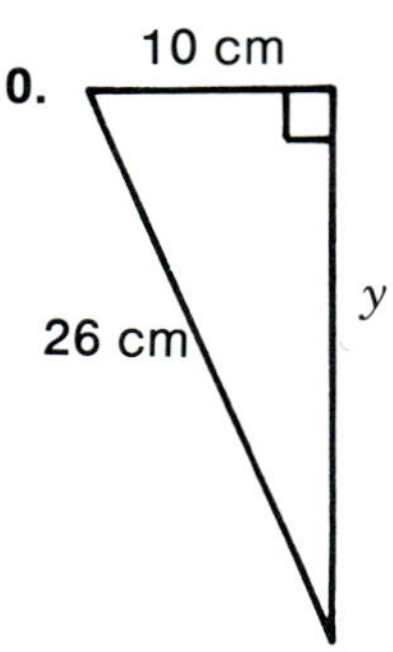

Chapter Challenge

USING A TABLE OF TANGENTS

The example shows how to find the approximate measure of an acute angle of a right triangle when you know the length of each leg.

Step 1. Write tan $\angle A$ correct to 4 decimal places.

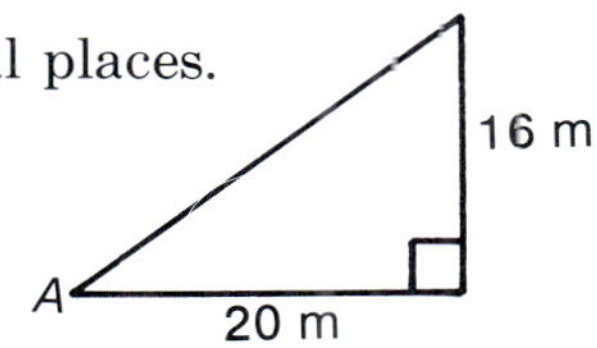

$$\tan \angle A = \frac{16}{20}$$
$$= 0.8000$$

Step 2. Locate 0.8000 in the table of tangents.

38°	0.7813
$\angle A$	0.8000
39°	0.8098

Step 3. By comparing the tangents, decide which angle measure is closer.

38°	0.7813	0.0187
$\angle A$	0.8000	
39°	0.8098	0.0098

Since 0.8000 is closer to 0.8098 than to 0.7813, $\angle A$ is closer to 39° than to 38°.

$\angle A \approx 39°$

Find the approximate measure of $\angle A$.

1.

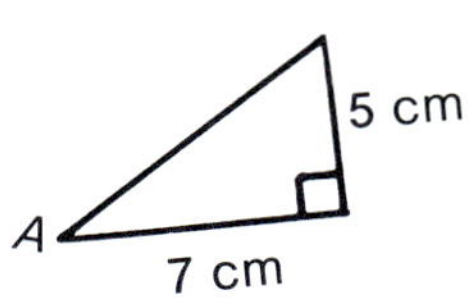

2.

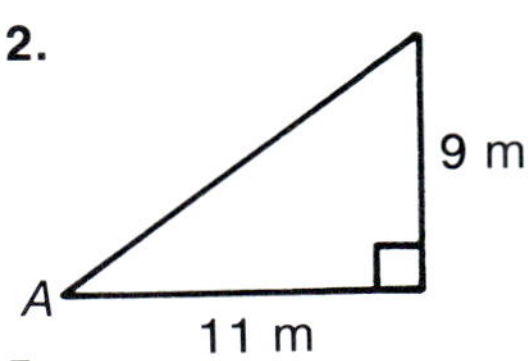

3.

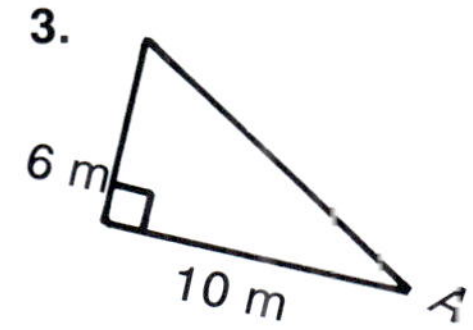

4.

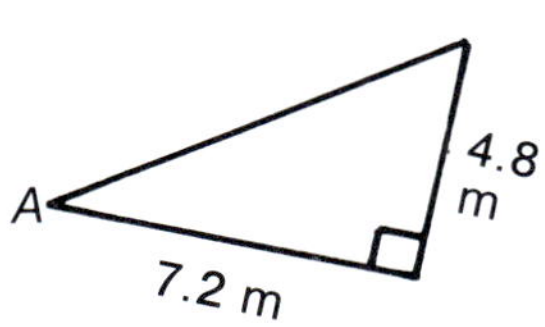

5.

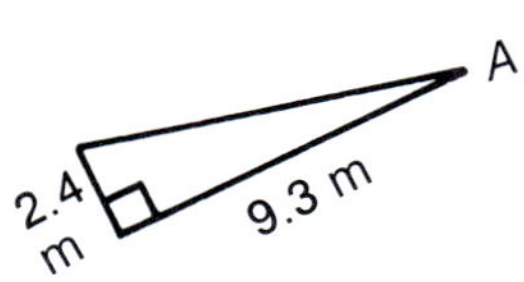

6.

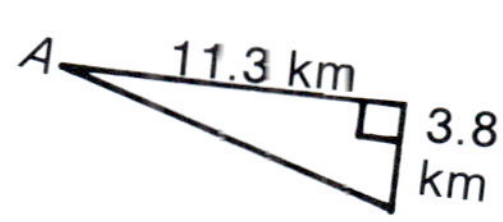

angle	*tangent*
1°	0.0175
2°	0.0349
3°	0.0524
4°	0.0699
5°	0.0875
6°	0.1051
7°	0.1228
8°	0.1405
9°	0.1584
10°	0.1763
11°	0.1944
12°	0.2126
13°	0.2309
14°	0.2493
15°	0.2679
16°	0.2867
17°	0.3057
18°	0.3249
19°	0.3443
20°	0.3640
21°	0.3839
22°	0.4040
23°	0.4245
24°	0.4452
25°	0.4663
26°	0.4877
27°	0.5095
28°	0.5317
29°	0.5543
30°	0.5774
31°	0.6009
32°	0.6249
33°	0.6494
34°	0.6745
35°	0.7002
36°	0.7265
37°	0.7536
38°	0.7813
39°	0.8098
40°	0.8391
41°	0.8693
42°	0.9004
43°	0.9325
44°	0.9657
45°	1.0000

Cumulative Checkup

STANDARDIZED FORMAT

Choose the correct letter.

1. $6.8 \div 10^3 = \underline{?}$

a. 0.68
b. 0.068
c. 0.0068
d. none of these

2. $\dfrac{\frac{2}{3}}{\frac{5}{9}} = \underline{?}$

a. $1\frac{1}{5}$
b. $2\frac{7}{10}$
c. $\frac{5}{6}$
d. none of these

3. Find the volume. Use 3.14 as an approximation for π.

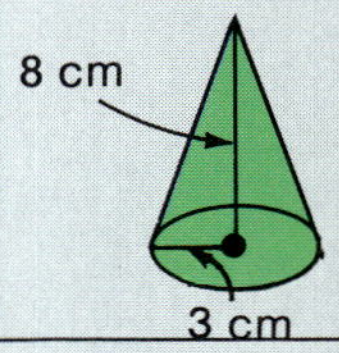

a. 75.36 cm^3
b. 28.26 cm^3
c. 226.08 cm^3
d. none of these

4. Solve. Round the answer to the nearest tenth.

$\frac{7}{2} = \frac{37}{n}$

a. 10.6
b. 10.4
c. 9.1
d. none of these

5. Find the installment price of a radio that can be bought for $19 down and monthly payments of $6.72 for 1 year.

a. $99.64
b. $234.72
c. $80.64
d. none of these

6. Solve.

$^-4n - 8 = 12$

a. 1
b. $^-1$
c. $^-5$
d. none of these

7. A parallelogram has how many lines of symmetry?

a. 1
b. 2
c. 4
d. none of these

8. *Event:* Picking a card from ten cards numbered 1 through 10.

P(prime number) = $\underline{?}$

a. $\frac{2}{5}$ **b.** $\frac{1}{2}$
c. $\frac{3}{5}$ **d.** $\frac{3}{10}$

9.

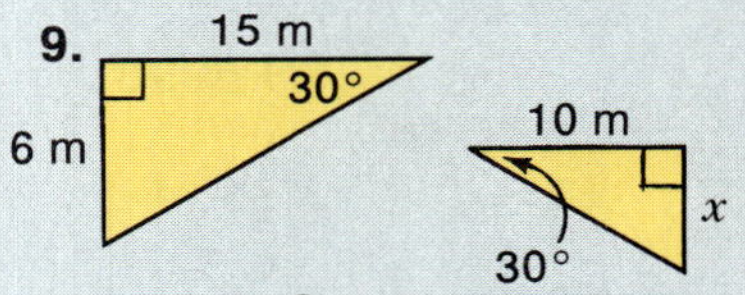

$x = \underline{?}$

a. 4 m
b. 5 m
c. 8 m
d. none of these

10.

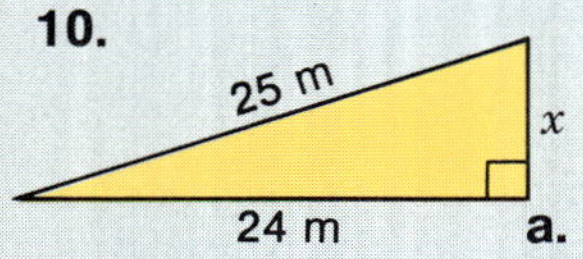

$x = \underline{?}$

a. 10 m
b. 7 m
c. 1 m
d. none of these

11. A recipe for 12 rolls uses $3\frac{1}{2}$ cups of flour. How much flour would be used for 8 rolls?

a. 28 cups **b.** $5\frac{1}{4}$ cups
c. $2\frac{1}{3}$ cups **d.** none of these

12. Kate spent $1.50 for 3 pens and a pad. Each pen cost 27¢. How much was the pad?

a. 69¢ **b.** 41¢
c. 23¢ **d.** none of these

RESOURCES

GETTING READY FOR COMPUTER PROGRAMMING

Using commands in programs

This bicycle factory has two robots. Robot 1 can leave Robot Base 1, pick up parts at any blue station, and take them to Store 1. Robot 2 can leave Base 2, pick up parts at any yellow station, and take them to Store 2.

When a robot is not working, it is at its Robot Base facing in the direction of the arrow.

You write a **program** to tell a robot where to move. The directions you write in a program are called **commands.**

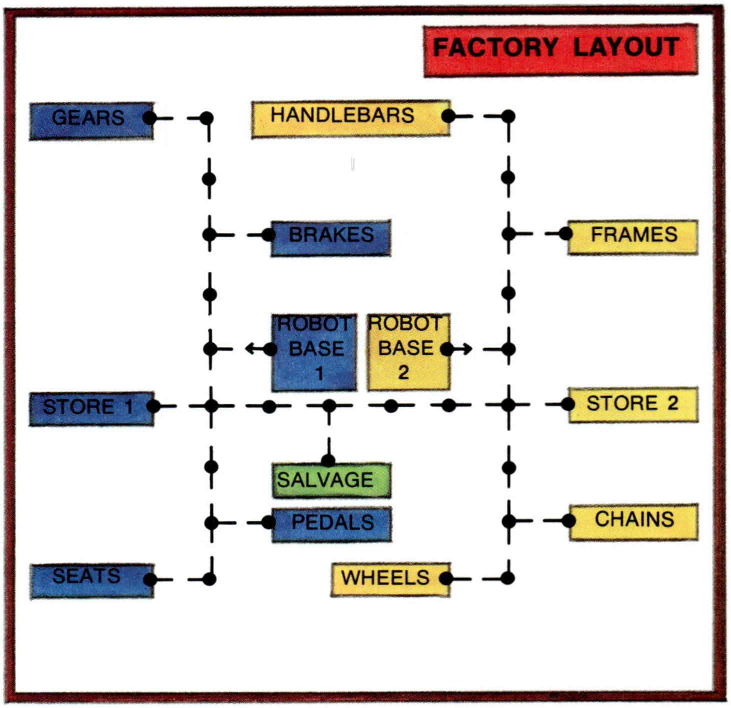

Follow the robot's path on the map.

1. PROGRAM BASE 1/SEATS
BEGIN ROBOT BASE 1
AHEAD 1 — *This command tells the robot to go ahead 1 space.*
TL $\frac{1}{4}$
AHEAD 4
TR $\frac{1}{4}$ — *This command tells the robot to turn right $\frac{1}{4}$ of a complete revolution.*
AHEAD 1
END SEATS

a. What is the name of the program?
b. Where did the robot begin?
c. Where did the robot end?

Complete the program.

2. PROGRAM BASE 2/____a____
BEGIN ROBOT BASE 2
AHEAD 1
TL $\frac{1}{4}$
AHEAD 3
TR $\frac{1}{2}$ — *This command tells the robot to turn right $\frac{1}{2}$ of a complete revolution.*
AHEAD 1
TL $\frac{1}{4}$
AHEAD 1
END ____b____

There are 2 mistakes in each program.
Rewrite each program correctly.

3. PROGRAM BASE 1/BRAKES
BEGIN ROBOT BASE 1
AHEAD 1
TL $\frac{1}{4}$
AHEAD 2
TR $\frac{1}{4}$
AHEAD 2
END BRAKES

4. PROGRAM BASE 2/CHAINS
BEGIN ROBOT BASE 2
AHEAD 1
TR $\frac{1}{4}$
AHEAD 2
TR $\frac{1}{4}$
AHEAD 1
END CHAINS

Complete each program.

5. PROGRAM BASE 1/GEARS
BEGIN ROBOT BASE 1
AHEAD 1
__a__
AHEAD 4
__b__
__c__
END GEARS

6. PROGRAM BASE 2/WHEELS
BEGIN ROBOT BASE 2
AHEAD 1
__a__
__b__
__c__
AHEAD 1
END __d__

Write these programs.

7. PROGRAM BASE 1/PEDALS

8. PROGRAM BASE 2/BARS

★ 9. Suppose that Robot 2 will only respond to these commands:

AHEAD 1, TR $\frac{1}{4}$, TL $\frac{1}{2}$

Write PROGRAM BASE 2/FRAMES.

New commands—TAKE and LEAVE

Each robot can be programmed to carry parts from one work station to another. To pick up or put down parts, the robot must face the work station.

To have the robot pick up some parts, you use the command TAKE PARTS. To have the robot put down the parts, you use the command LEAVE PARTS.

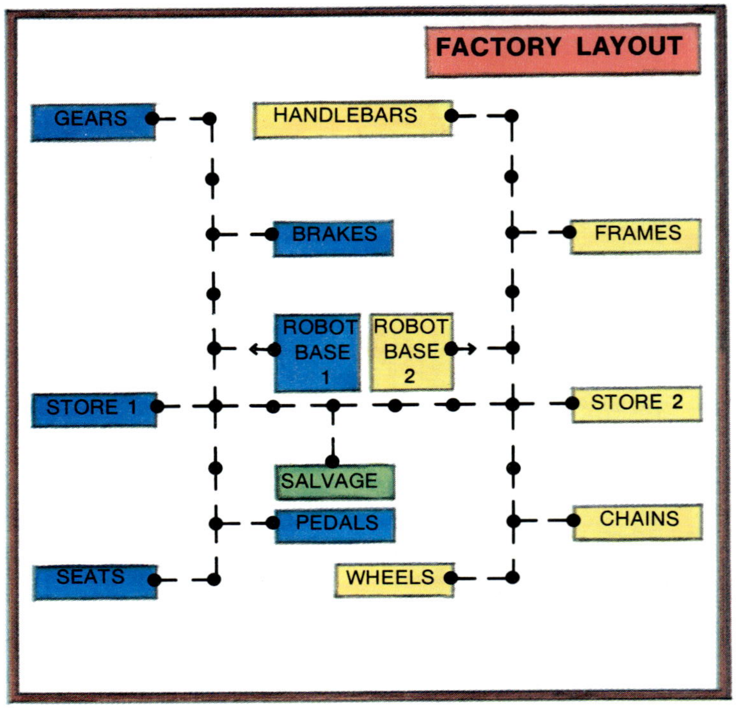

Study this program.

1. PROGRAM BASE 1/GEARS/STORE 1
BEGIN ROBOT BASE 1
AHEAD 1
TR $\frac{1}{4}$
AHEAD 4
TL $\frac{1}{4}$
AHEAD 1
TAKE PARTS ← *Use this command to pick up parts.*
TL $\frac{1}{2}$
AHEAD 1
TR $\frac{1}{4}$
AHEAD 5
TR $\frac{1}{4}$
AHEAD 1
LEAVE PARTS ← *Use this command to leave parts.*
END STORE 1

a. What is the name of the program?

b. Where did the robot begin?

c. Where did the robot pick up the parts?

d. Where did the robot leave the parts?

e. Where did the robot end?

Complete each program.

2. PROGRAM BASE 2/CHAINS/STORE 2
BEGIN ROBOT BASE 2
AHEAD 1
__a__
__b__
__c__
AHEAD 1
TAKE PARTS
__d__
AHEAD 1
__e__
AHEAD 2
__f__
AHEAD 1
____g____
END STORE 2

3. PROGRAM BASE 1/SEATS/STORE 1
BEGIN ROBOT BASE 1
AHEAD 1
__a__
__b__
__c__
AHEAD 1
____d____
__e__
AHEAD 1
__f__
__g__
TL $\frac{1}{4}$
AHEAD 1
LEAVE PARTS
END STORE 1

4. PROGRAM BASE 2/FRAMES/STORE 2
BEGIN ROBOT BASE 2
__a__
__b__
AHEAD 2
__c__
__d__
____e____
TR $\frac{1}{2}$
AHEAD 1
__f__
__g__
__h__
AHEAD 1
____i____
END STORE 2

★ **5.** This program was partly destroyed. Reconstruct it.

PROGRAM BAS
BEGIN BASE
AHEAD 1
TL $\frac{1}{4}$
AHEAD
TL $\frac{1}{4}$
AHEAD
TAKE
TL
AHEA
TR $\frac{1}{4}$
AHEAD 5
TL $\frac{1}{4}$
AHEA
LEAVE
END STOR

Procedures

A **procedure** is a list of commands for an action that is used often. You can use the name of a procedure as a command in a program.

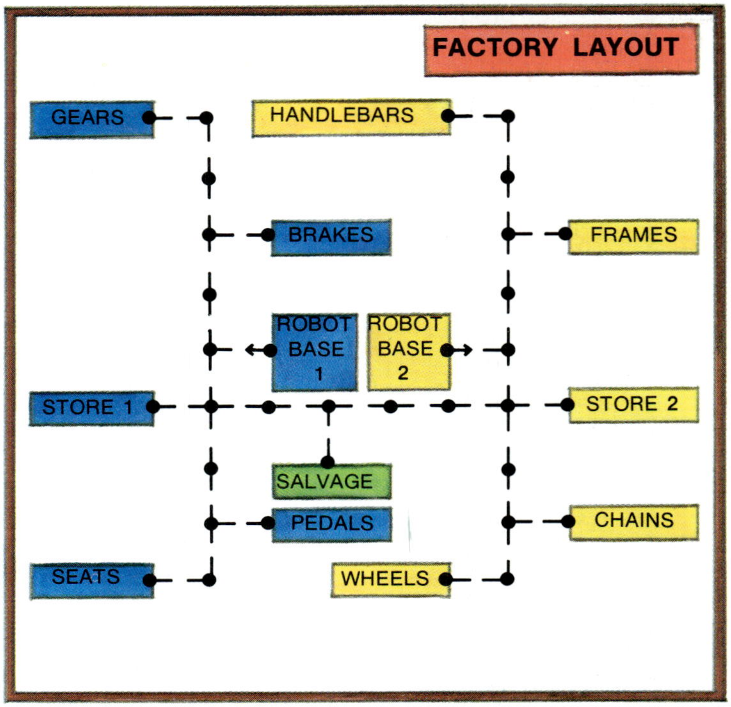

Complete each procedure.

1. After a robot picks up some parts at a work station, it always takes them to its storage area.

PROCEDURE GEARS/STORE 1
BEGIN GEARS
TR $\frac{1}{2}$
AHEAD 1
a ______
b ______
c ______
AHEAD 1

Remember: *The robot is facing the work station when it picks up parts.*

2. After a robot leaves some parts at a work station, it always returns to its robot base.

PROCEDURE STORE 1/BASE 1
BEGIN STORE 1
TR $\frac{1}{2}$
AHEAD 1
a ______
b ______
c ______
d ______
TR $\frac{1}{2}$

Remember: *The robot faces in the direction of the arrow.*

Write these procedures.

3. PROCEDURE FRAMES/STORE 2

4. PROCEDURE STORE 2/BASE 2

Assume from now on that you have procedures for moving a robot from any of its stations to any other station.

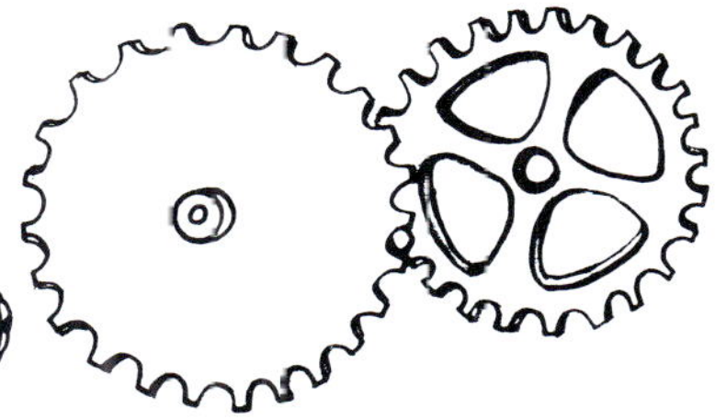

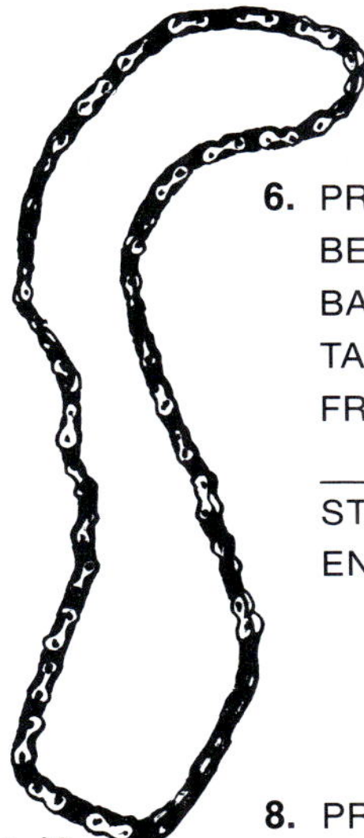

Complete each program.

5. PROGRAM PEDALS TO STORE 1
BEGIN ROBOT BASE 1
BASE 1/PEDALS
TAKE PARTS
PEDALS/____a____
LEAVE PARTS
STORE 1/____b____
END ROBOT BASE 1

6. PROGRAM FRAMES TO STORE 2
BEGIN ROBOT BASE 2
BASE 2/____a____
TAKE PARTS
FRAMES/____b____
____c____
STORE 2/____d____
END ROBOT BASE 2

7. PROGRAM SEATS TO STORE 1
BEGIN ROBOT BASE 1
____a____
TAKE PARTS
____b____
LEAVE PARTS
____c____
END ROBOT BASE 1

8. PROGRAM WHEELS TO STORE 2
BEGIN ROBOT BASE 2
____a____
TAKE PARTS
____b____
LEAVE PARTS
____c____
END ROBOT BASE 2

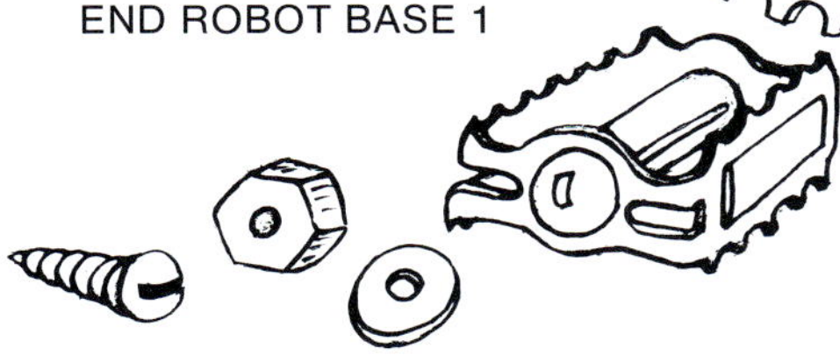

Write these programs using procedures.

9. PROGRAM GEARS TO STORE 1

10. PROGRAM CHAINS TO STORE 2

11. PROGRAM BRAKES TO STORE 1

12. PROGRAM BARS TO STORE 2

Decisions and loops

From now on you will restrict the amount of weight a robot can carry. To do this, your programs will have decisions in them.

First, if the box of parts weighs less than 80 pounds, the robot takes it directly to Store. Otherwise, parts are taken one at a time out of the box until the box weighs less than 80 pounds. This is an example of a loop.

A **loop** is a set of commands that describes an action that is repeated.

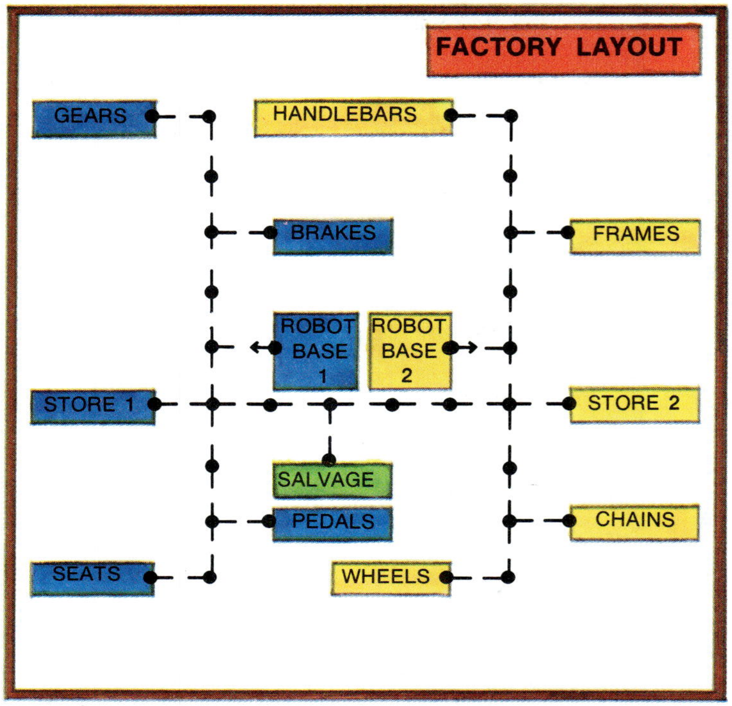

Study this program.

```
1. PROGRAM GEARS TO STORE 1
   BEGIN ROBOT BASE 1
   BASE 1/GEARS
   TAKE PARTS
   (A) IF PARTS WEIGH LESS THAN 80 POUNDS
         THEN GEARS/STORE 1
         OTHERWISE REMOVE 1 PART
                   REPEAT FROM (A)
   LEAVE PARTS
   STORE 1/BASE 1
   END ROBOT BASE 1
```

Parts are removed 1 at a time until the weight is less than 80 pounds.

The robot begins at _____a_____ and goes to _____b_____. If the parts weigh less than _____c_____ pounds, then the box is taken to _____d_____. Otherwise, parts are taken one at a time out of the box until the box of parts weighs less than 80 pounds. Then the box is taken to _____e_____. After leaving the parts at _____f_____, the robot returns to _____g_____.

Complete each program.

2. PROGRAM CHAINS TO STORE 2
   BEGIN ROBOT BASE 2
   ____a____
   TAKE PARTS
   Ⓐ IF PARTS WEIGH LESS THAN 80 POUNDS
   THEN ____b____
   OTHERWISE REMOVE 1 PART
   REPEAT FROM Ⓐ
   LEAVE PARTS
   ____c____
   END ROBOT BASE 2

3. PROGRAM PEDALS TO STORE 1
   BEGIN ROBOT BASE 1
   ____a____
   ____b____
   Ⓐ IF PARTS WEIGH LESS THAN 80 POUNDS
   THEN ____c____
   OTHERWISE ____d____
   ____e____
   LEAVE PARTS
   ____f____
   END ROBOT BASE 1

Write these programs. Restrict the weight the robot can carry to 80 pounds or less.

4. PROGRAM BRAKES TO STORE 1

5. PROGRAM FRAMES TO STORE 2

Each of these programs begins at a robot base, picks up parts that weigh less than 60 pounds, takes them to another work station, picks up parts that weigh less than 60 pounds, takes them to Store, and returns to Robot Base.

★ 6. PROGRAM BRAKES AND GEARS TO STORE 1

★ 7. PROGRAM CHAINS AND FRAMES TO STORE 2

More on decisions

In this lesson a robot will wait at its Store until the parts are inspected. Then if the parts pass inspection, the robot returns to its Robot Base. If the parts do not pass inspection, the robot takes the parts to Salvage and then returns to Robot Base.

Since both robots go to Salvage, you have to write another decision into each program to make sure the robots do not collide.

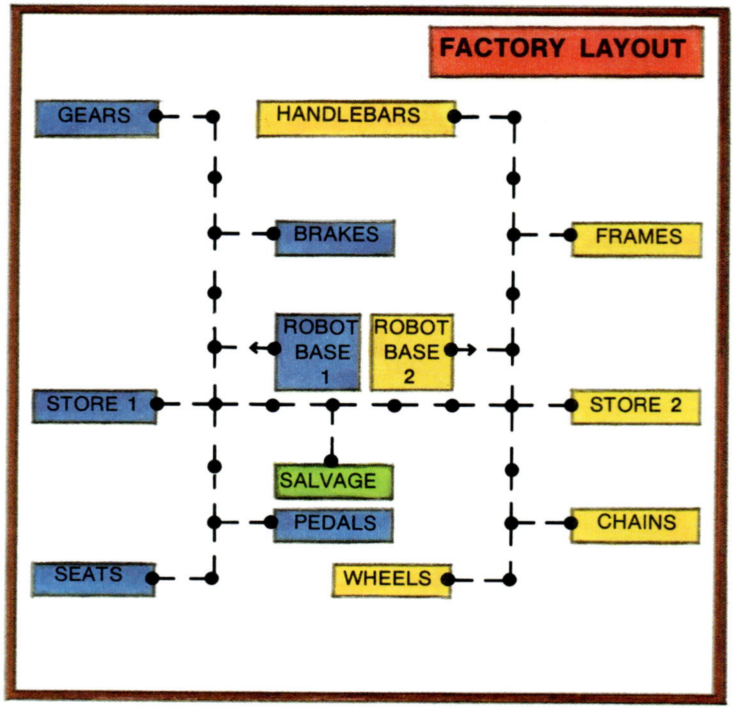

Study this program.

```
PROGRAM BRAKES TO STORE 1
BEGIN ROBOT BASE 1
BASE 1/BRAKES
TAKE PARTS
(A) IF PARTS WEIGH LESS THAN 80 POUNDS
      THEN BRAKES/STORE 1
      OTHERWISE REMOVE 1 PART
                REPEAT FROM (A)
LEAVE PARTS
WAIT
IF PARTS PASS INSPECTION
  THEN STORE 1/BASE 1
  OTHERWISE TAKE PARTS
            (B) IF PATH IS CLEAR
                  THEN STORE 1/SALVAGE
                  OTHERWISE WAIT 5 SECONDS
                            REPEAT FROM (B)
            LEAVE PARTS
            SALVAGE/BASE 1
END ROBOT BASE 1
```

1. The robot begins at ___a___ and goes to ___b___. If the parts weigh less than ___c___ pounds, then they are taken to ___d___. Otherwise, the parts are taken one at a time out of the box until the box weighs less than 80 pounds. Then the parts are taken to ___e___.

Complete each program.

2.

```
PROGRAM CHAINS TO STORE 2
BEGIN ROBOT BASE 2
____a____
TAKE PARTS
⟨A⟩ IF PARTS WEIGH LESS THAN 80 POUNDS
        THEN ____b____
        OTHERWISE REMOVE 1 PART
                          ____c____
LEAVE PARTS
WAIT
IF PARTS PASS INSPECTION
   THEN ____d____
   OTHERWISE TAKE PARTS
                  ⟨B⟩ IF PATH IS CLEAR TO SALVAGE
                          THEN ____e____
                          OTHERWISE WAIT 5 SECONDS
                                         REPEAT FROM ⟨B⟩
                  LEAVE PARTS
                  ____f____
END ROBOT BASE 2
```

Write these programs. Restrict the weight to less than 80 pounds.

3. PROGRAM PEDALS TO STORE 1

4. PROGRAM FRAMES TO STORE 2

★ **5.** Suppose that both robots can travel to all stations shown on the Factory Layout map. Robot 2 has to pick up a part at Gears, take it to Store 2, and then return to Base 2. Call your program PROGRAM GEARS/STORE 2. Be sure that the robots will not collide!

SKILL TEST

1	Add any two decimals	$3.68 + 2.59$ $47.3 + 9.86$	$42.8 + 7.6$ $1.509 + 0.94$		
2	Subtract any two decimals	$6.02 - 1.88$ $56.5 - 2.7$	$7.58 - 3.6$ $83.4 - 2.97$		
3	Multiply any two decimals	4.73×0.9	15.6×0.24	1.2×0.08	4.825×4.02
4	Divide a decimal by a whole number	$8\overline{)0.416}$	$9\overline{)21.33}$	$57\overline{)121.41}$	$34\overline{)28.39}$
5	Divide and round to the nearest hundredth	$0.6\overline{)43.1}$	$0.8\overline{)2.67}$	$0.43\overline{)0.87}$	$1.23\overline{)81.67}$
6	Multiply a decimal by a power of 10	4.63×10^2 5.7×10^2	0.526×10^3 0.975×10^4		
7	Divide a decimal by a power of 10	$35.8 \div 10^2$ $82.17 \div 10^3$	$267.0 \div 10^3$ $0.02 \div 10^2$		
8	Write a number in scientific notation	Write in scientific notation. 836,000 0.00045	 4,900,000 0.0000183		

9	Add two integers	$^{+}6 + ^{+}9$	$^{+}8 + ^{-}7$	$^{-}9 + ^{+}5$	$^{-}3 + ^{-}6$
10	Subtract two integers	$^{+}9 - ^{+}2$	$^{+}6 - ^{-}1$	$^{-}3 - ^{+}8$	$^{-}7 - ^{-}5$
11	Multiply two integers	$^{+}6 \times ^{+}3$	$^{+}7 \times ^{-}4$	$^{-}5 \times ^{+}9$	$^{-}8 \times ^{-}6$
12	Divide two integers	$^{+}36 \div ^{+}9$	$^{+}24 \div ^{-}8$	$^{-}32 \div ^{+}4$	$^{-}42 \div ^{-}6$
13	Find an equivalent fraction	$\frac{3}{4} = \frac{?}{12}$	$\frac{8}{5} = \frac{?}{40}$	$\frac{5}{7} = \frac{?}{35}$	$\frac{5}{8} = \frac{?}{40}$
14	Write a fraction in lowest terms	$\frac{12}{18} = \underline{?}$	$\frac{45}{81} = \underline{?}$	$\frac{15}{25} = \underline{?}$	$\frac{54}{72} = \underline{?}$
15	Change a fraction to a decimal	$\frac{2}{5} = \underline{?}$	$\frac{5}{8} = \underline{?}$	$\frac{3}{11} = \underline{?}$	$\frac{11}{9} = \underline{?}$
16	Change a decimal to a fraction	$0.6 = \underline{?}$	$0.25 = \underline{?}$	$1.5 = \underline{?}$	$0.625 = \underline{?}$
17	Compare fractions (< or >?)	$\frac{3}{8}$ ● $\frac{5}{8}$	$\frac{2}{3}$ ● $\frac{3}{5}$	$\frac{5}{7}$ ● $\frac{5}{9}$	$\frac{4}{9}$ ● $\frac{3}{7}$
18	Add fractions	$\frac{1}{4} + \frac{3}{8}$	$\frac{1}{5} + \frac{2}{3}$	$\frac{1}{9} + \frac{2}{3}$	$\frac{1}{4} + \frac{3}{10}$
19	Subtract fractions	$\frac{9}{10} - \frac{3}{5}$	$\frac{3}{4} - \frac{2}{3}$	$\frac{1}{2} - \frac{1}{8}$	$\frac{3}{4} - \frac{1}{6}$

20	Change a mixed number to a fraction	$3\frac{1}{2} = \underline{?}$	$2\frac{2}{3} = \underline{?}$	$4\frac{3}{4} = \underline{?}$	$5\frac{7}{8} = \underline{?}$
21	Change a mixed number to a decimal	$1\frac{2}{5} = \underline{?}$	$2\frac{3}{4} = \underline{?}$	$1\frac{9}{10} = \underline{?}$	$2\frac{3}{8} = \underline{?}$
22	Change a decimal to a mixed number	$4.6 = \underline{?}$	$6.75 = \underline{?}$	$3.125 = \underline{?}$	$2.08 = \underline{?}$
23	Change a fraction to a mixed number	$\frac{5}{2} = \underline{?}$	$\frac{10}{3} = \underline{?}$	$\frac{15}{4} = \underline{?}$	$\frac{35}{8} = \underline{?}$
24	Add mixed numbers	$13\frac{1}{4} + 5\frac{1}{3}$	$10\frac{2}{3} + 3\frac{1}{4}$	$5\frac{2}{3} + 7\frac{1}{3}$	$18\frac{2}{3} + 9\frac{3}{4}$
25	Subtract mixed numbers	$13\frac{1}{4} - 7\frac{1}{6}$	$10\frac{2}{3} - 6\frac{1}{5}$	$5\frac{1}{2} - 1\frac{3}{4}$	$12\frac{2}{3} - 7\frac{3}{4}$
26	Multiply fractions	$\frac{2}{3} \times \frac{4}{7}$	$\frac{6}{7} \times \frac{5}{3}$	$\frac{3}{5} \times \frac{5}{8}$	$\frac{12}{16} \times \frac{4}{3}$
27	Divide fractions	$\frac{3}{2} \div \frac{5}{4}$	$\frac{2}{3} \div \frac{4}{5}$	$\frac{3}{4} \div \frac{3}{8}$	$\frac{2}{5} \div \frac{3}{7}$
28	Multiply mixed numbers	$2\frac{1}{2} \times 3\frac{1}{4}$	$2\frac{4}{5} \times 3\frac{1}{8}$	$5\frac{2}{3} \times 2\frac{3}{4}$	$6\frac{5}{7} \times 2\frac{4}{5}$

29	Divide mixed numbers	$5\frac{3}{4} \div 1\frac{1}{4}$ $4\frac{1}{2} \div 1\frac{1}{2}$	$7\frac{4}{5} \div 2\frac{1}{4}$ $3\frac{2}{3} \div 2\frac{3}{5}$
30	Solve a proportion	$\frac{6}{5} = \frac{n}{25}$ $\frac{9}{n} = \frac{3}{2\frac{2}{3}}$	$\frac{n}{9} = \frac{6}{4\frac{1}{2}}$ $\frac{6}{3\frac{1}{2}} = \frac{8}{n}$
31	Change a percent to a fraction or a mixed number	25% = ? 87.5% = ?	180% = ? $66\frac{2}{3}\%$ = ?
32	Change a fraction to a percent	$\frac{1}{2}$ = ? $\frac{5}{8}$ = ?	$\frac{5}{6}$ = ? $\frac{7}{4}$ = ?
33	Change a decimal to a percent	0.08 = ? 0.75 = ?	3.21 = ? $0.06\frac{1}{4}$ = ?
34	Change a percent to a decimal	20% = ? $33\frac{1}{3}\%$ = ?	37.5% = ? 160% = ?
35	Find a percent of a number	20% of 65 = n 9.2% of 48 = n	$66\frac{2}{3}\%$ of 81 = n $6\frac{3}{4}\%$ of 125 = n
36	Find the number when a percent is given	40% of n = 24 8.5% of n = 3.23	$33\frac{1}{3}\%$ of n = 19 16.5% of n = 4.125

37	Find the percent	$n\%$ of 24 = 12		$n\%$ of 150 = 37.5	
		$n\%$ of 55 = 22		$n\%$ of 42 = 58	
38	Compare positive and negative numbers	< or >?			
		$^-9$ ● $^-4$	0 ● $^-\frac{2}{3}$	17 ● $^-23$	$^-\frac{5}{8}$ ● $^-\frac{3}{4}$
39	Add positive and negative numbers	$^+7 + {}^+4$	$^-\frac{2}{3} + {}^+\frac{1}{3}$	$^+9 + {}^-6$	$^-\frac{3}{4} + {}^-\frac{3}{2}$
40	Subtract positive and negative numbers	$^+8 - {}^+12$	$^-\frac{5}{8} - {}^+\frac{1}{2}$	$^+15 - {}^-8$	$^-\frac{3}{5} - {}^-\frac{3}{4}$
41	Multiply positive and negative numbers	$^+7 \times {}^+9$	$^+\frac{2}{3} \times {}^-\frac{3}{5}$	$^-6 \times {}^+7$	$^-\frac{3}{2} \times {}^-\frac{1}{2}$
42	Divide positive and negative numbers	$^+24 \div {}^+8$	$^-64 \div {}^+16$	$^+8 \div {}^-\frac{1}{2}$	$^-\frac{3}{4} \div {}^-\frac{5}{2}$
43	Find the square root of a number	**Give the square root of each number.**			
		36	81	169	1.44
44	Find a decimal approximation of a square root	**Give a decimal approximation rounded to the nearest tenth.**			
		$\sqrt{18}$	$\sqrt{50}$	$\sqrt{110}$	$\sqrt{275}$

EXTRA PRACTICE

Set 1 Write the standard numeral.

1. eight thousand 8000
2. five thousand one hundred twenty-four
3. forty-two thousand, nine hundred six
4. nineteen thousand, seven hundred fifty-three
5. three hundred twenty-eight thousand, six hundred seventy-six
6. five million, two hundred twelve thousand, four hundred thirty-seven
7. two hundred forty-two million, six thousand, two hundred eighty-three
8. five billion, two hundred twenty million, two hundred eleven thousand, sixty-six
9. twenty-four billion, nine hundred forty million, five thousand, eighty-two
10. seven trillion, one hundred thirty-four million, two hundred thousand, twenty

Set 2 Round each number to the nearest thousand.

1. 8500 9000	2. 9250	3. 16,743	4. 25,099
5. 291,842	6. 404,416	7. 598,540	8. 799,500

Round each number to the nearest ten thousand.

9. 36,721	10. 59,763	11. 85,000	12. 95,000
13. 374,821	14. 216,527	15. 534,083	16. 729,200

Round each number to the nearest million.

17. 8,834,527	18. 3,509,601	19. 29,498,937	20. 49,762,403
21. 312,291,742	22. 233,035,555	23. 456,621,384	24. 789,125,246

Set 3 Add.

1. 39 + 89 = 128	2. 356 + 84	3. 589 + 73	4. 7218 + 356	5. 9077 + 4227	6. 5881 + 3619
7. 42 + 68 + 68	8. 91 + 93 + 49	9. 78 + 86 + 35	10. 52 + 97 + 38	11. 23 + 65 + 59	12. 75 + 74 + 89
13. 338 + 729 + 74	14. 656 + 30 + 82	15. 529 + 476 + 54	16. 183 + 691 + 215	17. 706 + 938 + 455	18. 534 + 827 + 333

Set 4 Subtract.

1. $\begin{array}{r} 74 \\ -25 \\ \hline 49 \end{array}$	2. $\begin{array}{r} 62 \\ -37 \\ \hline \end{array}$	3. $\begin{array}{r} 80 \\ -43 \\ \hline \end{array}$	4. $\begin{array}{r} 759 \\ -163 \\ \hline \end{array}$	5. $\begin{array}{r} 842 \\ -256 \\ \hline \end{array}$	6. $\begin{array}{r} 673 \\ -459 \\ \hline \end{array}$
7. $\begin{array}{r} 7284 \\ -3162 \\ \hline \end{array}$	8. $\begin{array}{r} 5916 \\ -2448 \\ \hline \end{array}$	9. $\begin{array}{r} 6217 \\ -3485 \\ \hline \end{array}$	10. $\begin{array}{r} 801 \\ -345 \\ \hline \end{array}$	11. $\begin{array}{r} 705 \\ -277 \\ \hline \end{array}$	12. $\begin{array}{r} 903 \\ -658 \\ \hline \end{array}$
13. $\begin{array}{r} 600 \\ -158 \\ \hline \end{array}$	14. $\begin{array}{r} 900 \\ -219 \\ \hline \end{array}$	15. $\begin{array}{r} 700 \\ -307 \\ \hline \end{array}$	16. $\begin{array}{r} 8001 \\ -468 \\ \hline \end{array}$	17. $\begin{array}{r} 7002 \\ -3708 \\ \hline \end{array}$	18. $\begin{array}{r} 5000 \\ -2164 \\ \hline \end{array}$

Set 5 Multiply.

1. $\begin{array}{r} 398 \\ \times 6 \\ \hline 2388 \end{array}$	2. $\begin{array}{r} 805 \\ \times 7 \\ \hline \end{array}$	3. $\begin{array}{r} 1864 \\ \times 5 \\ \hline \end{array}$	4. $\begin{array}{r} 4780 \\ \times 9 \\ \hline \end{array}$	5. $\begin{array}{r} 65 \\ \times 53 \\ \hline \end{array}$	6. $\begin{array}{r} 34 \\ \times 48 \\ \hline \end{array}$
7. $\begin{array}{r} 583 \\ \times 29 \\ \hline \end{array}$	8. $\begin{array}{r} 714 \\ \times 51 \\ \hline \end{array}$	9. $\begin{array}{r} 927 \\ \times 64 \\ \hline \end{array}$	10. $\begin{array}{r} 521 \\ \times 40 \\ \hline \end{array}$	11. $\begin{array}{r} 438 \\ \times 30 \\ \hline \end{array}$	12. $\begin{array}{r} 652 \\ \times 60 \\ \hline \end{array}$
13. $\begin{array}{r} 153 \\ \times 417 \\ \hline \end{array}$	14. $\begin{array}{r} 916 \\ \times 628 \\ \hline \end{array}$	15. $\begin{array}{r} 791 \\ \times 240 \\ \hline \end{array}$	16. $\begin{array}{r} 654 \\ \times 320 \\ \hline \end{array}$	17. $\begin{array}{r} 882 \\ \times 207 \\ \hline \end{array}$	18. $\begin{array}{r} 439 \\ \times 406 \\ \hline \end{array}$
19. $\begin{array}{r} 5174 \\ \times 221 \\ \hline \end{array}$	20. $\begin{array}{r} 4382 \\ \times 325 \\ \hline \end{array}$	21. $\begin{array}{r} 2965 \\ \times 340 \\ \hline \end{array}$	22. $\begin{array}{r} 4708 \\ \times 506 \\ \hline \end{array}$	23. $\begin{array}{r} 9213 \\ \times 708 \\ \hline \end{array}$	24. $\begin{array}{r} 6458 \\ \times 950 \\ \hline \end{array}$

Set 6 Divide.

1. $7\overline{)562}$ = 80 R2 (-56, 2)	2. $8\overline{)301}$	3. $4\overline{)283}$	4. $9\overline{)418}$	5. $6\overline{)319}$
	6. $6\overline{)4352}$	7. $4\overline{)2413}$	8. $5\overline{)2089}$	9. $8\overline{)2963}$
10. $59\overline{)2130}$	11. $93\overline{)6790}$	12. $47\overline{)2360}$	13. $78\overline{)7111}$	14. $63\overline{)5057}$
15. $58\overline{)18,447}$	16. $41\overline{)16,665}$	17. $63\overline{)33,232}$	18. $82\overline{)23,013}$	19. $37\overline{)22,222}$
20. $71\overline{)39,914}$	21. $42\overline{)12,941}$	22. $83\overline{)39,095}$	23. $65\overline{)45,507}$	24. $34\overline{)28,909}$
25. $138\overline{)67,919}$	26. $219\overline{)23,752}$	27. $527\overline{)158,317}$	28. $603\overline{)503,855}$	29. $482\overline{)241,965}$

Set 7 **For 72.839501, give the digit that is in the**

1. ones place
2. tens place
3. tenths place
4. thousandths place
5. ten-thousandths place
6. millionths place
7. hundred-thousandths place
8. hundredths place

For 39.610857, give the digit that is in the

9. tens place
10. tenths place
11. ones place
12. thousandths place
13. millionths place
14. ten-thousandths place
15. hundredths place
16. hundred-thousandths place

Set 8 **Round each number to the nearest tenth.**

1. 73.86
2. 17.423
3. 36.551
4. 93.86
5. 63.845
6. 20.95
7. 36.284
8. 45.72
9. 79.36
10. 3.957

Round each number to the nearest hundredth.

11. 3.275
12. 4.062
13. 19.3085
14. 0.0850
15. 21.3416
16. 12.3689
17. 32.6517
18. 40.3965
19. 7.0965
20. 9.4022

Round each number to the nearest thousandth.

21. 34.8936
22. 56.4858
23. 0.9608
24. 26.75164
25. 17.0996
26. 5.45395
27. 39.84902
28. 9.97964
29. 34.4998
30. 4.3052

Set 9 $<$ **or** $>$**?**

1. 48.2 ● 48.3
2. 5.90 ● 5.39
3. 0.800 ● 0.799
4. 46.89 ● 47.31
5. 6.390 ● 6.382
6. 493.4 ● 489.8
7. 264.28 ● 272.22
8. 89.990 ● 10.099
9. 72.958 ● 73.094
10. 853.95 ● 835.95
11. 52.1638 ● 52.1368
12. 168.749 ● 168.579
13. 2984.31 ● 2983.41
14. 835.822 ● 843.262
15. 9873.42 ● 9783.42
16. 372.81 ● 372.79
17. 153.86 ● 153.865
18. 37.246 ● 372.46
19. 5.834 ● 5.843
20. 7.3965 ● 73.965
21. 4.005 ● 4.050
22. 15.064 ● 15.0064
23. 21.108 ● 21.180
24. 13.304 ● 13.340

Set 10 Add.

1. 7.65 + 3.82 = 11.47
2. 0.374 + 0.296
3. 5.08 + 0.39
4. 82.05 + 27.96
5. 58.48 + 6.97
6. 4.637 + 3.29
7. 37.5 + 64.7
8. 13.3 + 5.69
9. 2.76 + 52
10. 5.07 + 0.288
11. 36.8 + 8.37
12. 8.79 + 0.493
13. 283 + 0.762
14. 9.53 + 4.96
15. 0.265 + 5.86
16. 98.7 + 4.86
17. 0.254 + 6.48
18. 0.0584 + 0.737
19. 5 + 2.746
20. 3.98 + 17
21. 23 + 4.617
22. 5.06 + 118
23. 0.615 + 5.84
24. 9.99 + 3.7
25. 6.0821 + 0.377
26. 18.603 + 0.8155

Set 11 Subtract.

1. 72.1 − 50.9 = 21.2
2. 0.516 − 0.384
3. 8.26 − 4.19
4. 39.85 − 27.59
5. 7.098 − 5.65
6. 8.62 − 5.837
7. 37.5 − 8.3
8. 6.29 − 3.2
9. 538 − 4.75
10. 70.1 − 3.2
11. 83.7 − 4.47
12. 98 − 51.9
13. 6.4 − 6.27
14. 50 − 43.8
15. 0.4 − 0.377
16. 60.7 − 3.92
17. 5.8 − 0.387
18. 0.62 − 0.008
19. 9 − 3.5
20. 15 − 2.67
21. 24.9 − 15
22. 126.73 − 18
23. 56.7 − 2.99
24. 74.35 − 1.66
25. 81.05 − 19.9
26. 5.3 − 2.966

Set 12 Multiply.

1. 5.2 × 0.03 = 0.156
2. 0.48 × 0.02
3. 0.59 × 0.06
4. 3.4 × 0.05
5. 0.74 × 0.07
6. 2.9 × 0.09
7. 8.12 × 3.2
8. 67.4 × 2.8
9. 0.521 × 0.16
10. 0.380 × 0.24
11. 0.482 × 0.53
12. 39.5 × 0.64
13. 2.95 × 8.12
14. 0.596 × 2.38
15. 5.17 × 4.35
16. 0.462 × 34.2
17. 3.91 × 12.8
18. 73.8 × 42.7
19. 3.84 × 1.91
20. 56.5 × 5.32
21. 81.3 × 27.3
22. 7.02 × 5.82
23. 51.8 × 6.45
24. 6.24 × 9.03

Set 13 **Divide. Round each quotient to the nearest hundredth.**

1. $3\overline{)1.400}$ $0.466 \rightarrow 0.47$

$$\begin{array}{r} -12 \\ \hline 20 \\ -18 \\ \hline 20 \\ -18 \\ \hline 2 \end{array}$$

2. $6\overline{)5.3}$ 3. $8\overline{)3.1}$ 4. $9\overline{)4.4}$

5. $3\overline{)1.25}$ 6. $8\overline{)8.05}$ 7. $7\overline{)42.1}$ 8. $9\overline{)7.45}$

9. $6\overline{)0.555}$ 10. $7\overline{)8.791}$ 11. $9\overline{)2.803}$ 12. $3\overline{)16.07}$

13. $49\overline{)0.891}$ 14. $90\overline{)3.64}$ 15. $14\overline{)7.34}$ 16. $25\overline{)56.92}$

17. $38\overline{)7.53}$ 18. $68\overline{)4.82}$ 19. $26\overline{)2.96}$ 20. $53\overline{)1.834}$ 21. $69\overline{)1.78}$

22. $71\overline{)32.6}$ 23. $66\overline{)9.326}$ 24. $32\overline{)7.31}$ 25. $48\overline{)6.05}$ 26. $57\overline{)3.52}$

27. $85\overline{)53.4}$ 28. $94\overline{)89.1}$ 29. $83\overline{)2.729}$ 30. $79\overline{)3.92}$ 31. $42\overline{)5.11}$

Set 14 **Divide. Round each quotient to the nearest hundredth.**

1. $0.9\overline{)7.96}$ $8.844 \rightarrow 8.84$

$$\begin{array}{r} -72 \\ \hline 76 \\ -72 \\ \hline 40 \\ -36 \\ \hline 40 \\ -36 \\ \hline 4 \end{array}$$

2. $0.8\overline{)0.643}$ 3. $0.03\overline{)0.928}$ 4. $0.2\overline{)0.368}$

5. $0.09\overline{)3.826}$ 6. $0.004\overline{)0.8543}$ 7. $0.002\overline{)2.963}$ 8. $0.05\overline{)4.321}$

9. $5.6\overline{)2.007}$ 10. $0.64\overline{)741.8}$ 11. $2.8\overline{)0.4352}$ 12. $1.3\overline{)3.896}$

13. $7.38\overline{)5.961}$ 14. $92.4\overline{)43.75}$ 15. $3.86\overline{)517.4}$ 16. $1.24\overline{)29.63}$

17. $6.14\overline{)281.3}$ 18. $2.13\overline{)5.762}$ 19. $82.4\overline{)53.01}$ 20. $4.53\overline{)87.79}$

Set 15 **Write the standard numeral.**

1. 3^2 9   2. 2^3 3. 2^2 4. 6^1 5. 3^4 6. 2^5

7. 2^4 8. 3^5 9. 5^0 10. 3^3 11. 5^2 12. 4^1

13. 9^0 14. 4^2 15. 10^2 16. 10^1 17. 4^3 18. 6^2

19. 4^4 20. 8^1 21. 5^3 22. 7^2 23. 10^0 24. 4^5

25. 6^3 26. 5^4 27. 8^2 28. 10^4 29. 7^3 30. 3^6

Set 16 Multiply.

1. 5.62×10 56.2
2. 33.4×100
3. 0.5962×1000
4. 3.854×100
5. 0.326×100
6. 4.85×1000
7. 0.423×100
8. 52.6×10
9. 85.5×100
10. 74.3×10
11. 6.5811×1000
12. 0.3596×100
13. 5.046×10
14. 6.038×100
15. 4.295×1000
16. 8.1162×100

Divide.

17. $38.25 \div 10$
18. $51.06 \div 100$
19. $473.8 \div 1000$
20. $9.142 \div 10$
21. $826.4 \div 100$
22. $592.6 \div 10$
23. $378.2 \div 1000$
24. $714.2 \div 100$
25. $638.4 \div 10$
26. $109.8 \div 1000$
27. $8.164 \div 100$
28. $3.42 \div 1000$
29. $5463.7 \div 100$
30. $3801.6 \div 1000$
31. $74.235 \div 10$
32. $84.619 \div 100$

Set 17 Multiply.

1. 5.638×10^2 563.8
2. 2.917×10^1
3. 0.438×10^3
4. 2.635×10^4
5. 0.7189×10^3
6. 0.9544×10^5
7. 0.82936×10^4
8. 4.256×10^6
9. 87.2×10^4
10. 3.71×10^3
11. 0.6×10^3
12. 5.004×10^5
13. 5.36×10^2
14. 3.94×10^4
15. 2.836×10^3
16. 53.61×10^5

Divide.

17. $974.1 \div 10^2$
18. $83.59 \div 10^3$
19. $725.39 \div 10^1$
20. $2639 \div 10^4$
21. $3251 \div 10^5$
22. $846.57 \div 10^4$
23. $5.361 \div 10^3$
24. $72.81 \div 10^2$
25. $38 \div 10^3$
26. $4.16 \div 10^1$
27. $0.7 \div 10^2$
28. $0.02 \div 10^4$
29. $6.35 \div 10^2$
30. $241 \div 10^3$
31. $381.6 \div 10^4$
32. $37148.2 \div 10^5$

Set 18 Write in scientific notation.

1. 59,000 5.9×10^4
2. 87,000
3. 582,000
4. 356,000
5. 753,000
6. 6,054,000
7. 87,100,000
8. 597,000,000
9. 386,000
10. 695,000,000
11. 63,400,000
12. 71,650,000
13. 4,100,000
14. 73,000,000
15. 96,700,000
16. 5,836,000
17. 3,420,000,000
18. 5,617,000,000
19. 529,000,000
20. 66,500,000

Set 19 **Write in scientific notation.**

1. 0.048 4.8×10^{-2}
2. 0.071
3. 0.0056
4. 0.0083
5. 0.0516
6. 0.00055
7. 0.00112
8. 0.00062
9. 0.0941
10. 0.000753
11. 0.00664
12. 0.000017
13. 0.000833
14. 0.000056
15. 0.00161
16. 0.09154
17. 0.0000028
18. 0.0000138
19. 0.00077
20. 0.00000682

Set 20 < or >?

1. $^{+}8$ (>) $^{+}7$
2. $^{-}8$ ● $^{+}7$
3. $^{-}8$ ● $^{-}7$
4. $^{+}8$ ● $^{-}7$
5. $^{+}19$ ● $^{-}18$
6. $^{+}16$ ● $^{+}15$
7. $^{-}19$ ● $^{-}18$
8. $^{-}15$ ● $^{+}14$
9. $^{-}14$ ● $^{+}16$
10. $^{+}9$ ● $^{-}12$
11. $^{-}16$ ● $^{-}15$
12. 0 ● $^{-}16$
13. $^{+}12$ ● $^{+}10$
14. $^{+}12$ ● $^{-}10$
15. $^{-}12$ ● $^{+}10$
16. $^{-}12$ ● $^{-}10$
17. $^{+}16$ ● $^{-}13$
18. $^{-}16$ ● $^{+}13$
19. $^{+}16$ ● $^{+}13$
20. $^{-}16$ ● $^{-}13$
21. $^{+}15$ ● $^{+}20$
22. $^{-}15$ ● $^{-}20$
23. $^{+}15$ ● $^{-}20$
24. $^{-}15$ ● $^{+}20$
25. $^{-}18$ ● $^{+}24$
26. $^{+}18$ ● $^{-}24$
27. $^{-}18$ ● $^{-}24$
28. $^{+}18$ ● $^{+}24$

Set 21 **Add.**

1. $^{-}5 + {}^{-}6$ $^{-}11$
2. $^{-}8 + 0$
3. $^{-}5 + {}^{+}9$
4. $^{+}7 + {}^{+}8$
5. $^{-}6 + {}^{-}6$
6. $^{-}4 + {}^{-}7$
7. $^{+}3 + {}^{+}8$
8. $^{-}6 + {}^{-}4$
9. $^{-}3 + {}^{-}12$
10. $^{+}8 + {}^{+}11$
11. $^{-}10 + {}^{+}6$
12. $^{-}14 + {}^{-}6$
13. $^{-}7 + {}^{+}7$
14. $^{-}15 + 0$
15. $^{+}9 + {}^{-}9$
16. $^{+}16 + 0$
17. $^{+}18 + {}^{+}12$
18. $^{+}18 + {}^{-}12$
19. $^{-}18 + {}^{-}12$
20. $^{-}18 + {}^{+}12$
21. $^{-}16 + {}^{-}11$
22. $^{+}16 + {}^{+}11$
23. $^{-}16 + {}^{+}11$
24. $^{+}16 + {}^{-}11$
25. $^{+}20 + {}^{-}14$
26. $^{-}20 + {}^{+}14$
27. $^{+}20 + {}^{+}14$
28. $^{-}20 + {}^{-}14$
29. $^{-}25 + {}^{+}20$
30. $^{+}25 + {}^{+}20$
31. $^{+}25 + {}^{-}20$
32. $^{-}25 + {}^{-}20$
33. $^{+}26 + {}^{-}30$
34. $^{-}26 + {}^{-}30$
35. $^{+}26 + {}^{-}30$
36. $^{-}26 + {}^{+}30$

Set 22 Subtract.

1. $^{+}7 - {}^{-}5$ $^{+}12$
2. $^{+}7 - 0$
3. $^{+}6 - {}^{-}4$
4. $^{+}4 - {}^{+}6$
5. $^{+}12 - {}^{+}5$
6. $^{-}3 - {}^{+}9$
7. $^{-}5 - {}^{-}3$
8. $0 - {}^{+}6$
9. $^{-}8 - {}^{-}2$
10. $^{+}8 - {}^{+}3$
11. $^{-}4 - 0$
12. $^{-}8 - {}^{-}8$
13. $^{-}7 - {}^{+}2$
14. $0 - {}^{-}6$
15. $^{-}5 - {}^{-}9$
16. $^{+}8 - {}^{-}1$
17. $^{+}12 - {}^{+}8$
18. $^{+}12 - {}^{-}8$
19. $^{-}12 - {}^{+}8$
20. $^{-}12 - {}^{-}8$
21. $^{+}16 - {}^{-}10$
22. $^{-}16 - {}^{+}10$
23. $^{+}16 - {}^{+}10$
24. $^{-}16 - {}^{-}10$
25. $^{-}12 - {}^{+}19$
26. $^{+}12 - {}^{-}19$
27. $^{+}12 - {}^{+}19$
28. $^{-}12 - {}^{-}19$
29. $^{+}15 - {}^{-}13$
30. $^{-}15 - {}^{+}13$
31. $^{+}15 - {}^{+}13$
32. $^{-}15 - {}^{-}13$
33. $^{-}20 - {}^{-}25$
34. $^{+}20 - {}^{+}25$
35. $^{+}20 - {}^{-}25$
36. $^{-}20 - {}^{+}25$

Set 23 Multiply.

1. $^{-}6 \times {}^{+}6$ $^{-}36$
2. $^{-}8 \times {}^{-}8$
3. $^{+}7 \times {}^{-}3$
4. $^{+}5 \times {}^{+}5$
5. $^{-}4 \times {}^{+}3$
6. $^{-}9 \times {}^{+}9$
7. $^{-}3 \times {}^{-}9$
8. $^{-}9 \times {}^{-}8$
9. $^{-}9 \times {}^{+}5$
10. $^{+}4 \times {}^{+}6$
11. $0 \times {}^{+}8$
12. $^{-}6 \times 0$
13. $^{+}7 \times {}^{-}5$
14. $^{-}3 \times {}^{-}8$
15. $^{+}4 \times {}^{-}7$
16. $^{+}7 \times {}^{+}7$
17. $^{+}6 \times {}^{+}8$
18. $^{+}6 \times {}^{-}8$
19. $^{-}6 \times {}^{+}8$
20. $^{-}6 \times {}^{-}8$
21. $^{+}10 \times {}^{-}6$
22. $^{-}10 \times {}^{+}6$
23. $^{+}10 \times {}^{+}6$
24. $^{-}10 \times {}^{-}6$
25. $^{+}11 \times {}^{+}12$
26. $^{-}11 \times {}^{-}12$
27. $^{-}11 \times {}^{+}12$
28. $^{+}11 \times {}^{-}12$
29. $^{+}16 \times {}^{-}16$
30. $^{-}16 \times {}^{-}16$
31. $^{-}16 \times {}^{+}16$
32. $^{+}16 \times {}^{+}16$
33. $^{-}20 \times {}^{-}24$
34. $^{+}20 \times {}^{-}24$
35. $^{+}20 \times {}^{+}24$
36. $^{-}20 \times {}^{+}24$

Set 24 Divide.

1. $^{-}27 \div {}^{-}3$ $^{+}9$
2. $^{-}32 \div {}^{+}8$
3. $^{+}24 \div {}^{-}6$
4. $^{+}24 \div {}^{+}6$
5. $^{+}32 \div {}^{+}8$
6. $0 \div {}^{-}7$
7. $^{-}36 \div {}^{-}6$
8. $^{-}35 \div {}^{+}7$
9. $^{-}45 \div {}^{-}9$
10. $^{+}48 \div {}^{-}6$
11. $^{+}42 \div {}^{+}7$
12. $^{-}36 \div {}^{+}9$
13. $0 \div {}^{+}6$
14. $^{-}9 \div {}^{-}9$
15. $0 \div {}^{-}8$
16. $^{+}8 \div {}^{+}8$
17. $^{+}54 \div {}^{-}6$
18. $^{-}54 \div {}^{+}6$
19. $^{+}54 \div {}^{+}6$
20. $^{-}54 \div {}^{-}6$
21. $^{-}72 \div {}^{-}9$
22. $^{+}72 \div {}^{+}9$
23. $^{-}72 \div {}^{+}9$
24. $^{+}72 \div {}^{-}9$
25. $^{+}56 \div {}^{+}8$
26. $^{+}56 \div {}^{-}8$
27. $^{-}56 \div {}^{+}8$
28. $^{-}56 \div {}^{-}8$
29. $^{-}120 \div {}^{+}10$
30. $^{+}120 \div {}^{-}10$
31. $^{+}120 \div {}^{+}10$
32. $^{-}120 \div {}^{-}10$
33. $^{+}156 \div {}^{+}12$
34. $^{-}156 \div {}^{-}12$
35. $^{+}156 \div {}^{-}12$
36. $^{-}156 \div {}^{+}12$

Set 25 Simplify.

1. $|^-8|$
2. $|5|$
3. $|0|$
4. $|^-6|$
5. $|^-3| + |^-2|$
6. $|^-3 + ^-2|$
7. $|8| + |^-5|$
8. $|8 + ^-5|$
9. $|^-6| + |6|$
10. $|^-6 + 6|$
11. $|7| - |^-3|$
12. $|7 - ^-3|$
13. $|^-10| - |6|$
14. $|^-10 - 6|$
15. $|^-12| - |^-8|$
16. $|^-12 - ^-8|$
17. $|^-5| \times |^-3|$
18. $|^-5 \times ^-3|$
19. $|9| \times |^-7|$
20. $|9 \times ^-7|$
21. $|^-42| \div |6|$
22. $|^-42 \div 6|$
23. $|^-54| \div |^-6|$
24. $|^-54 \div ^-6|$
25. $|12 + ^-12|$
26. $|12| + |^-12|$
27. $|15 - ^-5|$
28. $|15| - |^-5|$
29. $|^-20 \times ^-5|$
30. $|^-20| \times |^-5|$
31. $|48 \div ^-16|$
32. $|48| \div |^-16|$

Set 26 Simplify.

1. $^-3 \times 9 - 2$ $^-29$
2. $12 + 8 \div ^-2$
3. $24 \div ^-6 - 2$
4. $^-35 - 10 - 3$
5. $3 \times (8 + ^-4)$
6. $(24 + 16) \div ^-8$
7. $12 + ^-6 \div 2 + 4$
8. $(12 + ^-6) \div 2 + 4$
9. $12 + ^-6 \div (2 + 4)$
10. $57 - (6 + ^-9) \times 3$
11. $57 - 6 + ^-9 \times 3$
12. $(57 - 6) + ^-9 \times 3$
13. $^-18 + 6 \div 3 + 3$
14. $^-18 + 6 \div (3 + 3)$
15. $(^-18 + 6) \div 3 + 3$
16. $^-24 - 8 \div 4 - 2$
17. $^-24 - 8 \div (4 - 2)$
18. $(-24 - 8) \div 4 - 2$
19. $48 + ^-6 \times 4 + 3$
20. $48 + ^-6 \times (4 + 3)$
21. $(48 + ^-6) \times 4 + 3$
22. $72 + ^-12 \div 3 - 1$
23. $72 + ^-12 \div (^-3 - 1)$
24. $(72 + ^-12) \div ^-3 - 1$

Set 27 Solve and check.

1. $n + 18 = ^-37$
   $n + 18 - 18 = ^-37 - 18$
   $n = ^-55$
   Check: $^-55 + 18 = ^-37$
2. $n + 52 = 28$
3. $n - 29 = 37$
4. $n - 34 = ^-19$
5. $19 = n - 11$
6. $36 = n + 42$
7. $^-25 = n - 17$
8. $^-16 = 12 + n$
9. $n + 35 = 15$
10. $27 + n = 19$
11. $n - 27 = 19$
12. $27 = n + 19$
13. $^-39 + n = ^-18$
14. $^-18 + n = ^-39$
15. $n - 42 = 0$
16. $0 = ^-17 + n$
17. $63 = n - 36$
18. $38 = n + 38$
19. $25 = n - 25$
20. $n + 16 = ^-16$
21. $n - 23 = ^-23$

Set 28 Solve and check.

1. $3n = {}^-27$

   $\frac{3n}{3} = \frac{^-27}{3}$

   $n = {}^-9$

   Check: $3(^-9) = {}^-27$

2. $5n = {}^-40$
3. $\frac{n}{9} = 5$
4. $\frac{n}{6} = {}^-8$
5. ${}^-42 = 7n$
6. $56 = 8n$
7. $56 = \frac{n}{8}$
8. $54 = \frac{n}{^-9}$
9. $54 = {}^-9n$
10. $6n = {}^-72$
11. $\frac{n}{6} = {}^-72$
12. $30n = 30$
13. $\frac{n}{8} = 8$
14. ${}^-10n = 130$
15. $\frac{n}{^-10} = 130$
16. ${}^-12 = \frac{n}{^-12}$
17. ${}^-12 = {}^-12n$
18. $\frac{n}{7} = 0$
19. $16n = 0$
20. ${}^-1 = \frac{n}{15}$
21. $20n = {}^-20$

Set 29 Solve and check.

1. ${}^-5n - 9 = 11$

   ${}^-5n - 9 + 9 = 11 + 9$

   ${}^-5n = 20$

   $\frac{^-5n}{^-5} = \frac{20}{^-5}$

   $n = {}^-4$

   Check: ${}^-5(^-4) - 9 = 11$

2. $\frac{n}{3} + 4 = {}^-20$
3. $9n - 15 = 21$
4. $15 + 2n = 33$
5. $22 = {}^-3n + 7$
6. ${}^-10 = \frac{n}{3} + 4$
7. $6n - 2 = 46$
8. ${}^-7n + 6 = 13$
9. $13 + \frac{n}{2} = 25$
10. ${}^-40 = 5 + 3n$
11. $12n - 24 = 0$
12. $\frac{n}{8} - 12 = {}^-44$
13. $\frac{n}{^-7} + 4 = {}^-8$
14. ${}^-16 = {}^-4n + 24$
15. $\frac{n}{^-3} - 18 = 22$
16. $0 = {}^-8n - 48$
17. $15 = \frac{n}{^-5} + 10$
18. ${}^-6 = 2 + \frac{n}{7}$
19. $\frac{n}{^-4} - 9 = 0$

Set 30 Complete.

1. 7.2 cm = <u>72</u> mm
2. 8.4 m = <u>?</u> dm
3. 3 mm = <u>?</u> cm
4. 6.8 dm = <u>?</u> m
5. 3.8 dm = <u>?</u> m
6. 6 mm = <u>?</u> cm
7. 5 m = <u>?</u> dm
8. 12.8 cm = <u>?</u> mm
9. 19.8 cm = <u>?</u> dm
10. 3.8 m = <u>?</u> dm
11. 14.9 cm = <u>?</u> dm
12. 14.3 cm = <u>?</u> mm
13. 7.4 dm = <u>?</u> m
14. 15.9 cm = <u>?</u> dm
15. 9.5 cm = <u>?</u> mm
16. 9 dm = <u>?</u> m
17. 9.25 dm = <u>?</u> cm
18. 5.3 m = <u>?</u> dm
19. 36.4 km = <u>?</u> m
20. 0.91 km = <u>?</u> m
21. 0.575 km = <u>?</u> m
22. 8.56 m = <u>?</u> cm
23. 0.3 m = <u>?</u> mm
24. 78.6 m = <u>?</u> km

Set 31 **Give the perimeter or circumference. Use 3.14 as an approximation for π.**

1. 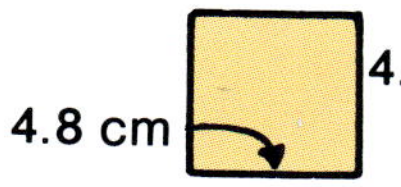

$P = 4s$
$= 4(4.8 \text{ cm})$
$= 19.2 \text{ cm}$

2. 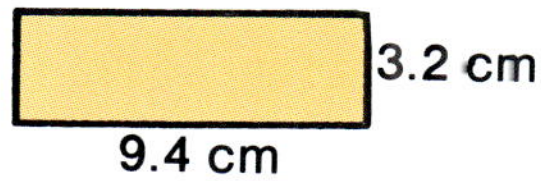

3. 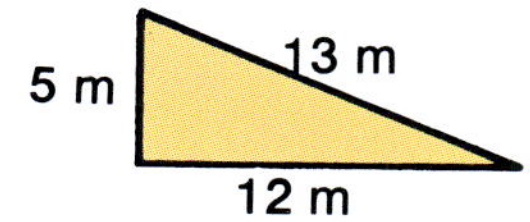

4. 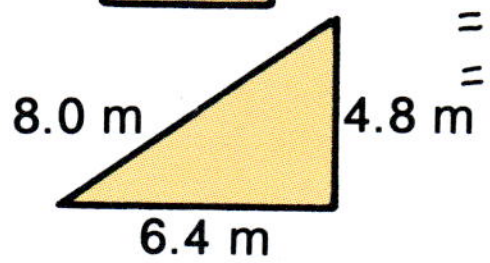

5. 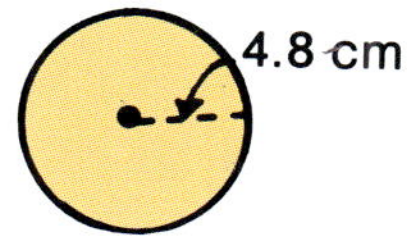

6. 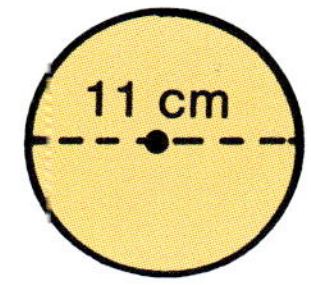

7. 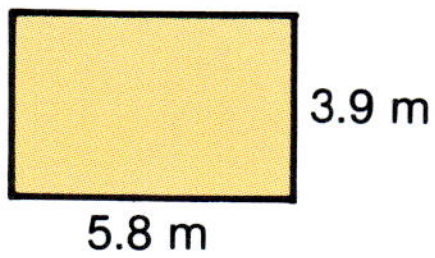

8. 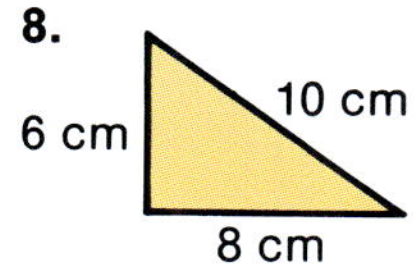

9. 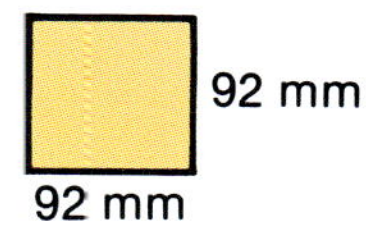

10. 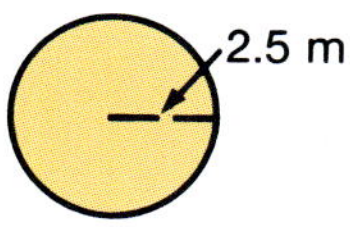

11. 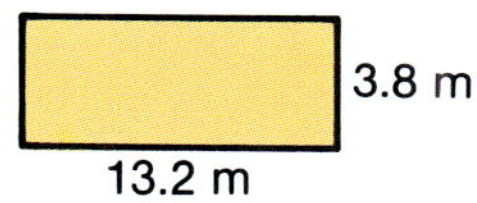

12. 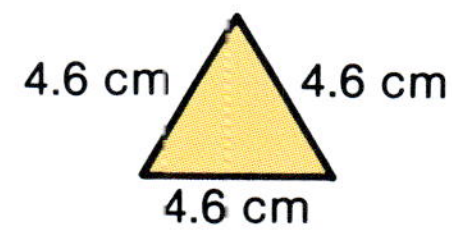

Set 32 **Complete.**

1. 2 L = 2000 mL
2. 3.4 L = ? mL
3. 0.26 L = ? mL
4. 5000 mL = ? L
5. 2500 mL = ? L
6. 568 mL = ? L
7. 3 kL = ? L
8. 1.6 kL = ? L
9. 0.9 kL = ? L
10. 3600 L = ? kL
11. 1250 L = ? kL
12. 796 L = ? kL
13. 4 g = ? mg
14. 5.8 g = ? mg
15. 0.74 g = ? mg
16. 3000 mg = ? g
17. 1600 mg = ? g
18. 725 mg = ? g
19. 5 kg = ? g
20. 2.4 kg = ? g
21. 0.8 kg = ? g
22. 4200 g = ? kg
23. 1640 g = ? kg
24. 598 g = ? kg

Set 33 **Complete.**

1. 5 ft = 60 in.
2. 4 yd = ? ft
3. 1 mi = ? yd
4. 1 mi = ? ft
5. 60 ft = ? yd
6. 60 in. = ? ft
7. 4 lb = ? oz
8. 2 T = ? lb
9. 144 oz = ? lb
10. 1000 lb = ? T
11. 8 c = ? pt
12. 12 pt = ? qt
13. 24 qt = ? gal
14. 12 qt = ? c
15. 6 gal = ? pt
16. 70 in. = 5 ft ? in.
17. 53 ft = 17 yd ? ft
18. 6000 ft = 1 mi. ? ft
19. 25 qt = 6 gal ? qt
20. 33 qt = ? gal 1 qt
21. 19 pt = ? qt 1 pt
22. 40 oz = 2 lb ? oz
23. 3 lb 7 oz = ? oz
24. 63 oz = ? lb 15 oz

Set 34 **Compute.**

1. 4 ft 5 in. + 2 ft 9 in.	**2.** 6 ft 9 in. + 3 ft 8 in.	**3.** 9 yd 2 ft + 2 yd 2 ft	**4.** 2 yd 2 ft 8 in. + 1 yd 1 ft 7 in.
5. 6 ft 2 in. − 2 ft 8 in.	**6.** 8 yd 1 ft − 3 yd 2 ft	**7.** 5 ft − 3 ft 7 in.	**8.** 9 yd 2 ft 6 in. − 3 yd 2 ft 8 in.
9. 1 ft 2 in. × 5	**10.** 3 ft 4 in. × 6	**11.** 4 yd 2 ft × 3	**12.** 1 yd 2 ft 10 in. × 4
13. 2)3 ft 4 in.	**14.** 3)4 yd	**15.** 4)10 ft 8 in.	**16.** 2)8 yd 1 ft 6 in.

Set 35 **Give the prime factorization.**

1. 18	**2.** 21	**3.** 14	**4.** 25	**5.** 6	**6.** 30
7. 40	**8.** 45	**9.** 8	**10.** 32	**11.** 10	**12.** 16
13. 27	**14.** 57	**15.** 64	**16.** 15	**17.** 52	**18.** 60
19. 24	**20.** 12	**21.** 50	**22.** 51	**23.** 68	**24.** 72
25. 54	**26.** 49	**27.** 63	**28.** 48	**29.** 56	**30.** 70
31. 42	**32.** 80	**33.** 44	**34.** 66	**35.** 75	**36.** 91
37. 65	**38.** 88	**39.** 94	**40.** 81	**41.** 90	**42.** 84
43. 100	**44.** 110	**45.** 108	**46.** 120	**47.** 132	**48.** 144

Set 36 **Give the greatest common factor (GCF).**

1. 4, 6	**2.** 4, 8	**3.** 3, 9	**4.** 5, 3	**5.** 2, 5	**6.** 4, 7
7. 3, 21	**8.** 3, 8	**9.** 7, 6	**10.** 4, 9	**11.** 6, 14	**12.** 2, 10
13. 18, 12	**14.** 12, 16	**15.** 16, 8	**16.** 8, 12	**17.** 10, 15	**18.** 9, 10
19. 24, 20	**20.** 15, 20	**21.** 16, 6	**22.** 25, 15	**23.** 18, 10	**24.** 36, 16
25. 16, 48	**26.** 12, 42	**27.** 36, 8	**28.** 12, 84	**29.** 36, 81	**30.** 32, 48
31. 20, 36	**32.** 40, 72	**33.** 35, 56	**34.** 48, 39	**35.** 15, 36	**36.** 32, 64
37. 27, 8	**38.** 24, 60	**39.** 30, 72	**40.** 42, 28	**41.** 81, 21	**42.** 48, 36
43. 48, 30	**44.** 40, 18	**45.** 18, 36	**46.** 24, 42	**47.** 30, 75	**48.** 48, 72

Set 37 **Write in lowest terms.**

1. $\frac{2}{4}$	2. $\frac{3}{12}$	3. $\frac{2}{6}$	4. $\frac{6}{16}$	5. $\frac{4}{10}$	6. $\frac{9}{6}$	7. $\frac{10}{16}$	8. $\frac{3}{9}$
9. $\frac{16}{18}$	10. $\frac{18}{12}$	11. $\frac{4}{8}$	12. $\frac{15}{12}$	13. $\frac{3}{6}$	14. $\frac{18}{16}$	15. $\frac{4}{12}$	16. $\frac{5}{20}$
17. $\frac{9}{24}$	18. $\frac{6}{12}$	19. $\frac{12}{15}$	20. $\frac{6}{24}$	21. $\frac{15}{10}$	22. $\frac{5}{10}$	23. $\frac{15}{24}$	24. $\frac{24}{36}$
25. $\frac{2}{8}$	26. $\frac{35}{40}$	27. $\frac{3}{15}$	28. $\frac{7}{14}$	29. $\frac{42}{48}$	30. $\frac{6}{18}$	31. $\frac{30}{45}$	32. $\frac{10}{20}$
33. $\frac{36}{48}$	34. $\frac{5}{15}$	35. $\frac{12}{32}$	36. $\frac{14}{35}$	37. $\frac{4}{16}$	38. $\frac{8}{16}$	39. $\frac{24}{32}$	40. $\frac{24}{40}$
41. $\frac{30}{36}$	42. $\frac{40}{48}$	43. $\frac{45}{54}$	44. $\frac{45}{50}$	45. $\frac{27}{63}$	46. $\frac{16}{64}$	47. $\frac{40}{60}$	48. $\frac{64}{72}$

Set 38 **Write as a decimal.**

1. $\frac{1}{3}$	2. $\frac{1}{2}$	3. $\frac{1}{4}$	4. $\frac{1}{8}$	5. $\frac{1}{6}$	6. $\frac{1}{7}$	7. $\frac{1}{9}$	8. $\frac{5}{6}$
9. $\frac{5}{8}$	10. $\frac{5}{3}$	11. $\frac{3}{4}$	12. $\frac{3}{2}$	13. $\frac{3}{3}$	14. $\frac{2}{3}$	15. $\frac{4}{7}$	16. $\frac{11}{8}$
17. $\frac{7}{4}$	18. $\frac{7}{2}$	19. $\frac{7}{9}$	20. $\frac{9}{8}$	21. $\frac{5}{4}$	22. $\frac{5}{2}$	23. $\frac{4}{3}$	24. $\frac{2}{9}$

Write as a fraction in lowest terms.

25. 0.75	26. 0.7	27. 0.5	28. 0.375	29. 0.08	30. 0.72	31. 0.1
32. 0.65	33. 0.2	34. 0.02	35. 0.4	36. 0.01	37. 0.125	38. 0.57
39. 1.25	40. 2.5	41. 1.6	42. 0.15	43. 0.9	44. 0.025	45. 0.09

Set 39 **Give the least common multiple (LCM).**

1. 4, 6	2. 4, 5	3. 4, 3	4. 3, 5	5. 2, 4	6. 4, 10
7. 6, 9	8. 6, 24	9. 3, 6	10. 6, 20	11. 9, 2	12. 8, 16
13. 5, 2	14. 8, 4	15. 6, 5	16. 7, 2	17. 6, 15	18. 8, 6
19. 12, 18	20. 3, 7	21. 6, 16	22. 12, 8	23. 18, 27	24. 10, 6
25. 5, 7	26. 9, 15	27. 16, 24	28. 9, 12	29. 8, 10	30. 8, 18

Set 40 $<$, $=$, or $>$?

1. $\frac{5}{12} \;(<)\; \frac{1}{2}$ 2. $\frac{1}{2} \bigcirc \frac{1}{3}$ 3. $\frac{2}{3} \bigcirc \frac{3}{2}$ 4. $\frac{1}{5} \bigcirc \frac{1}{4}$ 5. $\frac{5}{6} \bigcirc \frac{3}{4}$

6. $\frac{5}{8} \bigcirc \frac{2}{3}$ 7. $\frac{6}{16} \bigcirc \frac{3}{8}$ 8. $\frac{15}{18} \bigcirc \frac{5}{6}$ 9. $\frac{3}{4} \bigcirc \frac{6}{8}$ 10. $\frac{3}{8} \bigcirc \frac{1}{3}$

11. $\frac{5}{6} \bigcirc \frac{7}{8}$ 12. $\frac{1}{2} \bigcirc \frac{5}{9}$ 13. $\frac{15}{6} \bigcirc \frac{5}{2}$ 14. $\frac{6}{10} \bigcirc \frac{3}{5}$ 15. $\frac{5}{3} \bigcirc \frac{5}{4}$

16. $\frac{5}{9} \bigcirc \frac{2}{3}$ 17. $\frac{5}{6} \bigcirc \frac{5}{8}$ 18. $\frac{5}{2} \bigcirc \frac{7}{4}$ 19. $\frac{3}{4} \bigcirc \frac{4}{3}$ 20. $\frac{9}{5} \bigcirc \frac{5}{9}$

21. $\frac{3}{4} \bigcirc \frac{7}{10}$ 22. $\frac{5}{6} \bigcirc \frac{19}{24}$ 23. $\frac{13}{16} \bigcirc \frac{3}{4}$ 24. $\frac{11}{24} \bigcirc \frac{1}{2}$ 25. $\frac{7}{9} \bigcirc \frac{3}{4}$

26. $\frac{1}{4} \bigcirc \frac{5}{24}$ 27. $\frac{3}{2} \bigcirc \frac{23}{16}$ 28. $\frac{15}{32} \bigcirc \frac{7}{16}$ 29. $\frac{3}{5} \bigcirc \frac{7}{10}$ 30. $\frac{4}{5} \bigcirc \frac{5}{8}$

Set 41 **Give each sum in lowest terms.**

1. $\frac{2}{3} + \frac{1}{4} = \frac{8}{12} + \frac{3}{12}$ 2. $\frac{1}{2} + \frac{3}{8}$ 3. $\frac{2}{5} + \frac{1}{10}$ 4. $\frac{4}{9} + \frac{1}{6}$ 5. $\frac{2}{5} + \frac{1}{4}$

$= \frac{11}{12}$ 6. $\frac{3}{5} + \frac{1}{3}$ 7. $\frac{1}{2} + \frac{1}{6}$ 8. $\frac{5}{8} + \frac{1}{6}$ 9. $\frac{3}{8} + \frac{1}{3}$

10. $\frac{3}{5} + \frac{2}{15}$ 11. $\frac{3}{10} + \frac{1}{4}$ 12. $\frac{5}{12} + \frac{1}{4}$ 13. $\frac{5}{12} + \frac{3}{8}$ 14. $\frac{1}{2} + \frac{2}{9}$ 15. $\frac{3}{10} + \frac{1}{6}$

16. $\frac{1}{2} + \frac{1}{12}$ 17. $\frac{5}{9} + \frac{1}{9}$ 18. $\frac{7}{15} + \frac{1}{3}$ 19. $\frac{1}{10} + \frac{1}{3}$ 20. $\frac{1}{8} + \frac{3}{16}$ 21. $\frac{1}{5} + \frac{1}{2}$

22. $\frac{5}{16} + \frac{1}{4}$ 23. $\frac{1}{2} + \frac{3}{10}$ 24. $\frac{1}{5} + \frac{1}{6}$ 25. $\frac{7}{10} + \frac{1}{4}$ 26. $\frac{5}{24} + \frac{1}{8}$ 27. $\frac{1}{3} + \frac{7}{24}$

28. $\frac{9}{32} + \frac{1}{8}$ 29. $\frac{1}{6} + \frac{7}{10}$ 30. $\frac{5}{9} + \frac{1}{6}$ 31. $\frac{3}{20} + \frac{2}{5}$ 32. $\frac{1}{2} + \frac{1}{6}$ 33. $\frac{1}{2} + \frac{3}{10}$

Set 42 **Give each difference in lowest terms.**

1. $\frac{2}{3} - \frac{1}{4} = \frac{8}{12} - \frac{3}{12}$ 2. $\frac{1}{2} - \frac{3}{8}$ 3. $\frac{2}{5} - \frac{1}{10}$ 4. $\frac{4}{9} - \frac{1}{6}$ 5. $\frac{2}{5} - \frac{1}{4}$

$= \frac{5}{12}$ 6. $\frac{3}{5} - \frac{1}{3}$ 7. $\frac{1}{2} - \frac{1}{6}$ 8. $\frac{5}{8} - \frac{1}{6}$ 9. $\frac{3}{8} - \frac{1}{3}$

10. $\frac{3}{5} - \frac{2}{15}$ 11. $\frac{3}{10} - \frac{1}{4}$ 12. $\frac{5}{12} - \frac{1}{4}$ 13. $\frac{5}{12} - \frac{3}{8}$ 14. $\frac{1}{2} - \frac{2}{9}$ 15. $\frac{3}{10} - \frac{1}{6}$

16. $\frac{1}{2} - \frac{1}{12}$ 17. $\frac{5}{9} - \frac{1}{9}$ 18. $\frac{7}{15} - \frac{1}{3}$ 19. $\frac{7}{8} - \frac{1}{2}$ 20. $\frac{13}{16} - \frac{1}{8}$ 21. $\frac{1}{5} - \frac{1}{6}$

22. $\frac{3}{4} - \frac{1}{6}$ 23. $\frac{4}{5} - \frac{5}{8}$ 24. $\frac{5}{6} - \frac{5}{8}$ 25. $\frac{11}{16} - \frac{1}{4}$ 26. $\frac{11}{20} - \frac{2}{5}$ 27. $\frac{7}{8} - \frac{5}{6}$

28. $\frac{19}{32} - \frac{1}{2}$ 29. $\frac{5}{6} - \frac{1}{2}$ 30. $\frac{23}{32} - \frac{5}{16}$ 31. $\frac{7}{8} - \frac{3}{4}$ 32. $\frac{5}{6} - \frac{3}{8}$ 33. $\frac{3}{4} - \frac{9}{16}$

Set 43 **Write as a fraction.**

1. $2\frac{1}{3}$ $\frac{7}{3}$
2. $4\frac{3}{5}$
3. $3\frac{1}{8}$
4. $7\frac{2}{3}$
5. $3\frac{1}{5}$
6. $2\frac{5}{6}$
7. $1\frac{4}{9}$
8. $3\frac{7}{10}$
9. $2\frac{3}{8}$
10. $1\frac{1}{2}$
11. $3\frac{5}{12}$
12. $4\frac{4}{15}$
13. $5\frac{1}{4}$
14. $8\frac{7}{8}$
15. $3\frac{2}{7}$
16. $5\frac{3}{10}$
17. $2\frac{5}{9}$
18. $6\frac{3}{4}$
19. $2\frac{3}{16}$
20. $3\frac{7}{20}$
21. $7\frac{4}{5}$

Write as a decimal.

22. $2\frac{1}{4}$
23. $1\frac{4}{5}$
24. $3\frac{7}{10}$
25. $5\frac{3}{4}$
26. $2\frac{7}{8}$
27. $4\frac{1}{2}$
28. $3\frac{7}{25}$
29. $4\frac{9}{20}$
30. $7\frac{2}{3}$
31. $5\frac{7}{9}$
32. $3\frac{5}{6}$
33. $1\frac{5}{12}$
34. $2\frac{3}{16}$
35. $4\frac{2}{7}$
36. $2\frac{1}{16}$
37. $3\frac{7}{20}$
38. $1\frac{5}{16}$
39. $5\frac{3}{25}$
40. $4\frac{9}{20}$
41. $2\frac{1}{25}$
42. $6\frac{9}{16}$

Set 44 **Write in simplest form.**

1. $\frac{10}{8}$ $1\frac{1}{4}$
2. $\frac{9}{5}$
3. $\frac{10}{2}$
4. $3\frac{2}{4}$
5. $\frac{6}{9}$
6. $\frac{6}{4}$
7. $1\frac{12}{18}$
8. $\frac{18}{6}$
9. $\frac{15}{20}$
10. $\frac{15}{6}$
11. $\frac{18}{2}$
12. $\frac{17}{5}$
13. $5\frac{8}{10}$
14. $\frac{15}{3}$
15. $\frac{24}{15}$
16. $1\frac{4}{8}$
17. $\frac{12}{20}$
18. $\frac{10}{4}$
19. $3\frac{2}{6}$
20. $\frac{18}{24}$
21. $6\frac{10}{12}$
22. $\frac{24}{32}$
23. $\frac{18}{9}$
24. $\frac{15}{4}$
25. $\frac{36}{24}$
26. $\frac{30}{40}$
27. $\frac{10}{6}$
28. $\frac{42}{48}$
29. $7\frac{4}{12}$
30. $\frac{28}{8}$
31. $\frac{36}{12}$
32. $\frac{12}{7}$
33. $6\frac{10}{25}$
34. $\frac{18}{42}$
35. $\frac{48}{18}$

Set 45 **Add.**

1. $5\frac{1}{5} + 2\frac{1}{5}$ $7\frac{2}{5}$
2. $8\frac{1}{2} + 3\frac{1}{4}$
3. $9\frac{1}{2} + 1\frac{1}{2}$
4. $7\frac{1}{4} + 5\frac{3}{8}$
5. $8\frac{1}{3} + 4\frac{1}{4}$
6. $9\frac{1}{4} + 3\frac{1}{8}$
7. $15\frac{2}{3} + 8\frac{2}{3}$
8. $14 + 9\frac{3}{8}$
9. $13\frac{3}{4} + 6\frac{2}{5}$
10. $11\frac{1}{3} + 5\frac{1}{6}$
11. $14\frac{2}{3} + 7\frac{5}{9}$
12. $18\frac{3}{8} + 6\frac{7}{8}$
13. $23\frac{2}{9} + 15\frac{1}{3}$
14. $24\frac{1}{2} + 12\frac{3}{5}$
15. $32\frac{3}{4} + 14\frac{1}{2}$
16. $46\frac{2}{3} + 21\frac{3}{4}$
17. $40\frac{1}{4} + 35\frac{1}{8}$
18. $38\frac{3}{4} + 27\frac{5}{8}$

Set 46 **Subtract.**

1. $8\frac{2}{3} - 3\frac{1}{3}$
2. $9\frac{1}{2} - 4\frac{1}{2}$
3. $7\frac{3}{4} - 6\frac{1}{2}$
4. $8\frac{5}{9} - 6\frac{2}{9}$
5. $9\frac{3}{5} - 4\frac{1}{5}$
6. $8 - 3\frac{1}{2}$
7. $14\frac{1}{3} - 5\frac{2}{3}$
8. $15\frac{1}{8} - 6\frac{3}{4}$
9. $19\frac{3}{5} - 6\frac{7}{10}$
10. $18\frac{1}{2} - 5\frac{3}{4}$
11. $26 - 8\frac{3}{5}$
12. $27\frac{1}{4} - 6\frac{3}{5}$
13. $24\frac{4}{9} - 17\frac{2}{3}$
14. $36 - 15\frac{5}{8}$
15. $35\frac{1}{6} - 17\frac{1}{8}$
16. $25 - 14\frac{2}{3}$
17. $28\frac{1}{3} - 14\frac{1}{2}$
18. $27\frac{2}{3} - 18\frac{3}{4}$

Set 47 **Give each product in simplest form.**

1. $\frac{1}{2} \times \frac{1}{4}$
2. $\frac{4}{7} \times \frac{2}{5}$
3. $\frac{2}{3} \times \frac{3}{4}$
4. $\frac{5}{8} \times \frac{0}{6}$
5. $\frac{5}{9} \times \frac{3}{10}$
6. $\frac{3}{5} \times \frac{3}{5}$
7. $\frac{1}{7} \times \frac{1}{8}$
8. $\frac{1}{2} \times \frac{3}{8}$
9. $\frac{6}{4} \times \frac{2}{2}$
10. $\frac{3}{4} \times \frac{8}{3}$
11. $\frac{3}{4} \times \frac{6}{7}$
12. $\frac{5}{9} \times \frac{2}{3}$
13. $\frac{3}{4} \times \frac{12}{2}$
14. $\frac{3}{4} \times \frac{5}{3}$
15. $\frac{5}{8} \times \frac{7}{5}$
16. $\frac{2}{5} \times \frac{3}{4}$
17. $\frac{2}{5} \times \frac{3}{5}$
18. $\frac{7}{3} \times \frac{8}{3}$
19. $3 \times \frac{1}{2}$
20. $\frac{3}{4} \times \frac{3}{4}$
21. $\frac{5}{12} \times \frac{3}{2}$
22. $\frac{4}{5} \times \frac{5}{4}$
23. $4 \times \frac{3}{4}$
24. $\frac{9}{4} \times \frac{2}{3}$
25. $\frac{3}{5} \times 4$
26. $\frac{5}{8} \times \frac{8}{5}$
27. $\frac{12}{5} \times \frac{5}{16}$
28. $\frac{2}{3} \times 5$
29. $\frac{5}{6} \times \frac{5}{6}$
30. $\frac{5}{16} \times \frac{8}{3}$

Set 48 **Give each quotient in simplest form.**

1. $\frac{1}{2} \div \frac{2}{3}$
2. $\frac{4}{9} \div \frac{2}{3}$
3. $\frac{3}{2} \div \frac{3}{4}$
4. $\frac{3}{4} \div \frac{5}{8}$
5. $\frac{9}{5} \div \frac{3}{8}$
6. $\frac{3}{10} \div \frac{2}{5}$
7. $\frac{5}{6} \div \frac{1}{2}$
8. $\frac{1}{8} \div \frac{9}{5}$
9. $\frac{1}{5} \div \frac{5}{4}$
10. $\frac{7}{4} \div \frac{3}{4}$
11. $\frac{2}{3} \div \frac{2}{3}$
12. $\frac{7}{8} \div \frac{5}{4}$
13. $\frac{3}{5} \div \frac{3}{10}$
14. $\frac{5}{6} \div \frac{5}{8}$
15. $\frac{3}{4} \div \frac{2}{2}$
16. $\frac{3}{4} \div \frac{7}{7}$
17. $\frac{3}{10} \div \frac{7}{10}$
18. $\frac{9}{2} \div \frac{3}{4}$
19. $\frac{3}{10} \div \frac{1}{5}$
20. $\frac{1}{2} \div \frac{7}{2}$
21. $\frac{4}{5} \div \frac{2}{3}$
22. $\frac{8}{7} \div \frac{3}{2}$
23. $\frac{9}{5} \div \frac{2}{5}$
24. $\frac{5}{8} \div \frac{10}{2}$
25. $\frac{5}{12} \div \frac{3}{4}$
26. $8 \div \frac{3}{5}$
27. $\frac{9}{16} \div 12$

Set 49 Solve.

1. $\frac{3}{8}n = 12$

   $\frac{8}{3} \times \frac{3}{8}n = \frac{8}{3} \times 12$

   $n = \frac{96}{3}$

   $n = 32$

2. $\frac{3}{4}n = 24$
3. $\frac{3}{2}n = 9$
4. $\frac{5}{8}n = 40$
5. $\frac{5}{2}n = 10$
6. $\frac{4}{3}n = 24$
7. $\frac{4}{5}n = 40$
8. $\frac{3}{8}n = 21$
9. $\frac{1}{3}n = 14$
10. $\frac{3}{5}n = 18$
11. $\frac{5}{6}n = 20$
12. $\frac{5}{3}n = 30$
13. $\frac{1}{4}n = 24$
14. $\frac{3}{8}n = 24$
15. $\frac{3}{10}n = 18$
16. $\frac{7}{3}n = 42$
17. $\frac{5}{12}n = 60$
18. $\frac{5}{16}n = 5$
19. $\frac{7}{10}n = 35$
20. $\frac{11}{16}n = 44$
21. $\frac{2}{9}n = 36$
22. $\frac{6}{5}n = 60$

Set 50 Multiply.

1. $2\frac{1}{3} \times 7 = \frac{7}{3} \times 7$

   $= \frac{49}{3}$

   $= 16\frac{1}{3}$

2. $5 \times 6\frac{1}{3}$
3. $3 \times 4\frac{1}{2}$
4. $5\frac{1}{4} \times 5$
5. $2\frac{1}{5} \times 6$
6. $1\frac{1}{4} \times 2\frac{1}{5}$
7. $1\frac{1}{4} \times 1\frac{3}{8}$
8. $4\frac{1}{4} \times 3\frac{3}{4}$
9. $3\frac{1}{3} \times 3\frac{1}{4}$
10. $2\frac{1}{2} \times 2\frac{2}{3}$
11. $4\frac{2}{3} \times 1\frac{3}{4}$
12. $1\frac{1}{6} \times 1\frac{3}{4}$
13. $3\frac{1}{2} \times 4\frac{2}{3}$
14. $2\frac{1}{4} \times 2\frac{1}{9}$
15. $3\frac{1}{4} \times 2\frac{3}{8}$
16. $2\frac{3}{4} \times 2\frac{1}{4}$
17. $2\frac{3}{4} \times 1\frac{1}{5}$
18. $2\frac{1}{5} \times 3\frac{1}{3}$
19. $1\frac{1}{2} \times 1\frac{1}{2}$
20. $2\frac{1}{2} \times 3\frac{2}{3}$
21. $8 \times 3\frac{3}{4}$
22. $2\frac{1}{5} \times 1\frac{1}{4}$
23. $1\frac{2}{3} \times 10$
24. $2 \times 2\frac{5}{6}$
25. $3\frac{3}{4} \times 2\frac{2}{5}$
26. $3\frac{2}{5} \times 5$
27. $2\frac{5}{8} \times 2\frac{2}{7}$

Set 51 Divide.

1. $1\frac{1}{2} \div 1\frac{1}{4} = \frac{3}{2} \div \frac{5}{4}$

   $= \frac{3}{2} \times \frac{4}{5}$

   $= \frac{12}{10}$

   $= 1\frac{1}{5}$

2. $2\frac{2}{9} \div 1$
3. $3\frac{1}{2} \div 1\frac{3}{4}$
4. $3\frac{3}{4} \div 3$
5. $9\frac{1}{2} \div 3\frac{1}{2}$
6. $7\frac{2}{3} \div 4$
7. $5 \div 1\frac{1}{2}$
8. $0 \div 5\frac{1}{2}$
9. $7\frac{1}{3} \div 2\frac{1}{4}$
10. $2\frac{7}{8} \div 7\frac{2}{3}$
11. $6\frac{2}{3} \div 8\frac{3}{4}$
12. $8\frac{1}{4} \div 4\frac{2}{3}$
13. $3 \div 6\frac{2}{3}$
14. $6\frac{3}{5} \div 8\frac{2}{3}$
15. $6\frac{1}{3} \div 7\frac{1}{5}$
16. $9\frac{1}{3} \div 2\frac{3}{5}$
17. $4\frac{5}{8} \div 6\frac{3}{4}$
18. $5\frac{1}{4} \div 5\frac{4}{5}$
19. $8\frac{1}{2} \div 2\frac{1}{8}$
20. $4\frac{1}{2} \div 3\frac{1}{6}$
21. $2\frac{3}{4} \div 2\frac{3}{4}$
22. $10\frac{5}{6} \div 3\frac{1}{4}$
23. $16\frac{1}{3} \div 4\frac{2}{3}$
24. $7\frac{1}{3} \div 2\frac{1}{5}$
25. $9\frac{1}{6} \div 3\frac{2}{3}$

Set 52 **Give the area.**

1\.

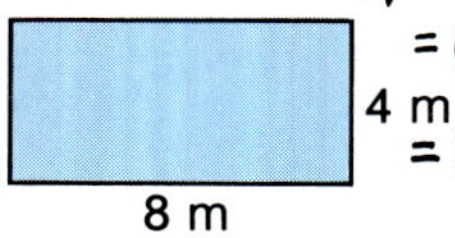

$A = lw$
$= (8m)(4m)$
$= 32m^2$

2\.

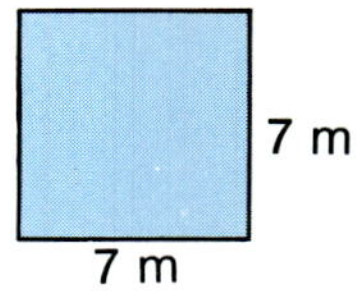

3\.

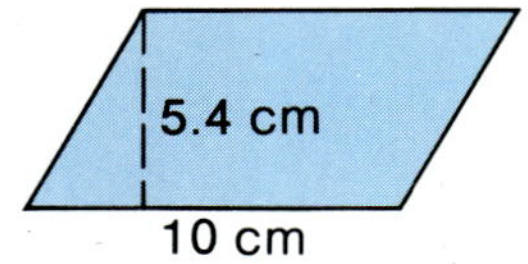

4\.

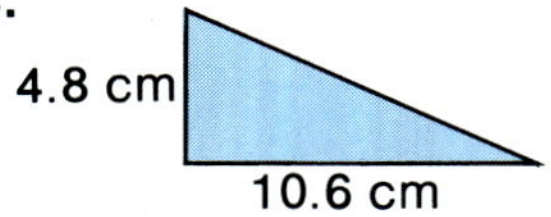

5\.

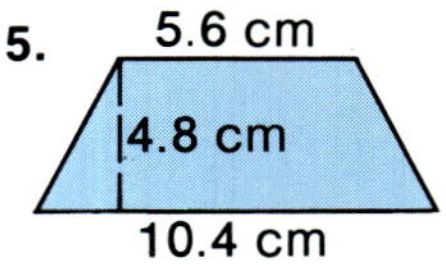

6\.

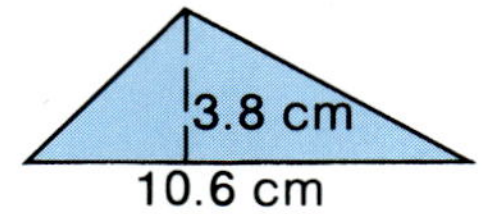

7\.

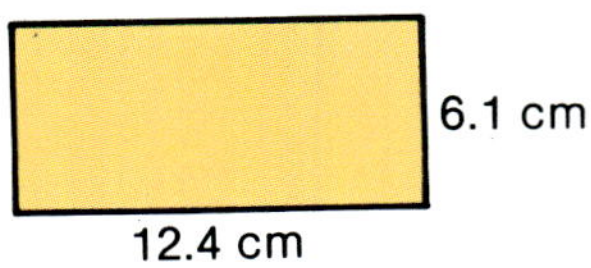

8\.

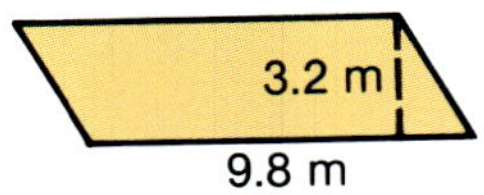

9\.

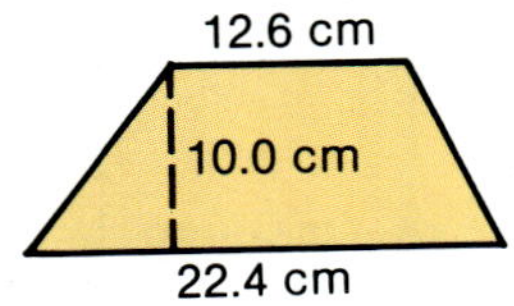

Set 53 **Give the area. Use 3.14 as an approximation for π.**

1\.

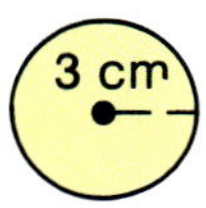

$A = \pi r^2$
$= 3.14(3cm)^2$
$= 3.14(9cm^2)$
$= 28.26 cm^2$

2\.

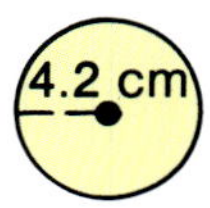

3\.

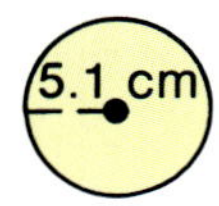

4\.

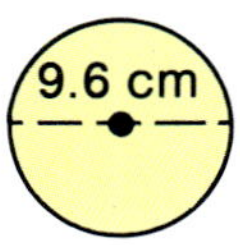

5\.

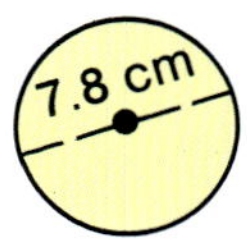

6\.

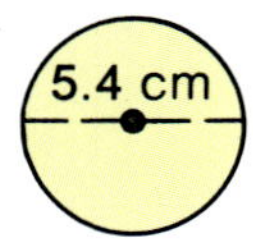

Set 54 **Give the surface area. Use 3.14 as an approximation for π.**

1\.

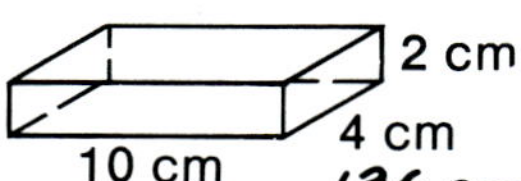

$136 cm^2$

2\.

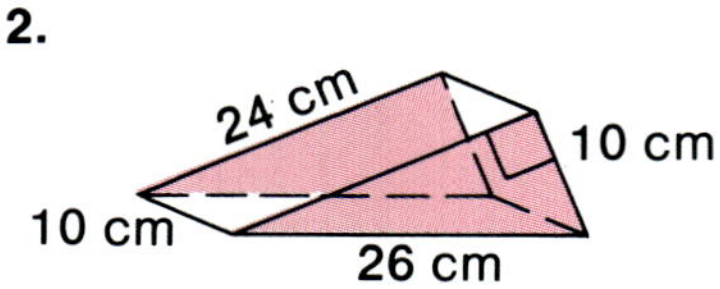

3\.

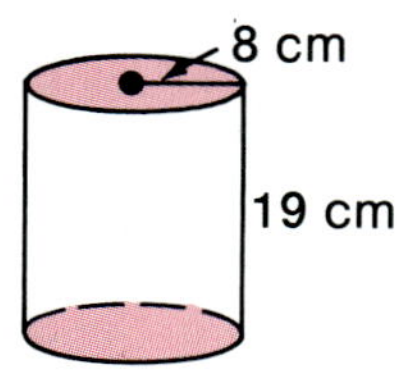

4\.

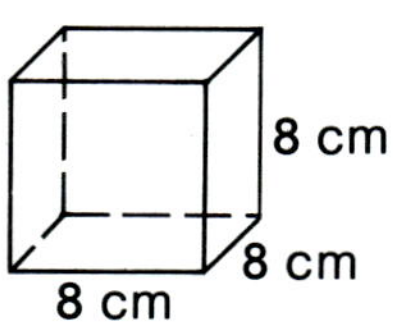

5\.

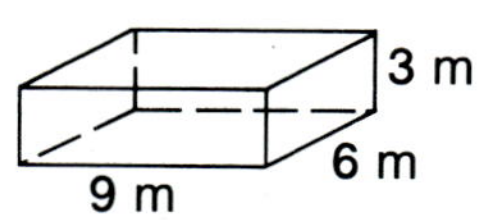

6\.

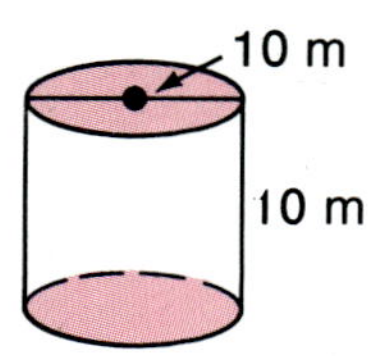

Set 55 **Give the volume. Use 3.14 as an approximation for π.**

1. 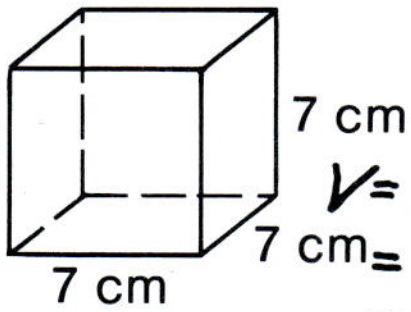

2. 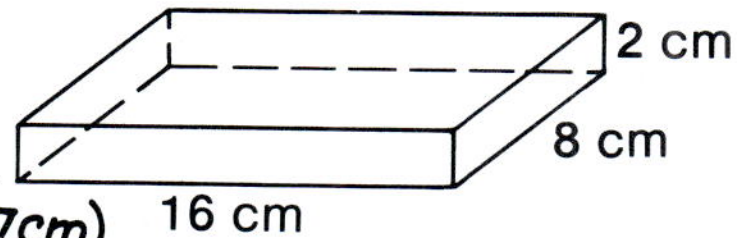

3. 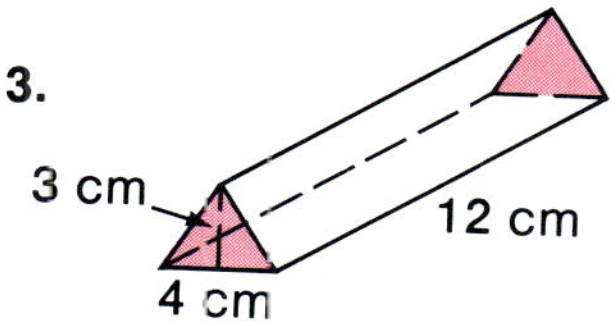

4. 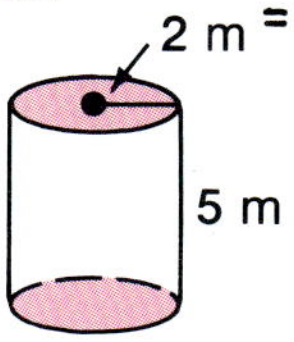

5. 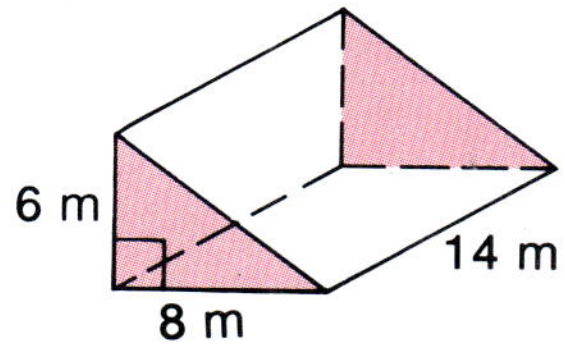

6. 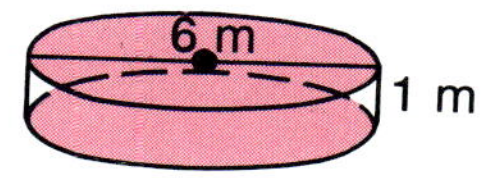

Set 56 **Give the volume. Use 3.14 as an approximation for π.**

1. 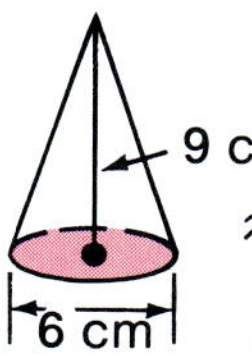

2. 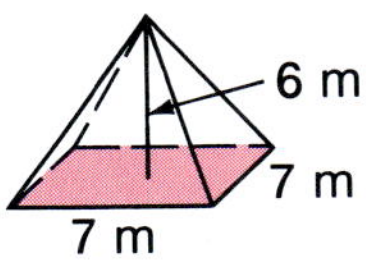

3. 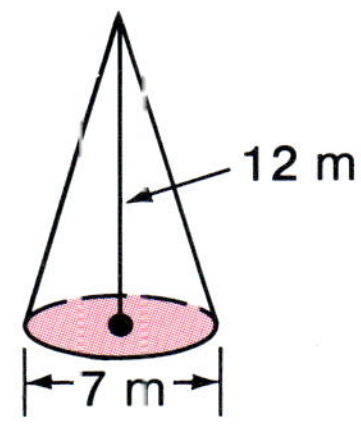

4. 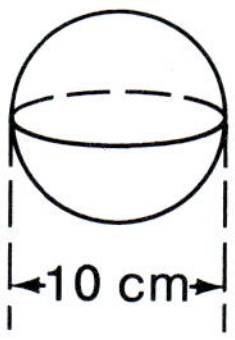

5. 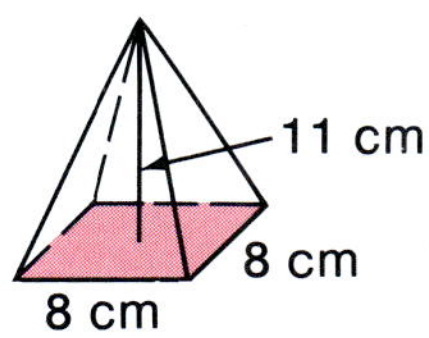

6. 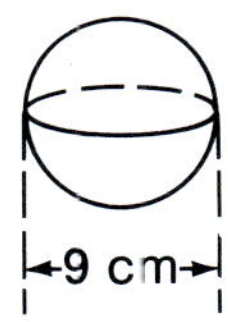

Set 57 **Find each area. Use 3.14 as an approximation for π.**

1. 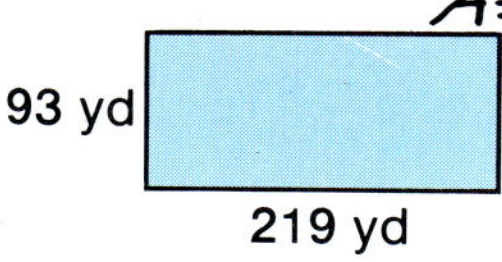

2. 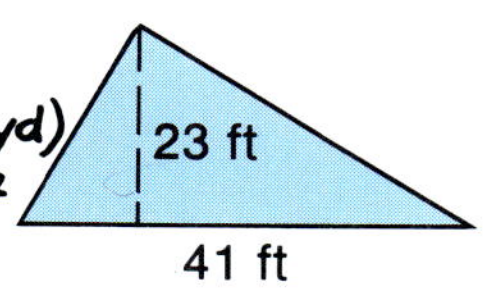

3. 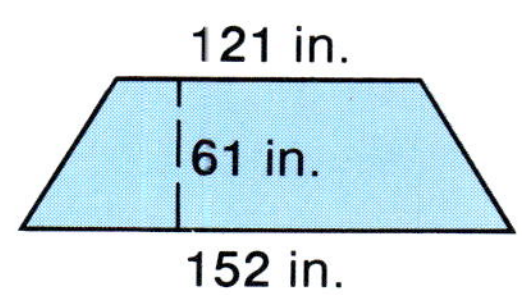

4. 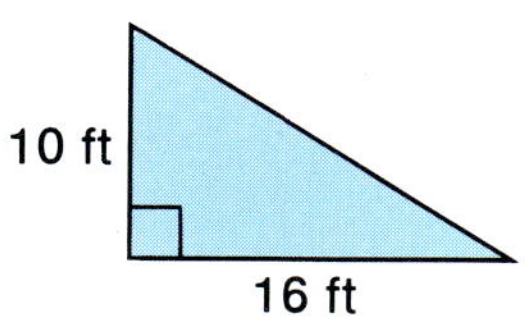

5. 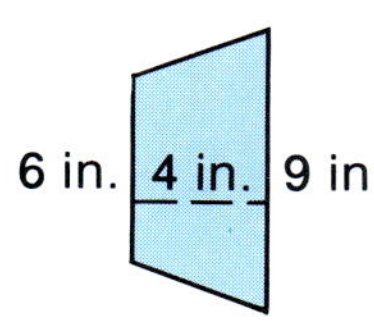

6. 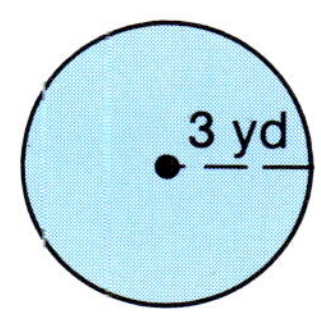

Set 58 **Find each volume. Use 3.14 as an approximation for π.**

1. 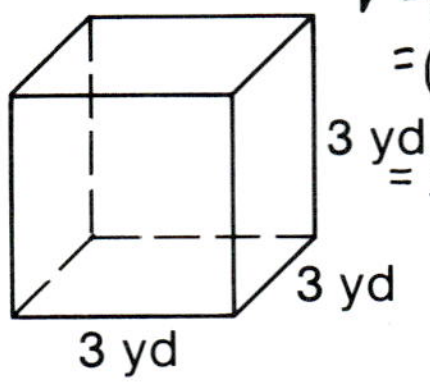

2. 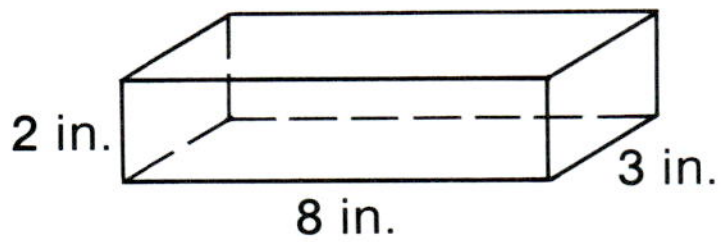

3. 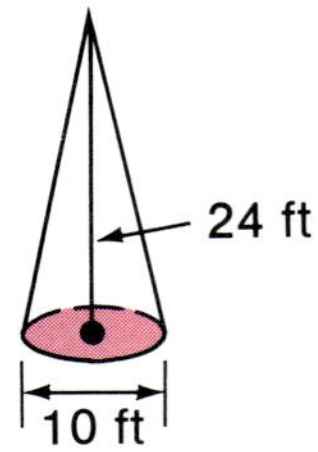

4. 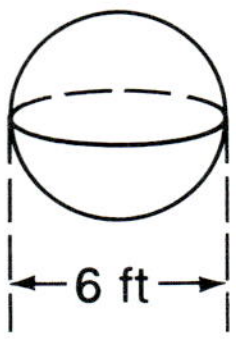

5. 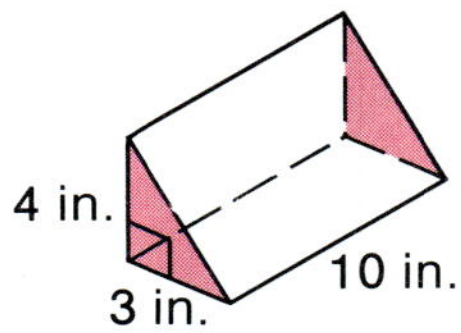

6. 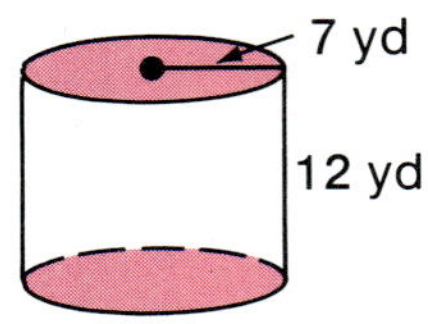

Set 59 **Solve each proportion.**

1. $\frac{2}{4} = \frac{n}{11}$ 2. $\frac{5}{6} = \frac{a}{12}$ 3. $\frac{f}{9} = \frac{4}{3}$ 4. $\frac{5}{6} = \frac{11}{a}$ 5. $\frac{8}{3} = \frac{13}{d}$

6. $\frac{3}{2} = \frac{5}{f}$ 7. $\frac{g}{12} = \frac{1}{4}$ 8. $\frac{h}{2} = \frac{9}{4}$ 9. $\frac{c}{11} = \frac{3}{8}$

10. $\frac{7}{8} = \frac{w}{4}$ 11. $\frac{z}{14} = \frac{7}{4}$ 12. $\frac{20}{24} = \frac{k}{12}$ 13. $\frac{6}{7} = \frac{9}{b}$

14. $\frac{a}{18} = \frac{5}{4}$ 15. $\frac{3}{y} = \frac{4}{7}$ 16. $\frac{5}{9} = \frac{8}{g}$ 17. $\frac{19}{j} = \frac{9}{5}$

18. $\frac{5}{9} = \frac{k}{5}$ 19. $\frac{9}{4} = \frac{9}{w}$ 20. $\frac{13}{x} = \frac{3}{5}$ 21. $\frac{7}{9} = \frac{4}{v}$ 22. $\frac{9}{t} = \frac{3}{5}$

23. $\frac{10}{f} = \frac{16}{9}$ 24. $\frac{12}{7} = \frac{9}{s}$ 25. $\frac{24}{11} = \frac{4}{y}$ 26. $\frac{r}{7} = \frac{10}{3}$ 27. $\frac{20}{13} = \frac{d}{5}$

Set 60 **Write as a fraction or mixed number in simplest form.**

1. 10% 2. 20% 3. 30% 4. 40% 5. 50% 6. 60% 7. 70%

8. 80% 9. 85% 10. 90% 11. 250% 12. 175% 13. $66\frac{2}{3}\%$ 14. $16\frac{2}{3}\%$

15. 1% 16. $\frac{1}{2}\%$ 17. $\frac{1}{4}\%$ 18. $87\frac{1}{2}\%$ 19. $83\frac{1}{3}\%$ 20. $12\frac{1}{2}\%$ 21. $11\frac{1}{9}\%$

22. 6.5% 23. 18.5% 24. 5.25% 25. 0.6% 26. 5.8% 27. 9.6% 28. 32.25%

Set 61 **Write as a percent.**

1. $\frac{1}{4}$ 25%
2. $\frac{2}{4}$
3. $\frac{3}{4}$
4. $\frac{4}{4}$
5. $\frac{5}{4}$
6. $\frac{6}{4}$
7. $\frac{1}{3}$
8. $\frac{2}{3}$
9. $\frac{3}{3}$
10. $\frac{4}{3}$
11. $\frac{1}{8}$
12. $\frac{3}{8}$
13. $\frac{5}{8}$
14. $\frac{7}{8}$
15. $\frac{1}{6}$
16. $\frac{5}{6}$
17. $\frac{1}{5}$
18. $\frac{2}{5}$
19. $\frac{3}{5}$
20. $\frac{4}{5}$
21. $\frac{5}{5}$
22. $\frac{6}{5}$
23. $\frac{7}{5}$
24. $\frac{8}{5}$
25. $\frac{5}{12}$
26. $\frac{5}{24}$
27. $\frac{19}{24}$
28. $\frac{7}{12}$
29. $\frac{5}{32}$
30. $\frac{21}{32}$
31. $\frac{25}{16}$
32. $\frac{35}{24}$
33. $\frac{5}{3}$
34. $\frac{13}{8}$
35. $\frac{19}{6}$
36. $\frac{11}{4}$
37. $\frac{13}{3}$
38. $\frac{11}{8}$
39. $\frac{15}{2}$
40. $\frac{19}{12}$

Set 62 **Write as a percent.**

1. 0.8 $= \frac{80}{100} = 80\%$
2. 1.46
3. 2.7
4. 0.06
5. 0.24
6. $0.62\frac{1}{2}$
7. 0.75
8. 1.05
9. 0.003
10. 0.024
11. $0.08\frac{1}{3}$
12. $0.41\frac{2}{3}$
13. $0.33\frac{1}{3}$
14. $0.16\frac{2}{3}$
15. 0.1
16. $0.66\frac{2}{3}$
17. 4.5
18. 2.375

Write as a decimal.

19. 15%
20. 130%
21. 2.6%
22. 18.3%
23. 40%
24. $66\frac{2}{3}\%$
25. $55\frac{5}{9}\%$
26. 32.5%
27. 275%
28. $188\frac{2}{5}\%$
29. 3%
30. $83\frac{1}{3}\%$
31. 200%
32. 36%
33. 0.8%
34. 3.45%
35. 304%
36. 0.25%

Set 63 **Solve.**

1. 75% of $164 = n$

   $\frac{3}{4}$ of $164 = \frac{3}{4} \times 164 = 123$
2. 50% of $386 = n$
3. 35% of $146 = n$
4. $12\frac{1}{2}\%$ of $136 = n$
5. 100% of $48 = n$
6. 40% of $437 = n$
7. $33\frac{1}{3}\%$ of $132 = n$
8. 6% of $80 = n$
9. 150% of $27 = n$
10. 300% of $48 = n$
11. 40% of $380 = n$
12. 30% of $480 = n$
13. 8% of $500 = n$
14. 20% of $125 = n$
15. $83\frac{1}{3}\%$ of $42 = n$
16. 500% of $17 = n$
17. $62\frac{1}{2}\%$ of $264 = n$
18. 10% of $521 = n$
19. 250% of $46 = n$

Set 64 Solve.

1. 10% of $n = 6$
   $0.1n = 6$
   $\frac{0.1n}{0.1} = \frac{6}{0.1}$
   $n = 60$
2. 50% of $n = 18$
3. 200% of $n = 38$
4. 150% of $n = 120$
5. 5% of $n = 8$
6. 25% of $n = 18$
7. $37\frac{1}{2}$% of $n = 27$
8. 175% of $n = 42$
9. $33\frac{1}{3}$% of $n = 30$
10. 1% of $n = 5$
11. 20% of $n = 6$
12. 60% of $n = 141$
13. $12\frac{1}{2}$% of $n = 42$
14. $83\frac{1}{3}$% of $n = 56$
15. $16\frac{2}{3}$% of $n = 27$
16. $87\frac{1}{2}$% of $n = 35$
17. 40% of $n = 48$
18. $133\frac{1}{3}$% of $n = 128$

Set 65 Solve.

1. n% of 8 = 4
   $\frac{4}{8} = \frac{1}{2}$
   $= 0.5$
   $= 50\%$
2. n% of 32 = 8
3. n% of 18 = 6
4. n% of 16 = 6
5. n% of 24 = 15
6. n% of 18 = 12
7. n% of 12 = 18
8. n% of 12 = 15
9. n% of 24 = 21
10. n% of 15 = 25
11. n% of 36 = 24
12. n% of 75 = 60
13. n% of 18 = 42
14. n% of 48 = 32
15. n% of 75 = 100
16. n% of 100 = 75
17. n% of 80 = 120
18. n% of 120 = 80
19. n% of 16 = 14
20. n% of 75 = 72
21. n% of 132 = 11

Set 66 The regular price is given. Compute the sale price.

1. \$4.80
   20% off \$3.84
2. \$6.64
   50% off
3. \$8.00
   25% off
4. \$9.60
   40% off
5. \$5.35
   40% off
6. \$6.29
   30% off
7. \$9.60
   $33\frac{1}{3}$% off
8. \$10.46
   28% off
9. \$20.48
   35% off
10. \$35.63
    $37\frac{1}{2}$% off
11. \$18.32
    45% off
12. \$43.49
    75% off
13. \$62.50
    30% off
14. \$49.98
    24% off
15. \$57.37
    32% off
16. \$73.65
    40% off

Set 67 **Complete.**

	Principal	*Yearly Rate*	*Time*	*Interest*
1.	\$1000	$12\frac{1}{2}\%$	3 months	\$31.25
2.	\$450	15%	8 months	?
3.	\$100	16%	1 year	?
4.	\$300	14%	2 years	?
5.	\$3600	$13\frac{1}{2}\%$	18 months	?

Set 68 **Complete.**

	Cash price	*Installment plan*	*Installment price*	*Difference between installment price and cash price*
1.	\$150	\$50 down and \$9 a month for 12 mo	\$158	\$8
2.	\$200	\$50 down and \$19.44 a month for 12 mo	?	?
3.	\$240	\$80 down and \$14.40 a month for 12 mo	?	?
4.	\$300	\$75 down and \$39.19 a month for 6 mo	?	?
5.	\$360	\$72 down and \$34.40 a month for 9 mo	?	?

Set 69 $<$, $=$, or $>$?

1. $^{-}3 > {}^{-}5$
2. 6 ● $^{-}8$
3. 0 ● $^{-}5$
4. 7 ● 9
5. 1.2 ● 3.5
6. $^{-}1.7$ ● 1.6
7. $^{-}3.25$ ● $^{-}3.52$
8. 4.10 ● 4.01
9. $^{-}0.8$ ● $^{-}0.80$
10. 0 ● 0.08
11. 1.5 ● 1.500
12. 0 ● $^{-}0.04$
13. $\frac{1}{3}$ ● $\frac{3}{8}$
14. $\frac{^{-}3}{4}$ ● $\frac{^{-}2}{3}$
15. $\frac{^{-}5}{6}$ ● 0
16. $\frac{3}{2}$ ● $\frac{9}{6}$

Set 70 **Give each sum or difference.**

1. $^{-}8 + 12$ 4
2. $^{-}9 + {}^{-}3$
3. $^{-}7 + 7$
4. $10 + {}^{-}6$
5. $^{-}12 - {}^{-}18$
6. $11 - 15$
7. $^{-}16 - 12$
8. $18 - {}^{-}16$
9. $\frac{3}{4} + \frac{1}{4}$
10. $\frac{^{-}3}{8} + \frac{1}{8}$
11. $\frac{1}{3} + \frac{^{-}5}{6}$
12. $\frac{^{-}1}{3} + \frac{^{-}1}{2}$
13. $\frac{5}{8} - \frac{3}{4}$
14. $\frac{7}{12} - \frac{^{-}2}{3}$
15. $\frac{^{-}4}{5} - \frac{3}{2}$
16. $\frac{^{-}4}{3} - \frac{^{-}1}{8}$
17. $3.6 + 4.2$
18. $^{-}5.7 + {}^{-}6.9$
19. $4.05 + {}^{-}6.11$
20. $^{-}8.32 + 6.08$
21. $2.9 - {}^{-}7.6$
22. $^{-}12.3 - {}^{-}14.7$
23. $^{-}6.28 - 5.37$
24. $10.33 - 15.46$
25. $6.3 + {}^{-}2.9$
26. $6.3 - {}^{-}2.9$
27. $^{-}11.74 + 15.06$
28. $^{-}11.74 - 15.06$

Set 71 **Give each product or quotient.**

1. $^{-}9 \times {}^{-}6$ 54
2. 7×8
3. $^{-}5 \times 0$
4. $^{-}9 \times 4$
5. $42 \div 7$
6. $^{-}32 \div 4$
7. $72 \div {}^{-}8$
8. $^{-}60 \div {}^{-}5$
9. $\frac{2}{3} \times \frac{1}{5}$
10. $\frac{^{-}3}{4} \times \frac{^{-}4}{5}$
11. $\frac{^{-}5}{2} \times \frac{5}{6}$
12. $\frac{3}{8} \times \frac{^{-}3}{8}$
13. $\frac{^{-}5}{6} \div \frac{2}{3}$
14. $\frac{9}{10} \div \frac{^{-}1}{2}$
15. $\frac{7}{8} \div \frac{4}{5}$
16. $\frac{^{-}11}{12} \div \frac{^{-}5}{3}$
17. $0.5 \times {}^{-}1.4$
18. $^{-}2.3 \times 6.8$
19. $^{-}4.52 \times {}^{-}2.6$
20. 7.01×5.55
21. $^{-}6.4 \div 0.8$
22. $7.2 \div {}^{-}1.2$
23. $9.6 \div 0.16$
24. $^{-}14.4 \div {}^{-}1.2$
25. $\frac{^{-}5}{4} \times \frac{^{-}3}{20}$
26. $\frac{7}{8} \div \frac{7}{2}$
27. $\frac{5}{2} \times \frac{^{-}5}{2}$
28. $\frac{^{-}9}{16} \div \frac{^{-}3}{4}$

Set 72 **Simplify.**

1. $9a - 3a$ $6a$
2. $6n + 9n$
3. $^{-}8k - 3k$
4. $4t + 6t + 2t$
5. $3r + 7r + 5r$
6. $6x + {}^{-}3x + {}^{-}2x$
7. $3c + 4c + 6$
8. $9 + 8f - 3f$
9. $5y + 6 + 9y$
10. $7w - 3w + 2w$
11. $10b - 3b - 3b$
12. $12x + 7x - 8x$
13. $m + m + m$
14. $3j + j + 4$
15. $11p - 3p + 9$
16. $15t - 7t + 4 + 5$
17. $16x + 8 + 3x - 5$
18. $20y - 6 + 10 - 11y$
19. $3s + t + 5s + 2t$
20. $4y - y + 7x - 3x$
21. $15b - 3 - 10b - 11$
22. $11c - 10d - 4c - d$
23. $13f - 6 - 4f - 10$
24. $18r - 6s - 15r + s$

Set 73 Solve and check.

1. $3t + t = {}^-32$
   $4t = {}^-32$
   $\frac{4t}{4} = \frac{{}^-32}{4}$
   $t = {}^-8$
   Check: $3({}^-8) + {}^-8 = {}^-32$
2. $2n + 3n = {}^-35$
3. $5x - 3x = 11$
4. $r + 7 = 29$
5. $t - 13 = 9$
6. $\frac{n}{8} = {}^-6$
7. $\frac{m}{3} = 12$
8. $6x = {}^-42$
9. ${}^-8v = 80$
10. ${}^-4 = \frac{n}{{}^-2}$
11. $16 = \frac{j}{{}^-3}$
12. $18 = 6k + 2k$
13. ${}^-24 = m - 3$
14. $36 = p + 27$
15. $3x + 2x = {}^-1.4$
16. $9w - 5w = 12.6$
17. $16n - n = {}^-4.5$
18. ${}^-6y = 9.6$
19. $\frac{y}{{}^-7} = 1.4$
20. ${}^-0.12 = 7y - y$

Set 74 Solve and check.

1. $8n + n - 6 = 39$
   $9n - 6 = 39$
   $9n - 6 + 6 = 39 + 6$
   $9n = 45$
   $\frac{9n}{9} = \frac{45}{9}$
   $n = 5$
   Check: $8(5) + 5 - 6 = 39$
2. $2j + 6 = {}^-8$
3. $3m - 27 = 0$
4. $15 = {}^-4t - 5$
5. ${}^-21 = 11w + 12$
6. $\frac{r}{{}^-3} + 4 = {}^-6$
7. $\frac{s}{5} - 8 = 15$
8. ${}^-16 = 7j - 9$
9. $24 = {}^-8k - 24$
10. $3x + 2x - 4 = 26$
11. $5y + y + 12 = {}^-12$
12. $4b - 3.6 = 9.4$
13. $10w + 8.3 = {}^-2.7$
14. $\frac{a}{0.5} + 2.4 = {}^-3.0$
15. $\frac{c}{0.2} - 3.6 = 7.2$
16. $3p + 7p - 2.8 = 7.1$
17. $11t - 7t + 6.1 = {}^-5.7$
18. $10.6 = 3r + 5r - 8.4$

Set 75 Match each inequality with its graph.

1. $x + 3 \geq 1$ d
2. $x + 3 < 1$
3. $x + 3 \neq 1$
4. $x + 3 > 1$
5. $x + 3 \leq 1$

a. 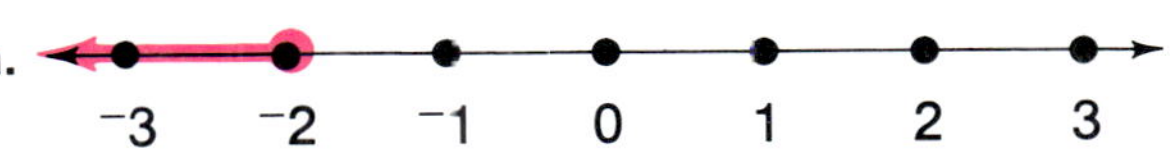

b. 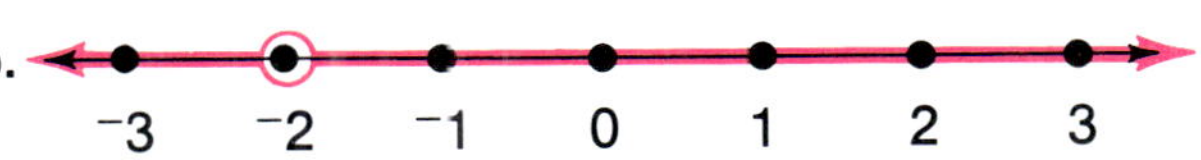

c. 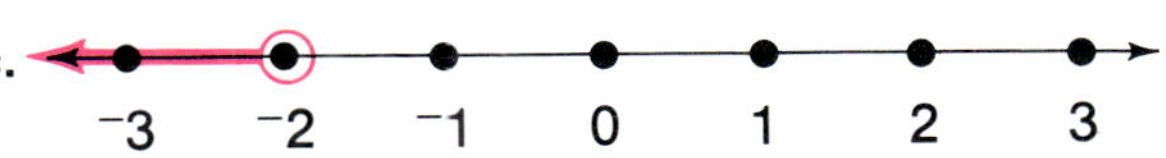

d. 

e. 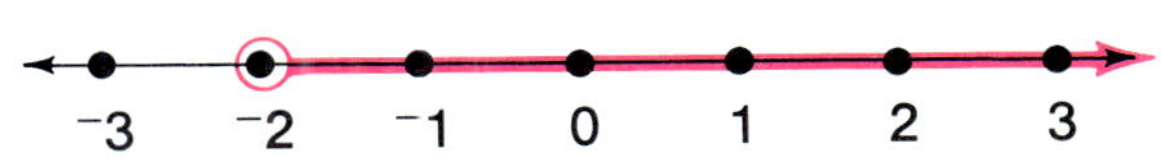

Set 76 **Draw a pair of axes on graph paper. Graph the following ordered pairs. Label each point with its letter.**

1. A:(0, 4)	**2.** W:(3, 0)	**3.** R:(1, 5)	**4.** T:(5, 1)	**5.** D:(5, $^-8$)	**6.** G:($^-6$, 0)
7. J:($^-3$, $^-5$)	**8.** K:(2, 7)	**9.** B:($^-3$, 5)	**10.** V:(6, 3)	**11.** U:($^-7$, $^-2$)	**12.** X:(5, $^-4$)
13. F:(1, 2)	**14.** C:($^-4$, 2)	**15.** S:(0, $^-8$)	**16.** E:(8, 5)	**17.** Z:($^-7$, 6)	**18.** H:($^-8$, $^-1$)

Set 77 **Give the coordinates.**

1. Point A (2,0)	**2.** Point B
3. Point C	**4.** Point D
5. Point E	**6.** Point F
7. Point G	**8.** Point H
9. Point I	**10.** Point J
11. Point K	**12.** Point L

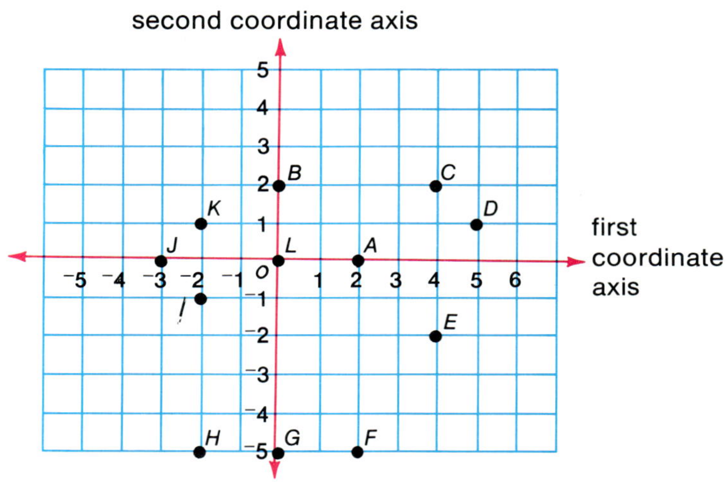

Set 78 **Complete each table of values.**

1. $2x + y = 12$

	x	y
a.	0	12
b.	1	?
c.	2	?
d.	3	?
e.	4	?
f.	5	?

2. $3x + 2y = 18$

	x	y
a.	6	?
b.	4	?
c.	2	?
d.	0	?
e.	$^-2$	?
f.	$^-4$	?

3. $\frac{x}{2} + y = {}^-8$

	x	y
a.	4	?
b.	2	?
c.	0	?
d.	$^-2$	?
e.	$^-4$	?
f.	$^-6$	?

Set 79 First tell how many outcomes. Then tell whether they are equally likely.

1. Toss die with sides numbered 1, 2, 3, 4. Look at the bottom number.

4
yes

2. Toss die with sides numbered 1–6. Look at the top number.

3. Spin this spinner. Look at the color where the arrow lands.

4. Toss coin.
Toss die.

5. Toss coin.
Toss die.

6. Toss die.
Spin spinner.

Set 80 Give each probability.

Event:
Toss a cube with sides numbered 1–6.

1. $P(1)$ $\frac{1}{6}$
2. P(even number)
3. $P(2)$
4. P(not 2)
5. P(odd number)
6. P(prime number)
7. P(less than 2)
8. P(greater than 3)
9. P(1 or 3)
10. P(1 and 3)

Set 81 Give each probability.

Event:
Toss a cube with sides numbered 1–6. Toss the cube again.

1. P(3 and then 4) $\frac{1}{36}$
2. P(even number and then even number)
3. P(1 and then odd number)
4. P(prime number and then composite number)
5. P(2 and then not 3)
6. P(not prime number and then not prime number)

Set 82 Give the odds.

Event:
Spin this spinner.

1. Odds in favor of 6
2. Odds against 6
3. Odds in favor of 5
4. Odds against 5
5. Odds in favor of even number
6. Odds against even number
7. Odds in favor of a number less than 2
8. Odds in favor of a prime number

Event:
Pick one marble.

9. Odds in favor of green.
10. Odds in favor of blue.
11. Odds in favor of yellow.
12. Odds against blue.
13. Odds against red.
14. Odds against green.

Set 83 Give the range, mean, median, and mode.

1. 3, 5, 8, 8, 10, 11, 14
2. 5, 6, 9, 9, 11, 12
3. 32, 38, 45, 49, 50
4. 26, 32, 34, 41, 42, 42, 46
5. 68, 69, 70, 71, 72
6. 39, 43, 56, 57, 65
7. 45, 45, 45, 45, 45, 45
8. 83, 86, 86, 89, 90, 91

Set 84 The two figures in each exercise are similar. Find the missing length.

1.

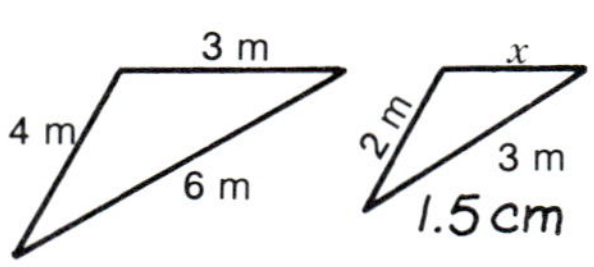

2.

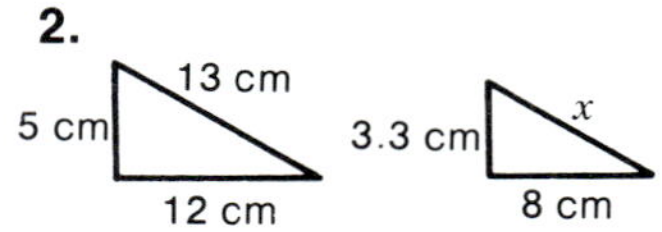

3.

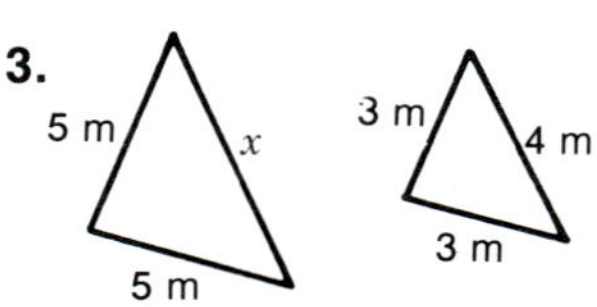

4.

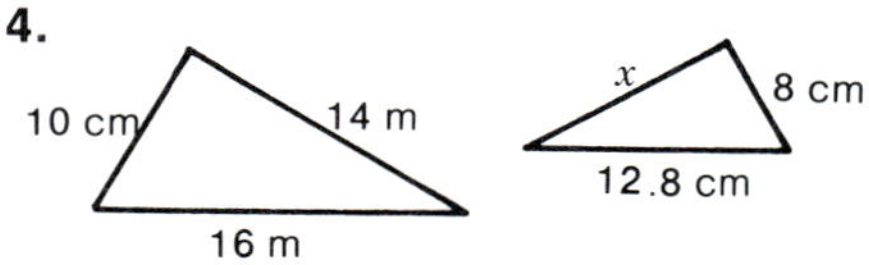

5.

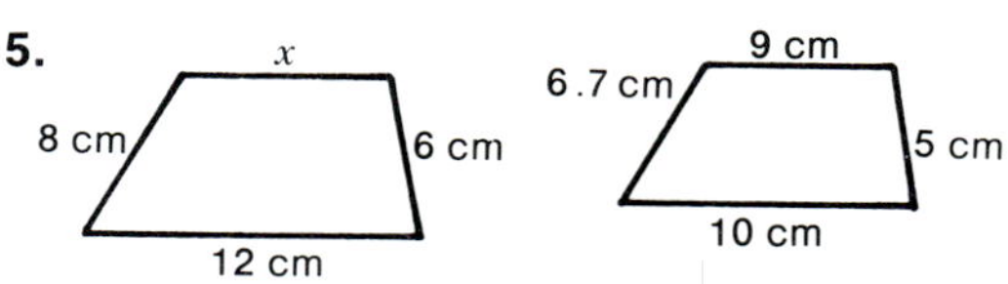

Set 85 **Find the missing lengths to the nearest 0.1 meter. Use the table of tangents.**

Tangent Ratios
tan 20° ≈ 0.36
tan 30° ≈ 0.58
tan 40° ≈ 0.84
tan 50° ≈ 1.19
tan 60° ≈ 1.73
tan 70° ≈ 2.75

1.

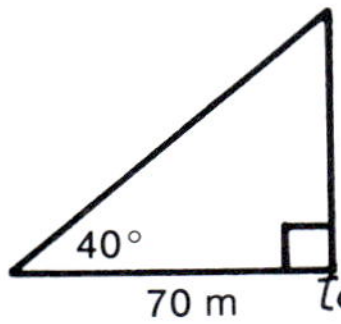

$\tan 40° = \frac{x}{70} \approx 0.84$

$x \approx 58.8$

2.

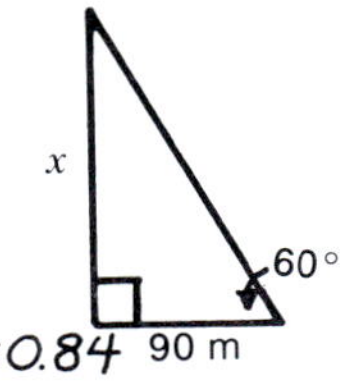

3.

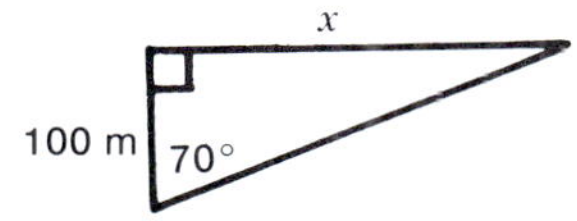

4.

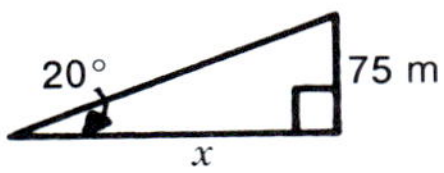

5.

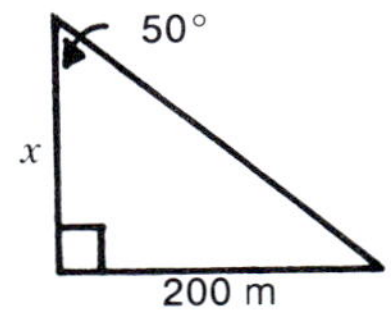

6.

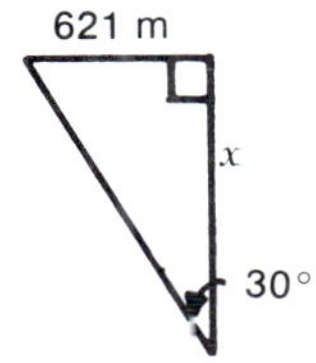

Set 86 **Simplify.**

1. $\sqrt{25}$ 5 **2.** $\sqrt{441}$ **3.** $\sqrt{100}$ **4.** $\sqrt{900}$ **5.** $\sqrt{784}$ **6.** $\sqrt{1600}$

7. $\sqrt{1.96}$ **8.** $\sqrt{5.29}$ **9.** $\sqrt{1.44}$ **10.** $\sqrt{10.89}$ **11.** $\sqrt{16.81}$ **12.** $\sqrt{0.49}$

Use the divide-and-average method.
Give an approximation to the nearest tenth.

13. $\sqrt{50}$ **14.** $\sqrt{60}$ **15.** $\sqrt{85}$ **16.** $\sqrt{90}$ **17.** $\sqrt{110}$ **18.** $\sqrt{150}$

19. $\sqrt{312}$ **20.** $\sqrt{456}$ **21.** $\sqrt{512}$ **22.** $\sqrt{675}$ **23.** $\sqrt{860}$ **24.** $\sqrt{958}$

Set 87 **Find the missing lengths to the nearest 0.1 meter.**

1.

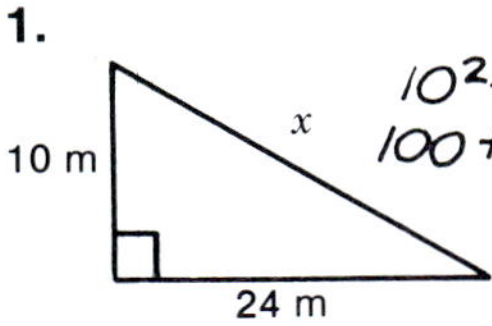

$10^2 + 24^2 = x^2$

$100 + 576 = x^2$

$676 = x^2$

$\sqrt{676} = x$

$26 = x$

2.

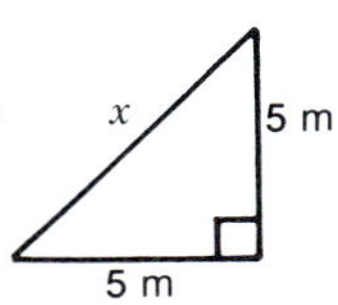

3.

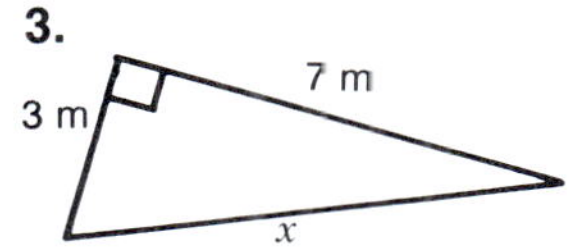

4.

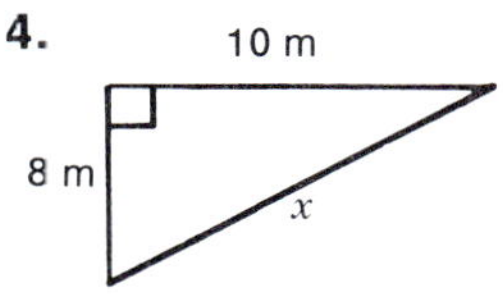

5.

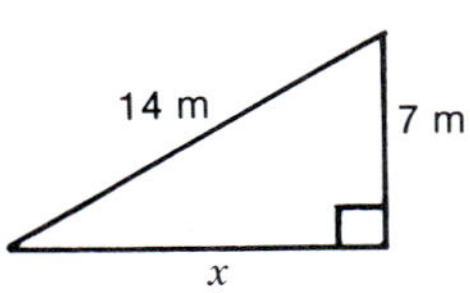

6.

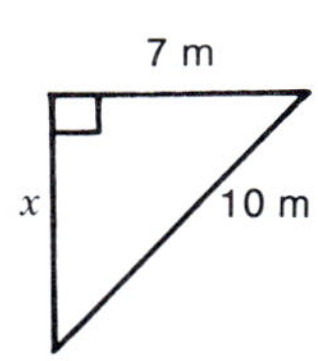

7.

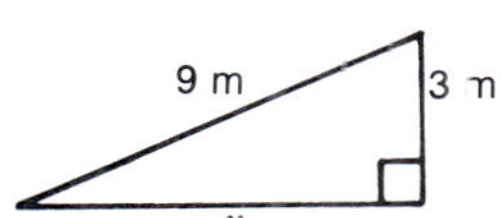

8.

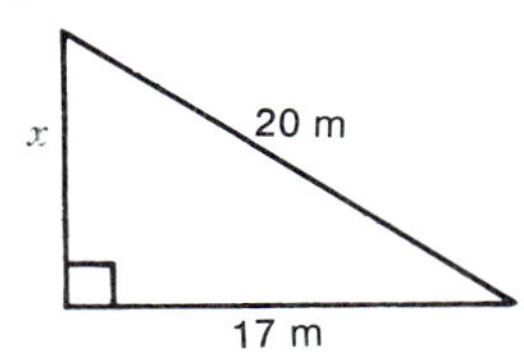

EXTRA PROBLEM SOLVING

Airplanes that fly beyond the speed of sound usually have their speeds given in *Mach* numbers. The Mach number is the quotient:

Speed of plane ÷ Speed of sound at plane's altitude

This means that the Mach number tells how many times the speed of sound a plane is flying.

PROBLEM SOLVING Set 1

1. Suppose that a plane flies at a speed of 2000 km per hour at an altitude where the speed of sound is 1000 km per hour. What is the Mach number?

2. Suppose that a plane is flying at a speed of Mach 1.5 at an altitude where the speed of sound is 1140 km per hour. How fast is the plane flying?

3. The speed of sound near sea level is 1190 km per hour. How many kilometers per hour is a plane near sea level flying if its speed is Mach 2.1?

4. At 14,000 m the speed of sound is 1060 km per hour. What is the speed in kilometers per hour of a plane at 14,000 m if its speed is Mach 1.3?

PROBLEM SOLVING Set 2

1. The first supersonic flight was by Charles Yeager on October 14, 1947. He flew a Bell XS-1 rocket plane at Mach 1.015 at an altitude of 12,800 m. This speed is 1079 kilometers per hour. What is the speed of sound at 12,800 m?

2. In 1962 the Russians held the jet speed record at 2680.99 km per hour. In 1976 a U.S. Air Force jet raised the record to 3529.56 km per hour. How much faster was that?

3. The Wright Brothers made their first flight on December 17, 1903. The longest flight, by Wilbur Wright, was 260 m long and lasted 59 seconds. What was the speed in kilometers per hour?

4. Speeds greater than Mach 5 are called *hypersonic*. On June 27, 1962, J. A. Walker flew an X-15 rocket plane at a speed of Mach 6.06, or 6696 km per hour. The plane was designed to fly at Mach 8 at that altitude. What is that speed in kilometers per hour?

PROBLEM SOLVING Set 3

The altitude record for a jet plane is 37.65 km.

1. The altitude record for a rocket plane was set in 1962 by Robert White of the U.S. His altitude was about 2.55 times the altitude record for a jet plane (set in 1977). What is the altitude record for a rocket plane?

2. The altitude record for a glider was set in 1961 by Paul Bikle of the U.S. His altitude was 4723 m less than half of the altitude record for a jet plane. What is the altitude record for a glider?

3. The altitude record for a jet plane is about 1.09 times the altitude record for a balloon. What is the altitude record for a balloon?

4. The altitude record for a turboprop plane is 15,549 meters. How many kilometers is that?

PROBLEM SOLVING Set 4

The distance record for a jet plane is 20,168.78 km.

1. The distance record for a piston-engine plane is 2086.79 km less than the distance record for a jet plane (set in 1962). What is the distance record for a piston-engine plane?

2. The distance record for a glider was set in 1972 by Hans Grosse of West Germany. The distance record for a jet plane is about 14 times this distance. How far did Grosse sail his glider?

3. The distance record for a helicopter is 3561.546 km. How much more is the distance record for a jet plane?

4. Charles A. Lindbergh was the first to fly alone across the Atlantic Ocean. He flew 5810 km in 33.5 hours. What was his average speed in km per hour?

PROBLEM SOLVING Set 5

1. An off-road vehicle has a tank that holds 22.2 gallons of gas. The vehicle can average 14.4 miles per gallon on the highway. How far can it travel on a tank of gas?

2. An off-road motorcycle has a 3.3-gallon gas tank. It can travel 90 miles on a full tank. How many miles per gallon does it get?

3. A motorbike can go 319 miles on a tank of gas. This distance is 29 miles farther than twice the distance it can go on a gallon of gas. How far can it go on a gallon of gas?

4. Carol drove 825 miles. Her car averaged 27.2 miles per gallon. Gas cost her \$1.429 per gallon. How much did she spend on gas for her trip?

5. Calvin filled the tank of his car with 17.2 gallons of gas. He drove 460.3 miles and then refilled the tank with 16.4 gallons. How many miles per gallon did his car average?

PROBLEM SOLVING Set 6

1. A Japanese 4-wheel-drive vehicle weighed 1701 pounds, 932 pounds of which was on the front axle. How many pounds of weight were on the rear axle?

2. The length of a Japanese car is 152 inches. The width is 60 inches. Its width is what fraction of its length?

3. A Japanese car can turn in a circle with a 33-foot diameter. What distance does the car travel in that circle?

4. Jim had tires with 30-inch diameters on his truck. How far would the truck move as the tires revolved once?

5. How many revolutions of the tires would there be in a mile of driving? (See exercise 4.)

PROBLEM SOLVING Set 7

1. Ken ordered a fiberglass top for his 1980 jeep for $700. There was a $28 sales tax, a $75 crating charge, and a $37.50 shipping charge. How much did the top cost in all?

2. Terry ordered a 3-inch roll bar for $199.95 and a power winch for $229.95. He had to pay a $23.64 sales tax and a $25 shipping charge. How much did his order cost him?

3. Dan drove his pickup 263.2 miles on 17.2 gallons of gas. Then he added a device that was supposed to increase his mileage (miles per gallon). He drove 299.9 miles on 16.3 gallons of gas. What was the increase in mileage?

4. An auto part was made from $\frac{3}{40}$-inch-thick steel. This was $\frac{1}{40}$ inch thicker than the steel usually used. What was the usual thickness of the steel?

5. Jed ordered special oversized wheels and tires for his truck. Each wheel and tire cost $189.95. How much did a set of 5 wheels and tires cost him?

PROBLEM SOLVING Set 8

1. The winner of the 1979 Baja International covered the 403 miles in 7 hours and 39 minutes. What was the average speed in miles per hour?

2. The entry fee for a 125-mile off-road race was $150. Then $\frac{7}{10}$ of the entry fees was returned to the drivers as prize money. What was the prize money if 38 drivers entered the race?

3. In the Indianapolis 500 a driver drove one lap of the 2.5-mile track at an average speed of 184.56 miles per hour. How many seconds did it take to drive that lap?

4. A land speed of 608 miles per hour was attained by the driver of a four-wheeled vehicle in 1965. The course was a mile long. About how many seconds did it take the driver to pass through the course?

5. In a race, the winner was 6 seconds ahead of the second-place finisher. If the vehicles were traveling at 60 mph at the finish, how many feet was the winner ahead of the second-place finisher?

PROBLEM SOLVING Set 9

Technology has produced many benefits.

1. Living standards are higher because of technology—and people live longer. Today people can expect to live to the age of 72 years. In 1900 few people lived past $\frac{2}{3}$ of that. What is $\frac{2}{3}$ of 72 years?

2. In 1850 a stagecoach was fast ground transportation—9 miles per hour. Today an electric rail line in Japan has trains that travel about $14\frac{1}{2}$ times as fast. How fast do the trains travel?

3. Today, Hovercraft in England travel on water at 75 miles per hour. This is $3\frac{3}{4}$ times the best speeds of the clipper ships that sailed the seas in 1850. How fast could the clipper ships sail?

4. Of course, there was no air travel at all in 1850. Today a Boeing 747 can fly 600 miles per hour and the supersonic *Concorde* can fly $2\frac{11}{12}$ times as fast. How fast is the *Concorde*?

PROBLEM SOLVING Set 10

Technology also has some bad side effects.

1. In the mid-1970's, 3 billion tons of solid waste were produced in the U.S. in a year. If the population was 210 million, what was the average number of pounds of solid waste produced per person?

2. A little over $\frac{3}{5}$ of the solid waste was mineral (mine waste, etc.). How much mineral waste was there in a year? (See exercise 1.)

3. 150 million tons of the solid waste was municipal waste (bottles, papers, etc.). What fraction of the solid waste was municipal waste? (See exercise 1.)

4. Technology has increased the use of fuels that cannot be replaced. 100,000 kilowatt hours of energy are generated in the U.S. per person per year. What is the total amount of energy generated in the U.S. in a year? (See exercise 1.)

PROBLEM SOLVING Set 11

1. At the Athlete's Foot Shop, jogging shoes sell for \$34.95 a pair. Soccer shoes cost $1\frac{1}{3}$ times as much. How much do soccer shoes cost?

2. A pair of tennis shoes costs $\frac{3}{4}$ of what the soccer shoes cost. How much do the tennis shoes cost? (See exercise 1.)

3. A sporting goods store was having a sale. The most expensive tennis racket regularly cost \$225 plus \$26.50 for stringing. The racket was on sale for \$195 including stringing. How much is saved by buying the racket at the sale?

4. Hal bought 2 pairs of sweat socks for \$2.95 each plus a sales tax of $\frac{1}{20}$ of the total. How much change did he get from a \$10 bill?

PROBLEM SOLVING Set 12

1. This ad appeared in a newspaper:

Cross-country Ski Package —only \$87	
cross-country skis	\$69.95 value
boots	\$19.95 value
ski poles	\$8.75 value

How much is saved by buying the package?

2. Jim bought a racket-ball racket for \$24.98 and a can of balls for \$1.98. He also paid a \$1.48 sales tax. How much change did he get from 2 \$20 bills?

3. Hockey skates were priced at \$45.50 on sale. They had been reduced to $\frac{7}{10}$ of the regular price. What was the regular price?

4. A sporting goods store received a shipment of skis. The ratio of the wholesale price to the retail price was the same for each pair of skis. One pair of skis was marked up from a wholesale price of \$90 to a retail price of \$210. What was the retail price on skis that cost \$105 wholesale?

5. A pair of roller skates was marked up \$60 from wholesale to retail and then marked down \$13.50 on sale. The sale price was \$121.50. What was the wholesale price?

PROBLEM SOLVING Set 13

1. A Senegal parrot was priced at \$200 and a male canary (guaranteed to sing) was priced at \$44.99. How much cheaper was the canary?

2. Suppose that the two birds were each reduced $\frac{1}{4}$ in price. How much cheaper than the parrot would the canary be? (See exercise 1.)

3. For someone who wanted an unusual pet, a pet shop had a Mexican red-leg tarantula. It cost $\frac{1}{8}$ as much as the Senegal parrot. How much did the tarantula cost? (See exercise 1.)

4. A 3-foot-long boa constrictor cost \$99.99, while a baby boa cost $\frac{3}{5}$ as much. What was the price of the baby boa?

5. A Florida king snake cost 1¢ less than $\frac{1}{2}$ of what the large boa cost. How much did the king snake cost? (See exercise 4.)

PROBLEM SOLVING Set 14

To get enough oxygen, fish need some water surface exposed to the air.

Kind of fish	*Minimum water surface needed*
swordtail	8 in.^{2}
barb	20 in.^{2}
angelfish	25 in.^{2}

1. How many swordtails can live in an aquarium with a 10-inch by 20-inch water surface?

2. How many barbs can live in an aquarium with a 10-inch by 20-inch water surface?

3. How much water surface is needed in an aquarium to keep 2 angelfish, 10 swordtails, and 6 barbs?

4. Bill has an aquarium with a 20-inch by 40-inch water surface. He wants to keep an equal number of swordtails and barbs in it. How many of each kind should he get?

PROBLEM SOLVING Set 15

1. A 20-gallon aquarium with all accessories was priced at $48.95. During a sale it was 20% off. What was the sale price?

2. Fancy fantail red guppies cost $2.99 a pair. How many guppies could you get with the money you would save by buying the aquarium on sale? (See exercise 1.)

3. Jim bought an aquarium regularly priced at $35.00. It was on sale for 25% off. There was a 5% sales tax in his state. How much did he pay in all?

4. Carol bought these fish for her aquarium:

1 gold-veil angel	$14.99
2 red-tail sharks	$3.99 each
6 green swordtails	$.79 each

She had to pay a 5% sales tax too. How much change did she get from 2 $20 bills?

PROBLEM SOLVING Set 16

1. A dachshund was reduced in price from $250 to $149. What was the percent of reduction?

2. A miniature schnauzer was reduced $101 to a sale price of $199. What was the percent of reduction?

3. Did the dachshund or the schnauzer have the greater price reduction? Which had the greater percent of reduction? (See exercises 1 and 2.)

4. A medium-sized dog eats $1\frac{1}{2}$ cups (1 c = 8 oz) of a dog food each day. The dog food is sold in 20-pound bags. How long would a bag of the food last a medium-sized dog?

5. A collie was priced at $280 before a 30% reduction. What was the sale price?

6. A German shepherd was priced at $240 after a $33\frac{1}{3}$% reduction. What was the original price?

PROBLEM SOLVING Set 17

1. In a men's downhill ski race the maximum vertical drop of the course is 1000 m. For women the maximum is 30% less than for men. What is the maximum vertical drop in a downhill race for women?

2. The highway speed limit for cars is 88 km per hour. Downhill racers have exceeded that by 16%. What speed is attained by downhill racers?

3. The maximum vertical drop in a men's slalom race is 122% of the maximum vertical drop in a women's slalom race. The drop for women is 180 m. What is it for men?

4. The maximum vertical drop in a women's giant slalom is 10% less than that for men, which is 500 m. What is the maximum drop for women?

5. There are 3 standard cross-country ski races for men. One is 30 km, one is 50% shorter, and one is $233\frac{1}{3}\%$ longer than the shortest race. What are the lengths of the 3 races?

PROBLEM SOLVING Set 18

1. Jenny took beginning downhill ski lessons using the Graduated Length Method. The first skis that she used were 0.9 m long. When she had completed her lessons, she was using skis that were 89% longer. How long were those last skis?

2. Jenny bought this equipment:

skis	\$125
bindings	\$48
boots	\$95
poles	\$15

How much did she pay in all if there was a 5.5% sales tax on the equipment?

3. Jenny could pay \$15 per hour for ski lessons. She could also buy a booklet of tickets for 8 one-hour lessons for \$100. What was the percent of savings on a lesson by using the tickets?

4. John bought cross-country ski equipment for \$89 and used it an average of 30 days per year for 2 years. He skied for a couple of hours each time in a park 3 blocks from his home. What was the average daily cost of his skiing?

PROBLEM SOLVING Set 19

1. Carol could buy skis, boots, bindings, and ski poles for $227, or she could rent the equipment for just under 4% of that. How many times would Carol have to use her ski equipment to make it cheaper to buy than to rent?

2. Carol bought the ski equipment and used it for 2 years. She averaged 20 days of skiing each year. A lift ticket cost her $14 each day, food cost her $3.75 each day, and round-trip transportation cost her $5 each day. What was the average daily expense related to skiing? (See exercise 1.)

3. Kevin wants to buy cross-country ski equipment that costs $129. He can buy it on the installment plan for $30 down and $22 per month for 5 months. How much extra does the installment plan cost?

4. Marna works in a ski shop and gets a 15% discount. How much would she pay for cross-country ski equipment that is regularly priced at $145.95?

PROBLEM SOLVING Set 20

1. Experts can ski cross-country at a rate of 16 km per hour. How long would it take an expert to ski 70 km?

2. The fastest that long-distance runners can run a marathon is about 42.2 km in 2 hours 10 minutes. How much faster is this running rate than the skiing rate? (See exercise 1.)

3. Years ago John Thompson, carrying mail, skied between two cities 140 km apart. Without mail, he could ski at an average rate of 9 km per hour. When he carried 45 kg of mail, his rate was reduced by 40%. How long did one trip take when Thompson was loaded with mail?

4. In a biathlon an athlete skis 20 km. He stops to shoot 5 times at each of 4 targets. He is penalized 2 minutes for each shot that he misses. What is the time of an athlete who skis the course in 83 minutes but misses 3 shots?

5. A man raced in a biathlon. His penalty time for missing shots was 5.4% of his skiing time. His total time was 78 minutes. How many shots did he miss? (See exercise 4.)

PROBLEM SOLVING Set 21

1. In a recent year 81.8 million kilograms of freshwater fish were caught by U.S. commercial fishermen. This was only 3% of the total fish caught commercially. How many kilograms of saltwater fish were caught that year?

2. In the same year 240.9 million kilograms of fish were caught in waters in and around New England. What percent of the total catch was caught in New England? (See exercise 1.)

3. The total fish catch for the world in 1972 was 65.5 billion kilograms. This was a 150% increase from 1945. What was the catch in 1945?

4. With proper management it is predicted that the total catch can be increased to 350% of the 1972 catch. What will be the total catch if this does occur? (See exercise 3.)

5. For 10 years the total yearly U.S. catch was 23 billion kg. But its share of the total world's catch dropped from 13% to 4%. Explain how this could happen.

PROBLEM SOLVING Set 22

1. Jerry bought a spin-casting rod and reel for $38.95, a tackle box for $4.49, and 6 lures for $1.89 each. There was a 5.5% sales tax on the fishing equipment. How much did he pay in all?

2. One day Jerry caught 3 walleye weighing 1.7 kg, 2.2 kg, and 0.9 kg. His 3 fish together were 58% lighter than the record walleye. How much did the record walleye weigh?

3. Jerry spent 18% of his savings for a boat and 12% of his savings for an outboard motor. This left him with $1200 in savings. How much did the boat cost?

4. After a year Jerry bought a cottage on his favorite fishing lake for $18,000. He paid 20% down and obtained a 9.5% per year loan for the rest. How much interest did he pay in the first month of his loan?

GLOSSARY

absolute value The number of units a digit is from zero on a number line.

acute angle An angle that measures between 0° and 90°.

addend A number used in an addition problem.

adding 0 property The sum of any number and 0 is the number.

adjacent angles Two angles that have a common side between them and a common vertex.

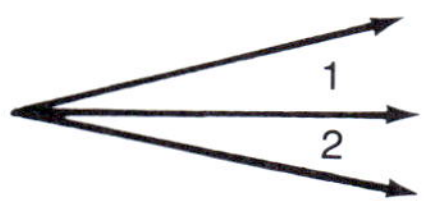

∠1 and ∠2 are adjacent angles.

angle A figure formed by two rays with the same endpoint.

arc Part of a circle.

area The number of unit squares that it takes to cover a region.

associative property of addition Changing the grouping of the addends does not change the sum.

associative property of multiplication Changing the grouping of the factors does not change the product.

average The sum of all the numbers divided by the number of numbers.

axes Two perpendicular lines used as a reference for graphing ordered pairs.

second coordinate axis

first coordinate axis

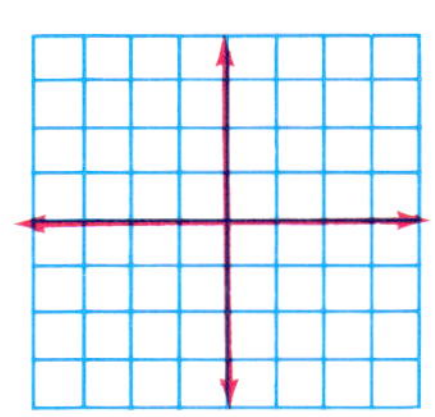

The horizontal line is the first coordinate axis and the vertical line is the second coordinate axis.

base (of an exponent) The number that is raised to a power.

base → 2^3 ← power

BASIC A language that a computer understands. It consists of specific symbols such as + and specific words such as PRINT.

basic counting principle If a first event has m outcomes and a second event has n outcomes, then the first event followed by the second event has $m \times n$ outcomes.

bisect To cut into halves.

This segment is bisected.

This angle is bisected.

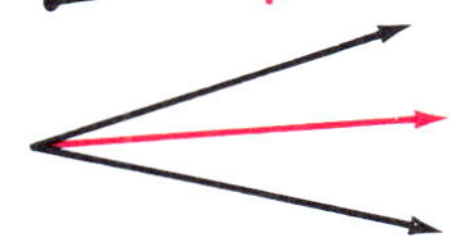

budget A plan for using your money.

canceling Dividing both the numerator and the denominator by a common factor before multiplying fractions.

$$\frac{\overset{4}{\cancel{8}}}{15} \times \frac{1}{\underset{3}{\cancel{6}}}$$

Celsius temperature scale (°C) The metric temperature scale, in which 0°C is the freezing point of water and 100°C is the boiling point of water.

centimeter A metric unit of length.
1 centimeter = 0.01 meter.

circle A curved plane figure with all points a given distance from the center.

circumference The distance around a circle.

command An instruction for the computer to follow.

commission The part of the total sales that goes to the salesperson.

common denominator A common denominator for $\frac{1}{2}$ and $\frac{1}{3}$ is 6, because $\frac{1}{2} = \frac{3}{6}$ and $\frac{1}{3} = \frac{2}{6}$. A common denominator is a common multiple of the denominators of two fractions.

common factor 2 is a common factor of 4 and 6, because 2 is a factor of both 4 and 6.

common multiple 30 is a common multiple of 5 and 6, because it is a multiple of both 5 and 6.

commutative property of addition Changing the order of the addends does not change the sum.

$$23 + 89 = 89 + 23$$

commutative property of multiplication Changing the order of the factors does not change the product.

$$19 \times 54 = 54 \times 19$$

compass An instrument used to draw circles.

complex fraction A fraction in which either the numerator or the denominator or both are fractions.

$$\frac{\frac{6}{7}}{\frac{2}{3}} \quad \frac{8}{\frac{3}{4}} \quad \frac{\frac{5}{2}}{6}$$

composite number A whole number other than 0 that has more than two factors.

compound interest Interest that is added to the principal at regular intervals. This makes the principal grow and earn more and more interest.

computer A machine that can count, store, and process information and give out the results.

computer program A detailed set of instructions given to a computer.

cone A three-dimensional figure with one flat surface (known as the base) that is a circle and with one other face that is curved.

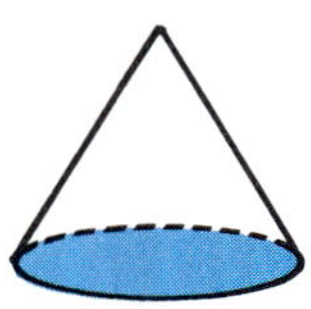

congruent figures Figures that have the same size and shape.

construct Draw a figure without measuring. Use only a compass and a straightedge.

conversion fraction A fraction equivalent to 1 that is multiplied by a measurement to convert the measurement from one unit to another unit.

$$2 \text{ ft} \times \frac{12 \text{ in.}}{1 \text{ ft}} = 24 \text{ in.}$$

coordinates An ordered pair of numbers that locates a point on a grid.

corresponding parts In congruent figures, the congruent angles and the congruent sides are corresponding parts. In similar figures, the congruent angles and the proportional sides are corresponding parts.

cube A rectangular prism whose six faces are squares.

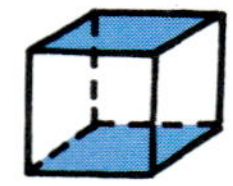

cube number The third power of a whole number.

$$64 = 4^3$$

customary system The system of measurement that uses foot, quart, pound, and Fahrenheit temperature.

cylinder A solid figure that has two circular bases that are the same size and are in parallel planes. It has one curved face.

data Numerical pieces of information.

decagon A polygon with ten sides.

decimal or decimal fraction A number written in our place-value system with a decimal point before the tenths place.

decimeter A metric unit of length.
1 decimeter = 0.1 meter.

degree A unit for measuring angles. This is a 1° (1-degree) angle:

dekameter A metric unit of length.
1 dekameter = 10 meters.

denominator In the fraction $\frac{2}{3}$, the denominator is 3.

dependent events Two events such that the outcome of the first affects the outcome of the second—for instance, drawing a first and a second card without replacing the first card.

diameter The distance across a circle through its center. The length of the diameter is twice the length of the radius.

difference The answer to a subtraction problem.

digits The basic symbols used to write numerals in a place-value system. In our base-ten system the digits are 0, 1, 2, 3, 4, 5, 6, 7, 8, and 9.

discount An amount subtracted from the regular price of an item.

disk A round magnetic object used to store computer information.

distributive property of multiplication A product can be written as the sum of two products.

$$3 \times (10 + 2) = (3 \times 10) + (3 \times 2)$$

divide-and-average method A method for finding square roots that uses division and averaging.

dividend The number that is divided.

$$3\overline{)18}^{\,6} \qquad 18 \div 3 = 6$$

dividend

divisible One number is divisible by another number if there is no remainder after division. 84 is divisible by 2, since 84 ÷ 2 leaves no remainder.

divisor The number that one divides by.

$$4\overline{)36}^{\,9} \qquad 36 \div 4 = 9$$

divisor

down payment The first amount paid when buying on an installment plan.

equally likely outcomes Outcomes such that each has the same chance of occurring.

equation A sentence with an equal sign, such as $3 \times 9 = 27$.

equilateral triangle A triangle with 3 lines of symmetry and 3 congruent sides.

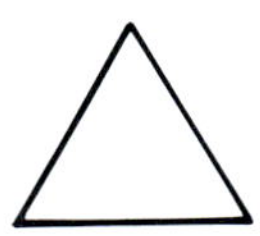

equivalent fractions Fractions that name the same number. $\frac{1}{2}$, $\frac{2}{4}$, and $\frac{3}{6}$ are equivalent fractions.

error of measurement The difference between the true measurement and the recorded measurement.

estimate To use rounded numbers to check whether an answer is correct. To estimate 47 + 32, you would add 50 + 30. The sum should be about 80.

even number Zero and the multiples of 2.

expanded numeral A numeral used to show the place value of each digit in a standard numeral.

$$248 = 200 + 40 + 8$$

exponent An exponent tells how many times a number is used as a factor.

exponent

$$2^3 = \underbrace{2 \times 2 \times 2}_{3 \text{ factors}}$$

factors Numbers used in a multiplication problem.

$$\begin{array}{r l} 8 & \leftarrow \text{factor} \\ \times 6 & \leftarrow \text{factor} \\ \hline 48 & \leftarrow \text{product} \end{array}$$

Fahrenheit temperature scale (°F) The customary temperature scale, in which 32°F is the freezing point of water and 212°F is the boiling point of water.

finance charge Buying an item on an installment plan costs more than paying cash. The difference is called the finance charge.

flow chart A general plan written before writing a program that gives step-by-step directions.

formula A general way of expressing a relationship using symbols.

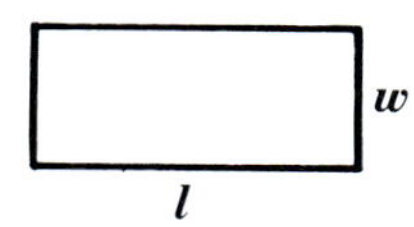

$$A = l \times w$$

fraction A numeral for part of a group or for part of a region. $\frac{1}{2}$, $\frac{4}{6}$, and $\frac{6}{5}$ are fractions.

frequency distribution A table showing the number of times different events occur.

function A set of number pairs of which no two first numbers are the same.

Function rule
$n \rightarrow n + 3$
(0, 3)
(1, 4)
(2, 5)

gram A metric unit of weight (mass). 1 gram = 0.001 kilogram.

graph A picture used to show numerical information. A graph can be a bar graph, a picture graph, a circle graph, or a line graph.

greatest common factor (GCF) The greatest number that is a factor of each of two or more numbers.

4 is the GCF of 8 and 12.

greatest possible error Half of the unit used in making the measurement.

hectometer A metric unit of length.

1 hectometer = 100 meters.

heptagon A polygon with seven sides.

hexagon A polygon with six sides.

histogram A special kind of bar graph that is used to show frequency.

hypotenuse The side of a right triangle that is opposite the right angle. It is the longest side of a right triangle.

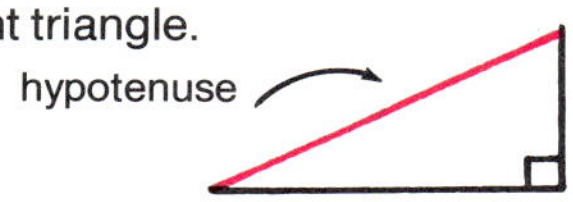

independent events Events such that the outcome of the first does not affect the outcome of the second.

Toss a coin.

Roll a die.

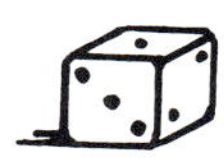

indirect measurement A measurement that is computed from other measurements rather than measured directly.

inequality A mathematical sentence that contains a symbol such as $\neq$, $<$, $>$, $\geq$, or $\leq$.

$$3 + 6 < 11 \qquad 2r + 5 \geq 6$$

input Information entered into the computer.

installment buying A way of buying expensive items. You pay part of the cost (the down payment) when you get the item and then agree to pay a certain amount each month for a certain number of months.

integers The numbers ..., $^-3$, $^-2$, $^-1$, 0, $^+1$, $^+2$, $^+3$,....

interest The amount a borrower pays for using the money.

intersecting lines Lines that meet at only one point.

intersecting segments (or rays) Segments (or rays) that are parts of intersecting lines.

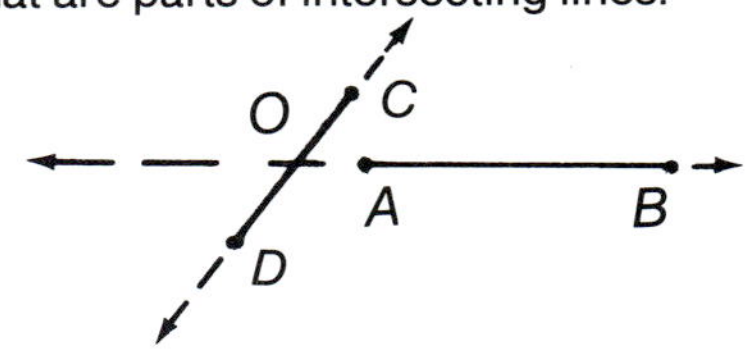

$\overline{AB}$ intersects $\overline{CD}$ at point O.

isosceles trapezoid A trapezoid with 1 line of symmetry through opposite sides and 2 congruent sides.

isosceles triangle A triangle with 1 line of symmetry and 2 congruent sides.

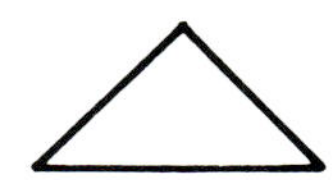

kilogram A metric unit of weight.

1 kilogram = 1000 grams.

kilometer A metric unit of length.

1 kilometer = 1000 meters.

kite A quadrilateral with a line of symmetry through opposite vertices and 2 pairs of congruent sides.

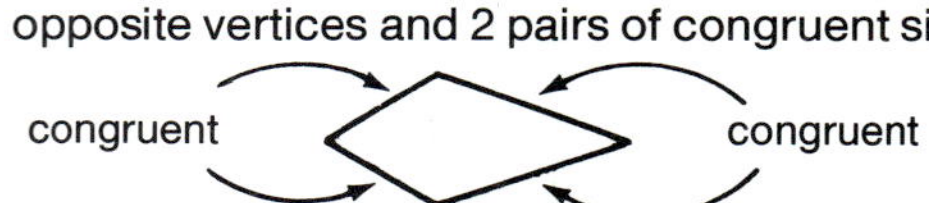

least common denominator The least common multiple of the denominators of two or more fractions.

least common multiple The least (smallest) common multiple of two or more numbers.

The least common multiple of 6 and 15 is 30.

leg of a right triangle Either of the two shorter sides of a right triangle.

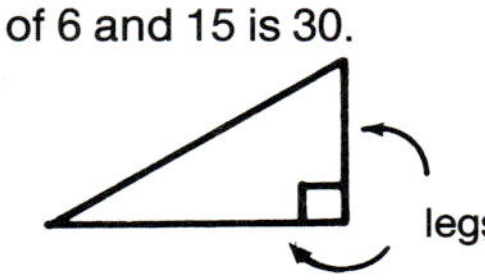

line of symmetry If a figure can be folded along a line so the two halves of the figure match, the fold line is a line of symmetry.

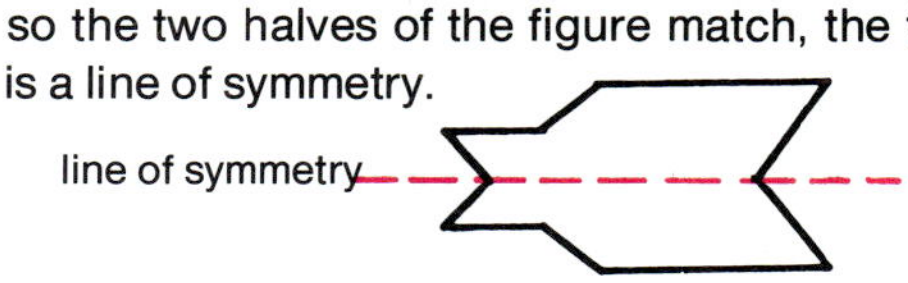

liter A metric unit of volume.

loop A set of commands that describes an action that is repeated.

lowest terms A fraction is in lowest terms if the greatest common factor of the numerator and the denominator is 1.

mean The average of all the numbers.

median If there is an odd number of numbers, the median is the number in the middle. If there is an even number of numbers, the median is the average of the two numbers in the middle.

meter A metric unit of length.
1 meter = 100 centimeters.

metric system An international system of measurement that uses meter, liter, gram, and Celsius temperature.

midpoint The point that bisects a segment.

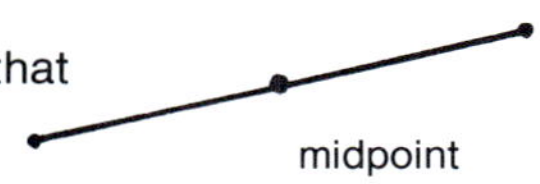

milligram A metric unit of weight (mass).
1 milligram = 0.001 gram.

milliliter A metric unit of volume.
1 milliliter = 0.001 liter.

millimeter A metric unit of length.
1 millimeter = 0.001 meter.

mixed decimal A number that has a decimal part and a fraction part. $0.25\frac{1}{2}$ is a mixed decimal.

mixed number A number that has a whole-number part and a fraction part. $2\frac{3}{4}$ is a mixed number.

mode The number that occurs most often.

multiple A product. 4, 8, 12, 16, 20, and so on, are multiples of 4.

multiplying by 1 property The product of any number and 1 is the number.

$$8 \times 1 = 8$$

negative number A number that is less than 0.

nonagon A polygon with nine sides.

number line A line with its points labeled with numbers.

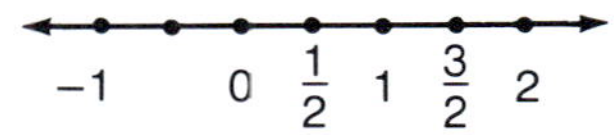

numeral A name or symbol for a number.
Numerals for ten: 10, 2×5, $12 - 2$, X

numerator In the fraction $\frac{2}{3}$, the numerator is 2.

obtuse angle An angle that measures between 90° and 180°

octagon A polygon with eight sides.

odd number A whole number that is not even. The numbers 1, 3, 5, 7, 9, 11, and so on, are odd.

odds The ratio of the number of ways that an outcome can occur to the number of ways that the outcome cannot occur.

opposites Two numbers are opposites if their sum is 0.

$$^-3 + {}^+3 = 0$$

↑ ↑
opposites

ordered pair A pair of numbers that give the coordinates of a point on a grid.

origin The point where axes intersect.

outcome A possible result.

output The results of a computer program.

parallel lines Lines in a plane that do not intersect.

parallelogram A quadrilateral with a point of symmetry and opposite sides parallel.

pentagon A polygon with five sides.

percent (%) *Percent* means "per hundred." 5% (5 percent) equals $\frac{5}{100}$.

perimeter The distance around a figure. The sum of the lengths of the sides.

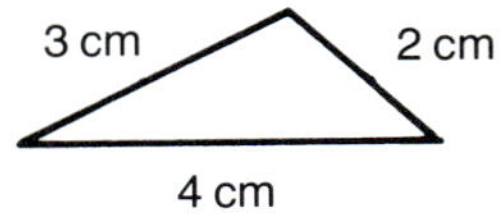

The perimeter is 9 cm.

perpendicular bisector The perpendicular bisector of a segment is the line that is perpendicular to the segment at the midpoint of the segment.

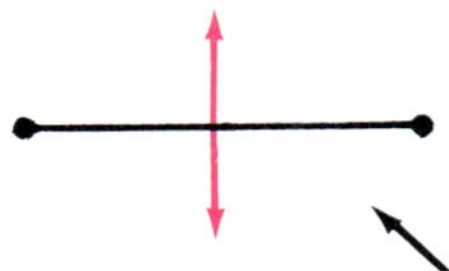

perpendicular lines Two lines that intersect to form right angles.

pi The number that is the ratio of the circumference of a circle to its diameter. It is represented by the Greek letter π and is approximately equal to 3.14.

place value A system for writing numbers in which the value of a digit is determined by its position.

plane A flat surface that extends endlessly in all directions.

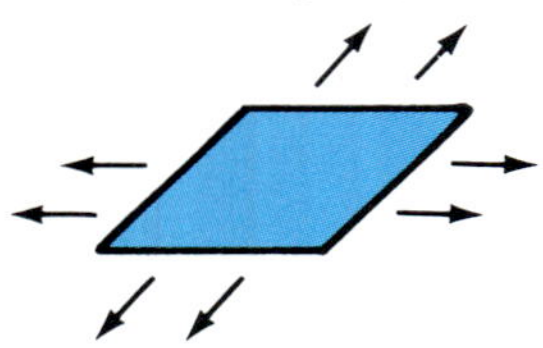

point of symmetry If a figure fits itself after a half-turn about a point, the point is a point of symmetry of the figure.

point of symmetry

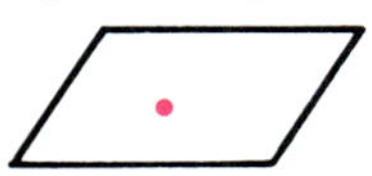

polygon A plane figure made up of segments.

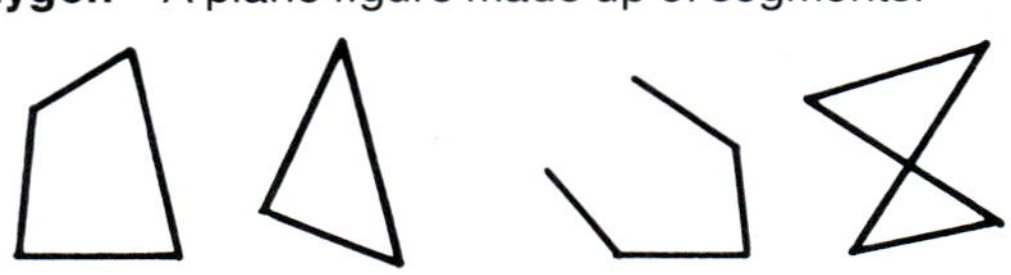

polygons not polygons

positive number A number greater than 0.

power An exponent. The number that tells how many times a number is used as a factor.

2^3 ← power

prime factorization Expression of a composite number as a product of prime numbers. The prime factorization of 18 is $2 \times 3 \times 3$.

prime number A whole number that has exactly two factors. 2, 3, 5, 7, 11, 13, and so on, are prime numbers.

principal An amount of money borrowed or deposited in an account.

prism A solid that has two bases that are the same size and shape and are in parallel planes. The other faces are all rectangles.

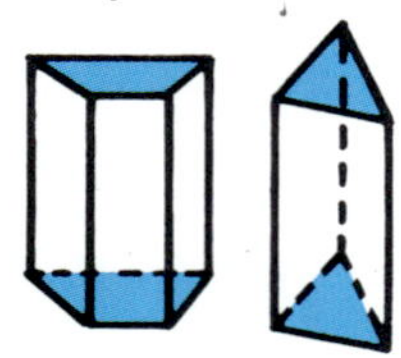

probability The ratio of the number of favorable outcomes to the total number of outcomes.

probability of picking black $= \frac{3}{5}$

procedure A list of computer commands for an action that is used often. You can use the name of a procedure as a command in another program.

product The answer to a multiplication problem.

proportion An equation stating that two ratios are equal.

$\frac{5}{8} = \frac{30}{48}$

protractor An instrument used to measure angles.

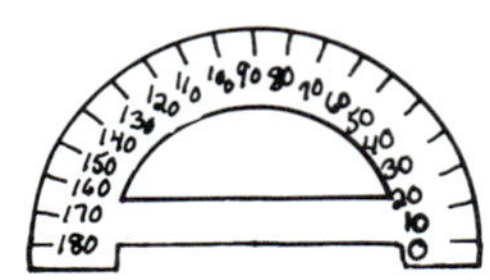

pyramid A solid figure with a face (known as the base) that is any polygon and with all other faces, which are triangles, sharing a common vertex.

quadrilateral A polygon with four sides.

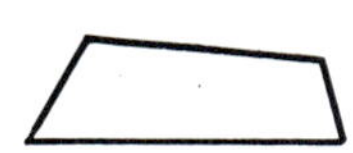

quotient The answer to a division problem.

radius The distance from the center of a circle to the circle. The radius is equal to one half the diameter.

range The difference between the least and the greatest numbers.

rate A comparison by division of two quantities. $\frac{\text{87 kilometers}}{\text{2 hours}}$

ratio A comparison of two quantities by division. Below, the ratio of squares to circles is 4 to 2, 4:2, or $\frac{4}{2}$.

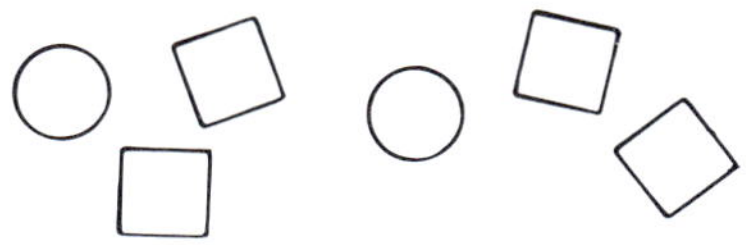

ray A part of a line that has one endpoint. This is ray *AB*:

reciprocal Two numbers are reciprocals when their product is 1.

$$\frac{3}{4} \times \frac{4}{3} = 1$$

↖ ↗ reciprocals

rectangle A parallelogram with a point of symmetry, a line of symmetry through opposite sides, and four right angles.

rectangular prism A prism whose six faces are rectangles.

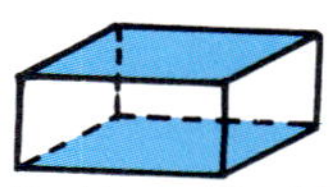

regular polygon A polygon that has all of its sides congruent and all of its angles congruent.

relative error The ratio of the greatest possible error of measurement to the actual measurement.

remainder In a division problem, the number that is "left over." When it is added to the product of the divisor and quotient, the sum is the dividend.

```
    3
8)29
 -24
   5 ← remainder
```

repeating decimal A decimal in which a digit or a group of digits repeats forever.

0.3333... 1.47474747...

rhombus A parallelogram with a point of symmetry, a line of symmetry through opposite vertices, and four congruent sides.

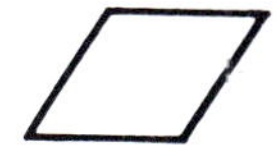

right angle An angle whose measure is 90°.

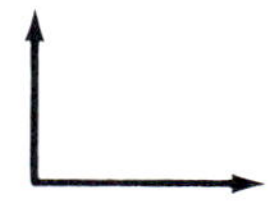

right triangle A triangle that has a 90° angle.

round a number To replace a number by another one that is easier to use. You round a number to the nearest ten by choosing the nearest multiple of 10. (5 is rounded up.)

13 → 10 27 → 30 45 → 50

You round a number to the nearest hundred by choosing the nearest multiple of 100.

487 → 500 1238 → 1200 550 → 600

Rule of Pythagoras The square of the hypotenuse of a right triangle is equal to the sum of the squares of the legs.

$$a^2 + b^2 = c^2$$

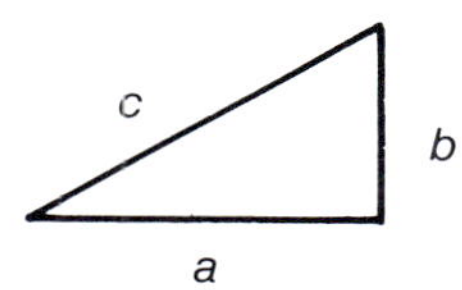

sales tax A tax paid when you make a purchase.

sample A small group, chosen from a larger group, that is examined carefully in order to make predictions about the larger group.

sample space The set of all possible outcomes of an event.

scale drawing A drawing of an object such that the ratio of a unit of length on the drawing to a unit of length on the object is fixed.

scalene triangle A triangle with no lines of symmetry and no congruent sides.

scientific notation A notation for writing a number as the product of a number between 1 and 10 and a power of ten.

$186.3 = 1.863 \times 10^2$

segment A part of a line that has two endpoints.

side of a plane figure One of the segments that make up a figure.

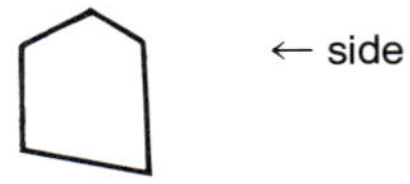

similar figures Two figures that have the same shape.

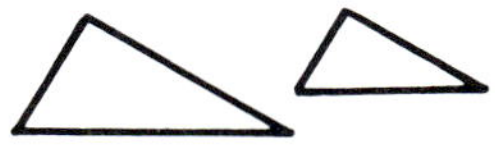

simple interest Interest that is computed by using the formula $I = prt$, where p is principal, r is rate, and t is time.

solution A number that makes an equation or inequality true.

solve Find all the numbers that make an equation or inequality true.

sphere A curved space figure with all points a given distance from the center.

square A rectangle with a point of symmetry, four lines of symmetry, and four congruent sides.

square number The second power of a whole number. $16 = 4^2$

square root The number that can be squared to get a given number. $\sqrt{49} = 7$

statistics A branch of mathematics that studies numerical facts as a basis for drawing general conclusions and making predictions.

substitute Replace a variable with a numeral.

$$7a + 3$$

$$7 \cdot 6 + 3$$

sum The answer to an addition problem.

supplementary angles Two angles are supplementary if the sum of their measures is 180°.

surface area The sum of the areas of all the surfaces of a solid figure.

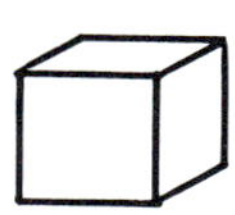

tangent ratio The ratio of the side opposite the given acute angle to the side adjacent to the given acute angle.
$\tan 30° = \frac{a}{b}$

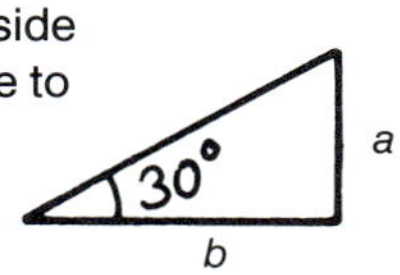

terminating decimal A decimal fraction, such as 0.5 or 1.47, which is not a repeating decimal.

terms (of a fraction) The numerator and denominator of a fraction. $\frac{3}{4}$ ⇐ terms

transversal A line that intersects two lines.

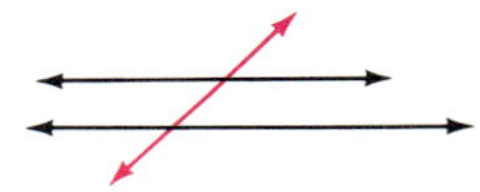

trapezoid A quadrilateral with two parallel sides.

tree diagram A diagram that shows all the possible outcomes of an event.

triangle A polygon with three sides.

triangular prism A prism with two triangular faces in parallel planes.

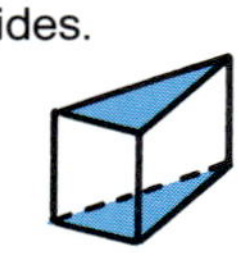

unbiased A sample of a given population is unbiased when each member of the population has an equally likely chance of being chosen for the sample.

unit price The cost per unit (weight, volume, etc.) of an item.

variable A symbol, usually a letter, that holds the place for a number.

$$8x + 19 = 23$$

vertex The point at the "corner" of an angle, plane figure, or solid figure.

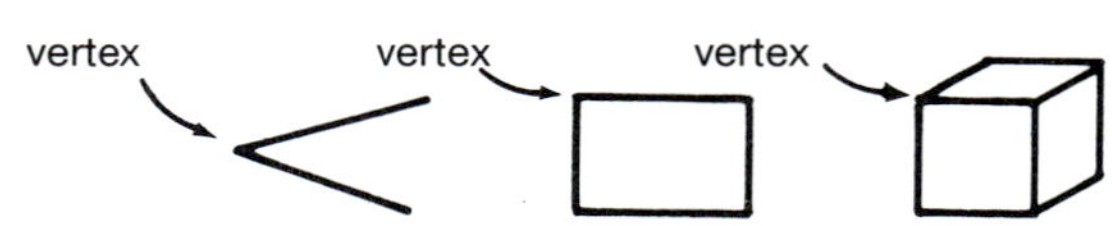

vertical angles Angles that are formed by two intersecting lines. They have a common vertex but no common side.

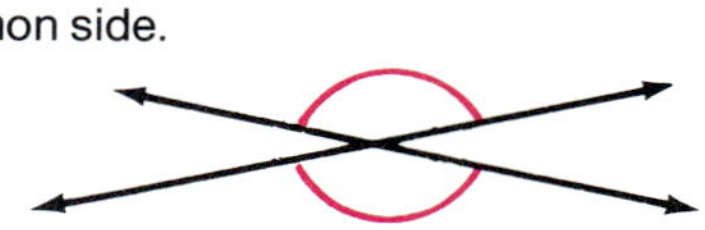

volume The measure of the space inside a solid figure.

whole number Any of the numbers 0, 1, 2, 3, 4, and so on.

Symbols

Symbol	Meaning	Symbol	Meaning
$<$	is less than	$\overline{AB}$	segment *AB*
$>$	is greater than	$\overrightarrow{AB}$	ray *AB*
$=$	is equal to	$\triangle$	triangle
$\approx$	is approximately equal to	$\parallel$	is parallel to
π	pi	$\perp$	is perpendicular to
$^\circ$	degree	$\leftrightarrow$	corresponds to
$'$	foot	$\cong$	is congruent to
$''$	inch	$^{+}5$	positive 5
$0.\overline{37}$	0.373737	$^{-}5$	negative 5
$a:b$	the ratio of a to b	$P(2)$	the probability of the outcome 2
$\neq$	is not equal to	$\sim$	is similar to
%	percent	tan 30°	tangent of 30°
$\angle$	angle	$\sqrt{\ }$	square root
*	multiply	$\lvert\ \rvert$	absolute value
/	divide		

Formulas

Formula	Description	Formula	Description
$P = 2(l + w)$	Perimeter of a rectangle	$V = lwh$	Volume of a rectangular prism
$C = \pi d$	Circumference of a circle	$V = Bh$	Volume of a prism
$A = lw$	Area of a rectangle	$V = \frac{1}{3}Bh$	Volume of a pyramid
$A = s^2$	Area of a square	$V = \frac{4}{3}\pi r^3$	Volume of a sphere
$A = bh$	Area of a parallelogram	$I = prt$	Interest
$A = \frac{1}{2}bh$	Area of a triangle	$F = \frac{9}{5}C + 32$	Temperature conversion to Fahrenheit
$A = \frac{1}{2}(b_1 + b_2)h$	Area of a trapezoid	$C = \frac{5}{9}(F - 32)$	Temperature conversion to Celsius
$A = \pi r^2$	Area of a circle	$a^2 + b^2 = c^2$	Rule of Pythagoras

Index